Nanomedicine

Nanomedicine

Fundamentals, Synthesis, and Applications

Edited by Yujun Song

WILEY-VCH

Editor

Prof. Yujun Song
University of Science and
Technology Beijing
Center for Modern Physics Technology
School of Mathematics and Physics
30 Xueyuan Road
Haidian District
Beijing 100083
China

Cover Image: © Olemedia/Getty Images

All books published by **WILEY-VCH** are carefully produced. Nevertheless, authors, editors, and publisher do not warrant the information contained in these books, including this book, to be free of errors. Readers are advised to keep in mind that statements, data, illustrations, procedural details or other items may inadvertently be inaccurate.

Library of Congress Card No.: applied for

British Library Cataloguing-in-Publication Data
A catalogue record for this book is available from the British Library.

Bibliographic information published by the Deutsche Nationalbibliothek The Deutsche Nationalbibliothek lists this publication in the Deutsche Nationalbibliografie; detailed bibliographic data are available on the Internet at <http://dnb.d-nb.de>.

© 2025 WILEY-VCH GmbH, Boschstraße 12, 69469 Weinheim, Germany

All rights reserved, including rights for text and data mining and training of artificial technologies or similar technologies (including those of translation into other languages). No part of this book may be reproduced in any form – by photoprinting, microfilm, or any other means – nor transmitted or translated into a machine language without written permission from the publishers. Registered names, trademarks, etc. used in this book, even when not specifically marked as such, are not to be considered unprotected by law.

Print ISBN: 978-3-527-34863-3
ePDF ISBN: 978-3-527-83038-1
ePub ISBN: 978-3-527-83039-8
oBook ISBN: 978-3-527-83040-4

Typesetting   Straive, Chennai, India
Printing and Binding   CPI Group (UK) Ltd, Croydon, CR0 4YY

Contents

Preface *xv*

1 State of the Art in Nanomedicine *1*
Yujun Song and Wei Hou
1.1 Intractable Diseases and Development of the Related Novel Therapy and Medicines *1*
1.2 Key Features of Nanomedicines *4*
1.3 Nanotechnology Translational Nanomedicine: Emergence and Progress *6*
1.4 Interdisciplinary Features of Nanomedicines: Multi-mode and Multi-function Features Promoting Nanomedicine-mediated Immunotherapy and/or Physical Field Ablation Therapy for Subversive Therapy *12*
1.5 Future Development of Nanomedicines by Coupling Advanced Biomedicines (Including Biochemistry and Biophysics), Modern Physicochemical Technologies, and Artificial Intelligence Technology *27*
References *29*

2 Fundamentals in Nanomedicine *49*
Xiangrong Song, Mengran Guo, Zhongshan He, Xing Duan, and Wen Xiao
2.1 Design Theory of Nanomedicines *49*
2.1.1 Passive Targeting Function Design of Nanomedicines *49*
2.1.1.1 Size-dependent Biological Functions for Nanomedicine Design *50*
2.1.1.2 Surface Charge and PEGylation of Nanoparticles for Nanomedicine Design *50*
2.1.1.3 Shape-dependent Biological Effect of Nanoparticles for Nanomedicine Design *50*
2.1.2 Active Targeting Function Design of Nanoparticles for Nanomedicine *51*
2.1.2.1 Small Molecule and Aptamer-directed Active Targeting Function Design *51*
2.1.2.2 Protein and Peptide-directed Active Targeting Function Design *52*
2.1.2.3 Antibody-directed Active Targeting Design *52*

2.1.3	Auto-targeting Design	53
2.1.3.1	Platelets Targeting Function Design	53
2.1.3.2	RBC Targeting Function Design	53
2.1.3.3	Macrophage Targeting Function Design	53
2.1.3.4	Exosomes and Supramolecular Cell Membrane Vesicle Targeting Design	54
2.1.4	In Vitro or In Vivo 3D Traceable Function	54
2.1.4.1	Gold Nanoparticles for Bioimaging	54
2.1.4.2	Quantum Dots for Bioimaging	54
2.2	Progress in the Controlled Synthesis of Nanomaterials	55
2.2.1	Ball Milling Through the Mechanical Method	55
2.2.2	Nanoprecipitation	56
2.2.3	Microfluidic Synthesis	56
2.2.4	Personalized Protein Nanoparticles	58
2.2.5	Biological Entities as Chemical Reactors	58
2.3	Progress in the Surface Modification and Functionalization of Nanomaterials for Nanomedicines	60
2.3.1	How Surface Modification Can Improve the Stability of Nanomedicine	60
2.3.1.1	Modification of Charged Materials	60
2.3.1.2	Increase the Steric Resistance Between Nanoparticles to Improve the Stability of Nanomedicine	62
2.3.1.3	Conformation Change of the Surface Polymer of Nanomaterials Improves the Stability of Nanomaterials	62
2.3.2	Surface Modification of Bioactive Groups Can Improve the Efficacy of Nanopreparations	63
2.3.2.1	Nanodrugs Modified by Single Targeting Ligand	63
2.3.2.2	Multiligand Modified Nanopreparations	66
	References	67
3	**Nanomedicine for Antitumors**	**73**
	Qiong Wu, Xinzhu Yang, Ruixue Zhu, and Yujun Song	
3.1	Introduction	73
3.2	Liposomal Nanoparticles	75
3.2.1	Synthesis Methods	76
3.2.1.1	Extrusion Technique	76
3.2.1.2	Sonication	76
3.2.1.3	Microfluidization	76
3.2.1.4	Heating Method	77
3.2.2	Application of Liposomes in Drug Delivery	77
3.3	Polymeric Nanoparticles	78
3.3.1	Techniques for Creating Polymeric Nanoparticles	79
3.3.1.1	Method of Evaporating Solvents	79
3.3.1.2	Method of Spontaneous Emulsification and Solvent Diffusion	79
3.3.1.3	Salting Out Method	79

3.3.1.4	Method of Solvent Displacement/Nanoprecipitation	*80*
3.3.1.5	Polymerization Methods	*81*
3.3.1.6	Nanoparticles Developed from Hydrophilic Polymers	*81*
3.3.2	Controlled Drug Release by Polymeric Nanoparticles	*81*
3.4	Inorganic Nanoparticles	*82*
3.4.1	Synthesis and Application of Colloidal Gold Nanoparticles	*83*
3.4.2	Synthesis and Application of Mesoporous Silica Nanoparticles	*84*
3.4.3	Synthesis and Application of Graphene	*85*
3.4.4	Synthesis and Application of Magnetic Nanoparticles	*86*
3.4.5	Quantum Dot Synthesis and Applications	*88*
3.4.6	Creation and Use of Stratified Double Hydroxides	*89*
3.4.7	Nanoparticles with Multifunctional Composite	*90*
3.5	Mixture (Hybrid) Nanoparticles	*90*
3.6	Cell Membrane Coating Nanotechnology	*93*
3.6.1	Synthesis of Cell Membrane–Nanoparticle Structures	*93*
3.6.1.1	Extrusion	*93*
3.6.1.2	Sonication	*93*
3.6.1.3	Electroporation	*94*
3.6.1.4	Graphene Nanoplatform-mediated Cell Membrane Coating	*94*
3.6.1.5	Encapsulation of In Situ Using Cell-derived Vesicles	*94*
3.7	Challenges and Current Limitations	*94*
3.7.1	Nanomaterials' Physicochemical Characterization	*95*
3.7.2	Safety Concerns	*95*
3.7.3	Regulatory Issues	*95*
3.7.4	Manufacturing Issues	*96*
3.8	Conclusions	*96*
	References	*97*
4	**Nanomedicine for the Treatment of Nervous System Diseases**	***113***
	Xiaojian Cui and Yujun Song	
4.1	Concepts and Types of Nanomedicines for Nervous System Diseases	*113*
4.2	Therapeutical Methods for Nervous System Diseases and Features of Current Nanomedicines	*114*
4.3	Synthesis Methods and Typical Examples (Including Clinical Trial) of Polymer-based Nanomedicines for Nervous System Diseases	*116*
4.3.1	Synthesis Methods of Polymer-based Nanomedicine	*116*
4.3.1.1	Solvent Evaporation	*116*
4.3.1.2	Emulsification Process	*116*
4.3.1.3	Reductive Reaction Reduction Method	*116*
4.3.2	Polymer-based Nanomedicine: Typical Application	*116*
4.3.2.1	Targeted Drug Delivery	*116*
4.3.2.2	Tissue Repair and Regenerative Medicine	*117*
4.3.2.3	Vaccines and Immunotherapy	*117*

4.3.2.4	Inclusion of Nanomaterials	*117*
4.4	Inorganic-based Nanomedicines: Synthesis Methods and Typical Application (Including Clinical)	*117*
4.4.1	Quantum Dots	*117*
4.4.2	Nanogold	*119*
4.4.3	Nanocarbon	*120*
4.4.4	Magnetic Nanoparticles	*120*
4.4.5	Nanomesoporous Silicon	*122*
4.4.6	Nanocalcium	*122*
4.5	Metallic-based Nanomedicines: Synthesis Methods and Typical Application (Including Clinical)	*122*
4.5.1	Preparation and Characterization of Metal-organic Frame Materials	*123*
4.5.2	Application of Functionalized Metal-organic Framework Materials in Tumor Therapy	*123*
4.6	Multifunctional Nanomedicines for Nervous System Diseases	*125*
4.6.1	Nanomedicine Therapy Strategy Combined with Chemotherapy	*125*
4.6.2	Nanomedicine Therapy Strategy Combined with Immunotherapy	*126*
4.6.3	Combination Treatment Strategy	*127*
4.6.4	Perspectives of Nanomedicines for Nervous System Diseases	*127*
	References	*128*

5 Nanomaterial Translational Nanomedicine for Anti-HIV and Anti-bacterial *131*
Hao Luo and Yujun Song

5.1	Concepts, Anti-HIV Theory, and Types and Features of Current Nanomedicines	*132*
5.1.1	Anti-HIV Reverse Transcriptase Inhibitors	*133*
5.1.2	Anti-HIV Protease Inhibitors	*133*
5.1.3	Characteristics of Polymer Nanocarriers	*141*
5.2	Polymer-based Nanomedicines: Synthesis Methods and Typical Application	*142*
5.3	Inorganic-based Nanomedicines: Synthesis Methods and Typical Application	*146*
5.4	Metallic-based Nanomedicines: Synthesis Methods and Typical Application	*148*
5.5	Multi-functional (Target) Nanomedicines	*150*
5.6	Future Development	*152*
	References	*153*

6 Nanomedicine for Next-generation Dermal Management *157*
Haibin Wu, Qian Chen, and Shen Hu

6.1	Introduction	*157*
6.2	Nano-biomaterials-based Therapeutics for Wound Healing	*159*
6.2.1	Nano-biomaterials for Effective Hemostasis	*160*

6.2.2	Antibacterial Nano-biomaterials *164*	
6.2.2.1	Inorganic Antibacterial Nano-biomaterials *164*	
6.2.2.2	Organic Antibacterial Nano-biomaterials *165*	
6.2.3	Engineering of the Wound Microenvironment *166*	
6.2.3.1	Redox Modulation of the Wound Microenvironment *167*	
6.2.3.2	Regulation of Microenvironmental MMP Activity *167*	
6.2.3.3	Targeting the Pro-inflammatory Mediators *169*	
6.2.4	Nano-biomaterials-enabled Biophysical Regulation of Wound Healing *169*	
6.2.4.1	Surface Nanotopographical Features *171*	
6.2.4.2	Mechanical Cues *171*	
6.2.4.3	Bio-electrical Stimulation *173*	
6.2.5	Angiogenic Nano-biomaterials *175*	
6.2.5.1	Nano-biomaterials as Delivery Vehicles for Angiogenic Therapeutics *175*	
6.2.5.2	Nano-biomaterials as Intrinsic Angiogenic Agents *176*	
6.3	Nano-biomaterials for Imaging and Monitoring of Cutaneous Wounds *176*	
6.3.1	Wound Infection Monitoring *177*	
6.3.1.1	Nanoprobes for Wound Infection Monitoring *177*	
6.3.1.2	Theranostic Nano-biomaterials for Wound Infection Control *177*	
6.3.2	Imaging of Wound Parameters and Markers *179*	
6.3.2.1	Imaging of Physiological and Pathological Wound Parameters *179*	
6.3.2.2	Imaging of Stem Cell-based Wound Therapy *180*	
6.3.2.3	Imaging of Wound Scarring Markers *180*	
6.4	Conclusion and Future Outlook *181*	
6.4.1	Standardization of the Preparation and Functionalization of Nano-biomaterials *182*	
6.4.2	Bio-safety of Nano-biomaterials *182*	
6.4.3	Computational Simulation and Machine Learning *183*	
	Acknowledgments *183*	
	References *184*	
7	**Nanomedicine for Targeting Delivery of Gene and Other DNA/RNA Therapies Based Viruses Engineering** *197*	
	Xiangrong Song, Mengran Guo, Zhongshan He, Xing Duan, and Wen Xiao	
7.1	Targeting Delivery of Gene Therapies to the Tumors *198*	
7.1.1	Passive Targeting *198*	
7.1.2	Active Targeting *199*	
7.2	Targeting Delivery of Gene-based Nanovaccines to the Spleen *199*	
7.3	Targeting Delivery of Gene-based Nanovaccines to the LNs *200*	
7.4	Targeting Delivery of Gene-based Nanomedicines to the Liver *200*	
7.5	Targeting Delivery of Gene-based Nanomedicines to the Lung *200*	
7.6	Targeting Delivery of Gene-based Nanomedicines to the Brain *202*	
	References *203*	

8	**Nanomedicine for Bio-imaging and Disease Diagnosis** *207*	
	Ziqi Wang and Yujun Song	
8.1	Concepts, Types, and Features of Current Nanoprobes *207*	
8.1.1	Optical Nanoprobe *208*	
8.1.2	Magnetic Nanoprobe *210*	
8.1.3	Photoacoustic Imaging Nanoprobe *212*	
8.1.4	CT Nanoprobe *213*	
8.1.5	Nuclide Imaging Nanoprobe *213*	
8.1.6	Multifunctional/Multimodal Nanoprobes *214*	
8.2	Synthesis Methods *214*	
8.2.1	Preparation and Synthesis of Noble Metal Nanoprobes *216*	
8.2.1.1	Chemical Synthesis Method *216*	
8.2.1.2	Photochemical Method *217*	
8.2.1.3	Template-based Method *217*	
8.2.1.4	Electrochemical Process *218*	
8.2.2	Preparation and Synthesis Method of Magnetic Nanoprobe *218*	
8.2.2.1	Coprecipitation Method *218*	
8.2.2.2	Thermal Decomposition Method *219*	
8.2.2.3	Hydrothermal Synthesis Method *219*	
8.3	Typical Application Examples in the Disease Diagnosis and Study of Biological Events *219*	
8.3.1	Application of Optical Nanoprobes *219*	
8.3.2	Application of Magnetic Nanoprobes *222*	
8.3.3	Application of Photoacoustic Imaging Nanoprobes *222*	
8.3.4	Application of CT Nanoprobe *223*	
8.4	Future Development *224*	
	References *225*	
9	**Magnetic Nanoparticles and Their Applications** *227*	
	Xiuyu Wang and Yuting Tang	
9.1	Introduction *227*	
9.2	Classification of Magnetic Nanoparticles *228*	
9.2.1	Magnetic Regulation *230*	
9.2.1.1	Size *230*	
9.2.1.2	Shape *232*	
9.2.1.3	Composition *233*	
9.2.2	Exchange–Coupling Interaction *235*	
9.3	Biomedical Applications *236*	
9.3.1	Magnetic Resonance Imaging *236*	
9.3.2	Magnetic Hyperthermia *238*	
9.3.3	Targeted Drug Delivery *239*	
9.3.4	Neuromodulation *240*	
9.4	Conclusion and Outlook *242*	
	References *242*	

10	**Nanomedicine-mediated Immunotherapy** *245*
	Wei Hou and Yujun Song
10.1	Immune System and Immune Response *245*
10.2	Mechanism of Tumor Immunotherapy *247*
10.3	Mechanism of Nanomedicine-enhancing Immunotherapy *248*
10.4	Immunological Applications of Nanomedicines *252*
10.4.1	Artificial Antibody *252*
10.4.2	Reprogrammed Immunity *253*
10.4.3	Nanomedicines as Agonists *255*
10.4.4	Nanomedicine-combined CAR-T Therapy *257*
10.4.5	Nanovaccines *258*
10.4.6	Nanomedicines Affect Cytokines *261*
10.5	Summary and Outlook *263*
	References *264*
11	**Nanomedicine-mediated Ultrasound Therapy** *269*
	Qingwei Liao, Yaoyao Liao, and Wei Si
11.1	Concept, Therapy Theory, and Devices *269*
11.1.1	The Concept of Ultrasound Therapy and Its Therapy Theory *269*
11.1.1.1	Thermal Effect *270*
11.1.1.2	Mechanical Effects *270*
11.1.1.3	Cavitation Effect *270*
11.1.1.4	Thixotropic Effect *271*
11.1.1.5	Acoustic Impulse and Acoustic Chemical Effects *271*
11.1.2	Concepts and Therapy Theory of Nanomedicines *272*
11.1.2.1	Concept of Nanomedicine *272*
11.1.2.2	Therapy Theory of Nanomedicine *272*
11.1.3	Equipment for Nanomedicine-mediated Ultrasound Therapy *273*
11.2	Nanomedicine Synthesis with Ultrasound Field Response *274*
11.2.1	Ultrasound Chemical Precipitation *274*
11.2.2	Ultrasonic Atomization Pyrolysis *274*
11.2.3	Ultrasonic Electrochemical Method *274*
11.2.4	Ultrasonic Reduction Method *275*
11.3	Application Examples *275*
	References *276*
12	**Nanomedicine-mediated Photodynamic and/or Photothermal Therapy** *279*
	Shangqing Jing and Yujun Song
12.1	Photodynamic and Photothermal Mechanism for Anti-cancer *281*
12.2	Nanomaterial-based PSs (Nano-PSs) for PDT *283*
12.2.1	Metal-based Nanomaterials *283*
12.2.1.1	Au NPs *283*
12.2.1.2	Ag NPs *284*
12.2.1.3	Metal Clusters *284*

12.2.2	Transition Metal Carbide Nanomaterials	285
12.2.3	Carbon-based Nanomaterials	285
12.2.4	Photothermal Agents for PTT	286
12.2.5	Noble Metal-based Nanomaterials	286
12.2.5.1	Au NPs	287
12.2.5.2	Ag NPs	287
12.2.6	Carbon-based Nanomaterials	288
12.2.6.1	Carbon Dots	288
12.2.6.2	Graphene Oxide and Reduced Graphene Oxide	288
12.2.6.3	Carbon Nanotubes	289
12.2.7	Semiconductor-based Nanomaterials	289
12.2.7.1	Cu-based Semiconductors	290
12.2.7.2	W-based Semiconductors	290
12.2.7.3	Mo-based Semiconductors	290
12.3	Photodynamic and Photothermal Synergistic Therapy	291
12.4	Opportunities and Challenges	292
12.4.1	Light Source	292
12.4.2	Hypoxia of Tumor Microenvironment	293
12.4.3	Combination with Other Therapy	293
12.5	Summary	293
	References	294

13 Nanomedicine-mediated Pulsed Electric Field Ablation Therapy 299
Xinzhu Yang, Qiong Wu, Ruixue Zhu, and Yujun Song

13.1	Introduction	299
13.2	Concept, Therapy Theory, and Devices	302
13.2.1	Concept, Therapy Theory	302
13.2.1.1	Enhanced Therapeutic Effects	303
13.2.1.2	Overcoming Drug Resistance	303
13.2.1.3	Reduce Side Effects	303
13.2.1.4	Promote Personalized Treatment	303
13.2.1.5	Pulse Generator	304
13.2.1.6	Electrode System	305
13.2.1.7	Pulse Parameter Adjustment	305
13.2.2	Devices	306
13.3	Synthesis of Nanomedicine with Electric Field Response	312
13.3.1	Synergistic Anticancer Research of Multimodal Iron-based Nanodrugs and Nanosecond Pulse Technology	312
13.3.1.1	Design and Synthesis of Multimodal Iron-based Nanodrugs	313
13.3.1.2	Characterization of Physical Properties and Structural Analysis	315
13.3.1.3	Research on Multimodal Physical Imaging Applications	318
13.3.1.4	Study on In Situ Anticancer Effects in Liver Cancer	320
13.3.2	Progresses of Innovative Combination Therapy	321
13.4	Application Examples	322

13.4.1	Cancer Treatment *322*	
13.4.2	Treatment of Neurological Diseases *323*	
13.4.3	Bacterial Infection Treatment *323*	
13.5	Conclusion *323*	
	References *324*	

14 Nanomedicine-mediated Magneto-dynamic and/or Magneto-thermal Therapy *329*
Yangfei Wang and Yujun Song

14.1	Concepts, Treatment Theories, and Methods *329*
14.1.1	Overview of Magnetic Hyperthermia Therapy *329*
14.2	Treatment Theories and Methods *330*
14.2.1	Magneto-thermal Conversion Mechanism *330*
14.2.2	The Killing Effect of Magnetic Hyperthermia on Tumor Cells *331*
14.2.3	Cytotoxicity *332*
14.3	Magnetic Field-responsive Nanomedicine Synthesis *333*
14.3.1	Fe_3O_4 Materials *333*
14.3.2	Ferrite Materials *334*
14.3.3	Alloy Materials *335*
14.3.4	Composite Materials *335*
14.3.5	Other Materials *336*
14.4	Applications *337*
14.4.1	Instance 1 *337*
14.4.2	Instance 2 *337*
14.4.3	Instance 3 *338*
	References *339*

15 Nanomedicine-mediated Radiotherapy for Cancer Treatment *341*
Ruixue Zhu and Yujun Song

15.1	Concept, Therapy Theory, and Devices *341*
15.2	Nanomaterials as Radiosensitizers for Radiation Therapy *347*
15.3	Nanomaterials Delivering Radioisotope for Internal Radioisotope Therapy *358*
15.4	Nanomedicine Synthesis with Radio/Nuclear Radiation Response *362*
15.5	Conclusion and Prospects *364*
	References *365*

16 Nanomedicine Conjugating with AI Technology and Genomics for Precise and Personalized Therapy *371*
Lin Liu, Siyu Chen, and Sen Zhang

16.1	Concept *371*
16.2	Genomics of Nanomedicine *374*
16.2.1	Genomic Modifications in Nanomedicine *376*
16.2.2	Genomic Toxicity of Nanomedicines *376*

16.2.3	Genomic Responses to Nanomedicines	*377*
16.3	Artificial Intelligence Technology in Nanomedicine Development	*377*
16.3.1	Machine Learning Techniques	*378*
16.3.2	Computer Vision Technology	*378*
16.3.3	Natural Language Processing Technology	*378*
16.4	Artificial Intelligence Facilitates Precise and Personalized Nanomedicine Based on Genomics	*379*
16.4.1	Cell Replacement Therapy for Diabetes	*380*
16.4.2	Precision Medicine in Cancer Treatment	*380*
	References	*382*
17	**Microfluidic Conjugating AI Platform for High-throughput Nanomedicine Screening**	*385*
	Xing Huang, Wenya Liao, Zhongbin Xu, and Yujun Song	
17.1	Introduction	*385*
17.2	Microfluidic Technologies for Medicine Development	*386*
17.2.1	Basics of Microfluidics	*386*
17.2.2	Fabrication of Microfluidic Chips	*387*
17.2.3	Representative Microfluidic Units	*390*
17.3	Microfluidic Preparation of Nanomedicines	*391*
17.3.1	Drug Nanoparticle Preparation	*392*
17.3.2	Nanocarrier Preparation	*393*
17.3.2.1	Polymer Nanoparticles	*393*
17.3.2.2	Liposomes	*394*
17.3.2.3	Inorganic Nanoparticles	*395*
17.4	Microfluidic High-throughput Drug Screening	*396*
17.4.1	Microfluidic Drug Screening Based on Cell Assays	*396*
17.4.2	Microfluidic Drug Screening Based on Organ-on-a-Chip	*398*
17.5	AI-assisted Microfluidic Development of Nanomedicine	*400*
17.6	Conclusions and Perspectives	*402*
	References	*404*

Index *413*

Preface

"The finest thoroughbred horse cannot travel ten paces in one leap, but the sorriest nag can go a ten days' journey. No accumulation of steps, can't lead to thousand miles" by Hsun-Tzu in *Encouraging to Learn*. Since the highly biocompatible liposome nanostructures and silicon polymer-based nanocarriers have been developed in 1960s for drug loading and safe transportation and for prolonged drug lasting, respectively, nanomaterials and nanotechnology translational medicine starts its long march. With unremitting efforts of millions of scientists, engineers, and graduates for almost half a century, nanomedicines ushered in a blowout in academic studies and clinical application after 2010s through the long-term interdisciplinary innovation of physics, chemistry, biology, medicine, life science, and information technology. Their biological effects and therapeutical mechanism of nanomaterials have been revealed gradually, such as enhanced permeability retention (EPR) effects, immunological effects, metaloptosis (e.g. ferroptosis, cuproptosis), enzyme-like function, and the enhanced therapeutical effects by the interaction with varieties of high-energy particles or physical fields.

Benefit from the progresses in physics, chemistry, biology, medicine, pathology, functional materials, biomedical engineering, life science, and nanotechnology, nanomedicines have been developed from single function as drug carriers or encapsulation to multifunctions (e.g. diagnosis, imaging, therapy, monitoring, and intelligent feedbacking) and multi-modality (e.g. chemotherapy, immunotherapy, physical field ablation, high energy particle, or electromagnetic wave radiation). These nanomedicines can be constructed from organics, inorganics, or composites with single components or multi-hierarchy microstructures to realize multi-targeting functions: detection, diagnosis, and therapy. Recently, nanomedicines of multifunction and multi-modality termed as nanotheranostics and nanotherapeutics have been further developed as smart nanomedicines, or intelligent nanomedicines by conjugating some biomarkers for specific active molecules or cells to these nanomedicines. Hitherto, varieties of nanomedicines have been designed and synthesized for intractable disease treatment, such as tumors or cancers, rheumatoid arthritis, cerebrovascular diseases (e.g. stroke, cerebral thrombosis, myocardial infarction), neural diseases (e.g. Parkinson, Alzheimer's, depression), recurrent and infectious skin diseases (particularly relating to blood, endocrine, and immunity), and highly contagious and lethal infectious diseases (e.g. HIV, COVID-19), and so on. Many of them have been approved by FDA, EMA, or the state food and drug administration of the relating countries for clinical trials and some of them have been commercialized to serve patients. Furthermore, as these multi-modality nanomedicines are mediating to varieties of new physical therapy (e.g. nanosecond pulsed electric fields, high-intensity focus ultrasound, pulsed or alternating magnetic field, ultrashort pulsed laser radiation, γ- or β-ray radiation, and fast neutron fragmentation irradiation) and immunotherapy, many innovative combined therapies can be developed for subversive therapeutical effects, particularly for these intractable diseases. Coupling with their diagnosis, imaging, monitoring, and intelligent feedbacking and expert consulting functions, smart nanotheronaustics have also been developed. In addition, the recently developed artificial intelligence generation content (AIGC) or machine learning technologies have been successfully used in the design and microstructure optimization of multi-mode nanomedicines. It shall be expected to address some fundamental and clinical application issues in the novel therapeutical effects of nanomedicines (e.g. quantum biomedical effects) and the coupling of nanomedicines with multi-physical fields for next generation of subversive therapies. The coupling of nanomedicines with AIGC will accelerate the discovery of novel nanomedicines or quantum medicines and their clinical translation, instrumentation of nanomedicines mediating multi-physical field coupling ablation, serving more and more patients and defending human life and health.

This monograph will try to summarize the above progresses in the nanomedicines systematically and comprehensively. Chapter 1 gives an introduction of the state of the art of nanomedicines by revealing their history, development, current state, and future prospect. Chapter 2 will give a brief fundamental on the design and fabrication of nanomedicines. From Chapters 3–6, some specific nanomedicines will be discussed for some intractable disease treatment based on some detailed examples. Chapters 7 and 10 will summarize the development of nanomedicines with some desired physicochemical or biological properties for unique diagnostic or therapeutical functions. Then some typical combined therapies via nanomedicines medicating physical field ablation will be discussed from Chapters 11–15. Finally, nanomedicine conjugating with AI technology and genomics for precise and personalized therapy and one creative technology for nanomedicine screening or microfluidic platform conjugating to AI will be summarized in Chapters 16 and 17, respectively.

We know that it is impossible to include all progresses and aspect of the rapid blooming and sunshine nanomedicines and the related fields in one monography. We hope that this book will contribute to the research and teaching of readers interesting in this vivid field. We also hope that this monograph can play a key role of tossing out a brick to get a jade gem, attracting more scientists, engineers, and students to join into this never-ending field full of perspectives in fundamental academic researches, applied technologies, and clinical trials. Best wishes that all scholars in this field can achieve more and more. Therefore, we will feel gratified if only this book can give readers some clues on learning and thinking of this interesting field and promote its scientific, technological, and clinical development.

Finally, I dedicate this monograph to my family (my father and mother: Mr Sigan Song and Ms Xiuyun Meng; and my daughter and son, Xinran Song and Haoran Song) for their great support and encouragement in editing and writing this book.

June 2024

Yujun Song
Beijing, China

1

State of the Art in Nanomedicine

Yujun Song[1,2,3] *and Wei Hou*[1]

[1]*University of Science and Technology Beijing, Center for Modern Physics Technology, School of Mathematics and Physics, 30 Xueyuan Road, Haidian District, Beijing 100083, China*
[2]*Zhengzhou Tianzhao Biomedical Technology Company Ltd., Zhengzhou New Technology Industrial Development Zone, 7B-1209 Dongqing Street, Zhengzhou 451450, China*
[3]*Key Laboratory of Pulsed Power Translational Medicine of Zhejiang Province, Hangzhou Ruidi Biotechnology Company Ltd., Room 803, Bldg. 4, 4959 Yuhangtang Road, Cangqian Street, Hangzhou 310023, China*

1.1 Intractable Diseases and Development of the Related Novel Therapy and Medicines

Lots of people get sick now and then in their lives, reducing their quality of life and even threatening their lives. Many diseases not only cause great suffering for people themselves but also for their families and then of the whole society, particularly for those intractable diseases, such as tumors or cancers, rheumatoid arthritis, cerebrovascular diseases (e.g. stroke, cerebral thrombosis, myocardial infarction), neural diseases (e.g. Parkinson's, Alzheimer's, depression), recurrent and infectious skin diseases (particularly relating to blood, endocrine, and immunity), and highly contagious and lethal infectious diseases (e.g. HIV, COVID-19), and so on. Among them, cerebrovascular diseases are the first killers of people, particularly for those more than 50 years old. There were about 1.79 million people in 2016 who died of these kinds of diseases all over the world, about 31% of global causes of death, according to the statistics of the World Health Organization (WHO). Cardiovascular outpatients in China are already more than 0.29 billion (B). Nearly three million people die from cardiovascular and cerebrovascular diseases in China every year, about 51% of the whole causes of death. Cerebrovascular diseases are the fifth cause of death in 2016, with about 373 deaths per 1 M people. In addition, these kinds of diseases preserve features of high suddenness, high disability rate (about 75% of the surviving patients have varying degrees of loss of labor ability and 40% are severely disabled), high recurrence rate, and more multiple complications (e.g. coronary heart disease, myocardial infarct, vascular dementia, subarachnoid hemorrhage, respiratory tract infection, and sudden deafness). China has entered an aging society like other developed countries (e.g. Japan) even though China is still a developing

Nanomedicine: Fundamentals, Synthesis, and Applications, First Edition. Edited by Yujun Song.
© 2025 WILEY-VCH GmbH. Published 2025 by WILEY-VCH GmbH.

country. The population of coronary artery disease (CAD) in China has increased from 2.27 M in 2016 to 2.53 M in 2020, with a compound annual growth rate of 2.7%. It was said that China's precision percutaneous coronary artery therapy (PCI) market has increased from ~$25.7 M in 2016 to ~101.4 M in 2020, with a compound annual growth rate of ~ 40.8%. The global market scale of PCI also shows a growth trend, which was $9.49 B in 2019 and was expected about $13.26 B in 2025 with a compound annual growth rate of ~ 5.4% (https://www.163.com/dy/article/HPGQBS2-R051481OF.html; https://wenku.baidu.com/view/933bb2d7f624ccbff121dd36a32d7375a417c6c4.html?_wkts_=1711608332102&bdQuery=2023%E7%BB%8F%E7%9A%AE%E5%86%A0%E7%8A%B6%E5%8A%A8%E8%84%89%E6%B2%BB%E7%96%97%E5%85%A8%E7%90%83%E5%B8%82%E5%9C%BA%E8%A7%84%E6%A8%A1).

Although cancers are the second killers of people, next to cerebrovascular diseases, the pain and burden of patients caused by cancers far outweigh the former due to their characteristics of chronic redundant diseases. It is said that there were about 19.29 M new cases of cancers, among which there were 10.06 M male cases and 9.23 M female cases according to the statistics in 2020. There were about 9.96 M death cases, including 5.53 M male cases and 4.43 M female cases. It is expected that there will be more than 21 M new cases in 2030 [1–3]. There are about 4.82 M and 2.37 M new cases of cancers, and about 3.21 M and 0.64 death cases of cancers, in China and the United States, respectively [1]. Partially thanks to innovative drugs and therapies promoted by medical technology, the overall trend of death cases of cancers in the United States is accelerated down since 1991 [1]. It is predicted that the new cases and death cases in 2023 will be continuously reduced by 410 K and 30 K compared with those in 2022. However, in China, the new cases and death cases of cancers in 2022 increased by 250 K and 210 K compared with those two years earlier (2020), and the 2022 death/new incidence rate in China is far more than that in United States (67% versus 27%) [1]. The death cases of lung cancers is the first among all death cases in China, and then the summed death cases of liver and pancreatic cancers. Particularly, the cases of liver cancers in China are almost half of the cases in the world. For many cases, cancers were found to be mostly the terminal stage [4–6]. While, thanks to vigorous anti-smoking measures in the United States, the first death case is breast cancer, not lung cancer, in the United States. The survival rates for some special cancers in China are lower than those in the United States, particularly for breast cancers and colorectal cancers. It is also said that cancer prognosis in China is much worse than that in the United States. China needs to make more efforts to provide effective cancer treatment and improve universal health coverage. New medicine and therapy of high anti-cancer efficiency are extremely urgent currently, especially for China. At the same time, the global market of anti-tumor medicine has increased to $192.2 B in 2022 with a compound annual growth rate of ~12.7% while in China, the sales of anti-tumor drugs have been showing a steady growth trend in recent years. The market size of anti-tumor drugs reached $28.2 B in 2020 and will have an estimated compound annual growth rate of ~16.1% from 2020 to 2025.

As for hyperuricemia and gout, there were about 1.03 billion (B) outpatients all over the world and about 0.18 B in China in 2022. The global market for gout medicine was about $3.0 B. The gout medicine market in China will grow rapidly in the future, which is expected to be about $1.54 B in 2030. China has been known as the country with the largest population of diabetes in the world. The total number of people related to diabetes exceeded 260 million in 2018, including 114.39 M diabetes outpatients and 148.70 M pre-diabetes population. It is forecasted that the number of diabetes people will reach 320 M, which will create a huge market for diabetes medicine, with a potential scale expected to reach $19.3 B.

There are four types of neural diseases: absence of symptoms, release of symptoms, irritation and shock, such as Parkinson's, Alzheimer's, Depression, and Huntington's diseases. There are more than 0.1 B people more than 15 years old with various mental disorders in China, among which there are about 16 M patients with severe mental disorders and most of the rest are people with mental or behavioral disorders such as depression or autism. These kinds of illnesses not only torture patients but also haunt their families for a long time. Particularly, they preserve some certain psychological infectivity (e.g. resulting in mass suicide groups), leading to great social harm. Developing these kinds of drugs for anti-mental disorders has to overcome the obstacle of passing the blood–brain barrier (BBB). It is more difficult for these drugs into nerve cells to break through the protective membranes of dendrites, myelin sheaths, axons, terminals, etc.

Clearly, fighting these intractable diseases is a long and arduous task for human beings, whose key is to develop diagnosis methods for early disease identification, innovative drugs, and subversive therapies. Starting from this century, medicine and health care entered into a rapid transformation period promoted by the interdisciplinary crossover of life science, biology, biophysics, biochemistry, nanotechnology, and information technology [7, 8]. As a result, many creative medicine or medical technologies sprout recently, such as personalized medicine, precise medicine, nanomedicine, and lots of innovative therapies, such as gene therapy (e.g. mRNA, DNA), targeting therapy (e.g. cell targeting, tumor microenvironment targeting), immunotherapy (e.g. PD1, PD-L1, vaccine, CAR-T,), new physical field ablation therapy (e.g. nanosecond pulsed electrical field (nsPEF) ablation), new physicochemical therapy (e.g. ferroptosis, cuproptosis, photothermal therapy, photodynamic therapy, magnetothermal therapy, magnetodynamic therapy), and heavy particle radiation therapy (e.g. proton beam radiation, neutron scattering radiation, boron neutron capture therapy) [7–21].

Particularly, nanomedicines, which are translated from some functional nanomaterials, deal with nanoscale matters that can be used in biomedicine or biomedical engineering as bioprobes for the detection of biomolecule, organelle, cells or tissues, or as biosensors for the diseases or pathological metabolism diagnosis, or special drugs for some disease treatment or life function regulation. Based on nanodrugs and nanomedical engineering, lots of disruptive solutions have been advanced for the treatment of intractable diseases, such as anti-tumor nanomedicines [22], nanodrugs for rheumatoid arthritis [23], efficient nanodrugs for nerve or brain diseases by overcoming brain–blood barriers [10, 24], and oral administration nanodrugs for anti-HIV at low dosage [25]. Why does nanomedicine have so many special and powerful functions in disease diagnosis and therapy?

1.2 Key Features of Nanomedicines

There are several critical features of nanomaterials for their translation into special and efficient drugs and therapies. First, as shown in Figure 1.1a,b, most nutrition molecules and key functional molecules in the cell microenvironment range from molecule size to nanoscale (e.g. H_2O, glucose, phospholipid, protein, antibody, antigen, DNA, RNA). The cross-membrane transportation sizes for small molecules are usually less than 10 nm, which are better if less than 6 nm, and the best ones for each component are 2–3 nm or less (Figure 1.1a, the red dotted circle) [26]. Nanoscale materials can be controlled and synthesized by matching to the nanoscale range of the key biological macromolecules (e.g. protein transport pathways, lysosome, centriole, ribosome) very well. Once their surfaces are modified similarly to those biomolecules (e.g. full of –OH ligands, amino acid side groups, glucose, lipids), they can preserve invisibility to the immune system and behave like zymogen during transportation. Second, sizes of organelles and majority of microstructures formed in cell membranes and organelles usually range from 30 nm (e.g. ribosome) to 10 μm (Figure 1.1a, the dotted pink circle, Figure 1.1c). Except that the nonmembrane structured ribosomes are 15–30 nm, the other organelles are generally ranging from 100 nm to 1.0 μm of mitochondria (the pink dotted circle in Figure 1.1a). As for the cells or bacteria for the motion space of nanomaterials and organelles, their sizes range from more than 1 μm for common bacterial to the smaller cells (i.e. red blood cells), and then at most up to less than 1 mm for the human eggs or frog cells (Figure 1.1a, the green dashed circle). The sizes for endocytosis and exocytosis for macromolecules, such as varieties of RNA, DNA, or proteins, range from several nanometers to several hundreds of nanometers or larger [27–29]. Three kinds of membrane transporting channels (i.e. voltage gates, ligand gates, and pressure activation channels) are all in the nanoscale [28, 29]. These size features of organelles and microstructures of cells provide enough free motion space for nanomaterials less than 10 nm and/or their aggregates less than 1.0 μm to exert their functions. As these nanomaterials are less than 10 nm, they can be surface-modified and functionalized easily by conjugating to some biomolecules and organelles, which facilitates their cross-membrane transportation and interaction with some certain organelles, and then targeting certain fine microstructures of organelles. Even they aggregate to several 10 nm or several 100 nm after biomolecule functionalization due to the strong interaction (e.g. coupling, crosslinking, salt bridges) or weak molecule interaction (e.g. van der Waals forces, hydrogen bonds), they can cross-membrane via endocytosis and exocytosis out or into cells and then lysis into nanometer or sub-nanometer effective components by special biomolecules or other cell microenvironment parameters (e.g. lysosomes, pH). Since they can be constructed with much similar surface properties and microstructures as those biomaterials in organisms, they can successfully avoid most of attacks from immunogenicity or autoimmunity, which can last their retention in organisms, leading to their unique enhanced permeability and retention (EPR) effect together with their high permeability [30–32].

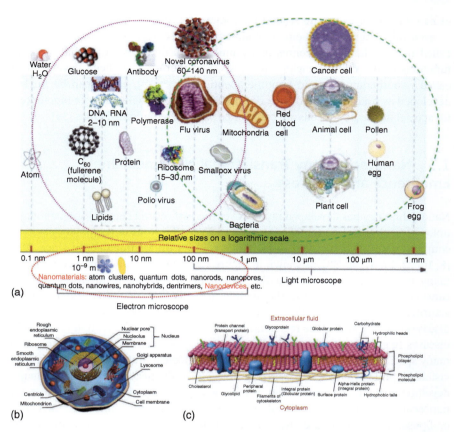

Figure 1.1 (a) Scale comparison of nanometer and some typical biomolecules and cells. Source: Adapted from Beijing Liuzhi Information Technology Co., Ltd./http://www.360doc.com/showweb/0/0/1102300150/last accessed December 28, 2023. The other small illustrations are original; (b) biomolecules and functional microstructures in cell membranes; and (c) organelles and microstructures in one single cell.

Clearly, due to their size effects and flexible surface modification and biomolecule functionalization, they show high biocompatibility and EPR effect as they interact with organs, tissues, cells, and organelles, which also benefit for them to overcome BBB for enhanced drug delivery to special focus in special organs or tissues and cells (e.g. brain or spinal, nerve cells, thrombus [preventing atherosclerosis]) [10, 33–35]. Particularly for those nanoparticles no more than 6 nm, better for 2–3 nm, they preserve much high bioactivity for efficiency-enhanced curative effect for treatment of tumors, cerebrovascular diseases, neural disease, etc. After they finish their bioactivity, they can be cleared via both urinary system and fecal system, which endows them high biosafety [10, 26, 36].

Nanomedicines can be constructed from organics, inorganics, or composites with single components or multi-hierarchy microstructures to realize some targeting functions: detection, and/or diagnosis, and/or therapy. Usually, their core parts can be ranged from 1.0 to 100 nm, which can be assembled into several hundreds of nanometers or even into several micrometers. Broadly, those with nanometers or even micrometers by assembly of subnanometer components can be generally called nanomedicines. Recently, many multi-mode nanomedicines that preserve functions of detection, imaging, diagnosis, and therapy have been developed or called nanotheranostics and nanotherapeutics [8, 37, 38], which can be further developed as smart nanomedicine, or intelligent nanomedicine if some biomarkers for special molecules or cells are conjugated with these nanomedicines [39].

1.3 Nanotechnology Translational Nanomedicine: Emergence and Progress

Nanotechnology has developed rapidly over the past several decades and nanotechnology translational nanomedicine entered into a blowout development stage by coupling with other biomedical technology and artificial intelligence since 2015 [8, 10, 22, 27, 37, 40–58], as shown in Figure 1.2. Up to now, they have constructed many exciting contributions to the treatment of intractable diseases, such as cancers [7, 22, 114, 118], the rheumatoid arthritis [23], or the collagen-induced arthritis [107], nerve or brain diseases [10, 24], anti-HIV [25], cerebrovascular diseases [34, 42], and tissue regeneration (e.g. spinal cord regeneration [128]), skin diseases (e.g. diabetic wound healing) [129–133], as well as precise diagnosis of many special diseases and tracking some key biological processes as ultrasensitive visible bioprobes [83, 98, 134–149]. Figure 1.2 gives the historical timeline of major developments of nanomedicines, revealing a gradually developing process for nanoscale materials translational medicine since the first nanoscale medicine, or liposome nanostructures was published in 1964, which were constructed by phospholipids nanoemulsion used as drug carriers to encapsulate readymade low molecule medicine or drugs not compatible with body liquids [59, 114]. In 1964, silicon polymer of high biocompatibility was also developed into nanocarriers for prolonged drug lasting time [70]. The first organic nanodrug entering the technical level was Gris (Griseofulvin)-PEG (polyethylene glycol) oral tablet contenting submicro griseofulvin particles with ultra-high absorption rate, which was applied and issued in 1970 [150]. Langer and Folkman reported the first polymer nano-system for sustained controlled release of ionic molecules and macromolecules in 1976 [60]. In 1986, the EPR effect of nanoparticles was revealed, which is a unique vascular phenomenon for selective concentration of nanoscale agents in tumor or lesion tissues and can greatly increase the utilization efficiency of drugs [32, 61, 62]. For this goal, drugs with long retention time during circulation was desired. However, some negative effects during the drug circulation, such as destruction of immune response and cellular microenvironment factors (e.g. phagocytosis of macrophages, protein corona), have to be addressed by surface modification and using materials of high biocompatibility, hydrophilicity, and biodegradability [118, 151–153]. Therefore, long-circulating poly(lactic acid)-*co*-poly(ethylene glycol) (PLGA-PEG) copolymers-based drugs (e.g. PLGA-PEG encapsulating RNAi or genes) were developed in 1994 by Langer et al. since PLGA-PEG copolymers

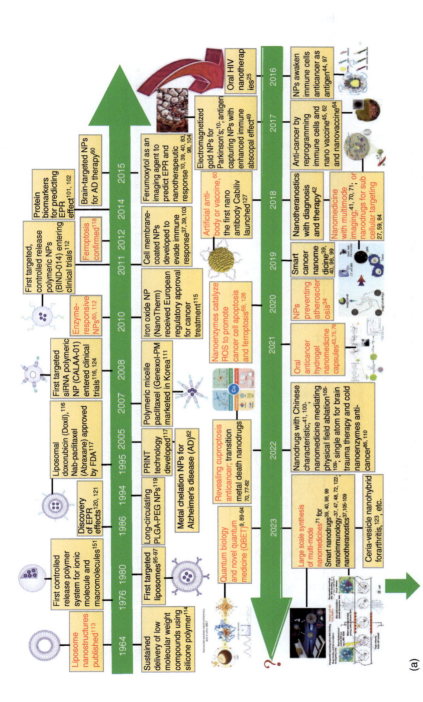

Figure 1.2 (a) Historical timeline of major developments of nanomedicine. EPR, enhanced permeability and retention; FDA, US Food and Drug Administration; nab, nanoparticle albumin-bound; NP, nanoparticle; PLGA-b-PEG: poly(D,L-lactic-co-glycolic acid)-b-poly(ethylene glycol); PRINT, particle replication in nonwetting template; siRNA, small interfering RNA; HIV, human immunodeficiency virus; QBET, quantum biological tunnelling for electron transfer. (Scheme modified from Figure 1 in the key reference [114] and literatures for the main progress published before 2015 and other related figures clipped from key references for the major progresses in nanoenzymes [66], ferroptosis [117], NPs awaken immune cells anti-cancer as antigen [56, 73], the immune abscopal effect [49], reprogramming immune cells [45, 115, 118], artificial anti-body or nanovaccines [46, 55, 116, 119–121], nanoenzymes catalyzing to produce ROS to promote cancer cell apoptosis [95, 122], nanomedicine with multi-mode imaging [41, 89, 98] or targeting and regulation of sub-cellular organelles [27] and tumor microenvironment [41, 101, 123, 124], oral anti-cancer hydrogel-encapsulating nanomedicines [43, 111, 112], cuproptosis anti-cancer mechanism reveal and other T (transition)-metalloptosis therapy confirmed [76, 89–94], smart nanodrug systems confirmed [10, 39, 40, 55, 82, 101, 110, 113, 125–127] and quantum effects in biology confirmed and forming quantum medicine [9, 83–88], etc., montage with permitted copyright.)

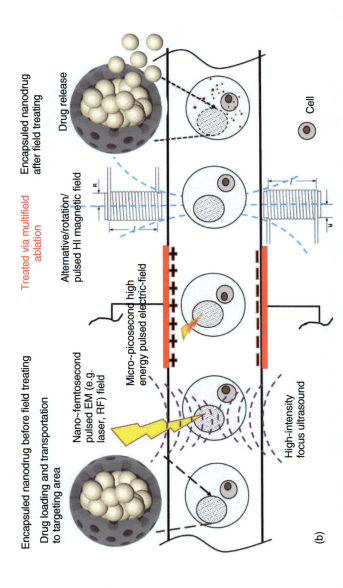

Figure 1.2 (*continued*). (b) The magnified scheme showing nanomedicine mediated multi-physical field (e.g. nanosecond–femtosecond electromagnetic (EM) field, high-intensity focus ultrasound (HIFU), microsecond–picosecond high-energy pulsed electric field, alternative/rotation/pulsed high-intensity magnetic field) ablation on single cells with controlled release and multi-modal therapeutical effects.

are of high biocompatibility and hydrophilicity [74]. However, active targeting is correspondingly of much importance when the tissue accumulation of drugs does not depend on EPR [154] or when the delivery of therapeutic agents requires active transcytosis of physiological barriers such as the intestinal mucosa or the BBB [155–157]. Therefore, many studies focused on the development of active targeting drugs and then the concept of active nanoparticle targeting was introduced in the 1970s [71, 72]. In 1976, some synthesized nanodrugs with active targeting functions made their way into clinical trials [30]. The liposomal doxorubicin (Doxil) nanodrugs of active targeting tumors were approved by FDA in 1995, showing greatly enhanced efficiency in cancer treatment [63, 114]. This is encouraging for the field of cancer nanomedicine. Then NP albumin-bound paclitaxel (nab-paclitaxel; Abraxane) became the second class of nanomedicines for long circulation to be approved by FDA in 2005 [63, 114] and commercialized, such as polymeric micelle paclitaxel (Genexol-PM) was successfully marketed in Korea in 2007 [158]. The nab platform enables formulation of hydrophobic drugs while largely mitigating the need to use toxic excipients, and then the regulatory filing for the approval of Vyxeos was projected in late 2016 [114]. Encouraged by the successful commercialization of Abraxane by addressing the enhanced retention time of drugs by the nanoscale strategy, the first targeted siRNA polymeric NPs (CALAA-01) were approved and entered into clinical trials in 2008 [50, 159]. Up to now, there are many targeting nanodrugs developed with controlled release including typical examples of targeted liposomes (for example, HER2) and single-chain variable fragment (scFv)-targeted liposome (MM-302) [160], the first targeted and controlled-release polymeric NP (BIND-014) [161], and the first targeted siRNA NPs (CALAA01) [50, 114, 162]. Simultaneously, the inorganic nanodrugs contenting 15 nm iron oxide particles were also approved as specific drugs for treatment of anemia in 1974 by FDA, following which many inorganic nanodrugs contenting gold, silver, iron oxides, silicon oxides, and titanium oxide nanoparticles have been gradually approved for clinical study by FDA [26, 114].

Besides the enhanced circulation time in body, the design and synthesis of nanodrugs have to consider how to avoid the immune response before they can be transported into the targeted lesion or cells. Many strategies were invented including the previous methods using materials of high biocompatibility, hydrophilicity, and biodegradability [118, 151]. Based on the direct bionics, cell membrane-coated NPs were further developed to construct cell membrane cloaking nanodrug aggregates by loading nanomedicine into the void cells (e.g. red blood cell) [79, 163, 164]. These cell membrane-coating nanodrug systems can efficiently evade immune response since surface characteristics of these cell membrane-coating nanodrugs are almost the same as those of healthy living cells [79, 163, 164]. During the development of nanodrugs, studies on EPR effects and the related biological mechanism of drugs continue to be paid attention to as their sizes are reduced to nanoscale. For these studies, many multi-mode nanobioprobes, such as protein biomarkers published in 2014 and the imaging agent of ferumoxytol in 2015, have been developed for predicting EPR effects and nanotherapeutic responses, which promotes the progress of nanodrugs with long retention time and active targeting. Together with the gradual realization of multi-functions of nanodrugs, these studies finally led to the emergence of new fields of precise nanomedicine [8, 27], smart nanomedicine [40], and nanotheranostics [37, 42, 165]. According to the therapy effect analysis on nearly 350 kinds of new drugs, contenting nanomaterials submitted for FDA certification from 1970 to 2015 by FDA Drug Evaluation and Research Center in the United States (CDER), the application cases of nanodrugs increased gradually in the past 20 years, among which some have been used to serve people for the treatment of many intractable diseases (e.g. cancers, stroke, Parkinson's, Alzheimer's) [166].

With the progress of nanomedicine and their processing technologies (e.g. synthesis, structures and function characterization, multi-functionalization and clinical practice), the related biological mechanism on their fundamental biomedical effects has been deeply revealed, particularly the discovery of nonapoptotic forms of cell death of iron-based nanoparticles, termed as ferroptosis that can significantly promote death of tumor cells, by Dixon et al. [117]. Ferroptosis is dependent upon intracellular iron, but not other metals, and is morphologically, biochemically, and genetically distinct from apoptosis, necrosis, and autophagy. It has been confirmed that the small molecule ferrostatin-1 is a potent inhibitor of ferroptosis in cancer cells and glutamate-induced cell death in organotypic rat brain slices, suggesting similarities between these two processes like glutamate, erastin inhibits cystine uptake by the cystine/glutamate antiporter, creating a void in the antioxidant defenses of the cell and ultimately leading to iron-dependent, oxidative death. Thus, activation of ferroptosis results in the nonapoptotic destruction of certain cancer cells, whereas inhibition of this process may protect organisms from neurodegeneration for some neurodisease treatments [117]. Since 2016, a research frenzy on ferroptosis in tumor treatment was sparked and many transition metal-based medicines or nanomedicines have been found preserving similar nonapoptotic forms of cell death [94, 122, 167–169].

In 2022, the biological mechanism of copper-induced cell death was successfully revealed by Tsvetkov et al. [94]. Copper is an essential cofactor for all organisms, and yet it becomes toxic if concentrations exceed a threshold maintained by evolutionarily conserved homeostatic mechanisms. In human cells, copper-dependent, regulated cell death is distinct from known death mechanisms and is dependent on mitochondrial respiration. Tsvetkov et al. found that copper-dependent death occurred by means of direct binding of copper to lipoylated components of the tricarboxylic acid (TCA) cycle. This results in lipoylated protein aggregation and subsequent iron–sulfur cluster protein loss, which leads to proteotoxic stress and ultimately cell death. These findings may explain the need for ancient copper homeostatic mechanisms. Cell death is an essential, finely tuned process that is critical for the removal of damaged and superfluous cells. Multiple forms of programmed and nonprogrammed cell death have been identified, including apoptosis, ferroptosis, and necroptosis. Using genetically modified cells and a mouse model of a copper overload disorder, the researchers report that excess copper promotes the aggregation of lipoylated proteins and links mitochondrial metabolism to copper-dependent death. Lipoylation determines sensitivity to copper-induced cell death. It can be proposed that Cu-based medicines preserve great promise in tumor cell treatment if they can be precisely targeted into the tumor cells.

With the investigation of the biomedical function of varieties of transition metal-based nanomedicines (e.g. Fe, Co, Cu, Au, Mn, and their compounds or alloys), their enzyme-like mechanism for disease treatment become more and more clear, which can catalyze many key molecule pathways (e.g. reactive ROS, glutamate) using nanomedicines based on the NPs of the transition metals, which can produce similar nonapoptotic forms of cell death in the cancer treatment as ferroptosis or cuproptosis [41, 94, 117, 170, 171], which can be reasonably termed as transition-metalloptosis [90]. Since almost all these nanomedicines preserve the enzyme-like catalyzing functions, which was further defined as a new research arena: nanoenzymes, recently [41, 58, 66, 95, 96, 104, 108–110, 172–175].

The research field of nanoenzymes has seen exponential growth over the past few years since the term was coined in 2016. This unique modality of cell death, driven by metal-dependent catalysis of some key biological reactions (e.g. ROS, glutamate, phospholipid peroxidation), is regulated by multiple cellular metabolic pathways, including produce of ROS; redox homeostasis; metal handling; mitochondrial activity; and metabolism of key amino acids, lipids, and sugars, in addition to various signaling pathways relevant to disease.

With the gradual reveal of the biomedical effects and their fundamental therapeutic mechanism, and the breakthrough in the controlled preparation methods and the administration technologies of nanomedicine, nanomedicine ushered in a blowout of development since 2016, as shown in Figure 1.2. Particularly, besides nanoenzyme functions [41, 58, 66, 95, 96, 104, 108–110, 172–175], many of them have played key roles in the overcome of multi-drug-resistant (MDR) during cancer treatment [176–180], in the development of innovative immunotherapy by the reveal of their immunological effects for nanoimmunotherapy [27, 37, 40, 47, 89, 110, 114, 118, 124, 168, 181–183].

MDR is a frequently encountered thorny issue as using traditional or even some innovative drugs to treat many diseases, particularly for some persistent infectious diseases and difficult miscellaneous diseases (e.g. cancers), which impedes the successful treatment of targeting diseases [176–180]. Developing novel long-circulating, self-assembled core–shell nanoscale coordination polymer (NCP) nanoparticles that efficiently deliver multiple therapeutics with different mechanisms of action to enhance synergistic therapeutic effects is an innovative strategy to overcome this multi-drug-resistant issue [179]. For example, Lin et al. invented NCPs code liver chemotherapeutics and siRNAs to eradicate tumors of cisplatin-resistant ovarian cancer in 2016 [178]. These NCP particles contain high payloads of chemotherapeutics cisplatin or cisplatin plus gemcitabine in the core and pooled siRNAs that target MDR genes in the shell. The NCP particles possess efficient endosomal escape via a novel carbon dioxide release mechanism without compromising the neutral surface charge required for long blood circulation and effectively downregulate MDR gene expression in vivo to enhance chemotherapeutic efficacy by several orders of magnitude. By silencing MDR genes in tumors, self-assembled core–shell nanoparticles suggest a more effective chemotherapeutic treatment for many challenging cancers [178].

1.4 Interdisciplinary Features of Nanomedicines: Multi-mode and Multi-function Features Promoting Nanomedicine-mediated Immunotherapy and/or Physical Field Ablation Therapy for Subversive Therapy

Tumor immunotherapy has become one of the key innovative methods in tumor treatment and many immunotherapy drugs (e.g. PD-1, PD-L1, CTLA-4) and therapy. (e.g. cart-T) have been developed recently. However, existing cancer immunotherapy drugs work in only 20%–30% of patients [44, 73], particularly for those patients with solid tumor. In some cases, even when the checkpoint molecules are blocked, there are too few active T cells around to sound the immune alarm, says Jedd Wolchok, a cancer immunotherapy expert at the Memorial Sloan Kettering Cancer Center in New York City [73]. Additionally, the key reason is that tumors do not display enough of the T cell's targets, so-called tumor antigens, on their surface. However, nanoparticles and their functionalized species can behave similar to antigens as they enter into bodies [44, 73, 184]. The immunological mechanism of nanomedicine has been studied intensively for the treatment of tumors in the past decade. Results indicate that nanomedicines preserve intensive immune abscopal effects and have great potential as artificial antigens to activate and train immune cells [185], and they can even reprogram cancer cells or immune cells to reshape the tumor immune microenvironment [37, 46–49, 51, 184, 186, 187]. It is said that tumor cells are usually produced every day in our body due to a variety of causes, which will not lead to cancer if they can die through their routine apoptosis themselves or be cleaned by our immune system [47, 48, 113, 188–192]. However, tumor cells are smart and can escape from our immune system since they can disguise themselves by releasing some signals or chemicals to let immune cells confirm that they are healthy cells, and then suppress the secretion function of immune cells not to release the corresponding cytokines killing cancer cells [47, 48, 54, 113, 189, 190, 193]. The immunological abscopal effect of nanomedicines was revealed in 2017 by Min et al. [49]. One function of nanomedicines on the immune system is to activate or awaken immune cells, called an immune agonist [44, 73, 178, 194], such as using the paclitaxel nanoparticles to awaken immune system to fight against cancer studied by Tang et al. [194]. In 2016, Lin et al. developed self-assembled core–shell NCP nanoparticles that efficiently delivered multiple therapeutics with different mechanisms of action to enhance synergistic therapeutic effects for ovarian cancer treatment [44]. These NCP NPs contained high payloads of chemotherapeutics cisplatin or cisplatin plus gemcitabine in the core and pooled siRNAs that target multi-drug-resistant (MDR) genes in the shell [44].

Some physical therapies, such as radiation therapy [44, 73] and electrical field therapy [13–15, 195, 196], can break tumor cells to expose some antigens. Therefore, the self-immune systems of some patients can be activated to produce immune response effects similar to immunotherapy after some physical therapy [15, 44, 178, 197, 198]. Based on this phenomenon, Lin et al. from the University of Chicago invented photosensitive ultra-small nanodrugs for tumor treatment, which can ignite the immune responses for some tumors insensitive to immunotherapy by coupling them with radiation therapy [44, 73]. The recent progress suggests that these nanodrug-mediating immunotherapies and/or physical field treatments are expected to enter into clinical trials, which have become the best partner in the field of immunotherapy [15, 37, 44, 73, 178, 197–199].

Another immunological function of nanomedicines was to reprogram immune cells to recognize cancer cells advanced in 2018 by Roth et al. [115] and Yang et al. [200], such as reprogramming the function and specificity of human T cells with nonviral genome targeting for anti-cancer therapy [115]. This strategy has been modified for the development of Parkinson's disease therapy using nanoparticles, such as electromagnetized gold nanoparticles mediating direct lineage reprogramming into induced dopamine neurons for the treatment of Parkinson's [10].

At the same time, some artificial antibodies (vaccines) have been developing, with active targeting functions [46]. In 2018, Cao and Wang developed a conformational engineering method to create an NP-based artificial antibody, denoted "Goldbody," through conformational reconstruction of the complementary-determining regions (CDRs) of natural antibodies on gold NPs (AuNPs) [46]. Upon anchoring both terminals of the free CDR loops on AuNPs, the "active" conformation of the CDR loops can be reconstructed by tuning the span between the two terminals, endowing these inorganic NPs the original specificity. Two Goldbodies have been created by this strategy to specifically bind with hen egg white lysozyme and epidermal growth factor receptor, with apparent affinities several orders of magnitude stronger than that of the original natural antibodies. As a result, it is possible to create protein-like functions on these much more stable metallic NPs in a protein-like way, namely by tuning flexible surface groups to the correct conformation, which will finally build up a category of Goldbodies that can target different antigens and thus be used as substitute for natural antibodies in various applications [46]. The first approved anti-body nanodrug Cabiliv (Caplacizumab-Yhdp) was approved by the European Union and then commercially launched for clinical use in 2018, which became the first nano anti-body drug for the treatment of allergic purpura (Henoch-Schonlein syndrome, HSS) in the world. (https://www.sohu.com/a/252250535_119250; https://www.vodjk.com/news/180904/1503187.shtml: Cabiliv™ [caplacizumab] approved in Europe for adults with acquired thrombotic thrombocytopenic purpura [aTTP]; Sanofi gets EU OK for Ablynx flagship drug Cabiliv). This drug was then approved by FDA in the United States in February 2019 (https://www.sohu.com/a/252250535_119250; https://www.drugs.com/history/cablivi.html). Since then, nanomedicines not only with more intimate to immunological effects but also with multiple-therapy functions and multi-targeting functions (e.g. cancer cells, organelles, or tumor microenvironment) have been developed forming a novel field of nano-immunotherapy up to now [47, 48, 51, 57, 101, 118, 182–184]. If readers need more details on the immunological effects of nanomedicines, Chapter 10 in this monograph can be referred to.

Simultaneously, coupling of nanomedicines to some advanced biophysical or biochemical methods for subversive combined therapies has been on the way for difficult miscellaneous diseases [27, 37, 40–42, 47, 58, 89, 110, 114, 118, 124, 165, 168, 181–183, 201–203]. Besides their biomedical functions (e.g. immunological effects, enzyme-like functions), nanomedicines constructed by nanohybrids conjugating to varieties of biochemical drugs or preparation, preserve unique physicochemical characteristics and can simultaneously interact with several physical fields (e.g. electrical field, magnetic field, optical field, ultrasound field, electric–magnetic field) [37, 46–48, 51, 186, 187]. Multi-mode therapy and diagnosis based on these multi-functional nanomedicine-mediated physical field ablation and/or the corresponding molecule imaging and bioprobe function have been developed recently into one novel medical methodology with both diagnosis and therapy functions, or nanotheranostics [37, 42]. Particularly, nanotheranostics as these multi-mode nanomedicines coupling with physical field ablation preserve greatly enhanced therapeutical effects and in situ precise diagnosis functions for treatment of varieties of intractable diseases by comparing with their counterparts (either physical field ablation or solely nanomedicine), many of which have been implemented in the clinical trial [37].

Besides the enhanced apparent therapeutic effect due to the synergistic effects among varieties of physical fields and nanomedicines, another amazing achievement of nanomedicine mediated physical field ablation in their biomedical application is the dramatically improved immunological effects that are far beyond that one single physical field or nanodrug can do. Table 1.1 summarizes the multi-mode of action, advantages and disadvantages, immunological effects, and clinical progress of multi-modal ablation by multiple physical field coupling.

Currently, there are five typical physical field coupling modes in the road of clinical trials as follows. (i) The coupling of high-intensity focused ultrasound (HIFU), with light irradiation (PR) or fluorescence excitation (CLE) produces dual mode and multi-modal therapeutic effects, such as tissue tearing, cell fragmentation, sonodynamic therapy (SDT), and photochemical kinetic ablation (PDT), leading to more active molecules (e.g. reactive oxygen species [ROS]) and high-temperature thermal effects [204–206]. These effects promote the release of tumor antigens to activate immune cells, induce inflammatory reactions, and enhance immune memory of related immune cells. To date, this kind of HIFU coupling to PP or CLE technology is still in preclinical stage. (ii) The coupling of HIFU and electrical field (EFT) produces dual mode or multi-modal effects, such as tissue tearing and cell fragmentation caused by ultrasound, and electrochemical (EDT) or electrostatic field polarization (ESFP) by electrical fields, and pore formation and polarization by nsPEF for cell internalization of drugs and surface electrical dipole regulation of sub-cellular structures [207], leading to ROS formation, reversible perforation, and electrical dipole interaction with organelles at the related microregions or biomolecules and their functional groups [208]. This kind of combined therapy by coupling HIFU and EFT preserves multi-immunological effects, such as immune cell activation, enhanced anti-tumor immune response, immune related gene expression regulation (pro-inflammatory and anti-inflammatory factors),

Table 1.1 Working modes of the currently developed multi-modal ablation therapy based on multi-physics coupling and their main features and key challenges, immunological effects, clinical progress, and main affiliation.

Modes of multi-physical field coupling	Energy/Momentum mode and/or biochemical reaction activation mode (modality)	Multi-functional modes at work	Main features	Key challenges	Immunological effects and mechanisms	Affiliation (company or institute)	Clinical stage	References
High-intensity focused ultrasound (HIFU), light irradiation (PR)/ fluorescence excitation (CL): dual mode multi-modal	Tissue tearing, cell fragmentation, sonochemical therapy (SDT), photochemical kinetic ablation (PDT)	Reactive oxygen species (ROS), high-temperature thermal effects	High tissue penetration, spatiotemporal control, and synergistic effects	Cause thermal damage to people	Promote the release of tumor antigens, activate immune cells, induce inflammatory reactions, and enhance immune memory	Ruiya biotechnology, Canada Covidien, United States	Preclinical	[204–206]
High-intensity focused ultrasound (HIFU) plus electrical field (EFT) or electrochemical field (EDT): dual mode multi-modal	Tissue tearing and cell fragmentation caused by ultrasound, electrochemical or electrostatic field polarization, nanosecond pulsed electric field (nsPEF) pore formation and polarization	ROS, reversible perforation, and interaction with organelles or biomolecular microregions and functional groups	High tissue penetration, oxygen independence, and synergistic effects	Thermal damage to human body, embedded electrode	Immune cell activation, enhanced anti-tumor immune response, immune related gene expression regulation (pro-inflammatory and anti-inflammatory factors), angiogenesis inhibition, and immunosuppressive cell regulation	MIT and SonaCare Medical Co, Shenzhen Maiwei Medical Company, University of Science and Technology Beijing	Preclinical, Clinical Phase I, II	[207, 208]

(Continued)

Table 1.1 (Continued)

Modes of multi-physical field coupling	Energy/Momentum mode and/or biochemical reaction activation mode (modality)	Multi-functional modes at work	Main features	Key challenges	Immunological effects and mechanisms	Affiliation (company or institute)	Clinical stage	References
Photoenergy irradiation (PR)/ chemical fluorescence excitation (CLE) plus electric field (EFT): dual mode multi-modal	Photodynamic therapy (PDT), electrochemical therapy (EDT), or electrostatic field polarization	ROS, high-temperature thermal effects, etc.	Temporal and spatial control, oxygen independence, and synergistic effects	Thermal damage to human body, embedded electrode	Activation and proliferation of immune cells and immune regulatory cells, enhancement of anti-tumor immune response and expression of immune related factors (pro-inflammatory and anti-inflammatory)	Griffin Adam holiday, Norway Photocure ASA, Hangzhou Ruidi Biotechnology Co., Ltd	Preclinical	[209, 210]
Electromagnetic field/magnetic field plus temperature field ablation: dual/triple mode multi-mode	Radiofrequency ablation (RFA)/ alternating magnetic field heat plus cryoablation (CryoA) cold plus radiofrequency chemical dynamic ablation	High temperature thermal effect, low temperature freezing effect, cell rupture	High tissue penetration, targeted therapy, noninvasive, synergistic effects	Depth limitation, uneven heat distribution, damage to people	Immune cell activation and enhancement (tumor-specific T cells and NK cells), anti-tumor immune response enhancement, immune-related gene expression regulation, immune regulatory cell regulation	Medtronic, United States; Immodulon Therapeutics Ltd, United Kingdom; Shanghai Meijie Medical Technology Co., Ltd.; Beijing Haijieya Medical Device Co., Ltd, Magforce AG, Germany; Lodespin Labs, Canada	Preclinical, Clinical Phase I, II	[211–214]

Electrical field plus radio-frequency electromagnetic field plus plasma physics: dual/three-mode multi-mode	Radio frequency ablation (RFA) plus electro-chemical therapy (EDT) plus electrostatic field polarization or pulsed electric field (PEF) ablation, or plus low-temperature plasma (CAP)	ROS, thermal effect, alternating electrical or electrostatic field, and pulse electric field polarization and activation, plasma activation	Targeted therapy, synergistic effects, noninvasive, and low toxicity side effects	Improved accuracy, parameter and dose optimization, tumor type free	Stimulate the activation and proliferation of immune cells (natural killer cells [NK cells] and cytotoxic T lymphocytes [CTLs]), inflammatory response, tumor microenvironment regulation, and promote tumor antigen expression	Oncosec Medical Incorporated, Korea Adtec Healthcare	Preclinical, Clinical Phase I, II	[215]

Source: Son et al. [217]/Royal Society of Chemistry.

angiogenesis inhibition and immunosuppressive cell regulation. Most of these therapies are in preclinical stage and some of them have been into clinical Phase I or II stage. (iii) The coupling of photo irradiation (PR) and/or chemical fluorescence excitation (CL) with EFT can also produce multi-modal therapeutic effects, such as photodynamic effect therapy (PDT) mode, electrochemical reaction therapy (EDT) mode and ESFP mode, which can ignite activation and proliferation of immune cells and immune regulatory cells, can enhance anti-tumor immune response and the related expression of immune-related factors (e.g. pro-inflammatory and anti-inflammatory) [209, 210]. This kind of multi-mode therapy is currently in preclinical stage. (iv) The coupling of EM field and/or magnetic field with temperature field ablation, which can produce the combination of dual or triple mode therapy, such as the combination of two or more ablation modes of radiofrequency ablation (RFA), alternating magnetic field heat, cryoablation (CryoA) cold ablation, and radiofrequency chemical dynamic ablation [211–214]. These multi-mode therapies can preserve the below immunological effects: immune cell activation and function enhancement (tumor-specific T cells and NK cells), enhanced anti-tumor immune response, and regulation of immune-related gene expression and immune regulatory cell. (v) The coupling of electrical field ablation, radiofrequency EM field ablation and plasma ablation, which can result in the multi-modal combination of RFA, EDT, ESFP, PEF, and low-temperature plasma (CAP). Studies on these multi-modal combinations indicate that they can activate the immune cells, stimulate the proliferation of immune cells (e.g. natural killer cells [NK cells] and cytotoxic T lymphocytes [CTLs]), promote tumor antigen expression, enhance the inflammatory response by recruiting more immune cells into lesions, and regulate the tumor microenvironment. Generally, most research suggests that multi-physical field coupling ablation can produce much more improved therapeutical effects, particularly in the activation of immune cells, enhancement of immune responses and the regulation of immune system, more than one single physical field mode due to their multi-field synergistic effects [215]. However, recent progress suggests that these immunological effects are generally not as great as desired when they are entering into the clinical trial stages, and most of these multi-modal therapies by the combination of different physical fields need agonists or multi-mode responsive nanomedicines to amplify their immunological effects and the final therapeutical effects or overcome some shortcomings [11, 37, 38, 42, 165, 207].

Typical cases of nanomedicines mediated physical-field ablation therapies were summarized in Table 1.2, which shows a brief description of their working modes, therapeutical effects, and anti-tumor applications. These multi-mode nanomedicines could also behave like agonists or amplifiers for their immunological effect by improving tri-interactions among medicines, physical fields, and lesions.

There are generally about nine types of multi-mode therapies based on nanomedicines mediated physical field ablation. (i) Nanomedicines (e.g. liposome nanodrugs) mediated HIFU by Indian Sun Pharmaceutical and Taiwan Toyo Pharmaceutical can result in tissue tearing and cell fragmentation by sonodynamic therapy (SDT) that increase the nanodrug cell internalization, utilization, and safety via overcoming BBB, which has entered into the clinical phases I and

Table 1.2 Working modes, therapeutical characteristics, applications, representative institutions, and clinical progress of nanomedicine-mediated physical field ablation therapies.

Physical ablation mode		Energy/Momentum or biochemical activity or working mode	Nano drug carrier	Conjugating drugs	Imaging mode	Therapeutical characteristics	Application fields	Representative institutions	Clinical stage	References
High-intensity focused ultrasound (HIFU)		Tissue tearing and cell fragmentation Sonochemical therapy (SDT)	Liposomes	Small molecules–macromolecules–nanomedicines	Ultrasonic	Enhances delivery safety and efficiency of drug utilization administration	Blood brain barrier, etc.	Indian sun pharmaceutical	Phase I/II	[217–221]
				Drug vesicles	Ultrasonic	Trigger drug release	Blood brain barrier, etc.	Taiwan Toyo Pharmaceutical	Phase I/II	[222–224]
Traditional electro-magnetic field	RF ablation (RFA) RF-CDT	High frequency electromagnetic wave	Polymeric nanoparticle	Biomacro-molecules	—	Greatly inhibiting tumor recurrence and preventing tumor metastasis	Breast cancer, etc.	Soochow University; Macau University of Science and Technology	Phase I/II	[105]
	Microwave ablation (MWT)	Microwave hyperthermia (MTT) Microwave-dynamic (MDT)	Liposomes	Small molecule drugs	—	Enhance ablation effect and prevent recurrence	Liver cancer, etc.	The Fifth Medical Center of Chinese People's Liberation Army General Hospital	Preclinical	[102]

(Continued)

Table 1.2 (Continued)

Physical ablation mode	Energy/Momentum or biochemical activity or working mode	Nano drug carrier	Conjugating drugs	Imaging mode	Therapeutical characteristics	Application fields	Representative institutions	Clinical stage	References
New electromagnetic field	Photo irradiation (PR) or CLE, pulsed laser photogenesis, photoshock								
	Photochemical photodynamic therapy (PDT)	Polymer, core-shell metal NPs	Macromolecular/small molecule drugs, porphyrins/saponins	Optical	Precision treatment	Breast cancer, liver cancer, etc.	Fudan Zhangjiang Bio-pharmaceuticals	Clinical approval, preclinical	[225–230]
	Photothermal ablation (PTA)	Inorganic metal NPs	Iron oxide nanomedicine	MRI	The curative effect is clear, and combined with other ways to enhance its curative effect	Prostate-cancer, glioblastoma, etc.	AMAG Pharmaceuticals; Advanced Magnetics, Inc	Phase I/II	[231–236]
	Chemiluminescence biochemical reaction excitation (CL PDT)	Inorganic NPs, liposomes, etc.	Photosensitizers, small molecule drugs		High tumor targeting and high efficiency	Lung cancer, etc.	Seoul National University; Shanghai Institute of Silicate, Chinese Academy of Sciences	Preclinical	[237–239]
X ray	X-ray photochemical dynamics (PDT)	Inorganic metal nanoparticles	Sustained luminescent nanoparticles	–	Low dose X-ray can activate	Breast cancer	Memorial Sloan Kettering Cancer Center; Fuzhou University	Preclinical	[240, 241]
	Radionuclide mediating photochemical dynamic therapy (CR PDT)	Polymers, inorganic NPs	Photosensitizer	–	Expanding the application scope of radiation therapy	Breast cancer, liver cancer, etc.	China Pharmaceutical University	Preclinical	[80, 237]

Category	Method	Mechanism	Material type	Specific materials	Imaging	Advantages	Cancer types	Institution/Company	Stage	References
	High energy radiation therapy (HERT)	Isotope decay emits high energy γ or β Isoray	Inorganic metal NPs	Au, Bi, Gd, Cs^{131} complexes	MRI, CT, nuclear imaging	Enhance the curative effect of radiotherapy	Liver cancer, pancreatic cancer, esophageal cancer, nonsmall cell lung cancer, etc.	Johnson&Johnson's subsidiary Janssen Pharmaceuticals, nanobiotix, IsoRay company	Preclinical, Phase I/II	[238, 242–245]
Electrical field	Conventional electric field	Electrochemical (EDT) or electrostatic field polarization	Inorganic metal nanoparticles	Pt nanoparticles, etc.	—	Minimal invasion and uniform ablation covering the relatively large tumors wholly	Breast cancer, etc.	Zhejiang University	Preclinical	[199]
		Millisecond to microsecond pulsed electric field (ms-μSPEF)	Liposomes	Small molecule drugs	—	Improving the transport of nanomedicines	Colon cancer, etc.	Wroclaw Medical University	Phase I/II/III	[246]
	New electrical field ablation	Nanosecond to picosecond pulsed electrical field ablation (ns-psPEF)	Inorganic NPs multi-mode nanomedicine	Small molecule drugs	MRI, CT	Improve drug delivery while improving curative effect and preventing recurrence	Liver cancers, etc.	AngioDynamics; Shanghai Ruidao Medical; Hangzhou Ruidi Biotechnology Parade Company; USTB	Preclinical	[247]
Magnetic field (alternating or pulsed amt/pmt)		Magnetocaloric (MHT) or magnetodynamics	Inorganic metal nanoparticles	Iron oxide nanoparticles	MRI	High tissue permeability	Nonsmall cell lung cancer, breast cancer	Xiamen University, National University of Singapore	Preclinical	[12]

RF, radiofrequency; CDT: chemodynamic therapy; CLE, chemical luminescent excitation; MRI, magnetic resonance imaging; CT, computed tomography; USTB, University of Science and Technology Beijing (USTB).
Source: Irvine and Dane [47] and Gawne et al. [37].

II [217–224]. (ii) Nanomedicines (e.g. some macromolecule or small molecule nanodrugs encapsulated by polymeric or liposome micelles) mediated traditional EM fields (e.g. RF ablation [RFA]: RF-CDT; microwave ablation [MWT]) can produce RF wave-induced chem-dynamic reaction and/or hyperthermia effects, which can greatly increase the ablation effects and prevent tumor recurrence or metastasis [102, 105]. Both Soochow University; Macau University of Science and Technology, and the fifth medical center of the Chinese People's Liberation Army General Hospital in China have invested in this method, and have paved this combination therapy into the clinical phase I or II in the treatment of breast cancers and the preclinical stage for liver cancers [102, 105]. (iii) There are many new combined therapies developed by new types of nanomedicines mediated new EM field ablation. One of these new therapies is polymeric or metal-based core–shell nanoparticles mediated photoirradiation (photochemical or photodynamics) therapy (PDT), photothermal ablation (PTA), chemiluminescence biochemical reaction excitation (CLE), and pulsed laser photogenesis or photoshock ablation, developed by Fudan Zhangjiang Biopharmaceuticals, AMAG Pharmaceuticals, Advanced Magnetics Inc, Seoul National University, Shanghai Institute of Silicate, Chinese Academy of Sciences, and so on [225–239]. Among them, iron oxide nanomedicines mediated PTA has been in the clinical phase I or II in the treatment of prostate cancer and glioblastoma, showing enhanced curative effects [234, 235]. The second combination is the inorganic or polymeric nanomedicines with sustained luminescence or photosensitivity mediated X-ray radiation for enhanced X-ray photochemical dynamics (PDT) or the radionuclide mediated photochemical dynamic therapy (CR PDT), which can greatly reduce the essential activation dosage of X-ray [240, 241] and/or expand the application scope of X-ray radiation therapy [80, 237]. X-ray radiation-based therapies have been used in the preclinical treatment of breast cancers and/or liver cancers by Memorial Sloan Kettering Cancer Center; Fuzhou University [240, 241] and China Pharmaceutical University [80, 237]. The third combination is the inorganic metal NPs (e.g. Au, Bi, Gd complexes or Cs^{131} like isotope NPs that also preserve MRI, CT, and nuclear resonance multi-mode imaging functions) mediated high energy radiation therapy (HERT: e.g. γ-rays or β-rays emitted by isotopes' decay), which have been in preclinical stage or in the clinical phase I/II stage for the treatment of liver cancer, pancreatic cancer, esophageal cancer, nonsmall cell lung cancer, etc., developed by Johnson&Johnson's subsidiary Janssen Pharmaceuticals, Nanobiotix Company, and Isoray Company [238, 242–245]. (iv) The inorganic NPs, liposome encapsulating small molecules, or inorganic and organic complicated NPs mediated electrical field ablation therapies have been developed from the conventional electrochemical (EDT), ESFP or high-energy millisecond to microsecond PEF (ms–μsPEF)-based therapies to ultrafast high-energy PEF (ns–psPEF). Among them, nanomedicines mediated ms–μsPEF therapies have been developed by Wroclaw Medical University entering into the clinical phase I/II/III for the treatment of colon cancers, which can improve the liposome-encapsulating small molecule nanodrugs [246]. Since ns–psPEF ablation therapies have no obvious thermal effect but significant electrical polarization effect, dramatically enhanced cell internalization and lesion

1.4 Interdisciplinary Features of Nanomedicines

penetration depth of drugs, their combination therapies have been paid much more attention rapidly recently, which have been used in the preclinical study for the treatment of varieties of cancers (e.g. liver cancers) by many company and institutes, such as AngioDynamics in the United States, Shanghai Ruidao Medical Company in China, Hangzhou Ruidi Biotechnology Company in China, Parade Company in China, and University of Science and Technology Beijing in China [207, 247]. (v) The inorganic NPs (e.g. iron oxide NPs) mediated magnetic fields (alternating or pulsed magnetic field) therapies can result in high tissue permeability of drugs due to the magnetocaloric (MHT) or magnetodynamic effects, which have been used in the treatment of nonsmall cell lung cancer and breast cancer by Xiamen University, National University of Singapore and currently in the preclinical stage [12].

Most of these multi-modal therapies by nanomedicines mediated a certain physical field ablation therapy are still in preclinical stages and some of them have been in clinical trials funded by some companies or institutes. Some of them have also shown greatly enhanced immunological effects during their treatment of intractable diseases (e.g. tumors) by comparing with the counterparts of the pure field ablation and/or only nanomedicines. Table 1.3 summarizes ten typical achievements obtained by the main leading institutes or companies in the physical field ablation and/or nanomedicine mediated physical field ablation to date, including types of multi-mode, their therapeutical characteristics and immunological effects, and their current clinical stages. Institutes or companies mainly include Covidien Company [72, 205–207, 217], Massachusetts Institute of Technology [209, 217] and Griffin Adam Holiday Company [88, 210, 211, 217] from the United States; King Saud University from Saudi Arabia [211–214, 216, 251, 252]; Immodulon Therapeutics Ltd. from the United Kingdom [88, 212–215, 217]; Shenzhen Maiwei Medical Company [254, 255], Hangzhou Ruidi Biotechnology Co. Ltd. (HZRD) [256–261], Shanghai Meijie Medical Technology Co. Ltd. [212, 213, 262], USTB and Zhenzhou Tianzhao Biomedical Company [41, 71, 89, 108, 111], and Peking University [268, 269] from China. Types of multi-mode are mostly focusing on the coupling effects from HIFU, electric fields, radiation by varieties of EM waves with broad wavelengths, radiofrequency radiation, temperature gradient induced thermal ablation and nanomedicine enhanced field ablation effects. All these combined therapies based on nanomedicines mediated multi-mode field ablation exhibit significant immunological effects (e.g. inducing inflammatory reactions and promoting the release of tumor antigens to recruit more immune cells and kill cancer cells; activating immune cells and immune regulatory cells; intensifying immune memory; enhancing anti-tumor immune response, and improving expression of related immune factors and immune genes) and the synergistic therapeutical effects (e.g. enhanced ablation effects to kill tumor cells; high penetration depth into lesion and cell internationalization of drugs to enhance the utilization of drugs; intensified dissolution and apoptosis of tumor cells). Particularly, the combined therapies by nanomedicines mediated ns–psPEF developed by USTB and HZRD have shown subversive therapeutic effects using HCC as the pathological model, which can co-stimulate the activation and proliferation of immune cells and promote the inflammatory response and expression of anti-tumor antigen [41, 89, 108, 123, 207, 270]. Based on the nanomedicine mediated ns–psPEF and high spatiotemporal resolution magneto-optic detection system, Song's group from USTB currently focuses on the development of series of nanomedicine mediated multi-physical-field ablation instruments by coupling rotating or pulsed magnetic fields, laser irradiation and pulsed ultrafast laser, HIFU and ultrafast high intensity PEF [271]. As investigation of the therapeutical effects via multi-physics coupling to nanomedicines using this edge tool, the synergistic immunological effects under different combination modes between multi-physics and multi-mode nanomedicines will be investigated systematically and comprehensively, such as activation of immune cells and immune regulatory cells, enhanced immune memory, reprogram of tumor immune microenvironment, regulation of immune factors and immune gene and their expression-related cytokines and chemokines, as well as the corresponding cell signal and biomolecule pathways.

It can be deduced that the future developed smart nanomedicines with multi-mode imaging, self-targeting transportation and controlled release can not only be used both in visible molecule imaging for the disease precise diagnosis (e.g. tumor classification and phase confirmation) and lesion localization but also their lesion tissue penetration depth and cell internalization can be improved dramatically by the empowerment from the coupling to the corresponding multi-physical fields, leading to smart nanotheranostics [37, 41, 47, 89, 94, 104, 108, 110, 123, 170]. Particularly, for the treatment of cancers, the synergistic effects between the nanomedicines and physical field ablation can endow the corresponding therapy with more flexible intelligent regulation ability on the TIME [127], and consequently, the elimination of cancers and prevention of their recurrence and metastasis will be possibly realized [104, 109, 272, 273].

If readers need more details on the nanomedicines mediated physical field ablation therapy, Chapters 11–15 ("Nanomedicine medicating ultrasound therapy," "Nanomedicine mediated photodynamic and/or photothermal therapy," "Nanomedicine mediated pulsed electric field therapy," "Nanomedicine mediated magneto-dynamic and/or magneto-thermal therapy," "Nanomedicine mediated radiofrequency or nuclear radiation therapy") in this monograph can be referred to.

In addition, there are lots of molecule-like features or discrete energy and momentum emerging in some nanomaterials with sizes less than 2 nm (i.e. superatom cluster with atom numbers from one to no more than several hundreds, such as $Au_{25}L_{18}$, $Ag_{44}L_{30}$, etc., where "L" denotes the ligand, such as thiolate) [274–277]. As they are translated into drugs, they can be called superatom cluster drugs or quantum drugs, which can also include the assemble of these superatom cluster drugs or even some assembly of single atom catalysts of special biological functions (e.g. molecule surgery) [53, 98, 174, 274, 277]. Together with many quantum effects that have also been found in many molecule signal pathways of biological processes (e.g. photosynthesis of chlorophyll) [9, 85, 88, 278]. These superatom cluster drugs will not only promote the progress of nanoenzymes and nanoimmunotherapy but also lead to a new arena in subversive medicines: quantum medicines or quantum drugs (Figure 1.2: one of the major breakthroughs or progresses in 2023) [9, 83–88], for the treatment of many complicated diseases precisely and efficiently, which

Table 1.3 Achievements of multiple physical field ablation and/or multi-mode nanomedicines mediated physical field ablation and the related instrumentation and clinical stages, obtained by 10 currently leading institutes or companies.

Institutes/Company	Multi-modal therapy via multi-physical field coupling	Typical therapeutical characteristics	Immunological effects	Clinical stage	References
Covidien, United States	High-intensity focused ultrasound, light irradiation/fluorescence excitation	High organizational penetration, spatiotemporal control, synergistic effects, and ROS generation	Promote the release of tumor antigens, activate immune cells, induce inflammatory reactions, and enhance immune memory	Preclinical, Clinical Phase I, II	[204–206, 216, 248]
Massachusetts Inst of Technology, United States	High-intensity focused ultrasound, electric field heating/polarization/perforation/molecular microstructure	High tissue penetration, oxygen independence, and synergistic effects	Dissolution and apoptosis of tumor cells, activation of immune cells, induction of immune memory, and regulation of immune suppression	Preclinical, Clinical Phase I, II	[208, 216, 249]
Griffin Adam Holiday, United States	Light energy irradiation/chemical fluorescence excitation, electric field heating/polarization/perforation/molecular microstructure	Spatiotemporal resolution control, oxygen independence, synergistic effects	Activation and proliferation of immune cells and immune regulatory cells, enhanced anti-tumor immune response, and improved expression of related immune factors	Preclinical	[209, 210, 216, 250]
King Saud University, Saudi Arabia	Nanomagnetic induction hyperthermia; magnetothermal/magnetodynamics; radiotherapy and chemotherapy	Enhancing sensitivity to radiotherapy or chemotherapy	Tissue penetration, targeted therapy, noninvasive, with synergistic effects of nanomedicines	Preclinical, Clinical Phase I, II	[211–214, 216, 251, 252]
Immodulon Therapeutics Ltd., United Kingdom	Low-temperature freezing, radio frequency high-temperature thermal/biochemical reaction activation	High temperature thermal effect, low temperature freezing effect, and cell rupture	Multiple immune cell activation and enhancement, enhanced anti-tumor immune response, and regulation of immune-related gene expression	Preclinical, Clinical Phase I, II	[211–214, 216, 253]

(Continued)

Table 1.3 (Continued)

Institutes/Company	Multi-modal therapy via multi-physical field coupling	Typical therapeutical characteristics	Immunological effects	Clinical stage	References
Shenzhen Maiwei Medical Company	Low-temperature freezing, pulsed electric field heating/polarization/perforation/molecular microstructure, RF high-temperature thermal/biochemical reaction activation	Efficient, minimally invasive, precise lesion localization, multi-modal ablation	Antigen release, immune cell activation, inflammatory response, immune memory	Preclinical, clinical phase I, II	[254, 255]
Hangzhou Ruidi Biotechnology Co., Ltd	Low temperature freezing, pulsed electric field heating/polarization/perforation/molecular microstructure, RF high-temperature thermal/biochemical reaction activation	Efficient, minimally invasive, precise lesion localization, and no heat sink	Stimulation of immune cell activation and proliferation, inflammatory response, regulation of tumor microenvironment, and promotion of tumor antigen expression	Preclinical, Clinical Phase I, II	[256–261]
Shanghai Meijie Medical Technology Co., Ltd	Low temperature freezing, radio frequency high temperature thermal/biochemical reaction activation	Accurate treatment of lesions	Multiple immune cell activation and enhancement, enhanced anti-tumor immune response	Clinical trial	[212, 213, 262]
University of Science and Technology Beijing, Zhenzhou Tianzhao Biomedical Company	Multi-mode nanomedicine mediated ns-psPEF ablation, alternating magnetic field, and laser irradiation; PEF or laser radiation producing thermal/polarization/perforation effects; magnetothermal/magnetodynamics; multi-function activation of nanomedicines	Noninvasive visible targeting lesion at nanoscale; overcoming multi-drug resistance; preventing recurrence and metastasis; regulating tumor cell microenvironment and cytokines; activating immune cells and their proliferation	Nanodrugs coupling multi-field ablation synergistically enhances the activation and the proliferation of immune cells, the inflammatory response, the regulation of tumor microenvironment, and the promotion of tumor antigen expression	Preclinical	[41, 89, 108, 111, 123, 247, 263–267]
Peking University	Light irradiation/fluorescence excitation; magnetothermal/magnetodynamics; radiotherapy and chemotherapy	Noninvasive, precise treatment of lesions, adjustable treatment effect	Heat stress effect, immune cell activation, inflammatory response, immune memory	Preclinical	[268, 269]

have been paid attention again recently [84, 85, 109, 174, 279]. These quantum medicines may also show great potential in mediating multi-physical field ablation for subversive therapies for intractable diseases.

1.5 Future Development of Nanomedicines by Coupling Advanced Biomedicines (Including Biochemistry and Biophysics), Modern Physicochemical Technologies, and Artificial Intelligence Technology

However, there are still many questions in the development of nanomedicines and the corresponding multi-mode therapies. One of the current key issues is how to obtain these smart nanomedicines with a desired multi-mode imaging function, zymogen-like immune stealth transport ability and special cell or cell microenvironment targeting releasing function. Usually, addressing all of these abilities or functions by one kind of materials is extremely difficult but varieties of materials are coupled together. Fortunately, Song et al. from USTB have developed a novel nanomedicine design strategy as below. The inorganic–organic nanocomposites are constructed based on onion-like nanohybrids of metal alloy or core–shell metal compounds as inorganic cores with multi-mode imaging functions, which can be surface modified and stabilized by organic shells forming inorganic and organic nanocomposites and then conjugating to some organic drugs (e.g. ginsenoside, Paclitaxel, Doxorubicin, antibodies [e.g. PD-L1], etc.) to construct high biocompatible nanomedicine [41, 89, 108]. Finally, these nanomedicines can be encapsulated by hydrogel (e.g. PEG-g-chitosan) or amphiphilic polymers (e.g. PEG-*b*-PLGA) forming nanocapsules or nanovesicles that can be simultaneously surface-linked with some cell or microenvironment specific targeting biomolecules [111]. Recently, Song's group invented a pilot microfluidic process (Figure 1.2) [98], an ultrasonic atomization coupling pyrolysis process, [258, 259, 280] a facile freeze-annealing process [281] and a solid-phase sintering and vapor–liquid–solid growth (SS–VLS-like) method [282] for the large-scale synthesis of these multi-mode nanoparticles, single atom clusters, N, P-doping 3D graphene nanodots supporting quantum dots and as precursors with flexible size (from single atoms to tens of nanometers), shape, composition, and supporter control as well as desired physicochemical properties (e.g. magnetic, optical, electronic, sonic, or mechanical properties) [41, 89, 98, 281, 283–290], for this kind of novel smart nanomedicines or nanoenzymes according to this strategy.

Another key issue is to develop varieties of multi-modal theranostics instruments based on the optimized coupling modes among varieties of nanomedicines and multi-modal ablation determined by varieties of multi-physical-field coupling ablation. Particularly, how to design the key devices or units realizing multi-mode nanomedicines mediated the suitable multi-physical-field coupling ablation instrument with high spatial-temporal resolution (for space time resolution: from μs to ns or even to fs; for space resolution: from micrometer to nanometer and even single molecule or ligand level) is crucial to optimize their synergistic effects to develop subversive smart therapies by comprehensively regulating the key immune systems (e.g. immune cells, immunological gene expression, and cell microenvironment) [37, 216, 291, 292]. Overcoming this issue is also essential to fulfill the clinical trials of nanomedicines themselves and these combination therapies [37, 41, 108, 187, 189]. However, this project is not a trivial work. Fortunately, many groups in institutes and companies have been on this road. For example, Song's group in USTB also developed the combined therapy by their multi-mode nanomedicines mediated the nsPEF, which has shown excellent therapeutic effects and great potential in the immune system regulation in the HCC treatment [207]. A novel type of instrument with many subversive features has been on the agenda based on multi-functional nanomedicines mediated multi-physical field coupling ablation by conjugating the rotating or pulsed magnetic fields, laser irradiation and ultrafast pulsed laser, HIFU to their instrument of multi-mode nanomedicines mediated nsPEF ablation [271].

In addition, the recently developed artificial intelligence generation content (AIGC) or machine learning technologies have been successfully used in the rapid and accurate development of varieties of novel specific drugs and therapy designs for intractable diseases [22, 293–296]. AIGC can be further used in the design and microstructure optimization of multi-mode nanomedicines and the assembly of multi-physical fields, as well as the optimization of the coupling modes and key devices, accelerating the instrument development of nanomedicines mediated multi-physical field coupling ablation and the key devices (e.g. tubing type microprobes assembling several physical fields for diagnosis and therapy of diseases). In the following 5–10 years, fundamental research and the key clinical trials of nanomedicines mediated multi-physical field coupling ablation may focus on the following aspects by the combination of progresses in fundamental biomedicines, modern physicochemical technologies, and artificial intelligence: (i) high-throughput design and synthesis methods (e.g. microfluidic processes for drug design and screening) for multi-functional nanohybrids and smart nanomedicines; (ii) basic studies on the interactions between nanomedicines and single cell, sub-cellular units, colossal cells and tissues (particularly for tumor cells and stem cells and neurocytes), the T-cells and B-cells or other immune cells for intractable disease therapy, and the related gene expression and molecular or signal pathway; (iii) nanomedicine database built-up and development assisted by AIGC; (iv) precise and personalized nanomedicine development assisted by AIGC. (v) database building of multi-physical field coupling ablation and their therapeutical effects and immunological effects assisted by AIGC; (vi) subversive therapy development and fundamental biomedical mechanism investigation of nanomedicines mediated multi-physical field ablation assisted by AIGC; and (vii) quantum medicine development for intractable diseases and their therapeutical mechanism study assisted by AIGC. For details on AIGC-assisted design of nanomedicines and combined therapies, readers can refer to Chapters 16 "Nanomedicine conjugating with AI Technology and Genomics for Precise and Personalized Therapy" and 17 "Microfluidic Conjugating AI Platform for High Throughput Nanomedicine Screening" of this monograph.

References

1 Xia, C., Dong, X., Li, H. et al. (2022). Cancer statistics in China and United States, 2022: profiles, trends, and determinants. *Chinese Medical Journal* 135: 584–590. https://doi.org/10.1097/CM9.0000000000002108.

2 Siegel, R.L., Miller, K.D., Wagle, N.S., and Jemal, A. (2023). Cancer statistics, 2023. *CA: A Cancer Journal for Clinicians* 73: 17–48. https://doi.org/10.3322/caac.21763.

3 https://www.iarc.fr/faq/latest-global-cancer-data-2020-qa.

4 Nielsen, S.R., Quaranta, V., Linford, A. et al. (2016). Macrophage-secreted granulin supports pancreatic cancer metastasis by inducing liver fibrosis. *Nature Cell Biology* 18: 549–560. https://doi.org/10.1038/ncb3340.

5 Manfredi, S., Lepage, C., Hatem, C. et al. (2006). Epidemiology and management of liver metastases from colorectal cancer. *Annals of Surgery* 244: 254–259.

6 Vidal-Vanaclocha, F. (2008). The prometastatic microenvironment of the liver. *Cancer Microenvironment* 1: 113–129. https://doi.org/10.1007/s12307-008-0011-6.

7 Kuncic, Z. (2015). Cancer nanomedicine: challenges and opportunities. *The Medical Journal of Australia* 203: 204–205. https://doi.org/10.5694/mja15.00681.

8 Chen, H., Gu, Z., An, H. et al. (2018). Precise nanomedicine for intelligent therapy of cancer. *Science China. Chemistry* 61: 1503–1552.

9 Jain, A., Gosling, J., Liu, S. et al. (2024). Wireless electrical–molecular quantum signalling for cancer cell apoptosis. *Nature Nanotechnology* https://doi.org/10.1038/s41565-023-01496-y.

10 Yoo, J., Lee, E., Kim, H.Y. et al. (2017). Electromagnetized gold nanoparticles mediate direct lineage reprogramming into induced dopamine neurons in vivo for Parkinson's disease therapy. *Nature Nanotechnology* 12: 1006–1014. https://doi.org/10.1038/nnano.2017.133.

11 Sun, W., Luo, L., Feng, Y. et al. (2020). Gadolinium–Rose Bengal coordination polymer nanodots for MR-/fluorescence-image-guided radiation and photodynamic therapy. *Advanced Materials* 32: 2000377. https://doi.org/10.1002/adma.202000377.

12 Zhang, Y., Wang, X., Chu, C. et al. (2020). Genetically engineered magnetic nanocages for cancer magneto-catalytic theranostics. *Nature Communications* 11: 5421. https://doi.org/10.1038/s41467-020-19061-9.

13 Breton, M. and Mir, L.M. (2012). Microsecond and nanosecond electric pulses in cancer treatments. *Bioelectromagnetics* 33: 106–123. https://doi.org/10.1002/bem.20692.

14 Ding, X., Stewart, M.P., Sharei, A. et al. (2017). High-throughput nuclear delivery and rapid expression of DNA via mechanical and electrical cell-membrane disruption. *Nature Biomedical Engineering* 1: 0039. https://doi.org/10.1038/s41551-017-0039.

15 Kotnik, T., Rems, L., Tarek, M., and Miklavčič, D. (2019). Membrane electroporation and electropermeabilization: mechanisms and models. *Annual Review of Biophysics* 48: 63–91. https://doi.org/10.1146/annurev-biophys-052118-115451.

16 Nuccitelli, R., McDaniel, A., Connolly, R. et al. (2020). Nano-pulse stimulation induces changes in the intracellular organelles in rat liver tumors treated in situ. *Lasers in Surgery and Medicine* 52: 882–889. https://doi.org/10.1002/lsm.23239.

17 Schoenbach, K.H. and Joshi, R.P. (2010). Bioelectric effects of intense ultrashort pulses. *Critical Reviews in Biomedical Engineering* 38: 255–304. https://doi.org/10.1615/CritRevBiomedEng.v38.i3.20.

18 Mohamed, N., Lee, A., and Lee, N.Y. (2022). Proton beam radiation therapy treatment for head and neck cancer. *Precision Radiation Oncology* 6: 59–68. https://doi.org/10.1002/pro6.1135.

19 Matsumoto, Y., Fukumitsu, N., Ishikawa, H. et al. (2021). A critical review of radiation therapy: from particle beam therapy (proton, carbon, and BNCT) to beyond. *Journal of Personalized Medicine* 11: 825. https://doi.org/10.3390/jpm11080825.

20 Casas, F., Abdel-Wahab, S., Filipovic, N., and Jeremic, B. (2017). *International Encyclopedia of Public Health*, 2ee (ed. S.R. Quah), 260–268. Academic Press.

21 Ferreira, T.H. and de Sousa, E.M.B. (2016). *Boron Nitride Nanotubes in Nanomedicine* (ed. G. Ciofani and V. Mattoli), 95–109. William Andrew Publishing.

22 Shamay, Y., Shah, J., Işık, M. et al. (2018). Quantitative self-assembly prediction yields targeted nanomedicines. *Nature Materials* 17: 361–368.

23 Li, R., He, Y., Zhu, Y. et al. (2019). Route to rheumatoid arthritis by macrophage-derived microvesicle-coated nanoparticles. *Nano Letters* 19: 124–134.

24 Furtado, D., Björnmalm, M., Ayton, S. et al. (2018). Overcoming the blood–brain barrier: the role of nanomaterials in treating neurological diseases. *Advanced Materials* 30: 1801362.

25 Giardiello, M., Liptrott, N.J., McDonald, T.O. et al. (2016). Accelerated oral nanomedicine discovery from miniaturized screening to clinical production exemplified by paediatric HIV nanotherapies. *Nature Communications* 7: 13184.

26 Bourquin, J., Milosevic, A., Hauser, D. et al. (2018). Biodistribution, clearance, and long-term fate of clinically relevant nanomaterials. *Advanced Materials* 30: 1704307.

27 Fu, X., Shi, Y., Qi, T. et al. (2020). Precise design strategies of nanomedicine for improving cancer therapeutic efficacy using subcellular targeting. *Signal Transduction and Targeted Therapy* 5: 262. https://doi.org/10.1038/s41392-020-00342-0.

28 Murray, R.K., Granner, D.K., Mayes, P.A., and Rodwell, V.W. (2003). *Harper's Illustrated Biochemistry*, 26ee, 693. Lange Medical Books/McGraw-Hill, Medical Publishing Division.

29 Glaser, R. (2001). *Biophysics*. 361 pp. Spring-Verlag.

30 Bertrand, N., Wu, J., Xu, X. et al. (2014). Cancer nanotechnology: the impact of passive and active targeting in the era of modern cancer biology. *Advanced Drug Delivery Reviews* 66: 2–25. https://doi.org/10.1016/j.addr.2013.11.009.

31 Greish, K. (2007). Enhanced permeability and retention of macromolecular drugs in solid tumors: a royal gate for targeted anticancer nanomedicines. *Journal of Drug Targeting* 15: 457–464. https://doi.org/10.1080/10611860701539584.

32 Maeda, H. (2015). Toward a full understanding of the EPR effect in primary and metastatic tumors as well as issues related to its heterogeneity. *Advanced Drug Delivery Reviews* 91: 3–6. https://doi.org/10.1016/j.addr.2015.01.002.

33 Elnaggar, Y.S.R., Etman, S.M., Abdelmonsif, D.A., and Abdallah, O.Y. (2015). Intranasal piperine-loaded chitosan nanoparticles as brain-targeted therapy in Alzheimer's disease: optimization, biological efficacy, and potential toxicity. *Journal of Pharmaceutical Sciences* 104: 3544–3556.

34 Flores, A.M., Hosseini-Nassab, N., Jarr, K.U. et al. (2020). Pro-efferocytic nanoparticles are specifically taken up by lesional macrophages and prevent atherosclerosis. *Nature Nanotechnology* 15: 154–161. https://doi.org/10.1038/s41565-019-0619-3.

35 Kumar, P., Wu, H., McBride, J.L. et al. (2007). Transvascular delivery of small interfering RNA to the central nervous system. *Nature* 448: 39–43. https://doi.org/10.1038/nature05901.

36 Yameen, B., Choi, W.I., Vilos, C. et al. (2014). Insight into nanoparticle cellular uptake and intracellular targeting. *Journal of Controlled Release* 190: 485–499. https://doi.org/10.1016/j.jconrel.2014.06.038.

37 Gawne, P.J., Ferreira, M., Papaluca, M. et al. (2023). New opportunities and old challenges in the clinical translation of nanotheranostics. *Nature Reviews Materials* https://doi.org/10.1038/s41578-023-00581-x.

38 Kunjachan, S., Ehling, J., Storm, G. et al. (2015). Noninvasive imaging of nanomedicines and nanotheranostics: principles, progress, and prospects. *Chemical Reviews* 115: 10907–10937. https://doi.org/10.1021/cr500314d.

39 Chen, X., Zhang, X., Guo, Y. et al. (2019). Smart supramolecular "Trojan Horse"-inspired nanogels for realizing light-triggered nuclear drug influx in drug-resistant cancer cells. *Advanced Functional Materials* 29 (13): 1807772. https://doi.org/10.1002/adfm.201807772.

40 van der Meel, R., Sulheim, E., Shi, Y. et al. (2019). Smart cancer nanomedicine. *Nature Nanotechnology* 14: 1007–1017. https://doi.org/10.1038/s41565-019-0567-y.

41 Zhao, X., Wu, J., Guo, D. et al. (2022). Dynamic ginsenoside-sheltered nanocatalysts for safe ferroptosis-apoptosis combined therapy. *Acta Biomaterialia* 151: 549–560. https://doi.org/10.1016/j.actbio.2022.08.026.

42 Xue, X., Huang, Y., Bo, R. et al. (2018). Trojan Horse nanotheranostics with dual transformability and multifunctionality for highly effective cancer treatment. *Nature Communications* 9: 3653. https://doi.org/10.1038/s41467-018-06093-5.

43 Lee, Y., Kamada, N., and Moon, J.J. (2021). Oral nanomedicine for modulating immunity, intestinal barrier functions, and gut microbiome. *Advanced Drug Delivery Reviews* 179: 114021. https://doi.org/10.1016/j.addr.2021.114021.

44 He, C., Duan, X., Guo, N. et al. (2016). Core–shell nanoscale coordination polymers combine chemotherapy and photodynamic therapy to potentiate checkpoint blockade cancer immunotherapy. *Nature Communications* 7: 12499. https://doi.org/10.1038/ncomms12499.

45 Smith, T.T., Stephan, S.B., Moffett, H.F. et al. (2017). In situ programming of leukaemia-specific T cells using synthetic DNA nanocarriers. *Nature Nanotechnology* 12: 813–820. https://doi.org/10.1038/nnano.2017.57.

46 Yan, G.-H., Wang, K., Shao, Z. et al. (2018). Artificial antibody created by conformational reconstruction of the complementary-determining region on gold nanoparticles. *Proceedings of the National Academy of Sciences* 115: E34. https://doi.org/10.1073/pnas.1713526115.

47 Irvine, D.J. and Dane, E.L. (2020). Enhancing cancer immunotherapy with nanomedicine. *Nature Reviews Immunology* 20: 321–334. https://doi.org/10.1038/s41577-019-0269-6.

48 Lakshmanan, V.-K., Jindal, S., Packirisamy, G. et al. (2021). Nanomedicine-based cancer immunotherapy: recent trends and future perspectives. *Cancer Gene Therapy* 28: 911–923. https://doi.org/10.1038/s41417-021-00299-4.

49 Min, Y., Roche, K.C., Tian, S. et al. (2017). Antigen-capturing nanoparticles improve the abscopal effect and cancer immunotherapy. *Nature Nanotechnology* 12: 877–882. https://doi.org/10.1038/nnano.2017.113.

50 D'Mello, S.R., Cruz, C.N., Chen, M.L. et al. (2017). The evolving landscape of drug products containing nanomaterials in the United States. *Nature Nanotechnology* https://doi.org/10.1038/NNANO.2017.67.

51 Conde, J., Oliva, N., Zhang, Y., and Artzi, N. (2016). Local triple-combination therapy results in tumour regression and prevents recurrence in a colon cancer model. *Nature Materials* 15: 1128–1138. https://doi.org/10.1038/nmat4707.

52 Singha, S., Shao, K., Yang, Y. et al. (2017). Peptide-MHC-based nanomedicines for autoimmunity function as T-cell receptor microclustering devices. *Nature Nanotechnology* 12: 701–710. https://doi.org/10.1038/nnano.2017.56.

53 Li, Q., Luo, T.Y., Taylor, M.G. et al. (2017). Molecular "surgery" on a 23-gold-atom nanoparticle. *Science Advances* 3: e1603193. https://doi.org/10.1126/sciadv.1603193.

54 Bai, S., Yang, L.L., Wang, Y. et al. (2020). Prodrug-based versatile nanomedicine for enhancing cancer immunotherapy by increasing immunogenic cell death. *Small* 16: 2000214. https://doi.org/10.1002/smll.202000214.

55 Cai, J., Wang, H., Wang, D., and Li, Y. (2019). Improving cancer vaccine efficiency by nanomedicine. *Advanced Biosystems* 3: 1800287. https://doi.org/10.1002/adbi.201800287.

56 Duan, X., He, C., Kron, S.J., and Lin, W. (2016). Nanoparticle formulations of cisplatin for cancer therapy. *Wiley Interdisciplinary Reviews. Nanomedicine and Nanobiotechnology* 8: 776–791. https://doi.org/10.1002/wnan.1390.

57 Nam, J., Son, S., Park, K.S. et al. (2019). Cancer nanomedicine for combination cancer immunotherapy. *Nature Reviews Materials* 4: 398–414. https://doi.org/10.1038/s41578-019-0108-1.

58 Zhao, X., Wang, J., Song, Y., and Chen, X. (2018). Synthesis of nanomedicines by nanohybrids conjugating ginsenosides with auto-targeting and enhanced MRI contrast for liver cancer therapy. *Drug Development and Industrial Pharmacy* 44: 1307–1316.

59 Bangham, A.D. and Horne, R.W. (1964). Negative staining of phospholipids and their structural modification by surface-active agents as observed in the electron microscope. *Journal of Molecular Biology* 8: 660–668.
60 Langer, R. and Folkman, J. (1976). Polymers for the sustained release of proteins and other macromolecules. *Nature* 263: 797–800.
61 Matsumura, Y. and Maeda, H. (1986). A new concept for macromolecular therapeutics in cancer chemotherapy: mechanism of tumoritropic accumulation of proteins and the antitumor agent Smancs. *Cancer Research* 46: 6387–6392.
62 Gerlowski, L.E. and Jain, R.K. (1986). Microvascular permeability of normal and neoplastic tissues. *Microvascular Research* 31: 288–305.
63 Smith, A.D. (2013). Big moment for nanotech: oncology therapeutics poised for a leap. *Oncology Live*® 14 (6) http://www.onclive.com/publications/Oncology-live/2013/June-2013/Big-Moment-for-Nanotech-Oncology-Therapeutics-Poised-for-a-Leap.
64 US National Library of Medicine. (2013). ClinicalTrials.gov: https://clinicaltrials.gov/ct2/show/NCT00689065?term.
65 US National Library of Medicine. (2016). *A study of BIND-014 given to patients with advanced or metastatic cancer*. http://clinicaltrials.gov/ct2/show/NCT01300533?term=NCT01300533&rank=1.
66 Ghadiali, J.E. and Stevens, M.M. (2010). Enzyme-responsive nanoparticle systems. *Advanced Materials* 20: 4359–4363.
67 Goto, Y., Yanagi, I., Matsui, K. et al. (2016). Integrated solid-state nanopore platform for nanopore fabrication via dielectric breakdown, DNA-speed deceleration and noise reduction. *Scientific Reports* 6: 31324. https://doi.org/10.1038/srep31324.
68 Yokoi, K., Kojic, M., Milosevic, M. et al. (2014). Capillary-wall collagen as a biophysical marker of nanotherapeutic permeability into the tumor microenvironment. *Cancer Research* 74: 4239–4246. https://doi.org/10.1158/0008-5472.Can-13-3494.
69 Yokoi, K., Tanei, T., Godin, B. et al. (2014). Serum biomarkers for personalization of nanotherapeutics-based therapy in different tumor and organ microenvironments. *Cancer Letters* 345: 48–55. https://doi.org/10.1016/j.canlet.2013.11.015.
70 Folkman, J. and Long, D.M. (1964). The use of silicone rubber as a carrier for prolonged drug therapy. *The Journal of Surgical Research* 4: 139–142.
71 Leserman, L.D., Barbet, J., Kourilsky, F., and Weinstein, J.N. (1980). Targeting to cells of fluorescent liposomes covalently coupled with monoclonal antibody or protein A. *Nature* 288: 602–604. https://doi.org/10.1038/288602a0.
72 Heath, T.D., Fraley, R.T., and Papahdjopoulos, D. (1980). Antibody targeting of liposomes: cell specificity obtained by conjugation of F(ab')$_2$ to vesicle surface. *Science* 210: 539–541.
73 Service, R.F. (2017). Nanoparticles awaken immune cells to fight cancer: in mice, new approach wipes out targeted tumor and metastases as well. *Science News* https://doi.org/10.1126/science.aal0581.

74 Gref, R., Minamitake, Y., Peracchia, M.T. et al. (1994). Biodegradable long-circulating polymeric nanospheres. *Science* 263: 1600–1603. https://doi.org/10.1126/science.8128245.

75 Rolland, J.P., Maynor, B.W., Euliss, L.E. et al. (2005). Direct fabrication and harvesting of monodisperse, shape-specific nanobiomaterials. *Journal of the American Chemical Society* 127: 10096–10100.

76 Liu, G., Garrett, M.R., Men, P. et al. (2005). Nanoparticle and other metal chelation therapeutics in Alzheimer disease. *Biochimica et Biophysica Acta* 1741: 246–252. https://doi.org/10.1016/j.bbadis.2005.06.006.

77 Biopharm, S. (2016). *Samyang Biopharm. History*.

78 MagForce. (2010). *MagForce. MagForce nanotechnologies AG receives European regulatory approval for its Nano Cancer® therapy*. magforce.de: http://www.magforce.de/en/presse-investoren/news-events/detail/article/magforce-nanotechnologies-ag-erhaelt-europaeische-zulassung-fuer-die-nano-krebsR-therapie.html

79 Hu, C.M.J., Aryal, S., Cheung, C. et al. (2011). Erythrocyte membrane-camouflaged polymeric nanoparticles as a biomimetic delivery platform. *Proceedings of the National Academy of Sciences of the United States of America* 108: 10980–10985.

80 Guo, J., Feng, K., Wu, W. et al. (2021). Smart ^{131}I labeled self-illuminating photosensitizers for deep tumor therapy. *Angewandte Chemie International Edition* 60: 21884–21889.

81 Miller, M.A., Gadde, S., Pfirschke, C. et al. (2015). Predicting therapeutic nanomedicine efficacy using a companion magnetic resonance imaging nanoparticle. *Science Translational Medicine* 7: 314ra183. https://doi.org/10.1126/scitranslmed.aac6522.

82 Sheng, Z., Hu, D., Zheng, M. et al. (2014). Smart human serum albumin-indocyanine green nanoparticles generated by programmed assembly for dual-modal imaging-guided cancer synergistic phototherapy. *ACS Nano* 8: 12310–12322.

83 Chi, X., Huang, D., Zhao, Z. et al. (2012). Nanoprobes for in vitro diagnostics of cancer and infectious diseases. *Biomaterials* 33: 189–206. https://doi.org/10.1016/j.biomaterials.2011.09.032.

84 Wang, Y. and Chen, L. (2011). Quantum dots, lighting up the research and development of nanomedicine. *Nanomedicine: Nanotechnology, Biology, and Medicine* 7: 385–402.

85 Bordonaro, M. (2019). Quantum biology and human carcinogenesis. *Biosystems* 178: 16–24. https://doi.org/10.1016/j.biosystems.2019.01.010.

86 Lambert, N., Chen, Y.N., Cheng, Y.C. et al. (2013). Quantum biology. *Nature Physics* 9: 10–18. https://doi.org/10.1038/nphys2474.

87 Zingsem, B.W., Feggeler, T., Terwey, A. et al. (2019). Biologically encoded magnonics. *Nature Communications* 10: 4345. https://doi.org/10.1038/s41467-019-12219-0.

88 Mangalhara, K.C., Varanasi, S.K., Johnson, M.A. et al. (2023). Manipulating mitochondrial electron flow enhances tumor immunogenicity. *Science* 381: 1316–1323. https://doi.org/10.1126/science.abq1053.

89 Zhang, W., Zhao, X., Yuan, Y. et al. (2020). Microfluidic synthesis of multi-mode Au@CoFeB–Rg3 nanomedicines and their cytotoxicity and anti-tumor effects. *Chemistry of Materials* 32: 5044–5056. https://doi.org/10.1021/acs.chemmater.0c00797.

90 Wang, W., Mo, W., Hang, Z. et al. (2023). Cuproptosis: harnessing transition metal for cancer therapy. *ACS Nano* 17: 19581–19599. https://doi.org/10.1021/acsnano.3c07775.

91 Gaetke, L. (2003). Copper toxicity, oxidative stress, and antioxidant nutrients. *Toxicology* 189: 147–163. https://doi.org/10.1016/s0300-483x(03)00159-8.

92 Liu, Z., Xiong, L., Ouyang, G. et al. (2017). Investigation of copper cysteamine nanoparticles as a new type of radiosensitizers for colorectal carcinoma treatment. *Scientific Reports* 7: 9290.

93 Pramanik, A., Pramanik, S., and Pramanik, P. (2017). Copper based nanoparticle: a way towards future cancer therapy. *Global Journal of Nanomedicine* 1: 555573.

94 Tsvetkov, P., Coy, S., Petrova, B. et al. (2022). Copper induces cell death by targeting lipoylated TCA cycle proteins. *Science* 375: 1254–1261. https://doi.org/10.1126/science.abf0529.

95 Yu, Z., Lou, R., Pan, W. et al. (2020). Nanoenzymes in disease diagnosis and therapy. *Chemical Communications* 56: 15513–15524. https://doi.org/10.1039/D0CC05427E.

96 Liu, J., Zhang, B., Luo, Z. et al. (2015). Enzyme responsive mesoporous silica nanoparticles for targeted tumor therapy in vitro and in vivo. *Nanoscale* 7: 3614–3626.

97 The first nano antibody Cabiliv launched after approved by European Marketing Association in Sept 2018 and then by FDA in Feb 2019: http://vdev.tip-lab.com/article/?uuid=ce20062cc1c44a5396f7eefce6d2a5fb (2019).

98 Wu, Q., Liu, R., Miao, F. et al. (2023). Multimodal ultra-small CoFe–WO$_x$ nanohybrids synthesized by a pilot microfluidic system. *Chemical Engineering Journal* 452: 139355. https://doi.org/10.1016/j.cej.2022.139355.

99 Nudelman, R., Alhmoud, H., Delalat, B. et al. (2019). Jellyfish-based smart wound dressing devices containing in situ synthesized antibacterial nanoparticles. *Advanced Functional Materials* 29: 1902783. https://doi.org/10.1002/adfm.201902783.

100 Xu, L., Wang, X., Wang, W. et al. (2022). Enantiomer-dependent immunological response to chiral nanoparticles. *Nature* 601: 366–373. https://doi.org/10.1038/s41586-021-04243-2.

101 Yang, J. and Zhang, C. (2020). Regulation of cancer-immunity cycle and tumor microenvironment by nanobiomaterials to enhance tumor immunotherapy. *WIREs Nanomedicine and Nanobiotechnology* 12: e1612. https://doi.org/10.1002/wnan.1612.

102 Hou, Q., Zhang, K., Chen, S. et al. (2022). Physical & chemical microwave ablation (MWA) enabled by nonionic MWA nanosensitizers repress incomplete MWA-arised liver tumor recurrence. *ACS Nano* 16 (4): 5704–5718.

103 Li, S., Chen, Z., Tan, L. et al. (2022). MOF@COF nanocapsule for the enhanced microwave thermal-dynamic therapy and anti-angiogenesis of colorectal cancer. *Biomaterials* 283: 121472. https://doi.org/10.1016/j.biomaterials.2022.121472.

104 Tao, N., Li, H., Deng, L. et al. (2022). A cascade nanozyme with amplified sonodynamic therapeutic effects through comodulation of hypoxia and immunosuppression against cancer. *ACS Nano* 16: 485–501. https://doi.org/10.1021/acsnano.1c07504.

105 Yang, Z., Zhu, Y., Dong, Z. et al. (2021). Tumor-killing nanoreactors fueled by tumor debris can enhance radiofrequency ablation therapy and boost antitumor immune responses. *Nature Communications* 12: 4299.

106 Yin, S., Chen, X., Hu, C. et al. (2014). Nanosecond pulsed electric field (nsPEF) treatment for hepatocellular carcinoma: a novel locoregional ablation decreasing lung metastasis. *Cancer Letters* 346: 285–291.

107 Koo, S., Sohn, H.S., Kim, T.H. et al. (2023). Ceria-vesicle nanohybrid therapeutic for modulation of innate and adaptive immunity in a collagen-induced arthritis model. *Nature Nanotechnology* https://doi.org/10.1038/s41565-023-01523-y.

108 Zhao, X., Wu, J., Zhang, K. et al. (2022). The synthesis of a nanodrug using metal-based nanozymes conjugated with ginsenoside Rg3 for pancreatic cancer therapy. *Nanoscale Advances* 4: 190–199. https://doi.org/10.1039/D1NA00697E.

109 Zhang, S., Li, Y., Sun, S. et al. (2022). Single-atom nanozymes catalytically surpassing naturally occurring enzymes as sustained stitching for brain trauma. *Nature Communications* 13: 4744. https://doi.org/10.1038/s41467-022-32411-z.

110 Zou, Y., Jin, B., Li, H. et al. (2022). Cold nanozyme for precise enzymatic antitumor immunity. *ACS Nano* 16: 21491–21504. https://doi.org/10.1021/acsnano.2c10057.

111 Liu, R., Wu, Q., Huang, X. et al. (2021). Synthesis of nanomedicine hydrogel microcapsules by droplet microfluidic process and their pH and temperature dependent release. *RSC Advances* 11: 37814–37823. https://doi.org/10.1039/D1RA05207A.

112 Miao, Y.-B., Lin, Y.J., Chen, K.H. et al. (2021). Engineering nano- and microparticles as oral delivery vehicles to promote intestinal lymphatic drug transport. *Advanced Materials* 33: 2104139. https://doi.org/10.1002/adma.202104139.

113 Huang, K.-W., Hsu, F.F., Qiu, J.T. et al. (2020). Highly efficient and tumor-selective nanoparticles for dual-targeted immunogene therapy against cancer. *Science Advances* 6: eaax5032. https://doi.org/10.1126/sciadv.aax5032.

114 Shi, J., Kantoff, P.W., Wooster, R., and Farokhzad, O.C. (2017). Cancer nanomedicine: progress, challenges and opportunities. *Nature Reviews. Cancer* 17: 20–37. https://doi.org/10.1038/nrc.2016.108.

115 Roth, T.L., Puig-Saus, C., Yu, R. et al. (2018). Reprogramming human T cell function and specificity with non-viral genome targeting. *Nature* 559: 405–409. https://doi.org/10.1038/s41586-018-0326-5.

116 Luo, M., Wang, H., Wang, Z. et al. (2017). A STING-activating nanovaccine for cancer immunotherapy. *Nature Nanotechnology* 12: 648–654. https://doi.org/10.1038/nnano.2017.52.

117 Sengupta, S. (2017). Cancer nanomedicine: lessons for immuno-oncology. *Trends in Cancer* 3: 551–560. https://doi.org/10.1016/j.trecan.2017.06.006.

118 Yuan, T., Wang, T., Zhang, J. et al. (2023). Robust and multifunctional nanoparticles assembled from natural polyphenols and metformin for efficient spinal cord regeneration. *ACS Nano* 17: 18562–18575. https://doi.org/10.1021/acsnano.3c06991.

119 Jin, G.R., Mao, D., Cai, P. et al. (2015). Conjugated polymer nanodots as ultrastable long-term trackers to understand mesenchymal stem cell therapy in skin regeneration. *Advanced Functional Materials* 25: 4263–4273.

120 Icli, B., Nabzdyk, C.S., Lujan-Hernandez, J. et al. (2016). Regulation of impaired angiogenesis in diabetic dermal wound healing by microRNA-26a. *Journal of Molecular and Cellular Cardiology* 2016: 151–159.

121 Achterberg, V.F., Buscemi, L., Diekmann, H. et al. (2014). The nano-scale mechanical properties of the extracellular matrix regulate dermal fibroblast function. *Journal of Investigative Dermatology* 134: 1862–1872.

122 Wu, H., Li, F., Shao, W. et al. (2019). Promoting angiogenesis in oxidative diabetic wound microenvironment using a nanozyme-reinforced self-protecting hydrogel. *ACS Central Science* 5: 477–485. https://doi.org/10.1021/acscentsci.8b00850.

123 Santos, F.K.D., Oyafuso, M.H., Kiill, C.P. et al. (2013). Nanotechnology-based drug delivery systems for treatment of hyperproliferative skin diseases – a review. *Current Nanoscience* 9: 159–167. https://doi.org/10.2174/157341313805117857.

124 Long, F., Gu, C., Gu, A.Z., and Shi, H. (2012). Quantum dot/carrier-protein/haptens conjugate as a detection nanobioprobe for FRET-based immunoassay of small analytes with all-fiber microfluidic biosensing platform. *Analytical Chemistry* 84: 3646–3653. https://doi.org/10.1021/ac3000495.

125 Agrawal, A., Sathe, T., and Nie, S. (2007). *New Frontiers in Ultrasensitive Bioanalysis. Vol. 172 Chemical Analysis: A Series of Monographs on Analysis Chemistry and Its Application* (ed. X.N. Xu), 71–89. John Wiley & Son, Inc.

126 Browning, L.M., Lee, K.J., Cherukuri, P.K. et al. (2016). Single nanoparticle plasmonic spectroscopy for study of the efflux function of multidrug ABC membrane transporters of single live cells. *RSC Advances* 6: 36794–36802. https://doi.org/10.1039/C6RA05895G.

127 Huang, T., Nallathamby, P.D., and Xu, X.-H.N. (2008). Photostable single-molecule nanoparticle optical biosensors for real-time sensing of single cytokine molecules and their binding reactions. *Journal of the American Chemical Society* 130: 17095–17105. https://doi.org/10.1021/ja8068853.

128 Kyriacou, S.V., Brownlow, W.J., and Xu, X.-H.N. (2004). Using nanoparticle optics assay for direct observation of function of antimicrobial agents in single live bacterial cells. *Biochemistry* 43: 140–147.

129 Song, Y., Nallathamby, P.D., Huang, T. et al. (2010). Correlation and characterization of 3D morphological dependent localized surface plasmon resonance spectra of single silver nanoparticles using dark-field optical microscopy and AFM. *Journal of Physical Chemistry C* 114: 74–81.

130 Choi, C.K., Li, J., Wei, K. et al. (2015). A gold@polydopamine core–shell nanoprobe for long-term intracellular detection of microRNAs in differentiating stem cells. *Journal of the American Chemical Society* 137: 7337–7346. https://doi.org/10.1021/jacs.5b01457.

131 Fan, Y. and Zhang, F. (2019). A new generation of NIR-II probes: lanthanide-based nanocrystals for bioimaging and biosensing. *Advanced Optical Materials* 7: 1801417. https://doi.org/10.1002/adom.201801417.

132 Liu, F., Le, W., Mei, T. et al. (2016). In vitro and in vivo targeting imaging of pancreatic cancer using a Fe_3O_4@SiO_2 nanoprobe modified with anti-mesothelin antibody. *International Journal of Nanomedicine* 11: 2195–2207.

133 Vo-Dinh, T., Wang, H.-N., and Scaffidi, J. (2009). Plasmonic nanoprobes for SERS biosensing and bioimaging. *Journal of Biophotonics* 3: 89–102.

134 Wabuyele, M. and Vo-Dinh, T. (2005). Detection of HIV type 1 DNA sequence using plasmonic nanoprobes. *Analytical Chemistry* 77: 7810–7815.

135 Wan, Y., Cheng, G., Liu, X. et al. (2017). Rapid magnetic isolation of extracellular vesicles via lipid-based nanoprobes. *Nature Biomedical Engineering* 1: https://doi.org/10.1038/s41551-017-0058.

136 Song, Y. (2011). *Recent Advances in Nanofabrication Techniques and Applications*, vol. 1 (ed. B. Cui), 505–532. In-Tech.

137 Song, Y. (2009). *Fabrication of high throughput biosensors based on single nanoparticles and nanoparticle arrays*. China patent.

138 Pinaud, F., Clarke, S., Sittner, A., and Dahan, M. (2010). Probing cellular events, one quantum dot at a time. *Nature Methods* 7: 275–285. https://doi.org/10.1038/nmeth.1444.

139 Xu, X.-H.N., Song, Y., and Nallathamby, P. (2007). *New Frontiers in Ultrasensitive Bioanalysis*, vol. 1 (ed. J.D. Winefordner) Ch. 3, 41–70. John Wiley & Son, Inc.

140 Letter, T.M. (1973). Griseofulvin: a new formulation and some old concerns. *The Medical Letter on Drugs and Therapeutics* 18: 17–18.

141 Rodriguez, P.L., Harada, T., Christian, D.A. et al. (2013). Minimal "self" peptides that inhibit phagocytic clearance and enhance delivery of nanoparticles. *Science* 339: 971–975. https://doi.org/10.1126/science.1229568.

142 Hajipour, M.J., Laurent, S., Aghaie, A. et al. (2014). Personalized protein coronas: a "key" factor at the nanobiointerface. *Biomaterials Science* 2: 1210–1221.

143 Sakulkhu, U., Maurizi, L., Mahmoudi, M. et al. (2014). Ex situ evaluation of the composition of protein corona of intravenously injected superparamagnetic nanoparticles in rats. *Nanoscale* 6: 11439–11450.

144 Howard, M., Zern, B.J., Anselmo, A.C. et al. (2014). Vascular targeting of nanocarriers: perplexing aspects of the seemingly straightforward paradigm. *ACS Nano* 8: 4100–4132. https://doi.org/10.1021/nn500136z.

145 Pridgen, E.M., Alexis, F., Kuo, T.T. et al. (2013). Transepithelial transport of Fc-targeted nanoparticles by the neonatal Fc receptor for oral delivery. *Science Translational Medicine* 5: 213ra167. https://doi.org/10.1126/scitranslmed.3007049.

146 Cheng, Y., Morshed, R.A., Auffinger, B. et al. (2014). Multifunctional nanoparticles for brain tumor imaging and therapy. *Advanced Drug Delivery Reviews* 66: 42–57. https://doi.org/10.1016/j.addr.2013.09.006.

147 Xi, Z., Wu, J., Shan, W. et al. (2016). Polymeric nanoparticles amenable to simultaneous installation of exterior targeting and interior therapeutic proteins. *Angewandte Chemie (International Ed. in English)* 55: 3309–3312. https://doi.org/10.1002/anie.201509183.

148 Samyang Biopharm. (2016). *History*. https://www.samyangbiopharm.com/eng/Aboutus/history.

149 US National Library of Medicine. (2013). ClinicalTrials.gov, h. c. g. c. s. N. t.

150 Espelin, C.W., Leonard, S.C., Geretti, E. et al. (2016). Dual HER2 targeting with trastuzumab and liposomal-encapsulated doxorubicin (MM-302) demonstrates synergistic antitumor activity in breast and gastric cancer. *Cancer Research* 76: 1517–1527.

151 Hrkach, J., Von Hoff, D., Ali, M.M. et al. (2012). Preclinical development and clinical translation of a PSMA-targeted docetaxel nanoparticle with a differentiated pharmacological profile. *Science Translational Medicine* 4: 128ra139. https://doi.org/10.1126/scitranslmed.3003651.

152 Davis, M.E., Zuckerman, J.E., Choi, C.H.J. et al. (2010). Evidence of RNAi in humans from systemically administered siRNA via targeted nanoparticles. *Nature* 464: 1067–1070. https://doi.org/10.1038/nature08956.

153 Jiang, Y., Krishnan, N., Zhou, J. et al. (2020). Engineered cell-membrane-coated nanoparticles directly present tumor antigens to promote anticancer immunity. *Advanced Materials* 32: 2001808. https://doi.org/10.1002/adma.202001808.

154 Malhotra, S., Dumoga, S., and Singh, N. (2022). Red blood cells membrane-derived nanoparticles: applications and key challenges in their clinical translation. *Wiley Interdisciplinary Reviews. Nanomedicine and Nanobiotechnology* 14: e1776.

155 Dai, Z. (2016). *Advances in Nanotheranostics II*, vol. 551. Springer.

156 D'Mello, S.R., Cruz, C.N., Chen, M.L. et al. (2017). The evolving landscape of drug products containing nanomaterials in the United States. *Nature Nanotechnology* 12: 523–530.

157 Dixon, S.J., Lemberg, K.M., Lamprecht, M.R. et al. (2012). Ferroptosis: an iron-dependent form of nonapoptotic cell death. *Cell* 149: 1060–1072.

158 Liu, T., Liu, W., Zhang, M. et al. (2018). Ferrous-supply-regeneration nanoengineering for cancer-cell-specific ferroptosis in combination with imaging-guided photodynamic therapy. *ACS Nano* 12: 12181–12192.

159 Chen, Q., Feng, L., Liu, J. et al. (2016). Intelligent albumin–MnO$_2$ nanoparticles as pH-/H$_2$O$_2$-responsive dissociable nanocarriers to modulate tumor hypoxia for effective combination therapy. *Advanced Materials* 28: 7129–7136. https://doi.org/10.1002/adma.201601902.

160 Cabral, H., Kinoh, H., and Kataoka, K. (2020). Tumor-targeted nanomedicine for immunotherapy. *Accounts of Chemical Research* 53: 2765–2776. https://doi.org/10.1021/acs.accounts.0c00518.

161 Dang, Q., Sun, Z., Wang, Y. et al. (2022). Ferroptosis: a double-edged sword mediating immune tolerance of cancer. *Cell Death & Disease* 13: 925. https://doi.org/10.1038/s41419-022-05384-6.

162 Liu, S., Jiang, Q., Zhao, X. et al. (2020). A DNA nanodevice-based vaccine for cancer immunotherapy. *Nature Materials* https://doi.org/10.1038/s41563-020-0793-6.

163 Pardi, N., Hogan, M.J., Porter, F.W., and Weissman, D. (2018). mRNA vaccines – a new era in vaccinology. *Nature Reviews Drug Discovery* 17: 261–279. https://doi.org/10.1038/nrd.2017.243.

164 Pouikli, A. and Frezza, C. (2023). Metabolic control of antitumor immunity. *Science* 381: 1287–1288. https://doi.org/10.1126/science.adk1785.

165 Ren, Z., Chen, X., Hong, L. et al. (2020). Nanoparticle conjugation of ginsenoside Rg3 inhibits hepatocellular carcinoma development and metastasis. *Small* 16: 1905233. https://doi.org/10.1002/smll.201905233.

166 Sun, B., Hyun, H., Li, L.-t., and Wang, A.Z. (2020). Harnessing nanomedicine to overcome the immunosuppressive tumor microenvironment. *Acta Pharmacologica Sinica* 41: 970–985. https://doi.org/10.1038/s41401-020-0424-4.

167 Oliveira, G. and Wu, C.J. (2023). Dynamics and specificities of T cells in cancer immunotherapy. *Nature Reviews Cancer* 23: 295–316. https://doi.org/10.1038/s41568-023-00560-y.

168 Wadhwa, A., Aljabbari, A., Lokras, A. et al. (2020). Opportunities and challenges in the delivery of mRNA-based vaccines. *Pharmaceutics* 12: 102.

169 Yan, W.-l., Lang, T.-q., Yuan, W.-h. et al. (2022). Nanosized drug delivery systems modulate the immunosuppressive microenvironment to improve cancer immunotherapy. *Acta Pharmacologica Sinica* 43: 3045–3054. https://doi.org/10.1038/s41401-022-00976-6.

170 Cao, F., Sang, Y., Liu, C. et al. (2022). Self-adaptive single-atom catalyst boosting selective ferroptosis in tumor cells. *ACS Nano* 16: 855–868. https://doi.org/10.1021/acsnano.1c08464.

171 Jiang, X., Stockwell, B.R., and Conrad, M. (2021). Ferroptosis: mechanisms, biology and role in disease. *Nature Reviews Molecular Cell Biology* 22: 266–282. https://doi.org/10.1038/s41580-020-00324-8.

172 Dale, B., Cheng, M., Park, K.S. et al. (2021). Advancing targeted protein degradation for cancer therapy. *Nature Reviews Cancer* https://doi.org/10.1038/s41568-021-00365-x.

173 Jiang, D., Ni, D., Rosenkrans, Z.T. et al. (2019). Nanozyme: new horizons for responsive biomedical applications. *Chemical Society Reviews* 48: 3683–3704. https://doi.org/10.1039/C8CS00718G.

174 Meng, F., Zhu, P., Yang, L. et al. (2023). Nanozymes with atomically dispersed metal centers: structure–activity relationships and biomedical applications. *Chemical Engineering Journal* 452: 139411. https://doi.org/10.1016/j.cej.2022.139411.

175 Mi, Y., Wolfram, J., Mu, C. et al. (2016). Enzyme-responsive multi-stage vector for drug delivery to tumor tissue. *Pharmacological Research* 113: 92–99.

176 Catalano, A., Iacopetta, D., Ceramella, J. et al. (2022). Multidrug resistance (MDR): a widespread phenomenon in pharmacological therapies. *Molecules* 27: 616.

177 Tanwar, J., Das, S., Fatima, Z., and Hameed, S. (2014). Multidrug resistance: an emerging crisis. *Interdisciplinary Perspectives on Infectious Diseases* 2014: 541340. https://doi.org/10.1155/2014/541340.

178 He, C., Poon, C., Chan, C. et al. (2016). Nanoscale coordination polymers code-liver chemotherapeutics and siRNAs to eradicate tumors of cisplatin-resistant ovarian cancer. *Journal of the American Chemical Society* 138: 6010–6019. https://doi.org/10.1021/jacs.6b02486.

179 Pacios, O., Blasco, L., Bleriot, I. et al. (2020). Strategies to combat multidrug-resistant and persistent infectious diseases. *Antibiotics (Basel)* 9: 65. https://doi.org/10.3390/antibiotics9020065.

180 Borisov, S.E., d'Ambrosio, L., Centis, R. et al. (2019). Outcomes of patients with drug-resistant-tuberculosis treated with bedaquiline-containing regimens and undergoing adjunctive surgery. *Journal of Infection* 78: 35–39. https://doi.org/10.1016/j.jinf.2018.08.003.

181 Lammers, T., Sofias, A.M., Van der Meel, R. et al. (2020). Dexamethasone nanomedicines for COVID-19. *Nature Nanotechnology* 15: 622–624. https://doi.org/10.1038/s41565-020-0752-z.

182 Sun, Q., Bai, X., Sofias, A.M. et al. (2020). Cancer nanomedicine meets immunotherapy: opportunities and challenges. *Acta Pharmacologica Sinica* 41: 954–958. https://doi.org/10.1038/s41401-020-0448-9.

183 Priem, B., van Leent, M.M., Teunissen, A.J. et al. (2020). Trained immunity-promoting nanobiologic therapy suppresses tumor growth and potentiates checkpoint inhibition. *Cell* 183: 786–801.e719. https://doi.org/10.1016/j.cell.2020.09.059.

184 Liu, J., Zhang, R., and Xu, Z.P. (2019). Nanoparticle-based nanomedicines to promote cancer immunotherapy: recent advances and future directions. *Small* 15: e1900262. https://doi.org/10.1002/smll.201900262.

185 Bird, L. (2019). Targeting trained immunity. *Nature Reviews Immunology* 19: 2–3. https://doi.org/10.1038/s41577-018-0097-0.

186 Mi, Y., Hagan, C.T. IV, Vincent, B.G., and Wang, A.Z. (2019). Emerging nano-/microapproaches for cancer immunotherapy. *Advancement of Science* 6: 1801847.

187 Mai, D., June, C.H., and Sheppard, N.C. (2022). In vivo gene immunotherapy for cancer. *Science Translational Medicine* 14: eabo3603. https://doi.org/10.1126/scitranslmed.abo3603.

188 Allen, G.M., Frankel, N.W., Reddy, N.R. et al. (2022). Synthetic cytokine circuits that drive T cells into immune-excluded tumors. *Science* 378: eaba1624. https://doi.org/10.1126/science.aba1624.

189 Bear, A.S., Vonderheide, R.H., and O'Hara, M.H. (2020). Challenges and opportunities for pancreatic cancer immunotherapy. *Cancer Cell* 38: 788–802. https://doi.org/10.1016/j.ccell.2020.08.004.

190 Dobrovolskaia, M.A., Aggarwal, P., Hall, J.B., and McNeil, S.E. (2008). Preclinical studies to understand nanoparticle interaction with the immune system and its potential effects on nanoparticle biodistribution. *Molecular Pharmaceutics* 5: 487–495. https://doi.org/10.1021/mp800032f.

191 Iwama, R.E. and Moran, Y. (2023). Origins and diversification of animal innate immune responses against viral infections. *Nature Ecology & Evolution* 7: 182–193. https://doi.org/10.1038/s41559-022-01951-4.

192 Manshian, B.B., Jimenez, J., Himmelreich, U., and Soenen, S.J. (2017). Presence of an immune system increases anti-tumor effect of Ag nanoparticle treated mice. *Advanced Healthcare Materials* 6: 1601099.

193 Yang, K., Han, W., Jiang, X. et al. (2022). Zinc cyclic di-AMP nanoparticles target and suppress tumours via endothelial STING activation and tumour-associated macrophage reinvigoration. *Nature Nanotechnology* 17: 1322–1331. https://doi.org/10.1038/s41565-022-01225-x.

194 Tang, W., Yang, J., Yuan, Y. et al. (2017). Paclitaxel nanoparticle awakens immune system to fight against cancer. *Nanoscale* 9: 6529–6536.

195 Garcia, P.A., Ge, Z., Moran, J.L., and Buie, C.R. (2016). Microfluidic screening of electric fields for electroporation. *Scientific Reports* 6: 21238. https://doi.org/10.1038/srep21238.

196 Weaver, J.C. and Chizmadzhev, Y.A. (1996). Theory of electroporation: a review. *Bioelectrochemistry and Bioenergetics* 41: 135–160. https://doi.org/10.1016/S0302-4598(96)05062-3.

197 Ni, K., Lan, G., Veroneau, S.S. et al. (2018). Nanoscale metal-organic frameworks for mitochondria-targeted radiotherapy-radiodynamic therapy. *Nature Communications* 9: 4321.

198 Jaffee, E.M., Van Dang, C., Agus, D.B. et al. (2017). Future cancer research priorities in the USA: a Lancet Oncology Commission. *The Lancet Oncology* 18: e653–e706. https://doi.org/10.1016/S1470-2045(17)30698-8.

199 Gu, T., Wang, Y., Lu, Y. et al. (2019). Platinum nanoparticles to enable electrodynamic therapy for effective cancer treatment. *Advanced Materials* 31: 1806803. https://doi.org/10.1002/adma.201806803.

200 Yang, H., Wang, Q., Li, Z. et al. (2018). Hydrophobicity-adaptive nanogels for programmed anticancer drug delivery. *Nano Letters* 18: 7909–7918.

201 Adir, O., Poley, M., Chen, G. et al. (2020). Integrating artificial intelligence and nanotechnology for precision cancer medicine. *Advanced Materials* 32: 1901989. https://doi.org/10.1002/adma.201901989.

202 Yang, B. and Shi, J. Chemistry of advanced nanomedicines in cancer cell metabolism regulation. *Advanced Science* 32: 2001388. https://doi.org/10.1002/advs.202001388.

203 Ho, D., Wang, P., and Kee, T. (2019). Artificial intelligence in nanomedicine. *Nanoscale Horizons* 4: 365–377. https://doi.org/10.1039/C8NH00233A.

204 堵建岗 & 激光生物学报 (2015). 劳. J. 光动力学结合声动力学治疗对小鼠鳞癌细胞超微结构的影响. 33-36+65.

205 李维娜 et al. (2010). 光动力学疗法和声动力学疗法. 27, 5.

206 Li, J.-H., Chen, Z.-Q., Huang, Z. et al. (2013). In vitro study of low intensity ultrasound combined with different doses of PDT: effects on C_6 glioma cells. *Oncology Letters* 5: 702–706.

207 赵枭雄 (2022). 多模铁基纳米药物联合纳秒脉冲抗肿瘤疗效及其机制研究 博士 thesis, 北京科技大学.

208 电场和超声波结合治疗癌症 %J 上海生物医学工程. 26, 1 (2005).

209 Melo, W.d.C.M.A.d., Celieiūt-Germanien, R., Imonis, P., and Virulence, A.S.J. (2021). Antimicrobial photodynamic therapy (aPDT) for biofilm treatments. Possible synergy between aPDT and pulsed electric fields. *Virulence* 12: 2247–2272.

210 de Cássia Martins Antunes Melo, W., Lee, A.N., Perussi, J.R., and Hamblin, M.R. (2013). Electroporation enhances antimicrobial photodynamic therapy mediated by the hydrophobic photosensitizer, hypericin. *Photodiagnosis and Photodynamic Therapy* 10: 647–650.

211 德国磁感应研究取得突破性进展 (2003). http://muchong.com/html/201108/3451682.html.

212 癌症治疗大势所趋!新型多模态冷冻消融技术+靶向治疗! (2020). https://www.innomd.org/article/5f6d53b0218ba11792991907.

213 "冷热"兼攻,"交复"联手,全球首发多模态肿瘤治疗系统,迎来抗癌新时代 (2023). https://new.qq.com/rain/a/20230715A06IMB00.

214 「初创」顶尖的医疗器械初创企业系列(四) (2023). https://baijiahao.baidu.com/s?id=1750792633833377119&wfr=spider&for=pc.

215 Wenjun, X., Xie, X., Wu, H. et al. (2022). Pulsed electromagnetic therapy in cancer treatment: progress and outlook. *Wiley Online Library* 3: 11–12.

216 赵枭雄 (2022). 多模铁基纳米药物联合纳秒脉冲抗肿瘤疗效及其机制研究, 北京科技大学.

217 Son, S., Kim, J., Kim, J. et al. (2022). Cancer therapeutics based on diverse energy sources. *Chemical Society Reviews* 51: 8201–8215. https://doi.org/10.1039/D2CS00102K.

218 Dimcevski, G., Kotopoulis, S., Bjånes, T. et al. (2016). A human clinical trial using ultrasound and microbubbles to enhance gemcitabine treatment of inoperable pancreatic cancer. *Journal of Controlled Release* 243: 172–181.

219 Duan, L., Yang, L., Jin, J. et al. (2020). Micro/nano-bubble-assisted ultrasound to enhance the EPR effect and potential theranostic applications. *Theranostics* 10: 462–483.

220 Morse, S.V., Mishra, A., Chan, T.G. et al. (2022). Liposome delivery to the brain with rapid short-pulses of focused ultrasound and microbubbles. *Journal of Controlled Release* 341: 605–615.

221 Morse, S.V., Pouliopoulos, A.N., Chan, T.G. et al. (2019). Rapid short-pulse ultrasound delivers drugs uniformly across the murine blood–brain barrier with negligible disruption. *Radiology* 291: 459–466.

222 Theek, B., Baues, M., Ojha, T. et al. (2016). Sonoporation enhances liposome accumulation and penetration in tumors with low EPR. *Journal of Controlled Release* 231: 77–85.

223 Centelles, M.N., Wright, M., So, P.W. et al. (2018). Image guided thermosensitive liposomes for focused ultrasound drug delivery: using NIRF labelled lipids and topotecan to visualise the effects of hyperthermia in tumours. *Journal of Controlled Release* 280: 87–98.

224 Fan, Z., Wang, Y., Li, L. et al. (2022). Tumor-homing and immune-reprogramming cellular nanovesicles for photoacoustic imaging-guided phototriggered precise chemoimmunotherapy. *ACS Nano* 16: 16177–16190.

225 Thebault, C.J., Ramniceanu, G., Boumati, S. et al. (2020). Theranostic MRI liposomes for magnetic targeting and ultrasound triggered release of the antivascular CA_4P. *Journal of Controlled Release* 322: 137–148. https://doi.org/10.1016/j.jconrel.2020.03.003.

226 Chen, L.J., Yang, C.X., and Yan, X.P.J.A.C. (2017). Liposome-coated persistent luminescence nanoparticles as luminescence trackable drug carrier for chemotherapy. *Analytical Chemistry* 89: 6936–6939.

227 Liu, J., He, S., Luo, Y. et al. (2022). Tumor-microenvironment-activatable polymer nano-immunomodulator for precision cancer photoimmunotherapy. *Advanced Materials* 34: e2106654. https://doi.org/10.1002/adma.202106654.

228 Lu, Y., Song, G., He, B. et al. (2020). Strengthened tumor photodynamic therapy based on a visible nanoscale covalent organic polymer engineered by microwave assisted synthesis. *Advanced Functional Materials* 30: 2004834. https://doi.org/10.1002/adfm.202004834.

229 Maldiney, T., Bessière, A., Seguin, J. et al. (2014). The in vivo activation of persistent nanophosphors for optical imaging of vascularization, tumours and grafted cells. *Nature Materials* 13: 418–426. https://doi.org/10.1038/nmat3908.

230 Wang, J., Ma, Q., Hu, X.X. et al. (2017). Autofluorescence-free targeted tumor imaging based on luminous nanoparticles with composition-dependent size and persistent luminescence. *ACS Nano* 11: 8010–8017.

231 Zheng, B., Bai, Y., Chen, H. et al. (2018). Near-infrared light-excited upconverting persistent nanophosphors in vivo for imaging-guided cell therapy. *ACS Applied Materials & Interfaces* 10: 19514–19522. https://doi.org/10.1021/acsami.8b05706.

232 Cho, M., Cervadoro, A., Ramirez, M.R. et al. (2017). Assembly of iron oxide nanocubes for enhanced cancer hyperthermia and magnetic resonance imaging. *Nanomaterials (Basel)* 7: 72.

233 Du, Y., Liu, X., Liang, Q. et al. (2019). Optimization and design of magnetic ferrite nanoparticles with uniform tumor distribution for highly sensitive MRI/MPI performance and improved magnetic hyperthermia therapy. *Nano Letters* 19: 3618–3626.

234 Hirsch, L.R., Stafford, R.J., Bankson, J.A. et al. (2003). Nanoshell-mediated near-infrared thermal therapy of tumors under magnetic resonance guidance. *Proceedings of the National Academy of Sciences of the United States of America* 100: 13549–13554.

235 Mai, B.T., Balakrishnan, P.B., Barthel, M.J. et al. (2019). Thermoresponsive iron oxide nanocubes for an effective clinical translation of magnetic hyperthermia and heat-mediated chemotherapy. *ACS Applied Materials & Interfaces* 11: 5727–5739.

236 Maier-Hauff, K., Ulrich, F., Nestler, D. et al. (2011). Efficacy and safety of intratumoral thermotherapy using magnetic iron-oxide nanoparticles combined with external beam radiotherapy on patients with recurrent glioblastoma multiforme. *Journal of Neuro-Oncology* 103: 317–324.

237 Rastinehad, A.R., Anastos, H., Wajswol, E. et al. (2019). Gold nanoshell-localized photothermal ablation of prostate tumors in a clinical pilot device study. *Proceedings of the National Academy of Sciences of the United States of America* 116: 18590–18596.

238 Lee, W., Jeon, M., Choi, J. et al. (2020). Europium-diethylenetriaminepentaacetic acid loaded radioluminescence liposome nanoplatform for effective radioisotope-mediated photodynamic therapy. *ACS Nano* 14: 13004–13015. https://doi.org/10.1021/acsnano.0c04324.

239 Liu, Y., Liu, Y., Bu, W. et al. (2015). Hypoxia induced by upconversion-based photodynamic therapy: towards highly effective synergistic bioreductive therapy in tumors. *Angewandte Chemie, International Edition* 54: 8105–8109. https://doi.org/10.1002/anie.201500478.

240 Yu, Z., Zhou, P., Pan, W. et al. (2018). A biomimetic nanoreactor for synergistic chemiexcited photodynamic therapy and starvation therapy against tumor metastasis. *Nature Communications* 9: 1–9.

241 Larson, S.M., Carrasquillo, J.A., Cheung, N.K.V., and Press, O.W. (2015). Radioimmunotherapy of human tumours. *Nature Reviews Cancer* 15: 509–509.

242 Song, L., Li, P.P., Yang, W. et al. (2018). Low-dose X-ray activation of W(VI)-doped persistent luminescence nanoparticles for deep-tissue photodynamic therapy. *Advanced Functional Materials* 28: 1707496.

243 Bort, G., Lux, F., Dufort, S. et al. (2020). EPR-mediated tumor targeting using ultrasmall-hybrid nanoparticles: from animal to human with theranostic AGuIX nanoparticles. *Theranostics* 10: 1319–1331.

244 Ma, M., Huang, Y., Chen, H. et al. (2015). Bi_2S_3-embedded mesoporous silica nanoparticles for efficient drug delivery and interstitial radiotherapy sensitization. *Biomaterials* 37: 447–455.

245 Marill, J., Anesary, N.M., Zhang, P. et al. (2014). Hafnium oxide nanoparticles: toward an in vitro predictive biological effect? *Radiation Oncology* 9: https://doi.org/10.1186/1748-717x-9-150.

246 IsoRay. http://www.maydeal.com/news/7829.html.

247 Kulbacka, J., Pucek, A., Wilk, K.A. et al. (2016). The effect of millisecond pulsed electric fields (msPEF) on intracellular drug transport with negatively charged large nanocarriers made of solid lipid nanoparticles (SLN): in vitro study. *The Journal of Membrane Biology* 249: 645–661.

248 Thiesen, B. and Jordan, A. (2008). Clinical applications of magnetic nanoparticles for hyperthermia. *International Journal of Hyperthermia* 24: 467–474. https://doi.org/10.1080/02656730802104757.

249 Faruq Mohammad, A. and Al-Lohedan, H.A. (2020). Multifunctional cancer targeting nanoparticles. US10561747B1.

250【首发】布局行业领先的多模态能量平台产品,迈微医疗完成数千万元 Pre-A 轮融资 (2023). http://news.sohu.com/a/650021191_133140.

251 Shenzhen Maiwei Medical Technology Co, L. (2022). *L. ablation catheter and multimodal ablation equipment.* Chinese invention patent, approval number: CN217907970U.

252 睿笛生物:以全球领先的纳秒脉冲电场技术, 突破肿瘤消融禁区, 布局心血管疾病 (2021). https://www.sohu.com/a/484907888_133140.

253 陈新华 & 等. 国家科技重大专项"原始创新型纳秒刀精准消融肝癌抗复发转移的研发及临床应用研究:项目编号:2018ZX10301201"结题报告 (2021).

254 Chen Yonggang (2021). *An adaptive pulse ablation instrument based on electrocardiogram waveform.* Chinese invention patent, approval number: ZL201910247941.8.

255 Chen Yonggang (2021). *A ablation electrode positioning system.* Chinese invention patent, approval number: ZL201811023704.5.

256 Yonggang, C. (2021). *A closed-loop control system for pulse ablation.* China invention patent, issued number: ZL202111466101.4.

257 Yonggang, C. (2022). *A miniaturized nanosecond pulse generation system for tumor ablation.* China Invention Patent Issued number: ZL202111495485.2.

258 Shanghai Meijie Medical Technology Co., L. (2023). *Multimodal tumor ablation probe system and its control method.* China patent.

259 Wang, Z., Zhang, F., Shao, D. et al. (2019). Janus nanobullets combine photodynamic therapy and magnetic hyperthermia to potentiate synergetic anti-metastatic immunotherapy. *Advanced Science (Weinh)* 6: 1901690.

260 University, P (2020). *Superconducting rotating frame for laser accelerated proton cancer treatment device.* Chinese invention patent, approval number: CN111790063A.

261 Yang, X., Wu, Q., Zhao, X. et al. (2023). Pulsed electric field technology translational medicine. *Small* submitted.

262 Song, Y. and Yang, X. (2023). *An instrument for the nanomedicine mediating multi-physics coupling ablation.* China patent.

263 *Autonomous ultrasound guided endoscope.* USA patent (2022).

264 MIT. (2000). *Effect of electric field and ultrasound for transdermal drug delivery.*

265 Holiday, G. A. (2012). *Pulsed power laser actuated catheter system for interventional oncology.* US Patent, approval number: US2012165808A1.

266 LTD, I. T (2013). *Immunogenic treatment of cancer.* Application number: WO2013079980A1.

267 宋玉军 and 等. 国家科技重大专项"原始创新型纳秒刀精准消融肝癌抗复发转移的研发及临床应用研究: 项目编号:2018ZX10301201"课题四"纳秒脉冲联合纳米药物精准消融肝癌及其机制研究"结题报告. (2021).

268 Yujun, S. (2009). *Preparation process of single nanoparticle and its array based biomolecular detectors.* China Invention Patent Issueed number: ZL 200910085973.9.

269 Wenqi, S., Yujun, S., and Wang, J (2015). *A anticancer alloy nanodrug with autonomous targeting and imaging functions and its preparation method.* China Invention Patent Issueed number: ZL 201510395518.4.

270 Yujun, S. (2018). *Construction of composite nanomedicine by coupling drug components with nanoparticles, preparation method and application.*

271 Yujun, S. and Jugang, M. (2020). *A device and method for large-scale continuous preparation of metal nanoparticles.* Chinese invention patent approval number: ZL 2020 11462174.1.

272 Ohki, Y., Munakata, K., Matsuoka, Y. et al. (2022). Nitrogen reduction by the Fe sites of synthetic [Mo_3S_4Fe] cubes. *Nature* 607: 86–90. https://doi.org/10.1038/s41586-022-04848-1.

273 Li, X., Khorsandi, S., Wang, Y. et al. (2022). Cancer immunotherapy based on image-guided STING activation by nucleotide nanocomplex-decorated ultrasound microbubbles. *Nature Nanotechnology* 17: 891–899. https://doi.org/10.1038/s41565-022-01134-z.

274 Mathew, A. and Pradeep, T. (2014). Noble metal clusters: applications in energy, environment, and biology. *Particle & Particle Systems Characterization* 31: 1017–1053. https://doi.org/10.1002/ppsc.201400033.

275 Chakraborty, P., Nag, A., Chakraborty, A., and Pradeep, T. (2019). Approaching materials with atomic precision using supramolecular cluster assemblies. *Accounts of Chemical Research* 52: 2–11. https://doi.org/10.1021/acs.accounts.8b00369.

276 Jena, P. and Sun, Q. (2018). Super atomic clusters: design rules and potential for building blocks of materials. *Chemical Reviews* 118: 5755–5870. https://doi.org/10.1021/acs.chemrev.7b00524.

277 Jena, P. (2022). Superatomic chemistry. *Journal of the Indian Chemical Society* 99: 100350. https://doi.org/10.1016/j.jics.2022.100350.

278 Devault, D., Parkes, J.H., and Chance, B. (1967). Electron tunnelling in cytochromes. *Nature* 215: 642–644.

279 O'Brien, E., Holt, M.E., Thompson, M.K. et al. (2017). The [4Fe4S] cluster of human DNA primase functions as a redox switch using DNA charge transport. *Science* 355: eaag1789. https://doi.org/10.1126/science.aag1789.

280 Wang, Y., Liu, R., Chen, T. et al. (2024). CoFe@[$CoFe_2O_4$/Fe(Co)Nx-C] nanoparticles prepared via an ultrasonic atomization coupling pyrolysis process and their magnetic and optical properties. *Journal of Alloys and Compounds* 970: 172358. https://doi.org/10.1016/j.jallcom.2023.172358.

281 Tong, X., Linong, L., Xiaona, Y. et al. (2021). N,P-codoped graphene dots supported on N-doped 3D graphene as metal-free catalysts for oxygen reduction. *ACS Applied Materials & Interfaces* 13: 30512–30523. https://doi.org/10.1021/acsami.1c03141.

282 Liao, Q., Hou, W., Liao, K. et al. (2022). Solid-phase sintering and vapor-liquid-solid growth of BP@MgO quantum dot crystals with a high piezoelectric response. *Journal of Advanced Ceramics* 11: 1725–1734. https://doi.org/10.1007/s40145-022-0643-x.

283 Ma, J., Tong, X., Wang, J. et al. (2019). Rare-earth metal oxide hybridized PtFe nanocrystals synthesized via microfluidic process for enhanced electrochemical catalytic performance. *Electrochimica Acta* 299: 80–88. https://doi.org/10.1016/j.electacta.2018.12.132.

284 Ma, J., Wang, L., Deng, Y. et al. (2021). Mass production of high performance single atomic FeNC electrocatalysts via a sequenced ultrasonic atomization, pyrolysis. *Science China Materials* 64: 631–641.

285 Ma, J., Wu, Q., Zhang, W. et al. (2022). Synergetic contribution of Co^{3+}/Co^{2+} and FeNC in CoFe@CoFe$_2$O$_4$ toward efficient electrocatalysts for oxygen reduction reaction. *Electrochimica Acta* 432: 141224. https://doi.org/10.1016/j.electacta.2022.141224.

286 Liang, H., Wang, Z., Junmei, W. et al. Synthesis of Fe$_{(1-x)}$Zn$_x$@Zn$_{(1-y)}$Fe$_y$O$_z$ nanocrystals via a simple programmed microfluidic process. *Materials Chemistry and Physics* 201: 156–164.

287 Wang, J., Wang, Z., Li, S. et al. (2017). Surface and interface engineering of FePt/C nanocatalysts for electro-catalytic methanol oxidation: enhanced activity and durability. *Nanoscale* 9: 4037–4312. https://doi.org/10.1039/c6nr09122a.

288 Wang, J., Zhao, H., Zhu, Y., and Song, Y. (2017). Shape-controlled synthesis of CdSe nanocrystals via a programmed microfluidic process. *Journal of Physical Chemistry C* 121: 3567–3572. https://doi.org/10.1021/acs.jpcc.6b10901.

289 Wang, R., Yang, W., Song, Y. et al. (2015). A general strategy for nanohybrids synthesis via coupled competitive reactions controlled in a hybrid process. *Scientific Reports* 5: 9189.

290 Zhao, X., Liang, H., Chen, Y. et al. (2021). Magnetic field coupling microfluidic synthesis of diluted magnetic semiconductor quantum dots: the case of Co doping ZnSe quantum dots. *Journal of Materials Chemistry C* 9: 4619–4627. https://doi.org/10.1039/D0TC06026G.

291 Huang, L., Li, Y., Du, Y. et al. (2019). Mild photothermal therapy potentiates anti-PD-L1 treatment for immunologically cold tumors via an all-in-one and all-in-control strategy. *Nature Communications* 10: 4871. https://doi.org/10.1038/s41467-019-12771-9.

292 Shi, L., Wang, J., Ding, N. et al. (2019). Inflammation induced by incomplete radiofrequency ablation accelerates tumor progression and hinders PD-1 immunotherapy. *Nature Communications* 10: 5421. https://doi.org/10.1038/s41467-019-13204-3.

293 Jumper, J., Evans, R., Pritzel, A. et al. (2021). Highly accurate protein structure prediction with AlphaFold. *Nature* 596: 583–589. https://doi.org/10.1038/s41586-021-03819-2.

294 Fan, K., Cheng, L., and Li, L. (2021). Artificial intelligence and machine learning methods in predicting anti-cancer drug combination effects. *Briefings in Bioinformatics* 22: https://doi.org/10.1093/bib/bbab271.

295 Decherchi, S., Berteotti, A., Bottegoni, G. et al. (2015). The ligand binding mechanism to purine nucleoside phosphorylase elucidated via molecular dynamics and machine learning. *Nature Communications* 6: 6155. https://doi.org/10.1038/ncomms7155.

296 Stokes, J.M., Yang, K., Swanson, K. et al. (2020). A deep learning approach to antibiotic discovery. *Cell* 180: 688–702.e613. https://doi.org/10.1016/j.cell.2020.01.021.

2

Fundamentals in Nanomedicine
Xiangrong Song, Mengran Guo, Zhongshan He, Xing Duan, and Wen Xiao

Sichuan University, State Key Laboratory of Biotherapy and Cancer Center, West China Hospital, Department of Critical Care Medicine, No. 37 Guoxue Lane, Wuhou District, Chengdu 610041, China

2.1 Design Theory of Nanomedicines

Nanomedicine is an interdisciplinary field by translating nanotechnology to specific drugs or therapies for the prevention and treatment of some intractable diseases. Comparatively, nanomedicines can break through the limitations of conventional disease therapies through improving many features of conventional therapeutics, including stability and solubility, circulation time, and biocompatibility. Most importantly, nanomedicines can overcome some in vivo biological barriers of cell microenvironment or tissues, targeted deliver drugs into specific lesion sites, and enhance cell internalization, thus highlighting the unique superiority of nanomedicines for targeted therapies. Therefore, it is of considerable interest and promise for disease management to design different targeted delivery strategies according to the characteristics of different lesion types. In this chapter, we will summarize the development and application of nanomedicines, including main design strategies, passive targeting and active targeting functions, auto-targeted and imaging abilities, and in vitro or in vivo 3D traceable treatment of diseases.

2.1.1 Passive Targeting Function Design of Nanomedicines

Passive targeted nanotherapeutics have the potential to increase the accumulation of drugs in the disease lesional areas by utilizing mechanisms such as endocytosis, fusion, adsorption, and material exchange of nanocarriers within the affected tissue [1]. Furthermore, the enhanced permeability and retention (EPR) effect serves as the predominant mechanism of passive targeting, attributed to the discontinuous arrangement of vascular endothelial cells in lesional sites. This facilitates the penetration of nanodelivery system more easily into cells compared to adjacent normal tissue (high permeability) [2]. It is worth noting that the size, surface charge, and shape of nanocarriers play an essential impact on the EPR effect [3].

Nanomedicine: Fundamentals, Synthesis, and Applications, First Edition. Edited by Yujun Song.
© 2025 WILEY-VCH GmbH. Published 2025 by WILEY-VCH GmbH.

2.1.1.1 Size-dependent Biological Functions for Nanomedicine Design

The EPR effect will be different based on the different disease or lesional tissue, and thus the size of the nanoparticles allowed to pass will also be different. Liu et al. researchers demonstrated that lesser than 50 nm and larger than 300 nm of liposomes were easily captured by the disease of liver tissue. Moreover, the liposomes with a diameter greater than 400 nm are easily eliminated by spleen. The liposomes (size of 90–200 nm) tended to be accumulated at the lesion site of tumor [4]. Additionally, the diameter of 100–140 nm PEG nanoparticles showed efficient accumulation capacity at tumor sites, while the slightly larger nanoparticles (size larger than 150 nm) were rapidly cleared by the liver [5]. On the other hand, different nanoparticle sizes influence the ability of blood vessels to exude, the penetration of tumor tissues, and the ability of tumor cells to phagocytose. In addition, designing and constructing a series of nanoparticles with adjustable sizes in the disease microenvironment can enhance the enrichment and tissue penetration of nanoparticles in the lesion sites. For example, in the tumor microenvironment, matrix metalloproteinases (MMPs)-responsive nanoparticles can become smaller nanoparticles, which enhances the penetration and retention of nanoparticles at lesion sites [6].

2.1.1.2 Surface Charge and PEGylation of Nanoparticles for Nanomedicine Design

The surface charge of nanoparticles plays an important impact on the blood compatibility, stability, and blood circulation of nanoparticles [7]. Specifically, the negative charge nanoparticles were more conducive to long-term blood circulation due to less absorption by the mononuclear phagocytosis system (MPS) [8]. On the contrary, the cationic surface nanoparticles were easily and quickly cleared by the MPS as they readily bind to plasma proteins and induce the formation of a protein crown [9]. Furthermore, the surface of nanoparticles was wrapped in PEGylated, which contributed to reducing opsonization increasing the circulation half-life, and effectively enhancing the accumulation of nanoparticles in the lesion site of tumor [10]. Moreover, Zentel et al. researchers demonstrated that higher PEG content of the nanoparticles can regulate the pharmacokinetic properties of nanoparticles, prolong persistent blood circulation time, and increase the accumulation of nanoparticles in tumors [11]. Besides, the above superior properties of PEGylation nanoparticles can be influenced by the modification density, molecular weight of modification density, and metabolic pathway of nanomaterials.

2.1.1.3 Shape-dependent Biological Effect of Nanoparticles for Nanomedicine Design

The shape of nanoparticles also plays important roles in their blood circulation and accumulation at lesional sites. Whether the spherical nanoparticles or other shapes, including rod, ellipse, filament, sheet, and disk can be used for the lesion site-targeted drug delivery. For instance, the rod-shaped nanoparticles exhibit an extended circulation period and enhanced accumulation at the tumor lesion site in comparison to spherical particles. Nevertheless, spherical nanoparticles still occupy a major position in nanomedicine research due to their preparation

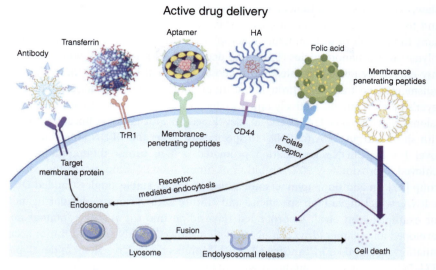

Figure 2.1 Drug-targeting strategies. Stimuli-responsive drug delivery.

method being more mature and easier to scale up in clinical conversion [12]. Furthermore, compared with spherical micelles, rod micelles significantly improve the half-life in blood, are more likely to be swallowed by target cells, and increase the accumulation of drugs in the lesion sites. Furthermore, the transition from a spherical to a nonspherical structure leads to an increased contact area and reduced drag resistance force when adhering to the endothelium, thereby enabling the particles to adhere more stably to the vascular wall surface [13]. Decuzzi et al. researchers confirmed that the disc nanoparticles be easily marginalized due to their high adhesion probability and propensity in laminar flow [14]. Therefore, regulating the shape of nanoparticles substantially promotes the marginalization of and the capture of nanoparticles in blood vessel walls, and enhances their potential to exude blood vessels.

2.1.2 Active Targeting Function Design of Nanoparticles for Nanomedicine

Active targeting uses a modified drug carrier as a "missile" to deliver drugs to the target area in a targeted manner for concentration and efficacy. Targeted nanoparticle drug delivery systems are designed to modify appropriate functional groups such as antibodies, glycoproteins, lipoproteins, transferrin, peptides, folic acid, nucleic acid, etc., on the surface of drug carriers, enabling them to bind specifically to receptors or antigens of target cells (Figure 2.1) [15].

2.1.2.1 Small Molecule and Aptamer-directed Active Targeting Function Design

Most of the small molecules as cell-targeting ligands are modified to the surface of the delivery carriers to enhance the accumulation in targeted cells or tissues, such as folate, αvβ3, and some aptamers. For instance, folic acid or vitamin B9 can efficiently bind to folate receptors that are overexpressed on the cell membranes (CMs) of many tumor cells or inflammatory macrophages. Guo et al. developed an oral lipid polymer-based nanoparticles (FA-LNPs) decorated with folic acid, which can not only effectively overcome intestinal mucosal–epithelial barrier by increasing the transmembrane transport through intestinal epithelial and the accumulation in Peyer's patches but also actively target to the aortic plaque sites and accumulate in lesional macrophages [16]. He et al. developed an FRα-targeted liposome loaded with short hairpin RNA (shRNA) to specifically target the CLDN3 molecule. The novel F-P-LP/CLDN3 formulation promoted benign tumor differentiation and achieved approximately 90% inhibition of tumor growth compared to the control group [17]. In addition, aptamers are a kind of highly stable single-stranded DNA or RNA, which have high specificity and affinity with various molecular ligands. For example, Ahn et al. incorporated floxuridine into the AS1411 aptamer and developed an aptamer–drug conjugate, which showed highly efficient tumor accumulation and uptake in cancer cells [18]. Moreover, some researchers developed PEG–PCL nanoparticles modified with the GMT8 aptamer to targeted therapy of glioblastoma. The GMT8 aptamer promoted glioblastoma penetration of drug and intracellular accumulation [19].

2.1.2.2 Protein and Peptide-directed Active Targeting Function Design

Some proteins or peptides can be modified to the surface of the delivery carriers to enhance the targeting efficiency for targeted cells or tissues. For example, cyclic RGD can be used as a ligand to target tumor tissue or cancer cells due to its high binding capacity with the overexpressed αVβ3 integrin receptor on endothelial cells or solid cancer cells [20]. Besides, the truncated bFGF peptide, which could effectively bind to FGFR1 overexpressed on endothelial cells of tumor neovasculature or tumor cells. Hence, the truncated bFGF peptide was selected to modify PLGA nanoparticles (D/P-NPs) simultaneously loaded with PEDF gene and paclitaxel in this study [21].

2.1.2.3 Antibody-directed Active Targeting Design

Vascular cell adhesion molecule-1 (VCAM-1), an adhesion molecule secreted by activated vascular endothelial cells at the site of atherosclerotic lesions, recruits inflammatory cells to the activated endothelial surface. Therefore, nanocarriers can target endothelial cells by carrying peptides on the surface of nanocarriers specifically identified by VCAM-1 in atherosclerotic plaques [22]. Furthermore, various vectors have been designed and synthesized to enhance mRNA delivery, such as LNPs, polymetric nanoparticles, and cationic nanoemulsions (CNEs) [23]. Kheirolomoom et al. developed a VHPK-CCL-anti-miR-712 NP cationic liposomes coating with the VHPK peptide on the surface of liposomes and containing anti-miR-712 molecules, which specifically targeted the aortic endothelial cells of mice. In addition, some studies showed that miRNA delivery was highly selective to the target organ, and selective and efficient silencing of inflammatory endothelial cells played an anti-AS role [24]. Tao and colleagues synthesize a macrophage-targeted peptide conjugated to the nanoparticle surface and loaded with siRNA against Camk2g (S2P-siCamk2g). CamKIIγ expression was significantly reduced in lesional macrophages in atherosclerosis-prone mice treated with S2P-siCamk2g nanoparticles compared with mice treated with no S2P targeting nanoparticles [25].

2.1.3 Auto-targeting Design

In recent years, the biomimetic drug delivery system based on CMs has attracted more and more attention because of its unique biological characteristics. Common bionic cell carriers, including platelets, red blood cells (RBCs), neutrophils, macrophages, exosomes, supramolecular CM vesicles, and stem cells have been used for drug delivery applications [26].

2.1.3.1 Platelets Targeting Function Design

The vascular endothelium provides a barrier between the subendothelial matrix and circulating cells such as hematocytes and platelets. Platelets can also accumulate and bind directly to injured endothelial cells. Various platelet surface molecules such as glycoprotein (GP)VI, GPIV, GPIb, GPIX, GPV, and GPIIb/IIIa are involved in platelet recruitment [27]. It has previously been reported that platelets could form co-aggregates with circulating CD34+ progenitors in patients with acute coronary syndromes, and these co-aggregates improve prognosis by promoting peripheral recruitment of CD34+ cells in the ischemic microcirculatory area and boosting their adhesion to the vascular lesion [28]. In the current study, to enhance pDNA delivery into the lungs, hybrid nanoparticles of DP and CM were developed from LA-4 lung epithelial cells. The CM components of the hybrid nanoparticles may interact with plasma membranes of target cells and facilitate intracellular uptake of pDNA [29].

2.1.3.2 RBC Targeting Function Design

The nanoparticle-coated RBC membrane has good biocompatibility and nonimmunogenicity, and can prolong the blood circulation time of nanocarriers. Yang et al. designed a bionic melanin nanocarrier of the red CM for photothermal treatment of tumors, which can significantly enhance tumor site and accumulation, and ultimately show good antitumor effects [30]. Although the lack of active targeting only the red cell membrane-modified vectors, erythrocyte membrane coating enhanced the accumulation of nanoparticles in tumors mainly due to effective escapes of RES recognition and improvement of circulation time.

2.1.3.3 Macrophage Targeting Function Design

Wang et al. reported a biomimetic drug delivery system derived from macrophage membrane-coated ROS-responsive nanoparticles. The macrophage membrane not only avoids the clearance of nanoparticles from the reticuloendothelial system but also leads nanoparticles to the inflammatory tissues. Moreover, the macrophage membrane sequesters proinflammatory cytokines to suppress local inflammation. The synergistic effects of pharmacotherapy and inflammatory cytokine sequestration from such a biomimetic drug delivery system led to improved therapeutic efficacy in atherosclerosis [31].

2.1.3.4 Exosomes and Supramolecular Cell Membrane Vesicle Targeting Design

Exosomes are nanoscale membrane vesicles secreted by cells, which can be used for drug delivery due to their unique biological activity. For example, Zhang and colleagues engineered an iRGD-targeted exosome to deliver DOX for targeting therapy of anaplastic thyroid carcinoma [32]. Moreover, Yu and colleagues developed a supramolecular CM vesicle, which efficiently loaded both indocyanine green and 1-methyl-tryptophan for combining PDT and immunotherapy, thereby showing high capability of tumor accumulation and antitumor efficiency [33].

These synthetic nanoparticles are not chemotactic and cannot be actively driven to the disease site. So, actively targeted modified nanoparticles are only absorbed after they are close to the CM. Recent years have seen a rise in research into auto-targeting because of the trend of automatic targeting, and the ability to guide nanoparticles to a disease site under the chemotaxis of CMs. At the same time, it has better biocompatibility and longer systemic circulation. Therefore, respective advantages of active targeting, passive targeting, and autologous targeting being combined, it may be possible to prepare nanoparticles with satisfactory targeting effects for the treatment of diseases.

2.1.4 In Vitro or In Vivo 3D Traceable Function

Nanobiotechnology and nanomedicine developed very rapidly and made remarkable achievements, especially, nanoprobes, as drug carriers or imaging tools, which ushered in a dawn and a new avenue for imaging and tracing. An important class of nanoparticles comprises those made of hybrid organic or inorganic nanoparticles such as semiconductors, metals, and metal oxides, as well as silica and rare earth minerals.

2.1.4.1 Gold Nanoparticles for Bioimaging

Gold nanosphere, as the most common type of gold nanoparticle, exhibits an intense ruby color in aqueous solutions. Gold nanoparticles are widely used in optical imaging because they excite scattered light at LSPR frequencies [34]. Furthermore, the LSPR stimulation of gold nanoparticles increases local electromagnetic fields, enhancing fluorophores and SERS reporters attached to a gold surface [35]. In addition, gold nanoparticles with near-infrared photothermal properties are ideal probes for photothermal therapy [36]. Recent studies have shown that gold nanoparticles can be heated by nonresonant shortwave radiofrequency fields, which allows the induction of targeted hyperthermia in deep tissue [37]. Last, gold nanospheres and gold nanorods can be used as superior CT contrast agents by virtue of their high material density and high number of subsequences.

2.1.4.2 Quantum Dots for Bioimaging

CdSe Qdots coated with peptides were shown to selectively target the vasculature of mouse tumors in vivo, providing a new approach to tumor vasculature imaging and monitoring the treatment for cancer. Although actively targeted nanoparticles show low toxicity to nontargeted tissues, there are still many reports of nanoparticles having some toxicity to some nontargeted tissue such as liver, spleen, kidneys, lymph nodes, heart, lungs, and bone marrow. They are deposited in those organs, penetrate CMs, lodge in mitochondria, and may trigger injurious responses. To minimize the risks brought by nanomaterials in future research, the following two basic ways can be achieved. The first is to develop new low-toxicity and high-biocompatibility nanomaterials, such as polylactic acid and silica nanoparticles. The other is surface modification of nanoparticles with biocompatible chemicals, such as polyethylene glycol (PEG), and chitosan.

2.2 Progress in the Controlled Synthesis of Nanomaterials

Nowadays, a number of nanomaterials are synthesized and applied in biomedical fields and clinical use, for example, liposomes, micelles, nanosuspension, emulsion, quantum dots, magnetic nanoparticles, and so on [38]. These synthesized nanoparticles always have unique features for application in sensing, detection, disease diagnosis, and medical equipment. Synthesis of these nanomaterials can be achieved either top-down or bottom-up. Top-down techniques are to crush large pieces of materials into nanoparticles, including mechanical milling, physical vapor deposition (PVD), lithography, and thermal evaporation pyrolysis [39]. Bottom-up methods can be divided into chemical and biological approaches [40]. Chemical approaches include chemical vapor deposition (CVD), microprecipitation, microemulsion, and supercritical methods. Biological approaches include biosynthesis in microorganisms and generating extracellular vesicle analogues through lipid self-assembly principles using plasma membrane fragments from cells. Compared with top-down methods, bottom-up methods have advantages of high product quality and high controllability. However, its lower production efficiency and higher preparation costs are also shortcomings that cannot be ignored. Nanoparticles are also synthesized from plant extracts, enzymes, agricultural waste, microorganisms, and actinomycetes [41].

2.2.1 Ball Milling Through the Mechanical Method

Ball milling is a top-down method. During the milling process, the carrier materials and milling balls made of glass, ceramics, and zirconia are grounded in a sealed container. The materials are milled to nanosize under the shear force between the milling balls, as well as the milling balls and the container wall [42]. According to the differences in production processes, this method has two modes: channel type and cycle type. In the channel mode, the coarse suspension enters from the feed tank, and the nanomaterial suspension formed by grinding is transferred to the discharge tank. To obtain uniformly sized nanocrystals, it is necessary to pour the suspension from the discharge tank into the feed tank or swap the positions of the feed tank and discharge tank for repeated grinding. In the recycling mode, the suspension only needs to be recycled and ground using one container, which improves the uniformity of the product and makes the final crystal size smaller. This method has the advantages of simple operation, low cost, and narrow particle size distribution, and has been widely used in laboratories and industry. However, the resulting nanomedicines often contain residual grinding media, which to some extent reduce the purity of the product.

2.2.2 Nanoprecipitation

An increasingly popular method for preparing nanoparticles is nanoprecipitation [43]. Poly(ethylene glycol)-poly(lactide-*co*-glycolic acid) (PEG-PLGA) is an amphiphilic block copolymer, which can be prepared into nanoparticles through self-assembly in solution using nanoprecipitation method. In an optimum case, high sheer homogenization techniques, ultracentrifugation, and surfactants are not needed for a polymer-based amphiphilic nanoparticle prepared using this method. However, this method also has some drawbacks, such as: (i) the concentration of the preparation is low, (ii) organic solvents may be introduced in some cases, and (iii) the cost of large-scale preparation is high.

2.2.3 Microfluidic Synthesis

In recent years, microfluidics have shown many advantages in the synthesis of nanomaterials and have received increasing attention. It is worth emphasizing that microfluidic technologies can be applied to different purposes, such as chemical reactions, nanostructure assembly, modification of nanoparticles, and even evaluation of biomedical functions [44]. For the application of nanomaterials, microfluidics have advantages in controlling different nanoparticle characteristics, for example, controlling multicomponent reactions, particle size, particle size distribution, morphology, and so on. The main issues to control the process are the progress in the design concept of microfluidics in chips or devices used for manufacturing nanomaterials, as well as the improved characteristics of nanomaterials generated on microfluidic devices.

Using microfluidics to precisely control the mixing effect of laminar liquid or droplet can realize the preparation of nanoparticles, which is difficult to complete by conventional liquid-mixing methods. Active small molecule compounds, protein macromolecules, vaccines, immune therapeutics, genes, and other drugs are wrapped or adsorbed in the functionalized nanocarrier structure to form a nanocarrier system, which can adjust the drug release speed, increase the permeability of biofilm, change the distribution in the body, improve bioavailability, etc., so as to improve the safety and effectiveness of drugs. The surface properties and drug-release behavior of nanodrug-carrying systems directly affect their bioactivity in vivo. The surface properties of nanocarriers can affect the interaction with physiological substances and the distribution effect of organs and tissues during drug delivery in vivo, and the controllable drug release behavior can effectively regulate the optimal therapeutic effect of drugs at the appropriate disease site and the appropriate release rate, and reduce the nonspecific damage or other toxic side effects of drugs in normal organs.

With the change of mixed liquid characteristics, channel structure size, and other factors, the flow in the microfluidic pipeline will form a variety of different flow patterns, such as plug flow, stratified flow, droplet flow, and ring flow. Thus, the fluid behavior between miscible and immiscible liquids makes it useful in the synthesis of different particles. Because the volume of fluid passing through the microfluidic channel is very small, the liquid flow rate and velocity are accurately regulated by the driving device, just like a microquantitative precision reactor. The particles of uniform size and good repeatability can be quickly obtained between the miscible liquids by controlling the fluid and intensifying the mixing. A uniform single or multiple emulsion drop structure was obtained between the immiscible liquids by preserving the interface and cutting the boundary fluid to the intermediate fluid during the mixing process.

In view of the characteristics of micro- and nanoscale fluid, the structure design of a microfluidic chip channel is particularly important. The structure of the channel (size, channel shape, wall surface, etc.) can significantly affect the fluid state. Microfluidic channel is generally composed of entrance, main channel, auxiliary channel, and exit. The main channel is the main part of the fluid mixing, separation, and related reactions. The auxiliary channel introduces fluid into the main channel for mixing, separation and reaction with the Y, T, and fan structures, and finally, the fluid flows out from the outlet after the physical or chemical reaction (Figure 2.2).

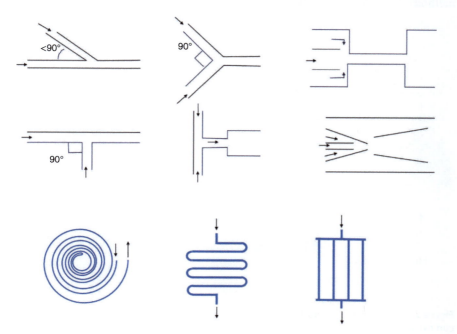

Figure 2.2 Different microfluidic flows of two-phase flow.

The precise control of fluid velocity and flow rate by microfluidic technology can make the liquid components introduced at different times fully mixed and highly uniform and orderly, and the prepared nanoparticles show obvious advantages in particle structure uniformity, batch-to-batch repeatability and drug-loading rate.

2.2.4 Personalized Protein Nanoparticles

Nanoparticles fabricated from metals, polymers, and different carbon allotropes may cause toxicity, inflammation, and immune activity after being administered into human bodies. These responses could be avoided by decorating nanoparticles with components derived from the host, such as proteins. Proteins are central to biological functions, ranging from DNA repair, and metabolic reaction catalysis to cell signaling. Personalized protein nanoparticles (PNPs) are size- and shape-tunable nanoparticles [45] made from patient-derived proteins (Figure 2.3). The PNP can be degraded in vivo and does not induce innate or adaptive immunity after single and repeated administrations. In order to obtain a nanoparticle with special in vivo properties, PNPs can be modified with specific protein cargos. Additionally, PNPs derived from different patients could generate unique molecular fingerprints.

2.2.5 Biological Entities as Chemical Reactors

A biosynthetic method uses different biological resources such as plants, fungi, and microorganisms to produce nanoparticles. Compared with chemical and physical methods, biological methods have many advantages, such as: (i) environmental

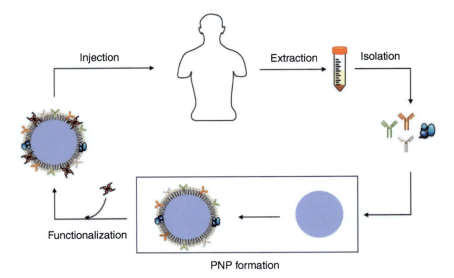

Figure 2.3 Schematic illustration of integrating patient-specific nanomaterials into a clinical setting.

Figure 2.4 Biological synthesis of NPs shows distinct advantages over physical and chemical methods.

protection, (ii) prevention of pollution and waste generation, (iii) efficient synthesis, and (iv) saving time and energy (Figure 2.4).

Fungi have been applied to synthesize nanoparticles. Compared with other biological methods, the use of fungal biomass filtrate containing different metabolites as a green synthesis method for the manufacture of metal nanoparticles is more economical and efficient. At present, many fungi have been widely used in the biosynthesis of nanoparticles, such as *Penicillium*, *Aspergillus nidus*, *Trichoderma*, *Claviculus*, *Fusarium oxysporum*, *Oryzolase*, and so on. Different fungi can synthesize different types of nanoparticles. The widespread use of fungi is due to their ability to secrete large amounts of proteins or enzymes, and they are easy to grow and reproduce, and easy to subculture renewable resources, with great potential to produce compounds that can be used in different directions [46]. The use of fungi as reducing agents and stabilizers in the biosynthesis of silver nanoparticles has the advantages of producing a large amount of protein, high yield, simple operation, and low toxicity.

Plant-mediated green synthesis of nanoparticles is the synthesis of nanoparticles using plant extracts. Compared to microorganisms, using plant extracts has cost-effectiveness and environmental advantages [47]. The synthesis of nanoparticles based on plant materials refers to the reduction of metal ions into nanoparticles using biomolecules (such as water-soluble plant metabolites and coenzymes) in extracts of stems, leaves, etc., at room temperature and atmospheric pressure. So far, plant parts such as leaves, stems, roots, flowers, seeds, bark, and their metabolites have been successfully used for the effective biosynthesis of nanoparticles.

The biosynthesis and practical application of nanoparticles mostly remain in the laboratory stage, and the synthesis mechanisms and exact metabolites involved in the bioreduction process are still unclear. The potential mechanisms still need to be further studied and explored.

2.3 Progress in the Surface Modification and Functionalization of Nanomaterials for Nanomedicines

Due to the small particles of nanomedical drugs, they have high surface energy, resulting in poor dispersion and easy interaction between nanoparticles and sediments. The modification of nanomedical drugs can improve their stability and prolong the storage time of drugs, which has great market value and application value. In addition, the distribution, absorption, metabolism, and excretion of nanomaterials in organisms are easily affected by the complex physiological environment in the body. Specific modification strategies for different barriers of nanomaterials in organisms can significantly change the systemic circulation time, organ and cell targeting, and effect of nanomaterials.

In view of the interaction between nanoparticles, many methods have been developed to avoid agglomeration. For example, antiflocculants, surfactants, and polymer chelating agents are added to the reaction system to increase the mutual repulsion between nanoparticles and avoid the agglomeration of nanoparticles. On the other hand, in order to better help nanomaterials cope with the complex physiological barriers in vivo, a variety of nanomaterial modification methods have been developed. For example, by coupling hydrophilic and hydrophobic components on the surface of nanomaterials, the systemic circulation time of nanoparticles can be improved, and the drugs loaded with nanoparticles can be controlled and slow-released. Another example is the modification of materials with different electrical properties on the surface of nanotherapeutic drugs, thus changing the organ selectivity of nanotherapeutic drugs. In addition, a large number of studies also focus on different ligands on the surface modification of nanotherapeutic drugs, so that they can specifically interact with different cells in the body and exert their effects in a targeted way.

2.3.1 How Surface Modification Can Improve the Stability of Nanomedicine

The DLVO theory of colloid stability was proposed by Deriagwin and Landau of the Soviet Union in 1941 and Verwey and Overbeek of the Netherlands in 1948 [48, 49]. The core idea is that the stability of the colloid depends on the total potential energy U of the system, which depends on the sum of the repulsive and gravitational potential energy. In other words, when the repulsion between particles is greater than the attraction, the sol tends to be stable. The repulsive force between particles is mainly related to two factors: charge and steric hindrance (Figure 2.5).

2.3.1.1 Modification of Charged Materials

The higher the absolute value of Zeta potential, the more stable the dispersion system is. The dividing line of dispersion stability of nanoparticles in aqueous phase is generally considered to be $+30$ mV or -30 mV. In other words, if the Zeta potential of nanoparticles in the water phase is higher than $+30$ mV or lower than -30 mV, the dispersion system is relatively stable. Therefore, cationic polymers such as

Figure 2.5 Classical DLVO theory of colloid stability.

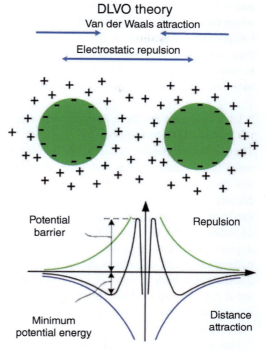

poly(L-lysine) (PLL), poly(ethylimine) (PEI), chitosan, and diethylaminoethyl glucan (DEAE-DEX) can significantly improve the stability of nanoparticles [50–54]. At the same time, since CMs are charged, nanoparticle surface charges play an important role in active cell interactions and cell uptake. Since the CM is negatively charged, the internalization level of positively charged nanoparticles modified with cationic polymers is higher than that of electronegative and neutral nanoparticles [51, 55]. At the same time, nanoparticles with positive charge are also effective in mucosal delivery. For example, increased uptake of chitosan-coated nanoparticles in vivo was achieved by enhancing gastrointestinal mucosal permeation [56]. The correlation of several studies had reported effective uptake of crosslinked chitosan nanoparticles by Caco-2 cells (e.g. particles with a zeta potential of +15 to +30 mV). Positively charged nanoparticles also showed increased penetration through the skin [57, 58].

Although the modification of positively charged materials has obvious advantages in cell uptake, the modification of negatively charged materials also has unique advantages in promoting the stability of nanoparticles, in vivo distribution, and reducing the toxicity of preparations. For example, the negatively charged nanoparticles prepared by modifying several anionic compounds such as hyaluronic acid, sodium alginate, and Dex-Aco onto cationic liposomes used for delivering mRNA drugs have the advantages of high safety and high blood stability [59, 60].

Electric charge also affects the distribution of nanomaterials in the body and cationic nanoparticles can cross the blood–brain barrier (BBB) to a greater extent, thereby increasing brain penetration [61]. However, high cationic and anionic nanoparticles have been found to cause charge-mediated disruption of the BBB, while neutral and slightly negative Nanoparticles are more effective in delivering therapeutic drugs through the brain, preserving BBB integrity [62]. At the same time, a number of research results also show that the positive charge nanoparticle is more easily distributed to the lung, while the negative charge nanoparticle has more obvious liver and spleen targeting [63].

2.3.1.2 Increase the Steric Resistance Between Nanoparticles to Improve the Stability of Nanomedicine

The surface properties of nanoparticles play the most important role in the storage stability of nanoparticles and their stability in blood circulation. The specially modified nanoparticles can increase the interaction between nanoparticles to a certain extent and avoid the aggregation and precipitation of particles caused by the interaction between particles. The different functional groups on the surface of nanoparticles are the main determinants of stability [64, 65]. Generally, the degree of hydrophobicity determines the binding degree of opsonin protein, and the hydrophobic surface induces the aggregation of nanoparticles through hydrophobic interaction, which minimizes the surface energy. The modified nanoparticles can avoid the adsorption of opsonin protein to a certain extent, avoid its recognition by the immune system, and thus be cleared by MPS (liver, spleen, and lung) and other organs [64].

The surface is coated with a biocompatible hydrophilic polymer/surfactant or a copolymer with hydrophilic properties. The hydrophilic polymer on the surface of the nanoparticle repels other molecules through spatial effects. Therefore, nanoparticles are not covered by opsonins and are not quickly cleared [66]. Surface-coating materials widely used to extend cycle times include PEG, polyethylene oxide (PEO), polyvinylpyrrolidone (PVP), polyacrylic acid, dextran, poloxam, and polysorbate (Tween-80). PEG is the most widely used material, which has effectively enhanced the stability of nanoparticles in many studies [67].

2.3.1.3 Conformation Change of the Surface Polymer of Nanomaterials Improves the Stability of Nanomaterials

The surface conformation of nanomaterials has a significant impact on the storage stability of nanomaterials and the avoidance or promotion of cell uptake. PEG, which is commonly used for modification of nanomaterials, has an obvious dosage mutation. PEG molecules with a brush structure on the surface of nanoparticles can reduce phagocytosis and complement activation, while PEG with a mushroom structure is an effective complement activator. It is beneficial for cell uptake [68, 69]. On the surface of nanoparticles, PEG density below 9% forms a mushroom-like structure, while PEG density higher than 9% forms a harder brushlike shape [70] (Figure 2.6).

In another study, researchers discovered that superparamagnetic ferric oxide nanoparticles (4–5 nm) were enveloped in a brush-like structure consisting of 20–30 glucan chains. This innovative coating mitigated the rapid elimination of nanoparticles from the blood, thereby prolonging their half-life ($t_{1/2}$) to 3–4 hours [71].

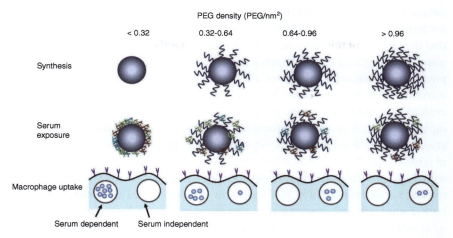

Figure 2.6 Influence of PEG density on serum protein adsorption to gold nanoparticles and their subsequent uptake by macrophages.

In addition to the use of neutral materials in the modification of nanocrafts, zwitterionic copolymers have demonstrated exceptional nanoparticle shielding in the bloodstream and have been found to maximize their interaction with target tissues. For instance, an amphoteric polymer nanoparticle, as developed by Yuan et al., maintains a neutral charge under physiological conditions, resulting in an extended circulation time (with a half-life of approximately 24.05 hours according to the three-chamber pharmacokinetic model). Upon leakage into the tumor site, the nanoparticle acquires a positive charge in the acidic extracellular environment, enabling effective uptake by the tumor cells [72]. Other reports have also indicated that silica nanoparticles modified by PEG-coupled polyethylenimide copolymer have a longer retention time in vivo and can effectively enter A549 lung cancer cells [73].

2.3.2 Surface Modification of Bioactive Groups Can Improve the Efficacy of Nanopreparations

2.3.2.1 Nanodrugs Modified by Single Targeting Ligand

The development of targeted nanomaterials has emerged as a key focus within the field of nanomaterials research and development. Various molecules, including folate, biotin, transferrin, peptides, enzyme substrates, monosaccharides, or oligosaccharides, have demonstrated successful utility as targeted components. These molecules interact specifically with membrane protein lectin and antibodies, underscoring their potential to advance the ongoing exploration and application of targeted nanomaterials [74]. Targeting ligand parts play a crucial role in specifically targeting body cells. When attached to nanoparticles in long chains, they enable the nanoparticles to extend beyond the dense polymer brush. This strategic design helps avoid steric hindrance and facilitates effective binding to the targeted receptors. Such precise targeting at the cellular level holds immense potential for advancing targeted drug delivery and precision medicine [75]. Despite the potential for targeting ligands to facilitate specific cell targeting, it is crucial to recognize that the efficacy of this approach does not unilaterally increase in proportion to the quantity of targeting ligands. Rather, it is imperative to undergo rigorous screening processes to determine the most effective targeting ligands. This precaution is warranted due to the propensity of nanoparticles to adsorb opsonins, and an escalation in the concentration of ligands on the nanoparticle surface may expedite rapid clearance. A pertinent illustration of this phenomenon is the discovery by Gu et al., revealing that low-density targeting ligands facilitated the effective targeting of prostate-specific antigens (PSMA) by nanoparticles [76], whereas nanomaterials with high surface ligand density were more likely to aggregate in the liver and spleen. Similarly, the study of Kirpotin et al. also proved this view. They used HER-2 antibody and observed that nanocapsules with a high density of ligands were more likely to adsorb serum proteins, which resulted in the faster removal of nanocapsules by the body and reduced the ability of the ligands to interact with the target receptor [77, 78]. Therefore, the number and density of targeting ligands on the surface of nanoparticles are important for effective targeting.

In addition to the number and density of targeting ligands on the surface of nanoparticles, the spatial orientation of targeting ligands also plays a crucial role in the targeting performance of nanopreparations. Lee et al. conducted a study on folic acid-modified tetrahedral oligonucleotide nanoparticles (28.6 nm) designed for siRNA delivery. They demonstrated that at least three molecules of folic acid are required per nanoparticle for optimal delivery to cancer cells. Increasing the number of folic acid molecules did not enhance cellular uptake, and the study revealed that gene silencing occurred only when the ligand was in the correct spatial orientation. Folic acid exhibits variations in density and location on the tetrahedron, and noteworthy gene silencing was observed when three folic acid molecules surrounded one face of the tetrahedron, forming the locally densest nanopreparation. However, no gene silencing was observed when the distance between the folic acid molecules increased. The researchers hypothesized that high local ligand density might impact the intracellular transport pathway of nanoparticles through cells, consequently influencing gene silencing [79]. So, the direction of targeting ligand should be taken into account for appropriate biological distribution and cellular uptake.

These studies underscore the pivotal role played by both the quantity and orientation of targeted ligands in achieving effective targeting of nanomaterials. The number of ligands employed inherently influences the spatial arrangement of these ligands. Consequently, in the pursuit of developing targeted nanomaterials, a holistic approach is imperative. This involves meticulous considerations in areas such as the judicious design of ligands, optimization of formulation proportions, and refinement of the preparation processes. A comprehensive understanding of these factors is essential for the development of efficient targeted nanomaterials, ensuring a synergistic integration of ligand properties, formulation characteristics, and preparation methodologies to enhance their targeting efficacy (Table 2.1).

Table 2.1 Different targeting moieties on nanoparticle surfaces and extent of their internalization.

Targeting moiety	Nanoparticle	Size (nm)	Surface charge (mV)	Targeted tissue	Result	Conjugation method	References
Folate	PLGA nanoparticles	108.9	−18.4	KB	6.7-fold higher uptake	Covalent binding (amide linkage)	[80]
	Iron oxide nanoparticles	60–80	−12.5	HeLa	>3-fold higher uptake	Covalent binding (amide linkage)	[81]
Transferrin	PLGA nanoparticles	220	−8	PC3	3-fold higher uptake	Linker chemistry	[82]
Biotin	Gold nanoparticles	20–40	–	HeLa, A549, MG63	>2-fold higher uptake	Covalent binding (amide linkage)	[83]
	Chitosan nanoparticles	165.3	+16.5	MCF-7 cells	>2-fold higher uptake	Covalent binding (amide linkage)	[84]
Monoclonal antibody	PLGA nanoparticles	124.2	+12.0	MDA-MB-231 cells	>2-fold higher uptake	Covalent binding (amide linkage)	[85]
RGD peptide	Albumin nanoparticles	130	−30.77	BxPC-3	>2-fold higher uptake (after 24 h)	Linker chemistry	[86]
Galactose	Gold nanoparticles	50	−30	Hepatocyte	16-fold higher uptake	Au–S binding	[87]

2.3.2.2 Multiligand Modified Nanopreparations

The distribution, absorption, metabolism, and excretion of nanomaterials in vivo are easily affected by the complex physiological environment in the body. These physiological obstacles include: (i) phagocytosis and clearance of macrophages; (ii) penetration and diffusion of nontarget organs; (iii) the natural barrier of cell membrane formation leads to the difficulty of efficient delivery of nanomaterials to target organs and/or target cells, and the ideal degree of high efficiency and low toxicity cannot be achieved. The existing delivery strategies based on single ligand modification can improve the delivery efficiency of nanopreparations in vivo to a certain extent, but there are still problems such as poor organ targeting and low expression of target cells, which make it difficult to play a good role. Therefore, specific modification strategies for different barriers of nanomaterials in vivo can significantly change the systemic cycle time, organ and cell targeting, and effect of nanomaterials. In this study, Song et al. adopted a new strategy of multiligand modification, and for the first time, adopted central composite design (CCD) in vivo to optimize the ratio of each ligand in the nanopreparation to obtain the optimal organ targeting mRNA delivery system. Researchers used self-peptide (SP) to "Don't eat me" macrophages and mannose to dendritic cells (dendritic cells, dendritic cells, dendritic cells, dendritic cells, and dendritic cells, respectively). At the same time, the cyclic transmembrane peptide W5R4K (WRK) was introduced to construct the triligand-modified mRNA delivery system SMW-LLNs. Furthermore, CCD was used to optimize the prescription ratio of SMW-LLNs in vivo, and Ol-LLNs, the optimal mRNA delivery system with liver targeting, was obtained. It is carried by the base editing system associated with mRNA and sgRNA, and is used in the treatment of the rare genetic disease phenylketonuria (PKU). The base editing efficiency of the multiligand-modified nanocatings reached 8.38% in the liver of adult PKU mice and reduced the concentration of phenylalanine in the blood of adult PKU mice.

Surface optimization of nanopreparations plays a pivotal role in enhancing the stability of nanodrugs and influencing various aspects such as drug absorption, distribution, metabolism, and excretion. Despite the identification of the impact of diverse surface characteristics and the development of various surface modification methods, achieving repeatable and effective surface modification necessitates the precise control and optimization of the conjugation of individual components at specific sites [88]. Simultaneously, the innovation in modifying materials emerges as a critical aspect. The exploration of novel targeting ligands and the judicious utilization of these ligands hold significant research value. In light of our current understanding of nanoparticle surfaces, we anticipate that advancements in surface functionalization will bring about transformative changes in the therapeutic application of nanoparticles, thereby revolutionizing the treatment modalities for various diseases in the near future.

References

1 Matsumura, Y. and Maeda, H. (1998). A new concept for macromolecular therapeutics in cancer chemotherapy: mechanism of tumoritropic accumulation of proteins and the antitumor agent smancs. *Cancer Research* 46: 6387–6392.

2 Kim, Y., Lobatto, M.E., Kawahara, T. et al. (2014). Probing nanoparticle translocation across the permeable endothelium in experimental atherosclerosis. *Proceedings of the National Academy of Sciences of the United States of America* 111: 1078–1083.

3 Lobatto, M.E., Calcagno, C., Millon, A. et al. (2015). Atherosclerotic plaque targeting mechanism of long-circulating nanoparticles established by multimodal imaging. *ACS Nano* 9: 1837–1847.

4 Liu, D., Mori, A., Huang, L. et al. (2002). Role of liposome size and RES blockade in controlling biodistribution and tumor uptake of GM1-containing liposomes. *Biomembranes* 1104: 95–101.

5 Schadlich, A., Caysa, H., Mueller, T. et al. (2021). Tumor accumulation of NIR fluorescent PEG–PLA nanoparticles: impact of particle size and human xenograft tumor model. *ACS Nano* 5: 8710–8720.

6 Shu, M., Tang, J., Chen, L. et al. (2021). Tumor microenvironment triple-responsive nanoparticles enable enhanced tumor penetration and synergetic chemo-photodynamic therapy. *Biomaterials* 9: 268–281.

7 Kankala, R.K., Han, Y.H., Na, J. et al. (2020). Nanoarchitectured structure and surface biofunctionality of mesoporous silica nanoparticles. *Advanced Materials* 32: e1907035.

8 Zahr, A.S., Davis, C.A., and Pishko, M.V. (2006). Macrophage uptake of core-shell nanoparticles surface modified with poly ethylene glycol. *Langmuir* 22: 8178–8185.

9 Campbell, R.B., Fukumura, D., Brown, E.B. et al. (2002). Cationic charge determines the distribution of liposomes between the vascular and extravascular compartments of tumors. *Cancer Research* 62: 6831–6836.

10 Sun, T., Zhang, Y.S., Pong, B. et al. (2014). Engineered nanoparticles for drug delivery in cancer therapy. *Angewandte Chemie, International Edition* 53: 12320–12364.

11 Allmeroth, M., Moderegger, D., Gundel, D. et al. (2013). PEGylation of HPMA-based block copolymers enhances tumor accumulation in vivo: a quantitative study using radiolabeling and positron emission tomography. *Journal of Controlled Release* 172: 77–85.

12 Geng, Y., Dalhaimer, P., Cai, S. et al. (2007). Shape effects of filaments versus spherical particles in flow and drug delivery. *Nature Nanotechnology* 2: 249–255.

13 Kinnear, C., Moore, T.L., Rodriguez-Lorenzo, L. et al. (2017). Form follows function: nanoparticle shape and its implications for nanomedicine. *Chemical Reviews* 117: 11476–11521.

14 Lee, S.Y., Ferrari, M., and Decuzzi, P. (2009). Shaping nano-micro-particles for enhanced vascular interaction in laminar flows. *Nanotechnology* 20: 203–241.

15 Bar-Zeev, M., Livney, Y.D., and Assaraf, Y.G. (2017). Targeted nanomedicine for cancer therapeutics: towards precision medicine overcoming drug resistance. *Drug Resistance Updates* 31: 15–30.

16 Guo, M.R., He, Z.S., Jin, Z. et al. (2023). Oral nanoparticles containing naringenin suppress atherosclerotic progression by targeting delivery to plaque macrophages. *Nano Research* 16: 925–937.

17 He, Z.Y., Wei, X.W. et al. (2013). Folate-linked lipoplexes for short hairpin RNA targeting claudin-3 delivery in ovarian cancer xenografts. *Journal of Controlled Release* 172: 679–689.

18 Tran, B.T., Kim, J., and Ahn, D.R. (2020). Systemic delivery of aptamer-drug conjugates for cancer therapy using enzymatically generated self-assembled DNA nanoparticles. *Nanoscale* 12: 22945–22951.

19 Gao, H., Qian, J., Yang, Z. *et al.* Whole-cell aptamer-functionalised nanoparticles for enhanced targeted glioblastoma therapy, *Biomaterials* 33 6264-6272 (2012).

20 Hai H., Yang X., Li H., *et al.* iRGD decorated liposomes: a novel actively penetrating topical ocular drug delivery strategy. *Nano Research* (2020) 13:3105–3109.

21 Xu, B., Jin, Q., Zeng, J. et al. (2016). Combined tumor and neovascular-"dual targeting" gene/chemo-therapy suppresses tumor growth and angiogenesis. *ACS Applied Materials & Interfaces* 8: 25753–25769.

22 Hu, C., Peng, K., Wu, Q. et al. (2021). HDAC1 and 2 regulate endothelial VCAM-1 expression and atherogenesis by suppressing methylation of the GATA6 promoter. *Theranostics* 20: 5605–5619.

23 Qin, S., Tang, X., Chen, Y. et al. (2022). mRNA-based therapeutics: powerful and versatile tools to combat diseases. *Signal Transduction and Targeted Therapy* 21: 166.

24 Kheirolomoom, A., Kim, C.W., Seo, J.W. et al. (2015). Multi-functional nanoparticles facilitate molecular targeting and miRNA delivery to inhibit atherosclerosis in ApoE$^{-/-}$ mice. *ACS Nano* 9: 8885–8897.

25 Huang, X., Liu, C., Kong, N. et al. (2022). Synthesis of siRNA nanoparticles to silence plaque-destabilizing gene in atherosclerotic lesional macrophages. *Nature Protocols* 17: 748–780.

26 Yang, L., Yang, Y., Chen, Y. et al. (2022). Cell-based drug delivery systems and their in vivo fate. *Advanced Drug Delivery Reviews* 187: 181–199.

27 Lippi, G., Franchini, M., and Targher, G. (2011). Arterial thrombus formation in cardiovascular disease. *Nature Reviews. Cardiology* 8: 502–512.

28 Stellos, K., Bigalke, B., Borst, O. et al. (2013). Circulating platelet-progenitor cell coaggregate formation is increased in patients with acute coronary syndromes and augments recruitment of CD34$^+$ cells in the ischaemic microcirculation. *European Heart Journal* 34: 2548–2556.

29 Zhuang, C., Piao, C., Kang, M. et al. (2023). Hybrid nanoparticles with cell membrane and dexamethasone-conjugated polymer for gene delivery into the lungs as therapy for acute lung injury. *Biomaterials Science* 11: 3354–3364.

30 Jiang, Z., Luo, Y., Men, Y. et al. (2017). Red blood cell membrane-camouflaged melanin nanoparticles for enhanced photothermal therapy. *Biomaterials* 143: 29–45.

31 Gao, C., Huang, Q., Liu, C. et al. (2020). Treatment of atherosclerosis by macrophage-biomimetic nanoparticles via targeted pharmacotherapy and sequestration of proinflammatory cytokines. *Nature Communications* 11: 2622.

32 Wang, C., Li, N., Li, Y. et al. (2022). Engineering a HEK-293T exosome-based delivery platform for efficient tumor-targeting chemotherapy/internal irradiation combination therapy. *Journal of Nanbiotechnology* 20: 247.

33 Qi, S., Zhang, H., Zhang, X. et al. (2022). Supramolecular engineering of cell membrane vesicles for cancer immunotherapy. *Scientific Bulletin* 67: 1898–1909.

34 Jain, P.K., Huang, X., El-Sayed, I.H., and El-Sayed, M.A. (2008). Noble metals on the nanoscale: optical and photothermal properties and some applications in imaging, sensing, biology, and medicine. *Accounts of Chemical Research* 41: 1578–1586.

35 Qian, X., Peng, X.H., Ansari, D.O. et al. (2008). In vivo tumor targeting and spectroscopic detection with surface-enhanced Raman nanoparticle tags. *Nature Biotechnology* 26: 83–90.

36 Hirsch, L.R., Stafford, R.J., Bankson, J.A. et al. (2003). Nanoshell-mediated near-infrared thermal therapy of tumors under magnetic resonance guidance. *Proceedings of the National Academy of Sciences of the United States of America* 11: 13549–13554.

37 Gannon, C.J., Patra, C.R., Bhattacharya, R. et al. (2008). Intracellular gold nanoparticles enhance non-invasive radiofrequency thermal destruction of human gastrointestinal cancer cells. *Journal of Nanobiotechnology* 6: 2–14.

38 Shaban, M. and Hasanzadeh, M. (2020). Biomedical applications of dendritic fibrous nanosilica (DFNS): recent progress and challenges. *RSC Advances* 10 (61): 37116–37133.

39 Baptista, A., Silva, F., Porteiro, J. et al. (2018). Sputtering physical vapour deposition (PVD) coatings: a critical review on process improvement and market trend demands. *Coatings* 8 (11).

40 Wani, I.A. (2014). *Nanomaterials, Novel Preparation Routes, and Characterizations*. American Institute of Physics.

41 Saxena, J. and Jyoti, A. (2020). Nanomaterials: novel preparation routes, characterizations, and applications. In: *Nanobiotechnology*, 23–33. Apple Academic Press.

42 Delogu, F., Gorrasi, G., and Sorrentino, A. (2017). Fabrication of polymer nanocomposites via ball milling: present status and future perspectives. *Progress in Materials Science* 86 (May): 75–126.

43 Almoustafa, H.A., Alshawsh, M.A., and Chik, Z. (2017). Technical aspects of preparing PEG-PLGA nanoparticles as carrier for chemotherapeutic agents by nanoprecipitation method. *International Journal of Pharmaceutics* 533 (1): 275–284.

44 Li, L.L., Li, X., and Hao, W. (2017). Microfluidic synthesis of nanomaterials for biomedical applications. *Small Methods* 1 (8): 1700140.

45 Lazarovits, J., Chen, Y.Y., Song, F. et al. (2019). Synthesis of patient-specific nanomaterials. *Nano Letters* 19 (1): 116–123.

46 Spagnoletti, F.N., Spedalieri, C., Kronberg, F., and Giacometti, R. (2019). Extracellular biosynthesis of bactericidal Ag/AgCl nanoparticles for crop protection using the fungus. *Journal of Environmental Management* 231: 457–466.

47 Ahn, E.Y., Jin, H., and Park, Y. (2019). Assessing the antioxidant, cytotoxic, apoptotic and wound healing properties of silver nanoparticles green-synthesized by plant extracts. *Materials Science and Engineering: C* 101: 204–216.

48 Kayes, J.B. (1977). Pharmaceutical suspensions: relation between zeta potential, sedimentation volume and suspension stability. *The Journal of Pharmacy and Pharmacology* 29: 199–204.

49 Nagata, T. and Melchers, G. (1978). Surface charge of protoplasts and their significance in cell–cell interaction. *Planta* 142: 235–238.

50 Zhang, Y., Yang, M., Portney, N.G. et al. (2008). Zeta potential: a surface electrical characteristic to probe the interaction of nanoparticles with normal and cancer human breast epithelial cells. *Biomedical Microdevices* 10: 321–328.

51 Verma, A. and Stellacci, F. (2010). Effect of surface properties on nanoparticle–cell interactions. *Small* 6: 12–21.

52 Ramos, J., Forcada, J., and Hidalgo-Alvarez, R. (2014). Cationic polymer nanoparticles and nanogels: from synthesis to biotechnological applications. *Chemical Reviews* 114: 367–428.

53 Hühn, D., Kantner, K., Geidel, C. et al. (2013). Polymer-coated nanoparticles interacting with proteins and cells: focusing on the sign of the net charge. *ACS Nano* 7: 3253–3263.

54 Bivas-Benita, M., Romeijn, S., Junginger, H.E. et al. (2004). PLGA-PEI nanoparticles for gene delivery to pulmonary epithelium. *European Journal of Pharmaceutics and Biopharmaceutics* 58: 1–6.

55 Chung, T.H., Wu, S.H., Yao, M. et al. (2007). The effect of surface charge on the uptake and biological function of mesoporous silica nanoparticles in 3T3-L1 cells and human mesenchymal stem cells. *Biomaterials* 28: 2959–2966.

56 Hillaireau, H. and Couvreur, P. (2009). Nanocarriers' entry into the cell: relevance to drug delivery. *Cellular and Molecular Life Sciences* 66: 2873–2896.

57 Wu, X., Landfester, K., Musyanovych, A. et al. (2010). Disposition of charged nanoparticles after their topical application to the skin. *Skin Pharmacology and Physiology* 23: 117–123.

58 Shanmugam, S., Song, C.K., Nagayya-Sriraman, S. et al. (2009). Physicochemical characterization and skin permeation of liposome formulations containing clindamycin phosphate. *Archives of Pharmacal Research* 32: 1067–1075.

59 Cui, P.F., Qi, L.Y., Wang, Y. et al. (2019). Dex-Aco coating simultaneously increase the biocompatibility and transfection efficiency of cationic polymeric gene vectors. *Journal of Controlled Release* 303: 253–262.

60 Duan, X., Zhang, Y., Guo, M. et al. (2023). Sodium alginate coating simultaneously increases the biosafety and immunotherapeutic activity of the cationic mRNA nanovaccine. *Acta Pharmaceutica Sinica B* 13: 942–954.

61 Jallouli, Y., Paillard, A., Chang, J. et al. (2007). Influence of surface charge and inner composition of porous nanoparticles to cross blood–brain barrier in vitro. *International Journal of Pharmaceutics* 344: 103–109.

62 Lockman, P.R., Koziara, J.M., Mumper, R.J. et al. (2004). Nanoparticle surface charges alter blood–brain barrier integrity and permeability. *Journal of Drug Targeting* 12: 635–641.

63 Cheng, Q., Wei, T., Farbiak, L. et al. (2020). Selective organ targeting (SORT) nanoparticles for tissue-specific mRNA delivery and CRISPR-Cas gene editing. *Nature Nanotechnology* 15: 313–320.

64 Sheng, Y., Liu, C., Yuan, Y. et al. (2009). Long-circulating polymeric nanoparticles bearing a combinatorial coating of PEG and water-soluble chitosan. *Biomaterials* 30: 2340–2348.

65 Wang, H., Zhao, P., Liang, X. et al. (2010). Folate-PEG coated cationic modified chitosan—cholesterol liposomes for tumor-targeted drug delivery. *Biomaterials* 31: 4129–4138.

66 Elsabahy, M. and Wooley, K.L. (2012). Design of polymeric nanoparticles for biomedical delivery applications. *Chemical Society Reviews* 41: 2545–2561.

67 Sun, T., Zhang, Y.S., Pang, B. et al. (2014). Engineered nanoparticles for drug delivery in cancer therapy. *Angewandte Chemie (International Ed. in English)* 53: 12320–12364.

68 Li, S.D. and Huang, L. (2010). Stealth nanoparticles: high density but sheddable PEG is a key for tumor targeting. *Journal of Controlled Release* 145: 178–181.

69 Wu, T.T. and Zhou, S.H. (2015). Nanoparticle-based targeted therapeutics in head-and-neck cancer. *International Journal of Medical Sciences* 12: 187–200.

70 Biswas, S. and Torchilin, V.P. (2014). Nanopreparations for organelle-specific delivery in cancer. *Advanced Drug Delivery Reviews* 66: 26–41.

71 Liu, S., Chiu-Lam, A., Rivera-Rodriguez, A. et al. (2021). Long circulating tracer tailored for magnetic particle imaging. *Nanotheranostics* 5: 348–361.

72 Yuan, Y.Y., Mao, C.Q., Du, X.J. et al. (2012). Surface charge switchable nanoparticles based on zwitterionic polymer for enhanced drug delivery to tumor. *Advanced Materials* 24: 5476–5480.

73 Zhang, T., Liu, H., Li, L. et al. (2021). Leukocyte/platelet hybrid membrane-camouflaged dendritic large pore mesoporous silica nanoparticles co-loaded with photo/chemotherapeutic agents for triple negative breast cancer combination treatment. *Bioactive Materials* 6: 3865–3878.

74 Byrne, J.D., Betancourt, T., and Brannon-Peppas, L. (2008). Active targeting schemes for nanoparticle systems in cancer therapeutics. *Advanced Drug Delivery Reviews* 60: 1615–1626.

75 Zhong, Y., Meng, F., Deng, C. et al. (2014). Ligand-directed active tumor-targeting polymeric nanoparticles for cancer chemotherapy. *Biomacromolecules* 15: 1955–1969.

76 Gu, F., Zhang, L., Teply, B.A. et al. (2008). Precise engineering of targeted nanoparticles by using self-assembled biointegrated block copolymers. *Proceedings of the National Academy of Sciences of the United States of America* 105: 2586–2591.

77 Kirpotin, D.B., Drummond, D.C., Shao, Y. et al. (2006). Antibody targeting of long-circulating lipidic nanoparticles does not increase tumor localization but does increase internalization in animal models. *Cancer Research* 66: 6732–6740.

78 Salvati, A., Pitek, A.S., Monopoli, M.P. et al. (2013). Transferrin-functionalized nanoparticles lose their targeting capabilities when a biomolecule corona adsorbs on the surface. *Nature Nanotechnology* 8: 137–143.

79 Lee, H., Lytton-Jean, A.K., Chen, Y. et al. (2012). Molecularly self-assembled nucleic acid nanoparticles for targeted in vivo siRNA delivery. *Nature Nanotechnology* 7: 389–393.

80 Kim, S.H., Jeong, J.H., Chun, K.W. et al. (2005). Target-specific cellular uptake of PLGA nanoparticles coated with poly(L-lysine)-poly(ethylene glycol)-folate conjugate. *Langmuir* 21: 8852–8857.

81 Sonvico, F., Mornet, S., Vasseur, S. et al. (2005). Folate-conjugated iron oxide nanoparticles for solid tumor targeting as potential specific magnetic hyperthermia mediators: synthesis, physicochemical characterization, and in vitro experiments. *Bioconjugate Chemistry* 16: 1181–1188.

82 Sahoo, S.K., Ma, W., and Labhasetwar, V. (2004). Efficacy of transferrin-conjugated paclitaxel-loaded nanoparticles in a murine model of prostate cancer. *International Journal of Cancer* 112: 335–340.

83 Heo, D.N., Yang, D.H., Moon, H.J. et al. (2012). Gold nanoparticles surface-functionalized with paclitaxel drug and biotin receptor as theranostic agents for cancer therapy. *Biomaterials* 33: 856–866.

84 Tian, X., Yin, H., Zhang, S. et al. (2014). Bufalin loaded biotinylated chitosan nanoparticles: an efficient drug delivery system for targeted chemotherapy against breast carcinoma. *European Journal of Pharmaceutics and Biopharmaceutics* 87: 445–453.

85 Chen, H., Gao, J., Lu, Y. et al. (2008). Preparation and characterization of PE38KDEL-loaded anti-HER2 nanoparticles for targeted cancer therapy. *Journal of Controlled Release* 128: 209–216.

86 Ji, S., Xu, J., Zhang, B. et al. (2012). RGD-conjugated albumin nanoparticles as a novel delivery vehicle in pancreatic cancer therapy. *Cancer Biology & Therapy* 13: 206–215.

87 Bergen, J.M., Von Recum, H.A., Goodman, T.T. et al. (2006). Gold nanoparticles as a versatile platform for optimizing physicochemical parameters for targeted drug delivery. *Macromolecular Bioscience* 6: 506–516.

88 Riehemann, K., Schneider, S.W., Luger, T.A. et al. (2009). Nanomedicine—challenge and perspectives. *Angewandte Chemie (International Ed. in English)* 48: 872–897.

3

Nanomedicine for Antitumors

Qiong Wu[1], Xinzhu Yang[1], Ruixue Zhu[1], and Yujun Song[1,2,3]

[1] University of Science and Technology Beijing, Center for Modern Physics Technology, School of Mathematics and Physics, 30 Xueyuan Road, Haidian District, Beijing 100083, China
[2] Zhengzhou Tianzhao Biomedical Technology Company Ltd., Zhengzhou New Technology Industrial Development Zone, 7B-1209 Dongqing Street, Zhengzhou 451450, China
[3] Key Laboratory of Pulsed Power Translational Medicine of Zhejiang Province, Hangzhou Ruidi Biotechnology Company Ltd., Room 803, Bldg. 4, 4959 Yuhangtang Road, Cangqian Street, Hangzhou 310023, China

3.1 Introduction

Cancer remains a major global health issue, with statistics indicating a consistent or rising number of cases worldwide [1–3]. By 2030, it is projected that over 21 million individuals will be diagnosed with cancer [4, 5]. Significant advancements have been made in nanotechnology in recent years, which have underscored potential strategies to mitigate toxicity, improve delivery, and address resistance in anticancer treatments [6, 7]. Nanotechnology is a rapidly developing discipline with a wide range of applications that span the environmental and medicinal domains. As a consequence of the promising outcomes of nanotechnology research, especially when it comes to treating long-term diseases like cancer, scientists are now thoroughly evaluating the toxicity of these novel products [8–12]. Nanocarriers present unique characteristics, such as their small scale, considerable surface area-to-volume ratio, and positive physical and chemical attributes [13, 14]. Drugs' pharmacokinetic and pharmacodynamic behaviors can be altered by nanocarriers, perhaps improving their therapeutic index. The encapsulation of pharmaceuticals in nanocarriers has the potential to improve the stability of the medications within the body, extend their bloodstream circulation, and provide controlled drug release [13]. As a result, medications can be dispersed differently throughout the body by using nanomedicine agents, with a greater concentration developing near the site of the tumor. The enhanced permeability and retention, or EPR), effect is the name given to this phenomenon [13]. One unique feature of the tumor microenvironment is the

---

Qiong Wu and Xinzhu Yang contributed equally to this chapter.

Nanomedicine: Fundamentals, Synthesis, and Applications, First Edition. Edited by Yujun Song.
© 2025 WILEY-VCH GmbH. Published 2025 by WILEY-VCH GmbH.

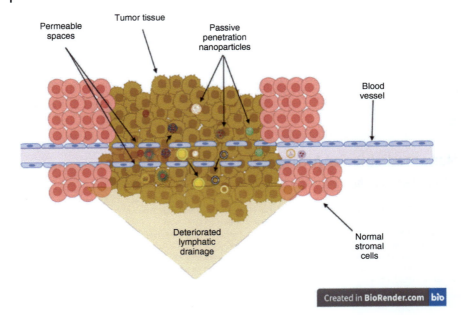

Figure 3.1 The enhanced permeability and retention (EPR) impact in the tissue of a tumor's microenvironment is schematically depicted. It illustrates how nanoparticles, with a size of less than 200 nm, are able to pass through the highly permeable blood vessel regions and are retained at the tumor site due to inadequate lymphatic drainage. Source: Al-Zoubi and Al-Zoubi [3]/with permission of Elsevier.

EPR effect, which occurs when the lymphatic system is unable to eliminate tumor fluids, leading to the accumulation of different materials, including nanoparticles, at the tumor site (refer to Figure 3.1) [3, 15, 16].

Recently, nanomedicine has emerged as a strategy to enhance clinical outcomes with minimal adverse effects. Research has focused on utilizing this for innovative diagnostic and therapeutic strategies in precise cancer therapies, exemplified by the approval of drugs like Abraxane, along with numerous other formulations advancing through clinical trials [4]. Nanoparticles have been purposefully constructed from various biocompatible materials, including organic molecules, proteins, lipids, polymers, and inorganic nanomaterials. This approach has enabled the creation of versatile, multifunctional nanoparticles with high-performance capabilities, tailored for the specific imaging and drug delivery needs in cancer treatment [4, 13].

The methods of delivering anticancer nanoparticles, whether through passive or active targeting, can be categorized into four primary types based on their structural features and the materials from which they are composed [17]. The categorization of anticancer nanoparticle delivery systems is based on their composition and structure, dividing them into four distinct groups: (i) organic nanoparticles, which encompass liposomes, polymeric nanoparticles, and dendrimers; (ii) viral nanoparticles; (iii) inorganic nanoparticles, including gold, carbon, silica, magnetic, and various metal nanoparticles; and (iv) hybrid nanoparticles, consisting of organic–inorganic nanoparticles, lipid–polymer nanoparticles, and those coated with cell membranes [7, 17].

Nanoparticles are at the forefront of advancement in pharmaceutical research [18]. They facilitate the mitigation of drawbacks associated with established delivery mechanisms, foster the creation of fresh diagnostic and therapeutic plans, and contribute to an enhanced comprehension of biological processes [18]. Accordingly, this review seeks to emphasize the primary methods of nanoparticle synthesis and their relevance in biomedical applications.

3.2 Liposomal Nanoparticles

It is anticipated that liposomal encapsulation will provide a safer and more beneficial anticancer treatment compared to traditional chemotherapy, which is often linked to severe side effects [3].

Cancer chemotherapy agents are nonspecific and may distribute widely in the body, often with less than ideal pharmacokinetic profiles [19, 20]. Because of this, several nanomaterials have been developed to solve the pharmacokinetic and nontargeted delivery issues with effective chemotherapy drugs [21, 22]. One major problem in the delivery of anticancer medicines is the effective transportation of encapsulated nanoparticles to the targeted tumor site, underscoring the advantageous structural properties of liposomal nanoparticles [3]. Liposomes serve as ideal analogues of biomembranes and cells in their structure and function [23, 24]. Their similarity to natural biological membranes earns liposomes the reputation of being an excellent model for investigating the origins, operations, and development of early cell membranes [24–26].

The fundamental composition of nanoliposomes involves a phospholipid double layer, which enhances the solubility of their payload within aquatic environments. This design, inspired by cell membranes, results in liposome-based nanoparticles exhibiting reduced toxicity and immunogenicity [27]. Furthermore, nanoliposomal encapsulation offers benefits like controlled drug delivery, reduced toxicity, enhanced shelf life, and minimized degradation [3, 27–31].

Nanoliposomes and liposomes typically exhibit comparable chemical composition, structural features, and thermodynamic behaviors. However, nanoliposomes, with their increased surface area, excel in enhancing solubility, refining the controlled release of payloads, boosting bioavailability, and enabling more accurate targeting of the encapsulated substances when contrasted with conventional liposomes [24, 32].

In some cases, formulations such as liposomal paclitaxel or liposomal doxorubicin nanoparticles showed better anticancer activity and less toxicity against breast and prostate cancer in comparison to the corresponding free medicines [3, 33–35]. Liposomal encapsulation also allows for drug combinations, which can intensify therapeutic effectiveness and lessen resistance to specific anticancer treatments [3, 36, 37].

Figure 3.2 Diagram showing the miniature extruder used to make nanoliposomes. Source: Panahi et al. [24]/with permission of Taylor & Francis.

3.2.1 Synthesis Methods

3.2.1.1 Extrusion Technique

One of the initial techniques for producing liposomes is extrusion, where the liposomes are altered in structure based on the pore size of the filters used, resulting in the formation of large unilamellar vesicles (LUVs) or nanoliposomes [24, 38, 39]. Application of high-pressure forces vesicles through polycarbonate filters, leading to extrusion (refer to Figure 3.2) [24]. A mini-extruder apparatus, fitted with 0.5 ml gas-tight syringes, is suitable for use in this method [24].

3.2.1.2 Sonication

Sonication is frequently considered the most commonly used approach to produce liposomes and nanoliposomes. It is a straightforward method that effectively reduces the size of liposomes and facilitates the creation of nanosized liposomes [24, 40, 41]. Certainly, this method has several limitations, including low encapsulation efficiency, size sorting limitations, thermal effects, and mechanical fragility [24, 42]. Two primary methods of sonication are commonly employed: probe sonication and bath sonication [24, 43, 44].

Upon completion of the product synthesis using either probe or bath sonication, the resulting material should be annealed. This entails raising the product's temperature over the critical temperature (Tc) for about an hour while it is in an inert atmosphere, such as argon or nitrogen gas. To obtain a purer nanoliposome suspension, centrifugation can be used to eliminate any remaining larger particles. Several factors significantly affect the size distribution and heterogeneity of the vesicles, including the temperature of annealing, duration of sonication, volume of the sample, settings on the sonicator, as well as the lipid type and concentration used [24, 45].

3.2.1.3 Microfluidization

The technique for producing liposomes involves a microfluidization process using a microfluidizer, which avoids the use of potentially harmful solvents [24, 46, 47]. The microfluidizer is utilized within the pharmaceutical industry for the creation of liposome-based products and pharmaceutical emulsions [24, 48]. The method employs a split pressure stream that directs each segment through a small opening, causing the streams to converge within the microfluidizer's internal chamber [24]. By guiding the flow stream via microchannels at high pressures (up to 12000 psi), cavitation and shear and impact forces within the interaction chamber are created, which cause the liposomes to shrink in size [24, 49, 50]. Advantages of this technology include high capture effectiveness (up to 70%), the ability to produce a large number of liposomes, and the flexibility to modify the average liposome size.

3.2.1.4 Heating Method
The application of heat can be used to synthesize hollow micro-liposomes (HM-liposomes) serving as drug delivery systems and gene transfer agents by allowing DNA to mix with preformed, empty HM-liposomes through incubation [24, 51–53]. By adding the medication to the reaction mixture with the liposomal constituents and glycerol, either above the lipids' critical temperature (Tc) or at room temperature following the synthesis of the HM-liposomes, the medicine can be encased within such particles. This approach is suitable for encapsulating various bioactive molecules, particularly those sensitive to temperature. Studies by Mozafari et al. indicate that nanoliposomes produced using the heating method are nontoxic to cultured cells, in contrast to those prepared by traditional methods that often involve the use of volatile solvents, which can be cytotoxic [24, 54]. The large-scale production of nanoliposomes is an additional significant advantage of this technique [24].

3.2.2 Application of Liposomes in Drug Delivery
A broad spectrum of liposome formulations containing diverse anticancer drugs has been demonstrated in numerous research studies to be less harmful than unconjugated drugs [24, 55]. Some medications, such as anthracyclines, insert themselves into DNA to target and damage rapidly proliferating cells in tumors, blood cells, hair, and the mucosa of the gastrointestinal tract, stopping the proliferation of these cells. These medications may thus be extremely harmful to the body. However, encapsulating drugs within liposomes can lead to a significant decrease in both acute and chronic toxicity by minimizing delivery to unintended tissues. While this may reduce the overall effectiveness due to the decreased bioavailability of the drug, it is particularly beneficial for tumors that do not actively engulf liposomes [24].

PEGylated liposomes appear to accumulate considerably in Kaposi's sarcoma and head and neck malignancies, according to experimental data. It is noted that their buildup is normal in lung cancer and comparatively reduced in breast cancer [24, 56]. Nonetheless, additional studies are necessary to precisely evaluate the link between the tumor's progression stage, its classification, the level of tissue angiogenesis, and the build-up of liposomes.

According to research on systemic lymphoma, the use of liposomes has been found to enhance treatment effectiveness through a sustained release mechanism and a prolonged presence of therapeutic levels in the blood. On the other hand, the therapeutic effect of the active medication is diminished when it enters the tissues of the mononuclear phagocytic system. In summary, the unique pharmacokinetics of liposomal medications can lead to increased bioavailability for targeted cells residing within the circulatory system [24, 57].

Intravenous injection is the primary method of liposome delivery, contrasting with the less favored oral route, which suffers from the carrier's degradation in the gastrointestinal tract, leading to reduced bioavailability of the drugs it contains [24, 58]. MarqiboVR stands out as the most recent addition to the liposome-based medications that have gained FDA approval. It employs a liposomal delivery system for vincristine, a treatment frequently chosen for grade 2 acute lymphoblastic leukemia. A primary challenge associated with liposome use is their limited retention in the body post-injection. To address this, enhancing the liposome's composition and modulating its size are key tactics for developing long-circulating liposomes [59]. Evidence has shown that smaller liposomes are more effective at avoiding the reticuloendothelial system (RES) compared to larger liposomes.

The modification of liposomes with polyethylene glycol (PEG) to create stealth liposomes is a beneficial technique that reduces their uptake by the RES, leading to a longer circulation time in the body. However, PEGylated liposomes can sometimes struggle to release their payload effectively at the intended target site and may also hinder the escape of drugs from endosomes after cellular uptake. One common strategy in the creation of targeted drug delivery systems (TDDS) is to target liposomes with certain ligands or antibodies. Numerous studies have been conducted to understand how ligand conjugation affects the efficacy of TDDS) [60]. Differentiating between target cells and normal cells is essential for the targeting ligand to minimize nonspecific binding and hazardous side effects. It is necessary for the target tissue to express the antigen or receptor selectively, thus in targeted drug delivery systems (TDDS), a longer circulation time is preferred. Experimental evidence generally suggests that liposome surface modification with antibodies is the most effective approach for creating TDDS [61].

Research done in vitro and in vivo has shown that, in comparison to nontargeted liposomes, targeted liposomes have improved binding to a variety of cancer cell types. Overall, the use of liposomes as drug carriers within the body has been associated with decreased toxicity and improved tolerability of the treatment [24].

3.3 Polymeric Nanoparticles

The investigation of polymer nanoparticles has arisen as a result of personalized drug delivery, enhanced bioavailability, regulated drug release from a single dose, and the ability to shield the drug until it reaches the intended location [18, 62, 63]. Various techniques are available for the production of polymeric nanoparticles. Selecting the most suitable synthesis method is crucial as it influences the nanoparticles' characteristics, including their size, the distribution of sizes within the population, and the manner in which the drug is encapsulated within the nanocarriers [18, 64, 65]. Synthetic polymers undergo intricate modifications that impart certain traits, including the enhancement of target specificity, the boost in bioavailability, the mitigation of toxicity, and the achievement of desired pharmacokinetic properties [18, 66, 67]. The relationship between the makeup and the surface characteristics of these polymers when interacting with biological systems continues to stimulate the development of fresh and enhanced technological designs [18].

When it comes to polymer selection, smart polymers stand out due to their ability to interact with biological processes, including alterations in pH and temperature,

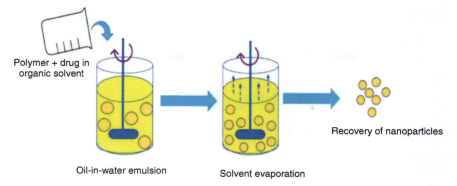

Figure 3.3 Solvent evaporation method. Source: Ahlawat et al. [69]/MDPI/Licensed under CC BY 4.0.

as well as external triggers, enabling them to perform targeted delivery. Hydrophilic polymers that are sensitive to acidic pH levels, such as vinyl esters, dual esters, and hydrazones, are extensively used to target inflamed and tumor tissues or lysosomal compartments, undergoing degradation to release the therapeutic agent [18, 65, 68].

3.3.1 Techniques for Creating Polymeric Nanoparticles

3.3.1.1 Method of Evaporating Solvents

The procedure entails combining the polymer with an organic solvent, such ethyl acetate or chloroform, and then dissolving the medication in the resultant polymeric solution. A surfactant, like polysorbate-80 or poloxamer-188, is then used to assist generate an oil-in-water emulsion. Various physical methods are employed to evaporate the organic solvent, such as increasing the temperature, applying pressure, or stirring continuously (refer to Figure 3.3) [69]. Considering these factors, this technique is favored for the production of water-soluble drug-encapsulated nanoparticles [70]. In essence, this method is intended for use solely at the laboratory scale for synthesis purposes.

3.3.1.2 Method of Spontaneous Emulsification and Solvent Diffusion

This methodology represents a variant of the solvent evaporation methodology. [71–73] In this instance, an organic solvent like dichloromethane is mixed with a water-soluble solvent like methanol or acetone to form the oil phase. Interfacial turbulence is produced by the spontaneous diffusion of methanol or acetone between the two phases, which results in the creation of nanoparticles (Figure 3.4) [69].

3.3.1.3 Salting Out Method

The purpose of this method was to do away with the requirement for organic solvents. To create nanoparticles, it involves dissolving the drug and polymer in a water-soluble solvent, like acetone, and then combining the mixture with an aqueous solution that contains a salting-out agent, like sucrose, magnesium

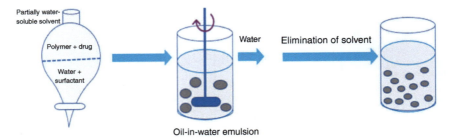

Figure 3.4 Solvent diffusion method. Source: Ahlawat et al. [69]/MDPI/Licensed under CC BY 4.0.

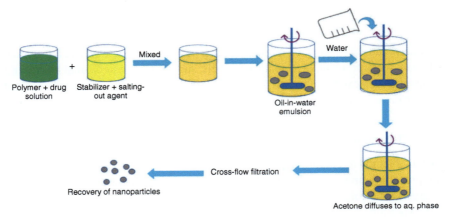

Figure 3.5 Salting out method. Source: Ahlawat et al. [69]/MDPI/Licensed under CC BY 4.0.

chloride, and calcium chloride, and a stabilizer, like hydroxyethylcellulose or PVP (Figure 3.5) [69]. Then, in order to promote the creation of nanoparticles through greater acetone penetration into the aqueous phase, the O/W emulsion is diluted [74]. In conclusion, because this method doesn't need raising the temperature, it is appropriate for materials that are sensitive to heat [75].

3.3.1.4 Method of Solvent Displacement/Nanoprecipitation

Water-soluble medications should not be encapsulated using this procedure; instead, lipophilic medicines should be used instead. The procedure comprises combining the medication and polymer with an organic solvent – dichloromethane, for example – with or without a surfactant, and then incorporating the combination into an aqueous solution that has stabilizer added to it. The polymer sticks to the interface due to the interfacial turbulence caused by the diffusion between the two phases (refer to Figure 3.6) [69]. The method is often used to a range of materials such as maleic acid (PVM/MA), PLGA, PLA, and polyvinylmethyl ether. Because of this method's high entrapment effectiveness of 98%, cyclosporin A was encapsulated using it [76–79].

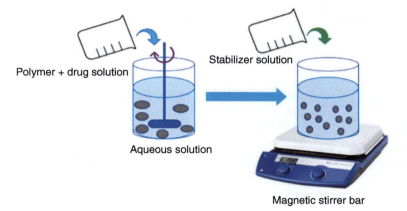

Figure 3.6 Nanoprecipitation method. Source: Ahlawat et al. [69]/MDPI/Licensed under CC BY 4.0.

3.3.1.5 Polymerization Methods

The synthesis of nanoparticles can be achieved through the polymerization of monomers. A research work by Couvreur and colleagues in 1998 described the polymerization of poly(alkyl cyanoacrylate) [80]. Firstly, cyanoacrylic monomers are introduced into a polymerization medium containing polysorbate-20, a surfactant, and are polymerized at room temperature with aggressive stirring. The drug is incorporated into the system either before or after the initiation of polymerization. Purification of the nanoparticles is typically achieved through ultracentrifugation. Controlling the pH (below 3.5), monomer content, stirring speed, and type and concentration of surfactant or stabilizer is critical for producing stable nanoparticles with a high molecular mass [69]. For example, a pH level of 3 or higher during polymerization can cause the nanoparticles to clump together [81].

3.3.1.6 Nanoparticles Developed from Hydrophilic Polymers

Researchers have employed various techniques to produce nanoparticles made from hydrophilic materials, such as chitosan and gelatin. A method for creating chitosan nanoparticles by ionic gelation was presented by Calvo and colleagues in 1996. This process required two aqueous phases interacting with each other: one having chitosan and EO (ethylene oxide), and the other containing polyanion sodium-TPP (tripolyphosphate) [82].

Subsequently, Mao and their collaborators employed the complex coacervation method to fabricate chitosan nanoparticles loaded with DNA for the purpose of oral gene delivery. Furthermore, nanoparticles composed of alginate were developed for the delivery of oligonucleotides (Figure 3.7) [81, 83].

3.3.2 Controlled Drug Release by Polymeric Nanoparticles

Polymeric nanoparticles exhibit greater potential than straightforward nanoparticles since they enable sustained and safe drug delivery with tunable pharmacokinetic characteristics [18, 84]. These nanoparticles provide a range of chemical

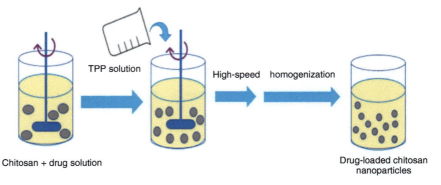

Figure 3.7 Ion gelation method. Source: Ahlawat et al. [69]/MDPI/Licensed under CC BY 4.0.

compositions, charges, and shapes, making them an adaptable carrier for regulated drug delivery. Additionally, they have modifiable release kinetics, which renders them appropriate for both active and passive cancer cell therapy applications [18, 85].

The release mechanisms within nanoparticles are determined by the pace at which the polymer degrades and the drug's capacity to pass through the polymeric substance. The modes of drug diffusion can be categorized into four types: (i) water uptake leading to matrix expansion, (ii) movement through the polymer matrix, (iii) erosion of the matrix, and (iv) osmotic pressure-driven release. In the first scenario, the polymeric matrix contains pores that expand with water uptake, facilitating drug exit. In the second scenario, the permeability and thickness of the drug influence its diffusion [18, 86]. The third scenario involves the polymer's degradation rate governing the release, providing important kinetic features and reliability. The fourth scenario is characterized by the medicament's release being powered by an osmotic pumping process [18, 87].

In a study exploring a potential prostate cancer treatment, nanoparticles encapsulated with resveratrol and made of PLGA demonstrated reduced cell viability and heightened cytotoxicity against LNCaP cells compared to free resveratrol. The cytotoxicity assessment revealed that the nanoparticles had IC_{50} and IC_{90} values approximately double those of resveratrol alone [88].

3.4 Inorganic Nanoparticles

Inorganic nanoparticles, despite their limited biocompatibility and propensity for degradation, are known for their straightforward chemical synthesis. They encompass a variety of forms, including but not limited to, AuNPs, AgNPs, and FeONPs [3, 7, 89]. Nanoparticles made of metal have proven beneficial in the treatment of cancer, utilizing methods such as drug transport or magnetic heat therapy. Nonetheless, it is necessary to coat them with a shielding coating to boost their robustness and therapeutic performance [3, 90]. Gold nanoparticles are often explored for cancer therapy because of their chemically passive and biocompatible characteristics, which improve their targeting efficacy on tumor cells. Furthermore, they have been utilized in some research to deliver various therapeutic agents, including photothermal therapy [3, 7]. Silica nanoparticles function as a potent drug delivery vehicle, featuring the capacity to generate extensively porous frameworks capable of accommodating large volumes of anticancer medications, along with reliable and effective drug dispersal [3, 91–94]. In addition, the use of carbon nanotubes is prevalent in systemic drug delivery platforms for cancer treatment that involve the incorporation of doxorubicin or paclitaxel, as well as in strategies for hyperthermia treatment targeting [95, 96].

The process of creating and altering inorganic nanoparticles is straightforward, allowing for precise control over their size, form, and surface characteristics. Furthermore, the inherent properties of different inorganic nanomaterials have enabled the development of integrated and multifunctional systems utilized in bioimaging, drug delivery, and therapeutic applications [97]. Despite the small number of FDA-approved nanomedicines based on inorganic nanoparticles that have made it to clinical use, their unique design and composition are already transforming conventional medical practices and hold promise for future use in diagnosis and/or therapy [98, 99].

3.4.1 Synthesis and Application of Colloidal Gold Nanoparticles

Gold nanoparticles), when dispersed colloidally, emerge as promising carriers for the transport of drugs and in biomedical applications [100]. The utility of colloidal gold nanoparticles became apparent with the advent of the citrate reduction process, which allowed for the creation of monodisperse gold nanoparticles. Their ease of production, extensive surface area, and tunable surface properties have positioned gold nanoparticles as viable carriers. Furthermore, the integration of smart polymers can enable the advancement of pharmacological transport mechanisms that are capable of releasing their contents in response to external cues [101, 102]. Additionally, on account of their substantial molar absorption coefficient within the visible-to-near-infrared (NIR) spectrum, gold nanoparticles are appropriate for acting as photothermal catalysts in the treatment of cancer via photothermal therapy. Furthermore, the SPR of GNPs has been investigated for its applications in a wide array of biological processes, including bimolecular detection and therapeutic activities [103].

To date, the Schiffrin–Brust biphasic method, which was established in 1994, remains the most favored synthetic route for generating gold nanoparticles because of its straightforward procedural steps and the simplicity of the chemicals required [104]. Following this, researchers such as El-Sayed and colleagues, as well as Murphy and his team, have successfully produced gold nanoparticles of varying sizes and forms, including nanorods, nanocages, and nanocubes, with the seed-mediated growth technique, achieving reliable reproducibility in the process [105, 106]. Given that the initially synthesized gold nanoparticles possess a restricted array of surface-capping ligands and functionalities, it is imperative to carry out ligand exchange and chemical alterations to enhance their utility for diverse applications in nanotechnology and biological fields [97]. Additionally, gold nanostructures with effective targeting abilities can be produced through meticulous surface engineering [98, 107]. Active targeting leverages the characteristics of cancer cells, which rapidly divide and overexpress specific receptors on their membranes. Regarding the cytotoxicity of gold nanoparticles, multiple studies have investigated their potential cellular toxicity [100, 108]. It is crucial to differentiate the toxicity associated with the core of gold nanoparticles from that related to their external ligands. Typically, gold nanoparticles with a positive charge are moderately toxic, whereas those coated with alkylthiolates containing carboxylate groups are comparatively nontoxic. Corroborating the importance of gold nanoparticle ligands, biomolecule-conjugated large gold nanoparticles demonstrated low toxicity toward human leukemia cells (K562) at concentrations reaching up to 250 mM. Contrary to this, solutions of $HAuCl_4$ demonstrated high toxicity, peaking at 90% toxicity levels [109].

3.4.2 Synthesis and Application of Mesoporous Silica Nanoparticles

Mesoporous silica nanoparticles (MSNPs) serve as a crucial class of inorganic carriers for delivery systems. Their suitability for biological applications is attributed to their adjustable sizes, biocompatible mesostructures, and straightforward decoration [110–112]. MSNPs are initially rendered hydrophilic owing to the abundance of silanol groups on their surfaces. This hydrophilicity, combined with their adaptability to modifications by diverse groups, facilitates the precise control of cargo molecule retention or release. Furthermore, the characteristics of porous materials enable a substantial loading of cargo molecules, minimizing their rapid dissolution in water and subsequent loss. This feature enhances the efficacy of the delivery system, enabling a greater amount of medication to reach its intended therapeutic site. Moreover, MSNPs capitalize on their substantial pore volume to enhance the transport of various drugs through the circulatory system. This is especially notable due to the reduced efficacy of these medications as a result of their inadequate water solubility [113].

MSNPs can be modified on both their external and internal surfaces, which enhances their utility as nanocarriers for drug delivery and endows them with various functions. Their mesoporous structure, with a substantial surface area relative to their volume, is essential for thorough functionalization [113, 114]. Organic compounds can be adsorbed onto the outer surface of MSNPs by binding to the silanol groups via covalent and electrostatic bonds. This allows for the attachment of targeting moieties that can enhance the targeting precision of medication distribution to the desired location and reduce injury to unaffected cells. Additionally, it has been observed that the interiors of MSNPs can also be tailored to carry particular cargo [113, 115, 116]. Prior to the widespread use of MSNPs in drug delivery systems, it is essential to assess their cellular uptake mechanisms and potential cytotoxic effects. Studies have demonstrated the internalization of MSNPs by cells and established their biocompatibility across various healthy and cancerous cell lines [117, 118]. Several studies have demonstrated that the uptake and cytotoxicity of MSNPs are related to the properties of the particles [111, 119]. Cytotoxicity is not observed at high concentrations for unmodified MSNPs with a diameter of 100 nm, which exceeds the concentrations typically needed for most therapeutic applications [120–123].

3.4.3 Synthesis and Application of Graphene

Graphene, GO, and RGO have garnered considerable attention in the biomedical field in recent times [124–127]. Single-layer graphene is extensively studied as a promising nanocarrier for drug and gene delivery due to its superior chemical and physical attributes. Graphene's inherent NIR light absorption properties make it suitable for photothermal therapy, which has shown effective antitumor effects. The ability to coat nanographene with various inorganic nanoparticles creates nanocomposites with enhanced traits, beneficial for applications like cancer treatment. Furthermore, the cytotoxicity of graphene materials has been a subject of investigation in both cellular and animal models [128, 129]. Research has highlighted that both the surface composition and the size of a particle are critical in regulating the distribution within the body, elimination, and toxicity of nanographene. Unfunctionalized graphene or graphene oxide (GO) prepared directly can be toxic, whereas GO derivatives with compatible biological coatings exhibit minimal adverse effects on cells at the concentrations tested. Nanographene with very small dimensions and biofriendly coverings may be effectively eliminated from the body post-intravenous injection, and at a tested dose of 20 mg kg^{-1}, it did not cause observable toxicity in mice [130].

Extensive research has focused on the production of graphene and its variants for a wide array of uses. Graphene can be synthesized using either from bottom to up strategies, such as CVD and chemical methods, or from top to down techniques involving mechanical, physical, and chemical exfoliation [131, 132]. GO is produced by subjecting graphite to powerful oxidizing agents, while reduced GO (RGO) is typically acquired through the process of graphite oxide exfoliation followed by chemical reduction [133]. Although GO is water-soluble, it has a propensity to clump together as a result of electrostatic charge screening and nonspecific protein adsorption on its surface. Hence, modifying the surface of GO is essential for boosting its compatibility with biological systems and controlling its interactions within them. GO can be further modified with a biocompatible polymer like PEG, a process known as PEGylation. The synthesized PEG-functionalized nanographene oxide (nGO-PEG), sized between 5 and 50 nm, exhibited remarkable stability within a variety of biological fluids, among them serum [134]. Graphene may also be modified noncovalently by associating with polymers or biomolecules via hydrophobic attractions, π–π interactions, or electrostatic forces, which bolsters its stability in water-based media [130].

The inherent attributes of graphene contribute to the potential of graphene-based nanomaterials as efficient vectors for drug delivery. The strategic coupling of functionalized GO or RGO with targeting moieties has enabled the selective targeted distribution of pharmaceuticals to specific cancerous cells. Variously modified GO surfaces have been utilized as nanocarriers for the encapsulation of multiple drugs, including DOX, CPT, and its analog SN38, employing either adsorption or covalent binding methods [134–137]. In 2008, Dai and colleagues demonstrated that GO could serve as a carrier for the loading and delivery of cancer drugs like SN38 through π–π stacking interplays. Remarkably, the resulting conveyance framework showed improved efficacy compared to irinotecan. Other teams have also chosen folic acid (FA) as a targeting agent for drug transport. Zhang et al. explored the controlled loading of DOX and CPT onto FA-conjugated GO, noting a direct relationship between the loading ratio and the concentration of the drugs [138].

3.4.4 Synthesis and Application of Magnetic Nanoparticles

Magnetic nanoparticles are attractive due to their superparamagnetism, adjustable sizes, and additional biological features [139, 140]. Magnetic nanoparticles smaller than the threshold for a single domain exhibit superparamagnetic behavior at ambient temperatures. Thanks to their distinctive magnetic characteristics and the ability to bind various biological and therapeutic molecules, MNPs have been extensively employed in biomedical and medicinal sciences. Applications include multimodal imaging, directed drug and gene delivery, cancer treatment through hyperthermia, bioseparation techniques, and promoting tissue regeneration [141–144].

Magnetic nanoparticles have often been considered safe for in vivo use due to the substantial natural presence of iron in the human body and their potential for degradation and elimination through endogenous processes. Toxic effects from iron overload in humans are only observed at very high concentrations (greater than 60 mg Fe/kg), which significantly exceeds the levels found as opposed to contrast agents, for instance, Endorem. However, coating materials have been reported to influence the ultimate cytotoxicity, highlighting the importance of their composition [145, 146]. Additionally, the size of hydrodynamic and surface charge of nanoparticles are parameters that appear to be associated with their potential toxicity. Numerous methods have been described for the synthesis of magnetic nanoparticles, including microemulsion techniques, solvothermal/hydrothermal processes, electrochemical methods, and laser pyrolysis, among others [147–149]. In the context of biomedical applications, the majority (over 95%) of reports regarding magnetic nanoparticle preparation involve two primary methods.

The initial step in creating targeted or therapeutic magnetic nanoparticles involves altering the organic coating around the magnetic core to produce a particle with reactive groups for additional customization. Ligand exchange is a common technique. Another viable method relies on molecules terminated with carboxylic groups [150]. The subsequent step in the process involves attaching targeting molecules to the nanoparticle surface to direct them to a specific location and/or loading them with drugs for cancer therapy. Furthermore, magnetic nanomedicines with therapeutic applications are intriguing due to their biocatalytic capabilities. For instance, Xiaoxiong Zhao and colleagues described a novel nanodrug. RNME combines the anticancer properties of NME and Rg3, mitigating their individual limitations and demonstrating synergistic effects. The core Fe@Fe$_3$O$_4$

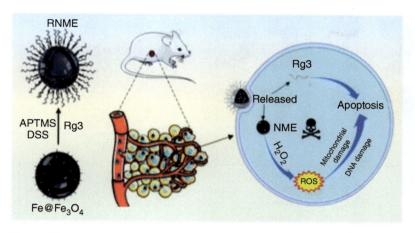

Figure 3.8 A novel metal-organic nanocomposite material serves as a synergistic approach for cancer treatment. Source: Adapted from Zhao et al. [151].

NME displays remarkable biocatalytic activity and imaging abilities. The most inner metal Fe(0) under conditions of weak acidity and the presence of Fe^{3+} in the tumor microenvironment continues to exhibit high Fe^{2+} catalytic activity. Consequently, these NMEs can promote cancer cell apoptosis (refer to Figure 3.8) [151]. Nanomedicine has shown promise in combating liver and pancreatic cancer, and it has also been found to substantially reduce morphological changes in the ileocecal region and alterations to the gut microbiota (see Figure 3.9) [151, 153].

MRI, an imaging technique that relies on computer-assisted detection of proton spin relaxation signals within human organs, stimulated by radiofrequency pulses and gradient fields, has emerged as a valuable diagnostic instrument in medical research [152]. Despite being a reliable imaging method, MRI has limitations in sensitivity and resolution, leading to efforts to integrate multiple imaging technologies to overcome these drawbacks. This has led to the development of hybrid imaging systems, which exploit the advantages of each modality. One example is the creation of biocompatible Fe_3O_4–TaO_x core–shell nanoparticles by Lee and Cho, which offer complementary information from both CT and MRI [154]. CT imaging can effectively visualize newly developed blood vessels within tumors, while MRI allows for the assessment of the tumor microenvironment, including both hypoxic and oxygenated areas. Labhasetwar and colleagues designed MNPs with optical imaging capabilities by incorporating NIR dyes, enabling the quantitative analysis of their biodistribution and tumor targeting efficiency in mice with breast tumor xenografts, both with and without the utilization of a magnetic field [155]. The application of high-sensitivity optical imaging can help assess how the properties of MNP formulations affect their tumor accumulation. Additionally, the clinical oncology potential of MRI/PET bimodal imaging is significant, offering enhanced soft-tissue contrast and better spatial alignment. Matsuda and colleagues demonstrate the benefits of PVE correction [156]. Desirable attributes for MRI/PET bimodal imaging include high sensitivity and superior soft-tissue contrast. With the burgeoning field

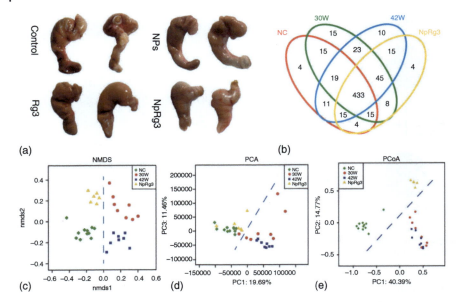

Figure 3.9 Administration of NpRg3 effectively prevents changes to the ileocecal morphology and gut microbiota in mice with hepatocellular carcinoma (HCC). Panel (a) compares the ileocecal morphology changes in the control, NPs, Rg3, and NpRg3 groups. Panel (b) illustrates the overlap and distinct operational taxonomic units (OTUs) among these groups using a Venn diagram. Panel (c) employs nonmetric multidimensional scaling (NMDS) to analyze the distribution and variation of gut microbiota among the four groups. Panel (d) uses principal component analysis (PCA) to demonstrate beta diversity and the degree of similarity among microbial communities. Panel (e) utilizes principal coordinate analysis (PCoA) to investigate the differences in gut microbiota among the four groups. Source: From Ren et al. [152]/John Wiley & Sons/CC BY 4.0.

of biomedical imaging, there is growing interest in multimodal imaging agents that combine optical imaging, PET, and MRI [157].

3.4.5 Quantum Dot Synthesis and Applications

Semiconductor nanoparticles known as quantum dots, with their unique photophysical properties, have become a popular choice for imaging probes and a flexible framework for the development of multipurpose nanodevices [158, 159].

The preparation of highly fluorescent quantum dots typically involves the use of organometallic methods and ligand exchange reactions, with surface modifications being essential for their use in biological systems. Coating the quantum dot surface with thiols is a flexible technique that not only facilitates the transfer of hydrophobic quantum dots into aqueous solutions but also allows for the introduction of functional groups necessary for biological conjugation. Additionally, depositing or attaching polymers onto the quantum dot surface serves as a strategy to improve biocompatibility and maintain stability against chemical degradation. Gao et al. have successfully coated CdSe/ZnS quantum dots with a triblock amphiphilic copolymer that provides resistance against enzymatic breakdown and hydrolysis [160]. Moreover, studies have shown that quantum dots are far more stable and compatible in aqueous conditions when they are encapsulated in silica shells [161].

Bioconjugated quantum dots are becoming commonplace instruments in biology research because of their use in medication and gene transport, sensing, cellular and biomolecular imaging, and sensing [162]. In contrast to mouse fibroblast cells (NIH_3T_3), Bentzen et al.'s study demonstrated that CdSe/ZnS quantum dots coated with amphiphilic poly(acrylic acid) (AMP) had a greater propensity to bind nonspecifically to human epithelial kidney (HEK) cells [163]. The intracellular distribution of rhodamine dextran and nuclear localization signal (NLS)-conjugated quantum dots was compared by Derfus et al. [164] Trifunctional polymer nanobeads were created by Pellegrino using a combination of magnetic nanoparticles, quantum dots, and an amphiphilic polymer. The surface of the beads was then further altered by adding FA [165]. Fan and Ding reported the encapsulation of quantum dots onto magnetite nanorings, resulting in a vortex core with exceptional luminosity and magnetic activity that serves as an innovative magnetic fluorescence nanoprobe [166]. Strategies to prevent in vivo quantum dots accumulation and degradation are essential to evaluate the clinical and biomedical nanotechnology potential of quantum dots.

3.4.6 Creation and Use of Stratified Double Hydroxides

$[M^{2+}_{1-x}M^{3+}_{x}(OH)_2](A^{n-})_{x/n} \cdot mH_2O$ is the formula) for layered double hydroxides in which the cations M^{2+} and M^{3+} are arranged inside brucite-like layers and M^{3+} serves as the interlayer anion to balance the charges [167]. With its use in gene and medication delivery systems, as well as in the creation of biological composite materials, layered double hydroxides have been thoroughly studied [168, 169]. Layered double hydroxide materials exhibit several benefits for drug and gene delivery applications in the biomedical field. Layered double hydroxides facilitate controlled release of therapeutics from within their layers, which is critical for the intended pharmaceutical effects [152].

Studies have indicated that layered double hydroxides exhibit a low even to negligible cytotoxicity toward mammalian cells [170, 171]. For instance, Tronto and associates intercalated a variety of pharmaceutical anions, such as aspartame, citrate, salicylate, and glutamate, into layered double hydroxides using two distinct synthesis techniques [172]. Layered double hydroxide hybrids containing myc antisense oligonucleotides were created by Kwak and his colleagues. When these hybrids were introduced to leukemia cells, HL-60 cell proliferation was inhibited [173]. One significant advantage of using layered double hydroxides to load drugs is their ability to provide sustained release [174]. Xu and colleagues investigated the cellular uptake mechanism of layered double hydroxide nanoparticles and identified that it involves clathrin-mediated endocytosis and escape from endosomes [175]. This modified endocytic pathway allows low-molecular-weight heparin-layered double hydroxide to achieve sustained release and enhanced pharmaceutical effects.

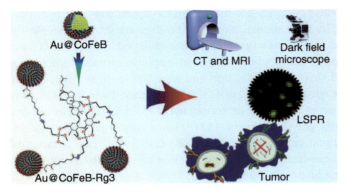

Figure 3.10 Au@CoFeB nanoparticles and Au@CoFeB–Rg3 nanomedicines show multimode imaging functions for medical diagnosis and exhibit a marked effect on the treatment of cancer cells. Source: Zhang et al. [177]/American Chemical Society.

3.4.7 Nanoparticles with Multifunctional Composite

As was previously indicated, different inorganic nanoparticles have distinct characteristics, including visible to NIR photoluminescence, photothermal properties, or the ability to load drugs. The integration of these nanoparticles can lead to multifunctional inorganic composites that are useful for both diagnostic and therapeutic applications in diseases [176]. The treatment approach of Au@SiO$_2$–DOX integrates chemotherapy and photothermal therapy, showing a synergistic effect that enhances the ability to kill cancer cells. Additionally, the MSNP platform has demonstrated the ability to encapsulate various components, including cancer drugs, superparamagnetic iron oxide nanocrystals, fluorescent markers, and targeting moieties. This platform enables simultaneous drug delivery, magnetic resonance imaging, fluorescence imaging, and cell targeting [152]. Recently, Weiwei Zhang and colleagues have described Au@CoFeB nanoparticles and Au@CoFeB–Rg3 s nanomedicines that display multimodal imaging capabilities for medical diagnosis and have shown a significant impact on the treatment of cancer cells, as illustrated in Figure 3.10 [177].

3.5 Mixture (Hybrid) Nanoparticles

The development of drug-delivery nanoparticles faces key challenges including efficacy, biocompatibility, degradation, stability, and toxicity. Organic and inorganic nanoparticles each come with their unique benefits and drawbacks. Consequently, the combination of these different types has been proposed as a strategy to address many of the hurdles associated with drug delivery [3, 178]. Chemodynamic therapy (CDT)-induced apoptosis is a promising cancer treatment approach. However, its clinical application is limited by significant side effects and suboptimal efficacy. Xiaoxiong Zhao and colleagues showed that altering the surface of ginsenoside Rg3 may drastically change the distribution of nanocatalysts in organs and their accumulation in tumors when supplied systemically. As seen in Figure 3.11, this strategy reduces toxicity and improves in vivo effectiveness while addressing common toxicological problems in nanomedicine [179]. Studies have demonstrated

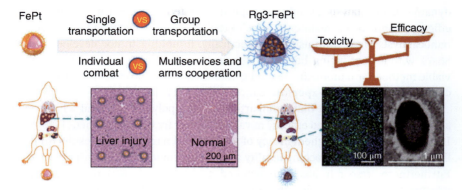

Figure 3.11 The safety and efficacy of ferroptosis-apoptosis combination treatment are enhanced by nanocatalysts after Rg3 surface engineering, offering a useful concept for clinical procedures. Source: Zhao et al. [179]/with permission of Elsevier.

that in contrast to solitary nanocatalysts, Rg3-protected dynamic nanocatalysts form hydrophilic nanoclusters that prolong their bloodstream circulation duration. This defense keeps the nanocatalysts from leaking and permits their precise release at the locations of tumors. Additionally, the Rg3 drug-loading platform provided by the nanoclusters increases the quantity of Rg3 that reaches the tumor and strengthens the synergistic effects with the nanocatalysts. The ability of the Rg3-protected

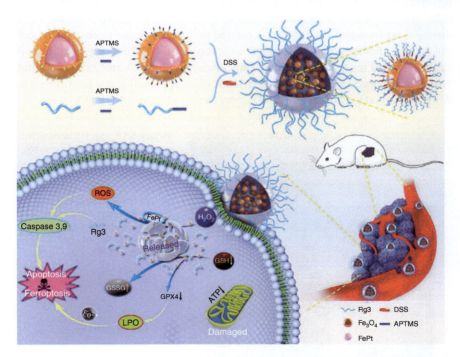

Figure 3.12 Diagram illustrating the structural component, the NFPR preparation procedure, and the mechanism of the combination ferroptosis-apoptosis therapy. Source: Zhao et al. [179]/with permission of Elsevier.

dynamic nanocatalysts to induce apoptosis and ferroptosis greatly improves the efficacy of anticancer treatments. Mice's lifespan was extended when Rg3-sheltered nanocatalysts were administered systemically, inhibiting tumor development by 86.6% without producing harm. As shown in Figure 3.12, our study presents a viable method for nanomedicine with great biosafety and provides new avenues for combination ferroptosis–apoptosis cancer therapy [179].

Studies have indicated that nanoparticles with a polymer core and a lipid outer shell can exhibit enhanced efficacy in various cancer treatment regimens [180, 181]. It is believed that the higher efficacy of these hybrid nanoparticles results from the lipid shell's improved biocompatibility and the polymer core's integrity [182, 183]. Furthermore, research has demonstrated that hybrid nanoparticles are more effectively taken up by cells and are subject to less clearance [184, 185]. Several investigations focused on breast, pancreatic, and prostate cancers have highlighted the effectiveness of silica–liposomal nanoparticles as a means of delivering anticancer agents [3, 182, 186, 187]. Additionally, hybrid nanoparticles like carbon nanotubes and chitosan nanoparticles have demonstrated improved anticancer capabilities [3, 188].

Moreover, the combination of chitosan-based hydrogels with inorganic nanomedicines has become a focus of research interest. In recent work, Ran Liu and associates created hydrogel microcapsules filled with nanomedicine using a polydimethylsiloxane-based microfluidic reactor. We evaluated the release characteristics of the nanomedicines from the hydrogel by simulating the target diseased cells' microenvironment and the pH and temperature of the gastrointestinal system during drug transport. The results indicated that the hydrogel microcapsule technology for nanomedicine development is appropriate for oral delivery. Furthermore, the medication functions within a highly biocompatible polymer interpenetrating network structure, opening the door to its possible application as a promising drug delivery method for the treatment of cancer (refer to Figure 3.13) [189].

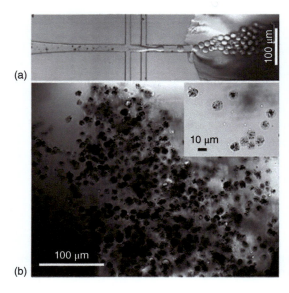

Figure 3.13 (a) Nanomedicine microcapsule production using a multijunction droplet microfluidic reactor and (b) synthetic nanomedicine microcapsules captured in a bright-field optical picture. Source: From Ref. Liu et al. [189]/Royal Society of Chemistry/CC BY 3.0.

3.6 Cell Membrane Coating Nanotechnology

Cell membrane coating nanotechnology allows scientists to impart certain cellular functions onto nanoparticles using simpler chemical methods, avoiding intricate procedures [190].

The replication of the natural structure is accomplished through a unique method involving the application of a natural cell membrane from various cellular sources, like red blood cells, onto the nanoparticles [3, 191–193].

3.6.1 Synthesis of Cell Membrane–Nanoparticle Structures

To create an effective nanomedicine, the cell membrane must be integrated with synthesized nanomaterials following extraction. Often, a strong and lasting bond is necessary, with attention paid to maintaining surface groups for improved same-type interactions. The prevalent techniques for cell membrane coating, as documented in literature, typically include repeated extrusion through porous filters and the use of sonication, along with the investigation of innovative methods. When establishing new techniques, it is crucial to verify that the cell membrane's original orientation is preserved, ensuring that surface groups remain externally displayed and retain their targeting or masking abilities [190].

3.6.1.1 Extrusion

The initial documented technique for cell membrane coating from 2011 entailed combining PLGA nanoparticles with membrane vesicles derived from red blood cells, which was then followed by seven sequential extrusions through 100 nm polycarbonate porous filters [194]. The technique has demonstrated its efficacy in employing mechanical pressure to distort cell membranes around a diverse array of nanomaterials, each with distinct physical characteristics and forms [195]. Challenges with the extrusion technique include the buildup of material on the filter, which leads to a reduction in pore size and the loss of substance, as well as the scalability limitations of this method [190].

3.6.1.2 Sonication

The sonication method addresses the issue of material loss and enables scaling up of production. Coating success is frequently shown after subjecting a mixture of cell membrane and nanoparticles to sonication, thereby avoiding the membrane loss encountered with extrusion [190, 196, 197]. Results obtained with the sonication approach are said to be the same or nearly equivalent to those obtained with the porous membrane extrusion method [191]. It is believed that the sonication process causes activation energy to attract the nanoparticles to the cell membrane, disrupting the bilayer of the membrane. Subsequently, the membrane tends to reform into its often spherical shape to achieve the minimum entropy, favoring the formation of the smallest structures [198]. The method of sonication can be incompatible with specific nanocore formations because of the alterations it induces in the size and stability of the substances [199].

3.6.1.3 Electroporation

In 2017, it was first shown that magnetic nanoparticles could be induced to enter cell membrane vesicles using a microfluidic electroporation method [200]. The process entailed combining the necessary elements within a microfluidic channel, followed by the application of electrical pulses through electrodes to facilitate the entry of the magnetic nanoparticle core into the cells. The research indicated that this method was preferable to traditional mechanical extrusion, as the nanocores generated using microfluidics exhibited better colloidal stability and more favorable in vivo responses, suggesting a more robust coating. Microfluidic electroporation could potentially improve upon the limited throughput of mechanical extrusion [190]. The study showed that it is possible to use a cell nuclear transfer device for the electroporation of gold nanorods into intact platelets. This finding is important for in vivo photothermal treatment (PTT) applications [201]. Interestingly, with each consecutive injection of the gold nanorod-platelet treatment, the accumulation of nanomedicine and the peak temperature at the tumor site increased. This phenomenon is believed to be caused by platelets' natural tendency to congregate in inflammatory areas, which is brought on by photothermal impacts [202].

3.6.1.4 Graphene Nanoplatform-mediated Cell Membrane Coating

A recently developed technique involving graphene has shown its efficiency in extracting leukocyte cell membranes and integrating them with a nanoparticle core in a single process [203]. When tested with cancer cells, the resulting nanoplatform achieved a cell-type specific capture efficiency of over 85% for circulating tumor cells, which remained viable for further culture. This innovation holds promise for the capture and isolation of circulating cancer cells, their propagation in culture, and the use of secondary biomimetic cancer therapy to specifically target and kill these cells in vivo [190].

3.6.1.5 Encapsulation of In Situ Using Cell-derived Vesicles

Vesicles containing different nanoparticles may be produced from live human umbilical vein endothelial cells by subjecting them to the nanoparticles in a serum-free environment [204]. The technique demonstrated the capability to encapsulate a variety of nanoparticles, such as iron oxide, gold, and quantum dots, within cell-derived vesicles. By employing a starvation medium, the cells were induced to form vesicles, which were then isolated containing the nanoparticles. These vesicles proved to be functional for applications including magnetic targeting, enhancing MRI contrast, and enabling hyperthermia. Since endogenous vesicles are known to be crucial for tumor migration and metastasis, repurposing them as drug delivery systems could capitalize on cancer cells' affinity for these vesicles, potentially enhancing treatment effectiveness [205].

3.7 Challenges and Current Limitations

Countless studies indicate that nanomedicine treatments are successful in combating cancer, both in laboratory settings and within living organisms. Nonetheless, a limited number of nanocarrier-based cancer therapies have progressed to clinical testing. Therefore, it is crucial to tackle the obstacles associated with creating optimized nanomedicine formulations that can be used in clinical stages [13, 206].

3.7.1 Nanomaterials' Physicochemical Characterization

In general, the structure, composition, size, surface features, porosity, charge, and aggregation tendencies of nanocarriers are their main physicochemical attributes [207, 208]. The diversity inherent in these properties complicates the characterization of nanomedicine products both before and after their introduction into the body. It is essential for quantitative analytical techniques to be capable of assessing all critical quality attributes of these nanosized substances [13].

Because minor variations in PD and physicochemical characteristics can result in considerable changes to secondary features including biocompatibility, toxicity, and therapeutic results in vivo, polydispersity plays a crucial role in the characterization of nanocarriers [13, 209, 210].

It might be challenging to evaluate the stability and storage qualities (shelf life) of goods containing nanomedicines [13, 211]. Nanocarriers may change their physicochemical characteristics, such as size, drug loading, and release kinetics, during polymer breakdown and entrance into biological fluids like serum. These changes may have an impact on the nanocarriers' in vivo performance. Moreover, storing nanocarriers in aqueous solutions, buffers, or as lyophilized powders can also change their characteristics [13, 212].

3.7.2 Safety Concerns

The extensive application of nanoparticles has prompted a critical need to evaluate their potential toxicity toward human health and the environment. Numerous investigations have shown that nanoparticles can induce adverse biological responses, giving rise to the field of nanotoxicology as a distinct area of scientific inquiry [13, 100, 213].

To mitigate risks in clinical development, the integration of sophisticated or predictive diagnostic methods with innovative targeting techniques can be beneficial. This approach can help identify "safe responders" and enable personalized cancer treatments. Theranostic strategies present significant potential in advancing this field [13, 214].

Beyond medical concerns, there are also environmental implications associated with the widespread use of nanomaterials. The large-scale manufacturing of nanomaterials makes it harder to track environmental and occupational exposure to nanoparticles. Yet it becomes increasingly crucial to assess and manage the potential impacts [13, 215].

3.7.3 Regulatory Issues

The authorization of nanomedicine therapies continues to face hurdles. The rise of more sophisticated and versatile nanocarriers is probably to complicate the approval process further. Regulatory agencies will need to enhance and harmonize the criteria for nanomedicine product approval to ensure patients can receive these innovative treatments in a timely manner [13].

3.7.4 Manufacturing Issues

The production of nanomedicines that meets technical challenges and adheres to good manufacturing practice (GMP) standards presents a significant challenge. Typically, preclinical and early clinical trials have been conducted using limited quantities of nanomaterials. When scaling up to larger production levels, variations in physical and chemical properties between batches can arise due to the nanomaterial's polydispersity [13, 216]. Thus, for the industrial-scale manufacturing of nanoparticle-based cancer therapies, it is essential to exercise strict control over the physicochemical properties to maintain consistency from one batch to another [13, 217]. A case in point illustrating the production hurdles for nanomedicine therapeutics is the shortage of Doxil® [218, 219]. In November 2011, production of Doxil® was suspended because of manufacturing and sterility problems. This caused a shortage that extended until 2014, leading to treatment interruptions for patients, higher drug expenses, and the necessity to develop a new manufacturing process for Doxil® [13].

Therefore, in order to achieve GMP-compliant, large-scale manufacturing of nanomedicines, it is vital to focus on reproducibility and thorough product analysis. Bioconjugation techniques, such as maleimide or succinimide reactions, are typically performed within a tight pH range to avoid hydrolysis. For the production of sophisticated, pH-responsive nanomedicines, it is crucial to keep the pH within this narrow interval throughout the manufacturing process, which calls for clearly established production procedures [13, 220].

The high cost of raw materials and the need for a multistep, intricate manufacturing process make the development of nanomedicine treatments unaffordable [13, 221]. Pharmaceutical businesses can be discouraged by the high costs involved in producing nanocarriers on a wide scale. Therefore, in order to justify their higher price compared to conventional medicines and to offset the costs of development and manufacture, the therapeutic benefits of nanomedicine products must be significant [13, 222].

3.8 Conclusions

A very promising avenue for enhanced cancer treatment is nanomedicine, a rapidly developing field of study. Numerous systems for nanomedicine have been developed and are now being applied in the therapeutic treatment of cancer. Phase 3 trials have demonstrated the benefits of anticancer drugs delivered through nanocarriers. However, challenges such as characterizing nanomaterials, addressing safety concerns, and dealing with regulatory and production issues are obstacles to the widespread use of nanomedicine in cancer therapy. For regulatory approval and large-scale production, precise standards and comprehensive instructions are absolutely required. It is best to start interacting with regulators as soon as possible to talk about new platforms for nanotechnology and to speed up review and approval. Collaboration among research labs, regulatory bodies, and the industry is vital to ensure patients have swift access to innovative, safe, and effective treatments in the realm of cancer nanomedicine [13].

References

1 Kulkarni, A., Kelkar, D.A., Parikh, N. et al. (2020). Meta-analysis of prevalence of triple-negative breast cancer and its clinical features at incidence in Indian patients with breast cancer. *JCO Global Oncology* 6: 1052–1062.
2 Sharma, R. (2021). Breast cancer burden in Africa: evidence from GLOBOCAN 2018. *Journal of Public Health* 43 (4): 763–771.
3 Al-Zoubi, M.S. and Al-Zoubi, R.M. (2022). Nanomedicine tactics in cancer treatment: challenge and hope. *Critical Reviews in Oncology/Hematology* 174: 103677.
4 Chen, H., Gu, Z., An, H. et al. (2018). Precise nanomedicine for intelligent therapy of cancer. *Science China Chemistry* 61 (12): 1503–1552.
5 Allemani, C., Matsuda, T., Carlo, V.D. et al. (2018). Global surveillance of trends in cancer survival 2000-14 (CONCORD-3): analysis of individual records for 37 513 025 patients diagnosed with one of 18 cancers from 322 population-based registries in 71 countries. *Lancet (London, England)* 391 (10125): 1023–1075.
6 de la Torre, P., Pérez-Lorenzo, M.J., Alcázar-Garrido, Á., and Flores, A.I. (2020). Cell-based nanoparticles delivery systems for targeted cancer therapy: lessons from anti-angiogenesis treatments. *Molecules* 25 (3): 715.
7 Yao, Y. and Fang, D. (2020). Design of automatic test system for multiplex digital array module. In: *2020 6th International Conference on Control, Automation and Robotics (ICCAR)*, 7–10.
8 Al-Trad, B., Alkhateeb, H., Alsmadi, W., and Al-Zoubi, M. (2019). Eugenol ameliorates insulin resistance, oxidative stress and inflammation in high fat-diet/streptozotocin-induced diabetic rat. *Life Sciences* 216: 183–188.
9 Aljabali, A.A.A., Hussein, E., Aljumaili, O. et al. (2018). Rapid magnetic nanobiosensor for the detection of Serratia marcescen. *IOP Conference Series: Materials Science and Engineering* 305 (1): 012005.
10 Aljabali, A.A.A., Zoubi, M.S.A., Al-Batanyeh, K.M. et al. (2019). Gold-coated plant virus as computed tomography imaging contrast agent. *Beilstein Journal of Nanotechnology* 10 (1): 1983–1993.
11 Aljabali, A.A.A., Al Zoubi, M.S., Alzoubi, L. et al. (2020). Chapter 16 – Chemical engineering of protein cages and nanoparticles for pharmaceutical applications. In: *Nanofabrication for Smart Nanosensor Applications. Micro and Nano Technologies* (ed. K. Pal and F. Gomes), 415–433. Elsevier.
12 Wang, Y., Deng, Y., Luo, H. et al. (2017). Light-responsive nanoparticles for highly efficient cytoplasmic delivery of anticancer agents. *ACS Nano* 11 (12): 12134–12144.

13 Andreas, W., Dominik, W., Vimalkumar, B., and Jrg, H. (2015). Nanomedicine in cancer therapy: challenges, opportunities, and clinical applications. *Journal of Controlled Release: Official Journal of the Controlled Release Society* 200: 138–157.

14 Ruth, D. and Rogerio, G. (2011). Nanomedicine(s) under the microscope. *Molecular Pharmaceutics* 8 (6): 2101–2141.

15 Kalyane, D., Raval, N., Maheshwari, R. et al. (2019). Employment of enhanced permeability and retention effect (EPR): nanoparticle-based precision tools for targeting of therapeutic and diagnostic agent in cancer. *Materials Science and Engineering C* 98: 1252–1276.

16 Rodallec, A., Benzekry, S., Lacarelle, B. et al. (2018). Pharmacokinetics variability: why nanoparticles are not just magic-bullets in oncology. *Critical Reviews in Oncology/Hematology* 129: 1–12.

17 Cho, K., Wang, X., Nie, S. et al. (2008). Therapeutic nanoparticles for drug delivery in cancer. *Clinical Cancer Research* 14 (5): 1310–1316.

18 Castro, K.C.D., Costa, J.M., and Campos, M.G.N. (2022). Drug-loaded polymeric nanoparticles: a review. *International Journal of Polymeric Materials and Polymeric Biomaterials* 71 (1): 1–13.

19 Danhier, F., Feron, O., and Préat, V. (2010). To exploit the tumor microenvironment: passive and active tumor targeting of nanocarriers for anti-cancer drug delivery. *Journal of Controlled Release* 148 (2): 135–146.

20 Olusanya, T.O.B., Haj Ahmad, R.R., Ibegbu, D.M. et al. (2018). Liposomal drug delivery systems and anticancer drugs. *Molecules* 23 (4): 907.

21 Balzus, B., Sahle, F.F., Hönzke, S. et al. (2017). Formulation and ex vivo evaluation of polymeric nanoparticles for controlled delivery of corticosteroids to the skin and the corneal epithelium. *European Journal of Pharmaceutics and Biopharmaceutics* 115: 122–130.

22 Kieler-Ferguson, H.M., Chan, D., Sockolosky, J. et al. (2017). Encapsulation, controlled release, and antitumor efficacy of cisplatin delivered in liposomes composed of sterol-modified phospholipids. *European Journal of Pharmaceutical Sciences* 103: 85–93.

23 Khosravi-Darani, K., Pardakhty, A., Honarpisheh, H. et al. (2007). The role of high-resolution imaging in the evaluation of nanosystems for bioactive encapsulation and targeted nanotherapy. *Micron* 38 (8): 804–818.

24 Panahi, Y., Farshbaf, M., and Mohammadhosseini, M. (2017). et al., Recent advances on liposomal nanoparticles: synthesis, characterization and biomedical applications. *Artificial Cells, Nanomedicine, and Biotechnology* 45 (4): 788–799.

25 Allison, S.D. (2007). Liposomal drug delivery. *Journal of Infusion Nursing* 30 (2): 89–95.

26 Mohammadreza, M., Parisa, E., Azim, A. et al. (2016). Efficacy of pegylated liposomal etoposide nanoparticles on breast cancer cell lines. *Turkish Journal of Medical Sciences* 46: 567–571.

27 Mohammed, A.R., Weston, N., Coombes, A.G.A. et al. (2004). Liposome formulation of poorly water soluble drugs: optimisation of drug loading and ESEM analysis of stability. *International Journal of Pharmaceutics* 285 (1): 23–34.

28 Allen, T.M., Cheng, W.W.K., Hare, J.I., and Laginha, K.M. (2006). Pharmacokinetics and pharmacodynamics of lipidic nano-particles in cancer. *Anti-Cancer Agents in Medicinal Chemistry* 6 (6): 513–523.

29 Allen, T.M. and Martin, F.J. (2004). Advantages of liposomal delivery systems for anthracyclines. *Seminars in Oncology* 31: 5–15.

30 Cristiano, M.C., Cosco, D., Celia, C. et al. (2017). Anticancer activity of all-trans retinoic acid-loaded liposomes on human thyroid carcinoma cells. *Colloids and Surfaces B: Biointerfaces* 150: 408–416.

31 Park, K., Kwon, I.C., and Park, K. (2011). Oral protein delivery: current status and future prospect. *Reactive and Functional Polymers* 71 (3): 280–287.

32 Mozafari, M.R., Flanagan, J., Matia-Merino, L. et al. (2010). Recent trends in the lipid-based nanoencapsulation of antioxidants and their role in foods. *Journal of the Science of Food and Agriculture* 86 (13): 2038–2045.

33 O'Brien, M.E.R., Wigler, N., Inbar, M. et al. (2004). Reduced cardiotoxicity and comparable efficacy in a phase III trial of pegylated liposomal doxorubicin HCl(CAELYXTM/Doxil®) versus conventional doxorubicin for first-line treatment of metastatic breast cancer. *Annals of Oncology* 15 (3): 440–449.

34 Satsangi, A., Roy, S.S., and Satsangi, R.K. (2015). et al., Synthesis of a novel, sequentially active-targeted drug delivery nanoplatform for breast cancer therapy. *Biomaterials* 59: 88–101.

35 Yari, H., Nkepang, G., and Awasthi, V. (2019). Surface modification of liposomes by a lipopolymer targeting prostate specific membrane antigen for theranostic delivery in prostate cancer. *Materials* 12 (5): 756.

36 Chen, X., Zhang, Y., Tang, C. et al. (2017). Co-delivery of paclitaxel and anti-survivin siRNA via redox-sensitive oligopeptide liposomes for the synergistic treatment of breast cancer and metastasis. *International Journal of Pharmaceutics* 529 (1): 102–115.

37 Meng, J., Guo, F., Xu, H. et al. (2016). Combination therapy using co-encapsulated resveratrol and paclitaxel in liposomes for drug resistance reversal in breast cancer cells in vivo. *Scientific Reports* 6: 22390.

38 Al-Remawi, M., Elsayed, A., Maghrabi, I. et al. (2017). Chitosan/lecithin liposomal nanovesicles as an oral insulin delivery system. *Pharmaceutical Development and Technology* 22 (3): 390–398.

39 Wang, S.X., Michiels, J., Ariën, K.K. et al. (2016). Inhibition of HIV virus by neutralizing Vhh attached to dual functional liposomes encapsulating Dapivirine. *Nanoscale Research Letters* 11 (1): 1–10.

40 Veneti, E., Tu, R.S., and Auguste, D.T. (2016). RGD-targeted liposome binding and uptake on breast cancer cells is dependent on elastin linker secondary structure. *Bioconjugate Chemistry* 27 (8): 1813–1821.

41 Eatemadi, A., Daraee, H., Zarghami, N. et al. (2016). Nanofiber: synthesis and biomedical applications. *Artificial Cells, Nanomedicine, and Biotechnology* 44 (1): 111–121.

42 Yu, F. and Tang, X. (2016). Novel long-circulating liposomes consisting of PEG modified beta-sitosterol for gambogic acid delivery. *Journal of Nanoscience and Nanotechnology* 16: 3115–3121.

43 Cao, H., Dan, Z., He, X. et al. (2016). Liposomes coated with isolated macrophage membrane can target lung metastasis of breast cancer. *ACS Nano* 10 (8): 7738–7748.

44 Davaran, S., Rezaei, A., Alimohammadi, S. et al. (2014). Synthesis and physicochemical characterization of biodegradable star-shaped poly lactide-*co*-glycolide-β-cyclodextrin copolymer nanoparticles containing albumin. *Advances in Nanoparticles* 03 (01): 14–22.

45 Daniele, P., Elena, T., Elena, R., and Gabriele, C. (2016). Characterization and investigation of redox-sensitive liposomes for gene delivery. *Methods in Molecular Biology* 1445: 217.

46 Tabatabaei Mirakabad, F.S., Akbarzadeh, A., Milani, M. et al. (2014). A comparison between the cytotoxic effects of pure curcumin and curcumin-loaded PLGA-PEG nanoparticles on the MCF-7 human breast cancer cell line. *Artificial Cells, Nanomedicine, and Biotechnology* 2014: 1–8.

47 Tong, Y., Ji, P., and Zhao, W. (2015). Preparation of baicalein liposome-lyophilized powder and its pharmacokinetics study. *Journal of Chinese Medicinal Materials* 38 (11): 2404–2507.

48 Emmanuel, S.A., Ivana, V., Samy, D. et al. (2016). A protocol for the systematic and quantitative measurement of protein–lipid interactions using the liposome-microarray-based assay. *Nature Protocols* 11 (6): 1021–1038.

49 Alizadeh, E., Akbarzadeh, A., Eslaminejad, M.B. et al. (2015). Upregulation of liver-enriched transcription factors HNF4a and HNF6 and liver-specific MicroRNA (miR-122) by inhibition of Let-7b in mesenchymal stem cells. *Chemical Biology & Drug Design* 85 (3): 268–279.

50 Davidson, E.M., Haroutounian, S., Kagan, L. et al. (2016). A novel proliposomal ropivacaine oil: pharmacokinetic–pharmacodynamic studies after subcutaneous administration in pigs. *Anesthesia & Analgesia* 122 (5): 1663.

51 Ying, Z., Mu-Qing, X., Lin, L. et al. (2016). Preparation of liposomal amiodarone and investigation of its cardiomyocyte-targeting ability in cardiac radiofrequency ablation rat model. *International Journal of Nanomedicine* 11: 2359–2367.

52 Endo-Takahashi, Y., Ooaku, K., Ishida, K. et al. (2016). Preparation of angiopep-2 peptide-modified bubble liposomes for delivery to the brain. *Biological & Pharmaceutical Bulletin* 39 (6): 977–983.

53 Meisel, J.W. and Gokel, G.W. (2016). A simplified direct lipid mixing lipoplex preparation: comparison of liposomal-, dimethylsulfoxide-, and ethanol-based methods. *Scientific Reports* 6: 27662.

54 Garello, F., Vibhute, S., Guenduez, S. et al. (2016). Innovative design of Ca-sensitive paramagnetic liposomes results in an unprecedented increase in longitudinal relaxivity. *Biomacromolecules* 17 (4): 1303–1311.

55 Patil, Y., Amitay, Y., Ohana, P. et al. (2016). Targeting of pegylated liposomal mitomycin-C prodrug to the folate receptor of cancer cells: intracellular activation and enhanced cytotoxicity. *Journal of Controlled Release* 225: 87–95.

56 Harrington, K.J., Mohammadtaghi, S., Uster, P.S. et al. (2001). Effective targeting of solid tumors in patients with locally advanced cancers by radiolabeled pegylated liposomes. *Clinical Cancer Research* 7 (2): 243–254.

57 Daraee, H., Etemadi, A., Kouhi, M. et al. (2014). Application of liposomes in medicine and drug delivery. *Artificial Cells, Nanomedicine, and Biotechnology* 44 (1): 381–391.

58 Bochot, A. and Fattal, E. (2012). Liposomes for intravitreal drug delivery: a state of the art. *Journal of Controlled Release* 161 (2): 628–634.

59 Kelly, C., Jefferies, C., and Cryan, S.A. (2011). Targeted liposomal drug delivery to monocytes and macrophages. *Journal of Drug Delivery* 2011: 727241.

60 Sawant, R.R. and Torchilin, V.P. (2012). Challenges in development of targeted liposomal therapeutics. *AAPS Journal* 14 (2): 303–315.

61 Sofou, S. and Sgouros, G. (2008). Antibody-targeted liposomes in cancer therapy and imaging. *Expert Opinion on Drug Delivery* 5: 189–204.

62 Mura, S., Nicolas, J., and Couvreur, P. (2013). Stimuli-responsive nanocarriers for drug delivery. *Nature Materials* 12 (11): 991–1003.

63 Rizvi, S.A.A. and Saleh, A.M. (2018). Applications of nanoparticle systems in drug delivery technology. *Saudi Pharmaceutical Journal* 26 (1): 64–70.

64 Jacob, J., Haponiuk, J.T., Thomas, S., and Gopi, S. (2018). Biopolymer based nanomaterials in drug delivery systems: a review. *Materials Today Chemistry* 9: 43–55.

65 Torchilin, P.V. (2006). Multifunctional nanocarriers. *Advanced Drug Delivery Reviews* 58 (14): 1532–1555.

66 Wong, P.T. and Choi, S.K. (2015). Mechanisms of drug release in nanotherapeutic delivery systems. *Chemical Reviews* 115 (9): 3388.

67 Kakkar, A., Traverso, G., Farokhzad, O.C. et al. (2017). Evolution of macromolecular complexity in drug delivery systems. *Nature Reviews Chemistry* 1 (8): 0063.

68 Torchilin, V.P. (2014). Multifunctional, stimuli-sensitive nanoparticulate systems for drug delivery. *Nature Reviews Drug Discovery* 13 (11): 813–827.

69 Ahlawat, J., Henriquez, G., and Narayan, M. (2018). Enhancing the delivery of chemotherapeutics: role of biodegradable polymeric nanoparticles. *Molecules* 23 (9): 2157.

70 Bénard, M., Schapman, D., Lebon, A. et al. (2015). Structural and functional analysis of tunneling nanotubes (TnTs) using gCW STED and gconfocal approaches. *Biology of the Cell* 107 (11): 419–425.

71 Wehrle, P., Magenheim, B., and Benita, S. (1995). The influence of process parameters on the PLA nanoparticle size distribution evaluated by means of factorial design. *European Journal of Pharmaceutics and Biopharmaceutics* 41: 19–26.

72 Niwa, T., Takeuchi, H., Hino, T. et al. (1993). Preparations of biodegradable nanospheres of water-soluble and insoluble drugs with D,L-lactide/glycolide copolymer by a novel spontaneous emulsification solvent diffusion method, and the drug release behavior. *Journal of Controlled Release* 25 (1): 89–98.

73 Murakami, H., Yoshino, H., Mizobe, M. et al. (1999). Preparation of poly(DL-lactide-*co*-glycolide) latex for surface-modifying material by a double coacervation method. *International Journal of Pharmaceutics* 187 (2): 143–152.

74 Anon (1988). Solvent selection in the preparation of poly(DL-lactide) microspheres prepared by the solvent evaporation method. *International Journal of Pharmaceutics* 43 (1–2): 179–186.

75 Chen-Yu, G., Chun-Fen, Y., Qi-Lu, L. et al. (2012). Development of a quercetin-loaded nanostructured lipid carrier formulation for topical delivery. *International Journal of Pharmaceutics* 430 (1–2): 292–298.

76 Austefjord, M.W., Gerdes, H.-H., and Wang, X. (2014). Tunneling nanotubes: diversity in morphology and structure. *Communicative & Integrative Biology* 7 (1): e27934.

77 Wu, T.-H., Yen, F.-L., Lin, L.-T. et al. (2008). Preparation, physicochemical characterization, and antioxidant effects of quercetin nanoparticles. *International Journal of Pharmaceutics* 346 (1–2): 160–168.

78 Min, K.H., Park, K., Kim, Y.-S. et al. (2008). Hydrophobically modified glycol chitosan nanoparticles-encapsulated camptothecin enhance the drug stability and tumor targeting in cancer therapy. *Journal of Controlled Release: Official Journal of the Controlled Release Society* 127 (3): 208–218.

79 Yen, F.-L., Wu, T.-H., Lin, L.-T. et al. (2009). Naringenin-loaded nanoparticles improve the physicochemical properties and the hepatoprotective effects of naringenin in orally-administered rats with CCl(4)-induced acute liver failure. *Pharmaceutical Research* 26 (4): 893–902.

80 Vauthier, C., Dubernet, C., Fattal, E. et al. (2003). Poly(alkylcyanoacrylates) as biodegradable materials for biomedical applications. *Advanced Drug Delivery Reviews* 55 (4): 519–548.

81 Soppimath, K.S., Aminabhavi, T.M., Kulkarni, A.R., and Rudzinski, W.E. (2001). Biodegradable polymeric nanoparticles as drug delivery devices. *Journal of Controlled Release* 70 (1): 1–20.

82 Calvo, P., Sanchez, A., Martinez, J. et al. (1996). Polyester nanocapsules as new topical ocular delivery systems for cyclosporin A. *Pharmaceutical Research* 13 (2): 311–315.

83 Mao, H.Q., Roy, K., Troung-Le, V.L. et al. (2001). Chitosan-DNA nanoparticles as gene carriers: synthesis, characterization and transfection efficiency. *Journal of Controlled Release: Official Journal of the Controlled Release Society* 70 (3): 399–421.

84 Beck-Broichsitter, M., Merkel, O.M., and Kissel, T. (2012). Controlled pulmonary drug and gene delivery using polymeric nano-carriers. *Journal of Controlled Release* 161 (2): 214–224.

85 Navya, P.N., Kaphle, A., Srinivas, S.P. et al. (2019). Current trends and challenges in cancer management and therapy using designer nanomaterials. *Nano Convergence* 6 (1): 23.

86 Fu, Y. and Kao, W.J. (2010). Drug release kinetics and transport mechanisms of non-degradable and degradable polymeric delivery systems. *Expert Opinion on Drug Delivery* 7 (4): 429.

87 Safdar, R., Omar, A.A., Arunagiri, A. et al. (2019). Potential of Chitosan and its derivatives for controlled drug release applications – A review. *Journal of Drug Delivery Science and Technology* 49642–49659.

88 Nassir, A.M., Shahzad, N., Ibrahim, I.A.A. et al. (2018). Resveratrol-loaded PLGA nanoparticles mediated programmed cell death in prostate cancer cells. *Saudi Pharmaceutical Journal* 26 (6): 876–885.

89 Al Zoubi, M.S., Aljabali, A.A.A., and Pal, K. (2021). Highly toxic nanomaterials for cancer treatment. In: *Bio-manufactured Nanomaterials: Perspectives and Promotion* (ed. K. Pal), 161–185. Cham: Springer International Publishing.

90 Basoglu, H., Goncu, B., and Akbas, F. (2018). Magnetic nanoparticle-mediated gene therapy to induce Fas apoptosis pathway in breast cancer. *Cancer Gene Therapy* 25 (5): 141–147.

91 Almeida, P.V., Shahbazi, M.-A., Mäkilä, E. et al. (2014). Amine-modified hyaluronic acid-functionalized porous silicon nanoparticles for targeting breast cancer tumors. *Nanoscale* 6 (17): 10377–10387.

92 Cheng, C.-A., Deng, T., Lin, F.-C. et al. (2019). Supramolecular nanomachines as stimuli-responsive gatekeepers on mesoporous silica nanoparticles for antibiotic and cancer drug delivery. *Theranostics* 9 (11): 3341–3364.

93 Xu, C.-Z. and Xu, G.Q. (2019). Saturated boundary feedback stabilization of a linear wave equation. *SIAM Journal on Control and Optimization* 57 (1): 290–309.

94 Zhang, F., Correia, A., Mäkilä, E. et al. (2017). Receptor-mediated surface charge inversion platform based on porous silicon nanoparticles for efficient cancer cell recognition and combination therapy. *ACS Applied Materials & Interfaces* 9 (11): 10034–10046.

95 Luo, E., Song, G., Li, Y. et al. (2013). The toxicity and pharmacokinetics of carbon nanotubes as an effective drug carrier. *Current Drug Metabolism* 14 (8): 879–890.

96 Madani, S.Y., Naderi, N., Dissanayake, O. et al. (2011). A new era of cancer treatment: carbon nanotubes as drug delivery tools. *International Journal of Nanomedicine* 6 (1): 2963–2979.

97 Liang, R., Wei, M., Evans, D.G., and Duan, X. (2014). Inorganic nanomaterials for bioimaging, targeted drug delivery and therapeutics. *Chemical Communications* 50 (91): 14071–14081.

98 Peer, D., Karp, J.M., Hong, S. et al. (2007). Nanocarriers as an emerging platform for cancer therapy. *Nature Nanotechnology* 2 (12): 751–760.

99 Kamaly, N., Xiao, Z., Valencia, P.M. et al. (2012). Targeted polymeric therapeutic nanoparticles: design, development and clinical translation. *Chemical Society Reviews* 41 (7): 2971–3010.

100 Lewinski, N., Colvin, V., and Drezek, R. (2008). Cytotoxicity of nanoparticles. *Small* 4: 1, 26–49.

101 Yavuz, M.S., Cheng, Y., Chen, J. et al. (2009). Gold nanocages covered by smart polymers for controlled release with near-infrared light. *Nature Materials* 8: 935–939.

102 Sershen, S.R., Westcott, S.L., Halas, N.J., and West, J.L. (2000). Temperature-sensitive polymer–nanoshell composites for photothermally modulated drug delivery. *Journal of Biomedical Materials Research* 51 (3): 293–298.

103 Krishnendu, S., Agasti, S.S., Chaekyu, K. et al. (2012). Gold nanoparticles in chemical and biological sensing. *Chemical Reviews* 112 (5): 2739–2779.

104 Brust, M., Fink, J., Bethell, D. et al. (1995). Synthesis and reactions of functionalised gold nanoparticles. *Journal of the Chemical Society, Chemical Communications* 16: 1655.

105 Dreaden, E., Alkilany, A., Huang, X. et al. (2012). The golden age: gold nanoparticles for biomedicine. *Chemical Society Reviews* 41 (7): 2740–2779.

106 Murphy, C.J., Gole, A.M., Hunyadi, S.E. et al. (2008). Chemical sensing and imaging with metallic nanorods. *Chemical Communications* 18 (5): 544–557.

107 Au, L., Zhang, Q., Cobley, C.M. et al. (2010). Quantifying the cellular uptake of antibody-conjugated Au nanocages by two-photon microscopy and inductively coupled plasma mass spectrometry. *ACS Nano* 4 (1): 35–42.

108 Murphy, C.J., Gole, A.M., Stone, J.W. et al. (2008). Gold nanoparticles in biology: beyond toxicity to cellular imaging. *Accounts of Chemical Research* 41 (12): 1721–1730.

109 Connor, E.E., Mwamuka, J., Gole, A. et al. (2005). Gold nanoparticles are taken up by human cells but do not cause acute cytotoxicity. *Small (Weinheim an Der Bergstrasse, Germany)* 1 (3): 325–327.

110 Popat, A., Hartono, S.B., Stahr, F. et al. (2011). Mesoporous silica nanoparticles for bioadsorption, enzyme immobilisation, and delivery carriers. *Nanoscale* 3 (7): 2801–2818.

111 Wu, S.-H., Hung, Y., and Mou, C.-Y. (2011). Mesoporous silica nanoparticles as nanocarriers. *Chemical Communications* 47 (36): 9972–9985.

112 Mai, W.X. and Meng, H. (2013). Mesoporous silica nanoparticles: a multifunctional nano therapeutic system. *Integrative Biology: Quantitative Biosciences from Nano to Macro* 5 (1): 19–28.

113 Li, Z., Barnes, J.C., Bosoy, A. et al. (2012). Mesoporous silica nanoparticles in biomedical applications. *Chemical Society Reviews* 41 (7): 2590–2605.

114 Lim, M.H. and Stein, A. (1999). Comparative studies of grafting and direct syntheses of inorganic–organic hybrid mesoporous materials. *Chemistry of Materials* 11 (11): 3285–3295.

115 Margolese, D., Melero, J.A., Christiansen, S.C. et al. (2000). Direct syntheses of ordered SBA-15 mesoporous silica containing sulfonic acid groups. *Chemistry of Materials* 12 (8): 2448–2459.

116 Solberg, S.M. and Landry, C.C. (2006). Adsorption of DNA into mesoporous silica. *The Journal of Physical Chemistry B* 110 (31): 15261–15268.

117 Lu, J., Liong, M., Zink, J.I., and Tamanoi, F. (2007). Mesoporous silica nanoparticles as a delivery system for hydrophobic anticancer drugs. *Small (Weinheim an Der Bergstrasse, Germany)* 3 (8): 1341–1346.

118 Radu, D.R., Lai, C.Y., Jeftinija, K. et al. (2004). A polyamidoamine dendrimer-capped mesoporous silica nanosphere-based gene transfection reagent. *Journal of the American Chemical Society* 126 (41): 13216–13217.

119 Vivero-Escoto, J.L., Slowing, I.I., Trewyn, B.G., and Lin, S.Y. (2010). Mesoporous silica nanoparticles for intracellular controlled drug delivery. *Small (Weinheim an der Bergstrasse, Germany)* 6 (18): 1952–1967.

120 Thomas, C.R., Ferris, D.P., Lee, J.-H. et al. (2010). Noninvasive remote-controlled release of drug molecules in vitro using magnetic actuation of mechanized nanoparticles. *Journal of the American Chemical Society* 132 (31): 10623–10625.

121 Meng, H., Liong, M., Xia, T. et al. (2010). Engineered design of mesoporous silica nanoparticles to deliver doxorubicin and P-glycoprotein siRNA to overcome drug resistance in a cancer cell line. *ACS Nano* 4 (8): 4539.

122 Lu, J., Choi, E., Tamanoi, F., and Zink, J.I. (2008). Light-activated nanoimpeller-controlled drug release in cancer cells. *Small* 4 (4): 421–426.

123 Lu, J., Liong, M., Li, Z. et al. (2010). Biocompatibility, biodistribution, and drug-delivery efficiency of mesoporous silica nanoparticles for cancer therapy in animals. *Small* 6 (16): 1794–1805.

124 Feng, L. and Liu, Z. (2011). Graphene in biomedicine: opportunities and challenges. *Nanomedicine* 2: 6.

125 Yang, K., Wan, J., Zhang, S. et al. (2012). The influence of surface chemistry and size of nanoscale graphene oxide on photothermal therapy of cancer using ultra-low laser power. *Biomaterials* 33 (7): 2206–2214.

126 Bo, T., Chao, W., Shuai, Z. et al. (2011). Photothermally enhanced photodynamic therapy delivered by nano-graphene oxide. *ACS Nano* 5 (9): 7000–7009.

127 Zhang, S., Yang, K., Feng, L., and Liu, Z. (2011). In vitro and in vivo behaviors of dextran functionalized graphene. *Carbon* 49 (12): 4040–4049.

128 Yang, K., Wan, J., Zhang, S. et al. (2011). In vivo pharmacokinetics, long-term biodistribution, and toxicology of PEGylated graphene in mice. *ACS Nano* 5 (1): 516–522.

129 Li, Y., Liu, Y., Fu, Y. et al. (2012). The triggering of apoptosis in macrophages by pristine graphene through the MAPK and TGF-beta signaling pathways. *Biomaterials* 33 (2): 402–411.

130 Yang, K., Feng, L., Shi, X., and Liu, Z. (2013). Nano-graphene in biomedicine: theranostic applications. *Chemical Society Reviews* 42 (2): 530–547.

131 Park, S. and Ruoff, R.S. (2009). Chemical methods for the production of graphenes. *Nature Nanotechnology* 4 (4): 217–224.

132 Dreyer, D.R., Park, S., Bielawski, C.W., and Ruoff, R.S. (2010). The chemistry of graphene oxide. *Chemical Society Reviews* 39 (1): 228–240.

133 Hummers, W.S. and Offeman, R.E. (1958). Preparation of graphitic oxide. *Journal of the American Chemical Society* 208: 1334–1339.

134 Liu, Z., Robinson, J.T., Sun, X., and Dai, H. (2008). PEGylated nanographene oxide for delivery of water-insoluble cancer drugs. *Journal of the American Chemical Society* 130 (33): 10876–10877.

135 Sahoo, N.G., Bao, H., Pan, Y. et al. (2011). Functionalized carbon nanomaterials as nanocarriers for loading and delivery of a poorly water-soluble anticancer drug: a comparative study. *Chemical Communications (Cambridge, England)* 47 (18): 5235–5237.

136 Kakran, M., Sahoo, N.G., Bao, H. et al. (2011). Functionalized graphene oxide as nanocarrier for loading and delivery of ellagic acid. *Current Medicinal Chemistry* 18 (29): 4503–4512.

137 Sun, X., Liu, Z., Welsher, K. et al. (2008). Nano-graphene oxide for cellular imaging and drug delivery. *Nano Research* 1 (3): 203–212.

138 Zhang, L., Xia, J., Zhao, Q. et al. (2010). Functional graphene oxide as a nanocarrier for controlled loading and targeted delivery of mixed anticancer drugs. *Small (Weinheim an Der Bergstrasse, Germany)* 6 (4): 537–544.

139 Qiao, R., Yang, C., and Gao, M. (2009). Superparamagnetic iron oxide nanoparticles: from preparations to in vivo MRI applications. *Journal of Materials Chemistry* 19: 6274.

140 Hao, R., Xing, R., Xu, Z. et al. (2010). Synthesis, functionalization, and biomedical applications of multifunctional magnetic nanoparticles. *Advanced Materials* 22 (25): 2729–2742.

141 Yang, H., Zhang, C., Shi, X. et al. (2010). Water-soluble superparamagnetic manganese ferrite nanoparticles for magnetic resonance imaging. *Biomaterials* 31 (13): 3667–3673.

142 Cherukuri, P., Glazer, E.S., and Curley, S.A. (2010). Targeted hyperthermia using metal nanoparticles. *Advanced Drug Delivery Reviews* 62 (3): 339–345.

143 Wang, W., Xu, Y., Wang, D.I.C., and Li, Z. (2009). Recyclable nanobiocatalyst for enantioselective sulfoxidation: facile fabrication and high performance of chloroperoxidase-coated magnetic nanoparticles with iron oxide core and polymer shell. *Journal of the American Chemical Society* 131 (36): 12892–12893.

144 Huang, D.M., Hsiao, J.K., Chen, Y.C. et al. (2009). The promotion of human mesenchymal stem cell proliferation by superparamagnetic iron oxide nanoparticles. *Biomaterials* 30 (22): 3645–3651.

145 Asati, A., Santra, S., Kaittanis, C., and Perez, J.M. (2010). Surface-charge-dependent cell localization and cytotoxicity of cerium oxide nanoparticles. *ACS Nano* 4 (9): 5321–5331.

146 Mahmoudi, M., Laurent, S., Shokrgozar, M.A., and Hosseinkhani, M. (2011). Toxicity evaluations of superparamagnetic iron oxide nanoparticles: cell "Vision" versus physicochemical properties of nanoparticles. *ACS Nano* 5 (9): 7263–7276.

147 Reddy, L.H., Arias, J.L., Nicolas, J., and Couvreur, P. (2012). Magnetic nanoparticles: design and characterization, toxicity and biocompatibility, pharmaceutical and biomedical applications. *Chemical Reviews* 112 (11): 5818–5878.

148 Laurent, S., Forge, D., Port, M. et al. (2008). Magnetic iron oxide nanoparticles: synthesis, stabilization, vectorization, physicochemical characterizations, and biological applications. *Chemical Reviews* 108 (6): 2064–2110.

149 Schladt, T.D., Schneider, K., Schild, H., and Tremel, W. (2011). Synthesis and bio-functionalization of magnetic nanoparticles for medical diagnosis and treatment. *Dalton Transactions* 40: 6315.

150 Lee, J.H. (2007). Artificially engineered magnetic nanoparticles for ultra-sensitive molecular imaging. *Nature Medicine* 13 (1): 95–99.

151 Biju, V., Itoh, T., and Ishikawa, M. (2010). Delivering quantum dots to cells: bioconjugated quantum dots for targeted and nonspecific extracellular and intracellular imaging. *ChemInform* 39 (8): 3031–3056.

152 Ren, Z., Chen, X., Hong, L. et al. (2020). Nanoparticle conjugation of ginsenoside Rg3 inhibits hepatocellular carcinoma development and metastasis. *Small* 16 (2): 1905233.

153 Zhao, X., Wu, J., Zhang, K. et al. (2022). The synthesis of a nanodrug using metal-based nanozymes conjugated with ginsenoside Rg3 for pancreatic cancer therapy. *Nanoscale Advances* 4 (1): 190–199.

154 Na, H.B., Song, I.C., and Hyeon, T. (2010). Inorganic nanoparticles for MRI contrast agents. *Advanced Materials* 21 (21): 2133–2148.

155 Anon (2012). Multifunctional Fe_3O_4/TaO(x) core/shell nanoparticles for simultaneous magnetic resonance imaging and X-ray computed tomography. *Journal of the American Chemical Society* 134 (25): 10309–10312.

156 Foy, S.P., Manthe, R.L., Foy, S.T. et al. (2010). Optical imaging and magnetic field targeting of magnetic nanoparticles in tumors. *ACS Nano* 4 (9): 5217–5224.

157 Matsuda, H., Ohnishi, T., Asada, T. et al. (2003). Correction for partial-volume effects on brain perfusion SPECT in healthy men. *Journal of Nuclear Medicine* 44 (8): 1243–1252.

158 Zhu, X., Zhou, J., Chen, M. et al. (2012). Core-shell Fe_3O_4@$NaLuF_4$:Yb,Er/Tm nanostructure for MRI, CT and upconversion luminescence tri-modality imaging. *Biomaterials* 33 (18): 4618–4627.

159 Zrazhevskiy, P., Sena, M., and Gao, X. (2010). Designing multifunctional quantum dots for bioimaging, detection, and drug delivery. *ChemInform* 39 (11): 4326–4354.

160 Gao, X., Cui, Y., Levenson, R.M. et al. (2004). In vivo cancer targeting and imaging with semiconductor quantum dots. *Nature Biotechnology* 22 (8): 969–976.

161 Correa-Duarte, M.A., Giersig, M., and Liz-Marzán, L.M. (1998). Stabilization of CdS semiconductor nanoparticles against photodegradation by a silica coating procedure. *Chemical Physics Letters* 286 (5–6): 497–501.

162 Michalet, X., Pinaud, F.F., Bentolila, L.A.B. et al. (2005). Quantum dots for live cells, in vivo imaging, and diagnostics. *Science* 307 (5709): 538–544.

163 Bentzen, E.L., Tomlinson, I.D., Mason, J. et al. (2005). Surface modification to reduce nonspecific binding of quantum dots in live cell assays. *Bioconjugate Chemistry* 16 (6): 1488–1494.

164 Derfus, A.M., Chan, W.C.W., and Bhatia, S.N. (2004). Intracellular delivery of quantum dots for live cell labeling and organelle tracking. *Advanced Materials* 16 (12): 961–966.

165 Riccardo, D.C., Bigall, N.C., Andrea, R., and Dirk, D. (2011). Multifunctional nanobeads based on quantum dots and magnetic nanoparticles: synthesis and cancer cell targeting and sorting. *ACS Nano* 5 (2): 1109–1121.

166 Lei, H.M.F. (2010). Quantum dot capped magnetite nanorings as high performance nanoprobe for multiphoton fluorescence and magnetic resonance imaging. *Journal of the American Chemical Society* 132 (42): 14803.

167 Hu, G. and O'Hare, D. (2005). Unique layered double hydroxide morphologies using reverse microemulsion synthesis. *Journal of the American Chemical Society* 127 (50): 17808–17813.

168 Darder, M., Aranda, P., and Ruiz-Hitzky, E. (2007). *Bionanocomposites: A New Concept of Ecological, Bioinspired, and Functional Hybrid Materials*, 10. WILEY-VCH Verlag.

169 Ruiz-Hitzky, E., Darder, M., Aranda, P., and Ariga, K. (2010). Advances in biomimetic and nanostructured biohybrid materials. *Advanced Materials* 22 (3): 323–336.

170 Kriven, W.M., Kwak, S.Y., Wallig, M.A., and Choy, J.H. (2004). Bio-resorbable nanoceramics for gene and drug delivery. *MRS Bulletin* 29 (1): 33–37.

171 Xu, Z.P., Walker, T.L., Liu, K.L. et al. (2007). Layered double hydroxide nanoparticles as cellular delivery vectors of supercoiled plasmid DNA. *International Journal of Nanomedicine* 2 (2): 163–174.

172 Tronto, J., José, M., dos Reis, F. et al. (2004). In vitro release of citrate anions intercalated in magnesium aluminium layered double hydroxides. *Journal of Physics and Chemistry of Solids* 65 (2): 475–480.

173 Kwak, S.Y., Jeong, Y.J., Park, J.S., and Choy, J.H. (2002). Bio-LDH nanohybrid for gene therapy. *Solid State Ionics* 151 (1–4): 229–234.

174 Gu, Z., Thomas, A.C., Xu, Z.P. et al. (2008). In vitro sustained release of LMWH from MgAl-layered double hydroxide nanohybrids. *Chemistry of Materials* 20 (11): 3715–3722.

175 Gu, Z., Rolfe, B.E., Thomas, A.C. et al. (2011). Cellular trafficking of low molecular weight heparin incorporated in layered double hydroxide nanoparticles in rat vascular smooth muscle cells. *Biomaterials* 32 (29): 7234–7240.

176 Zhang, Z., Wang, L., Wang, J. et al. (2012). Mesoporous silica-coated gold nanorods as a light-mediated multifunctional theranostic platform for cancer treatment. *Advanced Materials* 24 (11): 1418–1423.

177 Zhang, W., Zhao, X., Yuan, Y. et al. (2020). Microfluidic synthesis of multimode Au@CoFeB-Rg3 nanomedicines and their cytotoxicity and anti-tumor effects. *Chemistry of Materials* 32 (12): 5044–5056.

178 Mottaghitalab, F., Farokhi, M., Fatahi, Y. et al. (2019). New insights into designing hybrid nanoparticles for lung cancer: diagnosis and treatment. *Journal of Controlled Release* 295: 250–267.

179 Zhao, X., Wu, J., Guo, D. et al. (2022). Dynamic ginsenoside-sheltered nanocatalysts for safe ferroptosis-apoptosis combined therapy. *Acta Biomaterialia* 151: 549–560.

180 Fei, G., Jinming, Z., Chaomei, F. et al. (2017). iRGD-modified lipid–polymer hybrid nanoparticles loaded with isoliquiritigenin to enhance anti-breast cancer effect and tumor-targeting ability. *International Journal of Nanomedicine* 12: 4147–4162.

181 Hu, C.M.J., Kaushal, S., Cao, H.S.T. et al. (2010). Half-antibody functionalized lipid–polymer hybrid nanoparticles for targeted drug delivery to carcinoembryonic antigen presenting pancreatic cancer cells. *Molecular Pharmaceutics* 7 (3): 914–920.

182 Wang, Q., Alshaker, H., Bhler, T. et al. (2017). Core shell lipid–polymer hybrid nanoparticles with combined docetaxel and molecular targeted therapy for the treatment of metastatic prostate cancer. *Scientific Reports* 7 (1): 5901.

183 Zhang, R.X., Ahmed, T., Li, L.Y. et al. (2017). Design of nanocarriers for nanoscale drug delivery to enhance cancer treatment using hybrid polymer and lipid building blocks. *Nanoscale* 9 (4): 1334–1355.

184 Hu, Y., Hoerle, R., Ehrich, M., and Zhang, C. (2015). Engineering the lipid layer of lipid–PLGA hybrid nanoparticles for enhanced in vitro cellular uptake and improved stability. *Acta Biomaterialia* 28: 149–159.

185 Su, X., Wang, Z., Li, L. et al. (2013). Lipid–polymer nanoparticles encapsulating doxorubicin and 2′-deoxy-5-azacytidine enhance the sensitivity of cancer cells to chemical therapeutics. *Molecular Pharmaceutics* 10 (5): 1901–1909.

186 Colapicchioni, V., Palchetti, S., Pozzi, D. et al. (2015). Killing cancer cells using nanotechnology: novel poly(I:C) loaded liposome–silica hybrid nanoparticles. *Journal of Materials Chemistry B* 3 (37): 7408–7416.

187 Meng, H., Wang, M., Liu, H. et al. (2015). Use of a lipid-coated mesoporous silica nanoparticle platform for synergistic gemcitabine and paclitaxel delivery to human pancreatic cancer in mice. *ACS Nano* 9 (4): 3540–3557.

188 Cirillo, G., Vittorio, O., Kunhardt, D. et al. (2019). Combining carbon nanotubes and chitosan for the vectorization of methotrexate to lung cancer cells. *Materials* 12 (18): 2889.

189 Liu, R., Wu, Q., Huang, X. et al. (2021). Synthesis of nanomedicine hydrogel microcapsules by droplet microfluidic process and their pH and temperature dependent release. *RSC Advances* 11 (60): 37814–37823.

190 Sevencan, C., McCoy, R.S.A., Ravisankar, P. et al. (2020). Cell membrane nanotherapeutics: from synthesis to applications emerging tools for personalized cancer therapy. *Advanced Therapeutics* 3 (3): 1900201.

191 Fang, R.H., Kroll, A.V., Gao, W., and Zhang, L. (2018). Cell membrane coating nanotechnology. *Advanced Materials* 30 (23): 1706759.

192 Liu, C.-M., Chen, G.-B., Chen, H.-H. et al. (2019). Cancer cell membrane-cloaked mesoporous silica nanoparticles with a pH-sensitive gatekeeper for cancer treatment. *Colloids and Surfaces B: Biointerfaces* 175: 477–486.

193 Parodi, A., Quattrocchi, N., van de Ven, A.L. et al. (2013). Synthetic nanoparticles functionalized with biomimetic leukocyte membranes possess cell-like functions. *Nature Nanotechnology* 8 (1): 61–68.

194 Hu, C.M.J., Aryal, S., Cheung, C. et al. (2011). Erythrocyte membrane-camouflaged polymeric nanoparticles as a biomimetic delivery platform. *Proceedings of the National Academy of Sciences of the United States of America* 27: 108.

195 Zhu, J., Zhang, M., Zheng, D. et al. (2018). A universal approach to render nanomedicine with biological identity derived from cell membranes. *Biomacromolecules* 19 (6): 2043–2052.

196 Che-Ming, J., Fang, R.H., Wang, K.-C. et al. (2015). Nanoparticle biointerfacing by platelet membrane cloaking. *Nature* 526: 118–121.

197 Copp, J.A., Fang, R.H., Luk, B.T. et al. (2014). Clearance of pathological antibodies using biomimetic nanoparticles. *Proceedings of the National Academy of Sciences of the United States of America* 111: 13481–13486.

198 Israelachvili, J.N., Mitchell, D.J., and Ninham, B.W. (1976). Theory of self-assembly of hydrocarbon amphiphiles into micelles and bilayers. *Journal of the Chemical Society, Faraday Transactions 2: Molecular and Chemical Physics* 72: 1525–1568.

199 Pradhan, S., Hedberg, J., Blomberg, E. et al. (2016). Effect of sonication on particle dispersion, administered dose and metal release of non-functionalized, non-inert metal nanoparticles. *Journal of Nanoparticle Research* 18 (9): 285.

200 Rao, L., Cai, B., Bu, L.L. et al. (2017). Microfluidic electroporation-facilitated synthesis of erythrocyte membrane-coated magnetic nanoparticles for enhanced imaging-guided cancer therapy. *ACS Nano* 11 (4): 3496–3505.

201 Rao, L., Bu, L., Ma, L. et al. (2018). Platelet-facilitated photothermal therapy of head and neck squamous cell carcinoma. *Angewandte Chemie* 57 (4): 986–991.

202 Koupenova, M., Clancy, L., Corkrey, H.A. et al. (2018). Circulating platelets as mediators of immunity, inflammation, and thrombosis. *Circulation Research: A Journal of the American Heart Association* 122 (2): 337–351.

203 Zhou, X., Luo, B., Kang, K. et al. (2019). Leukocyte-repelling biomimetic immunomagnetic nanoplatform for high-performance circulating tumor cells isolation. *Small (Weinheim an Der Bergstrasse, Germany)* 15 (17): e1900558.

204 Silva, A.K.A., Corato, R.D., Pellegrino, T. et al. (2013). Cell-derived vesicles as a bioplatform for the encapsulation of theranostic nanomaterials. *Nanoscale* 5: 11374.

205 Muralidharan-Chari, V., Clancy, J.W., Sedgwick, A., and D'Souza-Schorey, C. (2010). Microvesicles: mediators of extracellular communication during cancer progression. *Journal of Cell Science* 123 (10): 1603–1611.

206 Gaspar, R.S. and Duncan, R. (2009). Polymeric carriers: preclinical safety and the regulatory implications for design and development of polymer therapeutics. *Advanced Drug Delivery Reviews* 61 (13): 1220–1231.

207 Joerg, H., Helene, K., Angela, S., and Peter, W. (2013). Engineered nanomaterial uptake and tissue distribution: from cell to organism. *International Journal of Nanomedicine* 2013: 3255–3269.

208 Fubini, B., Ghiazza, M., and Fenoglio, I. (2010). Physico-chemical features of engineered nanoparticles relevant to their toxicity. *Nanotoxicology* 4 (4): 347–363.

209 Aillon, K.L., Xie, Y., El-Gendy, N. et al. (2009). Effects of nanomaterial physicochemical properties on in vivo toxicity. *Advanced Drug Delivery Reviews* 61 (6): 457–466.

210 Paciotti, G.F., Myer, L., Weinreich, D. et al. (2004). Colloidal gold: a novel nanoparticle vector for tumor directed drug delivery. *Drug Delivery* 11 (3): 169–183.

211 Vasir, J.K. and Labhasetwar, V. (2007). Biodegradable nanoparticles for cytosolic delivery of therapeutics. *Advanced Drug Delivery Reviews* 59 (8): 718–728.

212 Drummond, D.C., Meyer, O., Hong, K.L. et al. (2000). Optimizing liposomes for delivery of chemotherapeutic agents to solid tumors. *Pharmacological Reviews* 51 (4): 691–743.

213 Dobrovolskaia, A.M. and McNeil, S.E. (2007). Immunological properties of engineered nanomaterials. *Nature Nanotechnology* 2 (8): 469–478.

214 Nie, S., Xing, Y., Kim, G.J., and Simons, J.W. (2007). Nanotechnology applications in cancer. *Annual Review of Biomedical Engineering* 9 (1): 257–288.

215 Tiede, K., Boxall, A., Tear, S.P. et al. (2008). Detection and characterization of engineered nanoparticles in food and the environment. *Food Additives & Contaminants Part A Chemistry Analysis Control Exposure & Risk Assessment* 25 (7): 795–821.

216 Zamboni, W.C., Torchilin, V., Patri, A.K. et al. (2012). Best practices in cancer nanotechnology: perspective from NCI nanotechnology alliance. *Clinical Cancer Research* 18 (12): 3229–3241.

217 Langer, K., Anhorn, M.G., Steinhauser, I. et al. (2008). Human serum albumin (HSA) nanoparticles: reproducibility of preparation process and kinetics of enzymatic degradation. *International Journal of Pharmaceutics* 347 (1–2): 109–117.

218 McBride, A., Holle, L.M., Westendorf, C. et al. (2013). National survey on the effect of oncology drug shortages on cancer care. *American Journal of Health-System Pharmacy* 70 (7): 609–617.

219 Berger, J.L., Smith, A., Zorn, K.K. et al. (2014). Outcomes analysis of an alternative formulation of PEGylated liposomal doxorubicin in recurrent epithelial ovarian carcinoma during the drug shortage era. *OncoTargets and Therapy* 7: 1409–1413.

220 Desai, N. (2012). Challenges in development of nanoparticle-based therapeutics. *The AAPS Journal* 14 (2): 282–295.

221 Resnik, D.B. and Tinkle, S.S. (2007). Ethics in nanomedicine. *Nanomedicine* 2 (3): 345–350.

222 Allen, T.M. and Cullis, P.R. (2013). Liposomal drug delivery systems: from concept to clinical applications. *Advanced Drug Delivery Reviews* 1: 65.

4

Nanomedicine for the Treatment of Nervous System Diseases

Xiaojian Cui[1] and Yujun Song[1,2,3]

[1] University of Science and Technology Beijing, Center for Modern Physics Technology, School of Mathematics and Physics, 30 Xueyuan Road, Haidian District, Beijing 100083, China
[2] Zhengzhou Tianzhao Biomedical Technology Company Ltd., Zhengzhou New Technology Industrial Development Zone, 7B-1209 Dongqing Street, Zhengzhou 451450, China
[3] Key Laboratory of Pulsed Power Translational Medicine of Zhejiang Province, Hangzhou Ruidi Biotechnology Company Ltd., Room 803, Bldg. 4, 4959 Yuhangtang Road, Cangqian Street, Hangzhou 310023, China

4.1 Concepts and Types of Nanomedicines for Nervous System Diseases

In pharmaceutics, the sizes of nanoparticles for nanomedicines are usually ranged from 1.0 nm to 1.0 μm. These NPs are commonly used as nanocarriers of some special drugs to improve their biocompatibility, transportation ability, and safety. Nanocarriers refer to various nanoparticles that dissolve or disperse some specific drugs for enhanced transportation and permeability into lesions and/or cells. Nanomedicine refers to the direct processing of raw materials into nanoparticles. Types of nanoparticles used are classified as follows. (i) Lipid nanoparticles. Liposomes are novel drug preparations designed for targeted drug delivery, consisting of vesicles formed by a phospholipid bilayer membrane that possesses both hydrophilic and hydrophobic properties. Lipid nanoparticles, on the other hand, are a specific type of nanoparticle composed of lipids. With a particle size of about 100 nm and surface modification with hydrophilic materials, such as polyethylene glycol, lipid nanoparticles have the characteristics of "long circulation" and "invisible" or "stereoscopic stability" after intravenous injection, which plays an important role in reducing the phagocytosis of drugs by liver macrophages, improving drug targeting, hindering the binding of blood protein components to phospholipids, and prolonging the circulation time in vivo. Lipid nanoparticles are also used as carriers to improve the oral absorption of biomacromolecular drugs and other routes of administration, such as transdermal nanoflexible liposomes and insulin nanoliposomes. (ii) The structure of solid lipid nanoparticles is different from that of liposome bilayer with phospholipid as the main component. Solid lipid nanoparticles are solid particles formed by a variety of lipid materials such as fatty acids, fatty alcohols, and phospholipids. They are stable in nature, easy to

Nanomedicine: Fundamentals, Synthesis, and Applications, First Edition. Edited by Yujun Song.
© 2025 WILEY-VCH GmbH. Published 2025 by WILEY-VCH GmbH.

prepare, and have a certain sustained release effect. They are mainly suitable for the encapsulation of insoluble drugs and are used as carriers for intravenous injection or local administration to achieve targeted positioning and controlled release. (iii) Nanocapsules and nanospheres are mainly prepared by biodegradable polymer materials such as polylactic acid, polylactide–glycolide, chitosan, and gelatin. It can be used to encapsulate hydrophilic drugs or hydrophobic drugs. According to the properties of the material, it is suitable for different routes of administration, such as the targeting effect of intravenous injection, and the slow and controlled release effect of intramuscular or subcutaneous injection. Nanocapsules and nanospheres for oral administration can also be used as nondegradable materials, such as ethyl cellulose and acrylic resin. (iv) Polymeric micelles. This is a new type of nanocarrier that has been developed in recent years. Water-soluble block copolymers or graft copolymers with both hydrophilic and hydrophobic groups were synthesized. After dissolving in water, they spontaneously form polymeric micelles to complete the solubilization and encapsulation of drugs. Because of the hydrophilic shell and hydrophobic core, they are suitable for carrying drugs of different properties [1].

4.2 Therapeutical Methods for Nervous System Diseases and Features of Current Nanomedicines

The initial stage of Parkinson's disease is typically managed pharmacologically, aiming for "low-dose drug maintenance to achieve satisfactory clinical outcomes" while minimizing drug-related side effects and complications. Treatment initially involves a single medication at a low dosage. At present, the clinical drugs for the treatment of Parkinson's disease are compound levodopa preparations, anticholinergic drugs, amantadine, dopamine receptor agonists, monoamine oxidase B (MAO-B) inhibitors, and catechol. Oxygen-methyltransferase (COMT) inhibitors, and neuroprotective agents. The development of drug therapy, neurosciences, and other related sciences has brought new opportunities and hopes for the drug treatment of Parkinson's disease. However, there are still no drugs to cure Parkinson's disease, and the disease gradually worsens with the prolongation of the disease. Therefore, the treatment of Parkinson's disease requires long-term administration, and in the long run, follows the principle of "not seeking full effect, fine water flow." In clinical treatment, the titration scheme should be adopted in combination with drug efficacy and adverse reactions, that is, starting from a small dose and gradually increasing the dose. Within the dose range of tolerable adverse reactions, a reasonable dose to achieve the best efficacy should be sought, and then the dose should be maintained for treatment. In addition, the treatment of Parkinson's disease should follow the principle of individualization, according to the patient's clinical manifestations, disease severity, age, physical condition, and economic conditions, such as comprehensive consideration, according to the person. The "honeymoon period" of levodopa treatment is generally five years. After that, patients often have complications such as motor fluctuations or dyskinesia. For complications such as motor fluctuations, it is mainly caused by two reasons. One is that the half-life of levodopa is short, and the concentration fluctuation in blood and brain tissue is obvious; the second is that with the progression of the disease, dopaminergic neurons continue to degenerate and die, resulting in a decrease in their buffering capacity. The treatment strategy of excitant (CDS) is the latest development in the treatment concept of Parkinson's disease in recent years, and it is expected to solve problems such as central movement fluctuation. Continuous dopaminergic stimulation can be achieved through the following ways: (i) changing the dosage form of dopa, the route or method of administration. For example, switching to sustained-release agents, continuous intravenous or intestinal dosing, increasing the frequency of dosing, etc. (ii) Select a long-acting type of barr receptor agonist. (iii) Application of CONT inhibitors. (iv) Short-acting apomorphine intravenous or subcutaneous injection. At present, the COMT preparation has been listed. Entacapone, combined with the compound levodopa preparation, is a simple and ideal scheme to achieve the retention of dopaminergic stimulation. In the process of use, due to the price of entacapone, clinicians must be combined with the actual situation of patients to formulate corresponding countermeasures.

For advanced Parkinson's disease, drug treatment is usually ineffective, and surgery may be chosen in this case, such as in patients with severe movement complications, symptoms are more obvious, and in the case of drugs that cannot be controlled, surgery can be chosen. At present, deep brain stimulation (DBS) is one of the latest advances in the treatment of Parkinson's disease. This technique is also known as brain pacemaker surgery. Stereotactic method is used to implant electrodes in a special position in the brain to release weak electrical pulses, stimulate the related nerve nuclei that control movement, and inhibit the release and conduction of abnormal electrical activity, so as to reduce the clinical symptoms of Parkinson's disease.

The field of cell and tissue transplantation therapy holds promising potential. It is transplanted into the host brain tissue by selecting appropriate cells or tissue groups to replace damaged neurons to achieve the purpose of reconstructing or restoring neurological function. Due to the selectivity of the lesion site, Parkinson's disease patients are more suitable for dry transplantation treatment. It is easier to achieve through the current stereotactic technique, and there are easy-to-establish and reliable animal models for the determination and evaluation of transplantation efficacy.

Cell and tissue transplantation is an important direction for the treatment of Parkinson's disease, but it is invasive and has the risk of graft rejection. Considering this, scientists are exploring another treatment approach-gene therapy. The primary problem of gene therapy for Parkinson's disease is the selection of target genes. In the past 10 years, deep research has been carried out on the following three categories of target genes. The first category: the target genes are a variety of enzymes involved in dopamine metabolism, including tyrosine kinase (TH), guanosine triphosphate cyclohydrolase I (GCHI), and L-aromatic amino acid decarboxylase (AADC). The selection of the above target genes aims to increase the dopamine content in the brain tissue and improve the symptoms of Parkinson's disease. The second category: target genes are various neurotrophic factors, including brain-derived neurotrophic factor (BDNF), glial cell line-derived neurotrophic factor (GDNF), etc. The selection of the above target genes is based on the nutritional protection of these neurotrophic factors. The third category: the target gene is gene transduction to interfere with apoptosis, which can be achieved through multiple links. The choice of transplantation pathway is also an important issue in gene therapy for Parkinson's disease. There are usually two methods: ex vivo and in vivo. The former is to replant human brain tissue after transfection of target cells with expression vectors carrying therapeutic genes in vitro. At present, the focus of attention is to transfect the target gene into stem cells before transplantation, and it is expected that both gene therapy and stem cell therapy will play a role; the latter is to directly transplant the expression vector carrying the target gene into human brain tissue to achieve the purpose of treating Parkinson's disease [2].

4.3 Synthesis Methods and Typical Examples (Including Clinical Trial) of Polymer-based Nanomedicines for Nervous System Diseases

4.3.1 Synthesis Methods of Polymer-based Nanomedicine

There are many methods for the preparation of polymer nanoparticles, such as solvent evaporation method, emulsification method, and reaction reduction method [3].

4.3.1.1 Solvent Evaporation
The solvent evaporation method first dissolves the mixture of polymers and drugs in an organic solvent and then volatilizes the organic solvent so that molecules such as polymers and drugs can self-assemble into nanoparticles. This method is simple and low-cost, but it is easy to cause drug loss and difficult to control the formation of hollow particles.

4.3.1.2 Emulsification Process
The emulsification method is to add a surfactant to two immiscible solutions, and then to emulsify the two liquids by stirring. The polymer and drug were then emulsified into the oil phase, and the self-assembled nanoparticles were then packaged. The nanoparticles prepared by this method are small and much regular in shape.

4.3.1.3 Reductive Reaction Reduction Method
The reduction reaction is a method of reducing the polymer composition to nanoparticles by reducing the reducing agent under mild reaction conditions. This method can obtain polymer nanoparticles with very regular morphology and uniform size. This method is more flexible for the control of particles, but the operation is more complicated and the cost is higher.

4.3.2 Polymer-based Nanomedicine: Typical Application

4.3.2.1 Targeted Drug Delivery
Polymer nanoparticles can encapsulate drugs and secrete them in vivo. Recent studies have shown that polymer nanoparticles can be targeted to release drugs and treat patients due to their receptor-mediated telomere and endocytosis. These targeted

drugs can reduce the side effects of drug distribution in other tissues so as to improve the efficacy of drugs.

4.3.2.2 Tissue Repair and Regenerative Medicine

Polymer nanoparticles can serve as an effective substrate to support tissue engineering and stem cell transplantation. They promote the growth and differentiation of cells in interstitial fluid and can be used to repair damage to the central nervous system.

4.3.2.3 Vaccines and Immunotherapy

Polymer nanoparticles can also be used as vaccine and immunotherapy carriers. They can do this by binding to antigen proteins and thereby triggering an immune response. They can also be targeted to destroy tumors and cancer cells by formulating drugs and monoclonal antibodies.

4.3.2.4 Inclusion of Nanomaterials

Polymer nanoparticles are not the only nanocarriers that can be used to incorporate drugs and nanomaterials. Other nanocarriers include liposomes, metal nanoparticles, carbon nanotubes, etc. However, due to the controllability and durability of polymer materials, they are more widely used in the field of nanomedicine than other kinds of materials.

4.4 Inorganic-based Nanomedicines: Synthesis Methods and Typical Application (Including Clinical)

Nanomedicine integrates drugs and imaging reagents into nanoparticles, and uses the small size effect, surface effect, and quantum effect of nanoparticles to make them have unique light, sound, heat, magnetic, electric, and other special properties to deliver drugs to diseased tissues in a targeted manner. Based on these special properties of nanomaterials, nanomedicine diagnostic agents can realize one or more therapeutic means, including photokinetic, photothermal, sonodynamic and drug therapy, and combine with tumor site imaging to accurately treat tumor lesion tissues [4]. Inorganic nanomaterials include quantum dots (QDs), nanogold, nanocarbon, nanomesoporous silicon, magnetic nano, and nanocalcium materials.

4.4.1 Quantum Dots

QDs are nanoparticles composed of group II–VI or III–V elements, such as sulfides (ZnS, CuS, PbS, CdS), selenides, tellurides, InP, GaN, GaAs, and carbon (silicon) QDs. They are about 2–10 nm in diameter and can emit light signals of a certain wavelength after being irradiated by the laser. Compared with traditional organic fluorescent dyes, QDs have the following advantages: good light stability, long fluorescence life, wide and continuous absorption spectrum, narrow and symmetrical emission spectrum, and a single light source can be used to detect QDs of different sizes synchronously. However, the actual synthesis and application of QDs also have some shortcomings, such as high toxicity, easy accumulation, poor stability, and difficult surface modification. Based on these advantages and disadvantages,

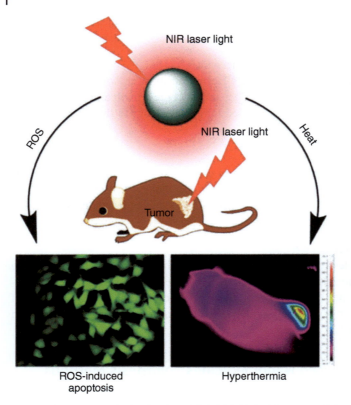

Figure 4.1 Schematic illustration of DOX@PEG-Ag2S nanomedicine and real-time visualization of tumor therapy. Source: Reproduced from Ref. [6] with permission from American Chemical Society.

many researchers modify the surface of QDs to make them widely used in biomedical fields such as immunofluorescence diagnosis, drug screening, in vivo imaging, and biomacromolecular interaction [5].

Using Ag_2S QDs as the core, Wang et al. [6] designed and synthesized DOX@PEG-Ag_2S nanodrugs for visualization of tumor therapy, as shown in Figure 4.1. Through modulation, the particle size of Ag_2S QDs can be adjusted to emit optical signals in the near-infrared II region. The Ag_2S core of surface-modified DT (dodecyl mercaptan) can be self-assembled with C18PMH/PEG to form nanoparticles. Finally, the anticancer drug doxorubicin (DOX) is loaded into its interior, thus avoiding the accumulation of QDs and making the nanomedical drugs have good dispersion.

As a nanomedicine carrier, QDs can be used to improve the therapeutic effect. The design strategy of QDs can be started from two aspects: precise surface modification target and material structure innovation. The former can promote the enrichment of QDs in tumor sites, and then give full play to the fluorescence, photokinetic, and photothermal advantages of QDs. The latter can prolong the circulation time of QDs in vivo, promote the permeability of tumor sites, and broaden the long-wavelength fluorescence properties of QDs.

4.4.2 Nanogold

Nanogold material refers to the synthesis of gold nanoparticles with different morphologies under different reaction conditions using gold as the substrate, the size of which is between 1 and 100 nm. At present, researchers have successfully synthesized gold nanospheres, nanogold rods, nano-Venus, and nanogold cages [7]. The advantages of nanogold are: high electron density, dielectric properties, and catalysis, which can be combined with a variety of biological macromolecules for biomedicine and detection; surface-enhanced Raman scattering properties can be used to detect the level of biomolecules in vivo and accurately diagnose the lesion site. By increasing the longitudinal surface plasmon resonance ratio, the gold nanoparticle can be expanded into the near-infrared region and can be converted into heat energy or singlet oxygen to exert the near-infrared thermal effect. Its disadvantages are that the surface modification means are not abundant, the cost is high, and the biosafety of nanogold is still questioned. In recent years, based on the excellent performance of nanogold as a nanoparticle carrier, the expansion of nanogold surface modification methods and the design and preparation of efficient tumor diagnosis and treatment reagents have become a research [8–11].

In 2015, Song et al. [12] reported a case of nanodrug rGO-AuNRVe, which was self-assembled with nanogold rods and loaded with the anticancer drug DOX on graphene oxide. As shown in Figure 4.2, the nanodrug particle size of (65 ± 12.6) nm, near-infrared laser irradiation can produce photothermal effect and obvious photoacoustic effect. After the cells ingested the nanomedicine through endocytosis,

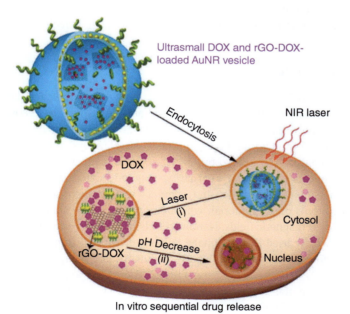

Figure 4.2 Schematic diagram of rGO-AuNRVe in vivo. Source: Reproduced from Ref. [12] with permission from American Chemical Society.

the laser triggered the nanomedicine gold rod to release the DOX-loaded graphene oxide; due to the slightly acidic biological environment in cancer cell tissue, the chemical bond between DOX and graphene oxide can be broken so as to accurately release anticancer drugs, reducing the toxic side effects on normal cells.

In recent years, a large amount of research work has focused on the development of new surface modification means of gold nanoparticles as drug carriers, as well as the collaborative treatment of tumors by various therapeutic means, which has achieved major breakthroughs, but it still cannot solve the biosafety problems of gold nanoparticles and the complex and costly synthesis. So, the economic and safe gold nanodrug carriers are still the hot spot of future research.

4.4.3 Nanocarbon

Carbon nanomaterials are carbon materials with a dispersion scale of at least one dimension less than 100 nm, including graphene, carbon nanotubes, and fullerenes. Graphene is a two-dimensional material formed from sheets of elemental carbon atoms. The hexagonal carbon atom network structure of carbon nanotubes is bent to a certain extent, forming a tubular structure. Both carbon nanotubes and graphene have excellent electrical, optical, and thermal properties and are widely used in biological detection and analysis [13]. In the application of tumor medicine, the abundant lumen on carbon nanotubes can load antitumor drugs, and the drugs can also be loaded on the surface of graphene through π–π stacking to realize the delivery of antitumor drugs. In addition, most carbon materials can efficiently absorb infrared light, and they are ideal materials for the synthesis of photothermal diagnostic reagents. In recent years, nanometer drug carriers designed and synthesized based on the excellent properties of carbon nanomaterials have attracted wide attention [14–16].

4.4.4 Magnetic Nanoparticles

Magnetic material is an ancient and widely used functional material. Magnetic nanomaterials were gradually produced, developed, and expanded after the 1980s, and have now become the most vitality and broad application prospects of new magnetic materials, mainly including Fe_3O_4, garnets, metal alloys, perovskite compounds, magnetic composites, and so on. At large sizes, magnetic materials can produce thermal effects only under alternating magnetic fields, while they show superparamagnetism at nanoscale, and there is a permanent magnetic moment between the atoms of the material, which can produce obvious thermal effects under the action of magnetic fields. Nanomaterials with superparamagnetic properties have been widely used in MRI image enhancement and drug-targeted conduction due to the absence of hysteresis [17, 18]. In recent years, a large number of studies based on Fe_3O_4 nanocarriers have greatly promoted the application of magnetic nanomaterials in the biological field, and proposed design strategies to reduce the aggregation of Fe_3O_4 nanomaterials and improve the targeting of nanomaterials [19–21].

In 2015, Zhang et al. [22] designed and prepared DOX-loaded nanomedical drugs by using Fe_3O_4 as the core and modified polymer on the outer layer to

4.4 Inorganic-based Nanomedicines: Synthesis Methods and Typical Application (Including Clinical)

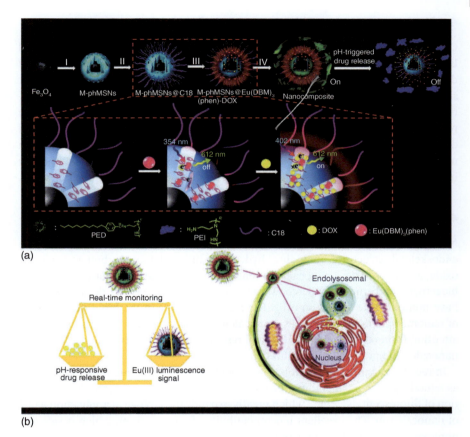

Figure 4.3 (a) Synthesis route for nanocomposites. (b) Selectively trigger the delivery of DOX by nanocomposites at lower pH. Source: Reproduced from Ref. [22] with permission from Royal Society of Chemistry.

wrap DOX drugs. As shown in Figure 4.3, the outer layer of Fe_3O_4 is coated with M-phMSNs (a magnetic benzene-based polyporous silicon material), and the long carbon chain C^{18} and $Eu(DBM)_3$ are further cross-linked on the surface through hydrogen bonding and π–π interaction and loaded with DOX. Finally, the porous material was sealed by acid-sensitive PED, which realized the function of releasing the slightly acidic environment of cancer cells and the fluorescently labeled drug. Researchers have successfully applied multimode combination therapy to magnetic nanomaterials. They use different release mechanisms to control drug release, and realize the three-mode combination therapy of chitosan encapsulated nanoporous, in vivo photoacoustic imaging, photokinetic photothermal, and chemotherapy, which makes magnetic nanomaterials show excellent performance in reducing toxic side effects and improving therapeutic effects. In recent years, magnetic nano Fe_3O_4 has been widely used as a nanometer drug carrier in the design of chemotherapy, photokinetic photothermal therapy, photoacoustic imaging, and other nanomaterials, which has promoted the biological application of magnetic nanomaterials for tumor therapy.

4.4.5 Nanomesoporous Silicon

Compared with traditional silica with high density, small specific surface area and limited application, mesoporous silica has a specific pore structure, hollow, small density, and large specific surface area, showing unique permeability, molecular sieve ability, optical properties, and adsorption, so that it has great application potential in biological materials, catalysis, new assembly materials and other aspects. In recent years, with the adjustable size of mesoporous pores and the development of drug-filling technology in mesoporous pores, more and more mesoporous silicon nanomaterials have been selected as nanomedical drug carriers, which have greatly improved the therapeutic effect of cancer and have great biological application value.

4.4.6 Nanocalcium

Nanometer calcium is a bionic nanomaterial designed according to human endoskeletal structure, including hydroxyapatite, calcium carbonate, calcium oxide, and so on. These nanocalcium materials have excellent biocompatibility and bioactivity, and under acidic conditions (pH 3.5–6), are easy to decompose, and have nontoxic side effects on cells. In recent years, due to the rapid development of nanomedical design strategies, nanocalcium drug materials have received more attention from researchers, and the acid response and excellent biodegradability of nanocalcium have been fully utilized.

In recent years, nanocalcium has been developed by more and more researchers as nanodrug carriers, including the composite with other nanomaterials, modification of fluorescent dyes, etc., which greatly expands the biological application value of nanocalcium. The excellent biocompatibility of nanocalcium, which is safe and nontoxic, can ensure the safety of nanomaterials and provide a guarantee for future clinical applications and biological applications, but the shortcomings of nanocalcium easy to gather need to be solved urgently.

4.5 Metallic-based Nanomedicines: Synthesis Methods and Typical Application (Including Clinical)

Metal-organic frameworks (MOFs) are a kind of crystalline porous materials with periodic frame structures formed by the self-assembly of organic ligands and metal ions (clusters) through coordination bonds. MOFs are developed on the basis of coordination chemistry, which combines the structural characteristics of coordination polymers and crystalline materials, and is obviously different from traditional organic polymers and inorganic polymers in nature. MOFs have the characteristics of large specific surface area, orderly and controllable structure, easy functionalization, and biodegradation, and have been widely used in the fields of gas adsorption and separation, synthetic catalysis, luminescent materials, biological imaging, drug slow release, and chemical sensing, etc. MOFs are multifunctional new hybrid materials with important application prospects [23]. With the improvement of modern synthesis technology and the cross between multiple disciplines, MOFs have been rapidly developed, and have been widely concerned by various fields in recent years.

4.5.1 Preparation and Characterization of Metal-organic Frame Materials

With the development of modern synthesis technology, various preparation methods for MOFs are also emerging, including solvent volatilization method, diffusion method, hydrothermal method, solvothermal method, microwave synthesis method, etc. These preparation methods have their own characteristics and advantages. You can choose the appropriate preparation method according to your own needs. Of course, in the process of synthesizing MOFs, it is also necessary to provide suitable reaction conditions, such as reaction temperature, pH, reaction time, and solvent. Choosing the right reaction temperature is very important for crystal growth and avoiding side reactions. Factors such as the polarity and solubility of the solvent also affect the synthesis of MOF crystals, and the solvent provides an environment for the growth of MOFs. Some organic ligands can be deprotonated effectively by choosing the right pH, and at the same time, the reaction can avoid precipitating too fast. The formation of MOFs is also affected by other factors, such as the ratio of reactants and the properties of the external ions. Generally, the structure and properties of MOFs are characterized in the following ways: (i) the crystal structure is determined by single crystal X-ray diffraction; (ii) the crystallinity and phase purity of MOFs were characterized by powder X-ray diffraction; (iii) the pore size distribution and specific surface area of MOFs were determined by gas adsorption test; (iv) the surface morphology and size distribution of MOFs were observed by electron microscopy. (v) Its structural composition was determined by ultraviolet, infrared, and solid-state nuclear magnetism and other characterization methods.

4.5.2 Application of Functionalized Metal-organic Framework Materials in Tumor Therapy

As a new kind of crystalline porous material, MOFs have attracted much attention in the field of materials research. Their ordered porous structure is formed by linking metal ions with organic ligands through coordination bonds. With the deepening of the research on MOFs, the types of MOFs are increasing, and their applications in various fields are becoming more and more abundant. Nanodrug carriers based on MOFS have been continuously developed in the basic research fields of tumor chemotherapy, photodynamic therapy, photothermal therapy, immunotherapy, and combination therapy. However, the development of biocompatible nanomaterials based on biofunctionalization of MOFs is still slow. How to bio-functionalize MOFs has also been a puzzle for researchers. At present, the main strategies for biofunctionalization of MOFs are as follows: (i) covalently modifying macromolecules or biomolecules on the surface of MOFs; (ii) the enzyme, protein, and other biological macromolecules are fixed in the cavity frame structure of MOFs; and (iii) modify or coat the biofilm structure on the surface of MOFs.

The new biomaterials combined with MOFs and polymers have great application potential, achieving a clever combination of the advantages of MOFs (high porosity, diverse frame topologies, and precisely controlled structures) and the advantages of polymers (flexibility, good biocompatibility, and stability). By covalently modifying the polymer on the surface of MOFs, the surface-modified polymer gel is produced, and the soft nanoporous material is obtained. The surface polymer modification of MOFs can regulate the ability to adhere to cells, enabling the delivery of drug molecules as well as bioactive molecules. It has a variety of topological structures and a high specific surface area and is suitable for immobilization and transfer of biomolecules. In recent years, the construction of nanomaterials based on MOFs and biomacromolecules and their application in the field of biocatalysis has been reported more and more. The internal framework structure of MOFs can provide a suitable microenvironment for biomolecules [12]. MOFs can generate hollow nanomaterials with large cavity structures. In the cavity, the biomolecules exist in the free state, and in the case of biological enzymes, the free-state enzyme molecules can show higher catalytic activity. At the same time, the thin-walled spherical shell structure of the hollow nanomaterial can slow down the loss of biomolecules compared with the open pore structure. Xianzheng Zhang's team [24] reported that a nanosystem based on reactive oxygen species response of porphyrin MOFs encapsulated bioactive molecules (L-Arg@PCN @Mem) could be used for nitric oxide therapy and PDT in cancer (Figure 4.4). The system uses porphyrin MOFs (PPCN-224) as the carrier and adsorbs arginine (L-Arg) in the form of coordination. Under light, the nanosystem can produce a large amount of reactive oxygen species. On the one hand, reactive oxygen species can directly cause oxidative damage to tumor cells. On the other hand, reactive oxygen species can oxidize L-Arg to produce nitric oxide, and a high concentration of nitric oxide will further cause nitrification damage to

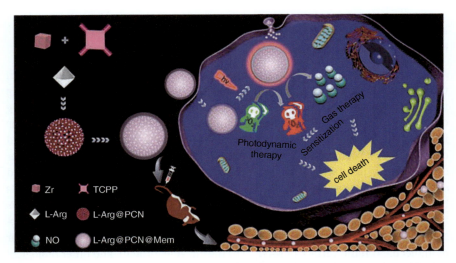

Figure 4.4 Schematic illustration of L-Arg@PCN@Mem for gas therapy and sensitized PDT. Source: Reproduced from Ref. [24], © 2018, with permission from Elsevier.

tumor cells. Compared with reactive oxygen species, nitric oxide has a long life and a large diffusion radius. So, it can penetrate the oxygen-deficient areas of tumors to sensitize PDT and improve the poor efficacy of PDT in the hypoxic areas. Cell membrane systems have been used by more and more researchers to develop functional biomedical materials. In recent years, researchers have carried out in-depth research in the field of using biofilm materials as targeting materials and carriers of drugs and nanoparticles [25]. In order to achieve efficient point-to-point delivery of drugs and nanoparticles, a series of nanodrug carriers modified by biofilm materials have been used in tumor therapy; for example, the direct insertion or modification of drug molecules on the tumor cell membrane (chemotherapy drug camptothecin, photosensitizer protoporphyrin, imaging agent cy-5, etc.). In addition, biofilm modification was performed on nanoparticles containing drugs (DOX, Tirazamine, etc.) and enzymes (glucose oxidase, catalase, etc.) (MOFs, such as ZIF-8 and PCN-224) to improve their homologous targeting and immune escape performance.

4.6 Multifunctional Nanomedicines for Nervous System Diseases

In the central nervous system (CNS) of the brain, there are different types of tumors in system, which are mainly divided into primary and secondary categories [26]. Due to the particularity of the occurrence site, brain tumor has become one of the important tumors endangering human life and health, and the disease burden is increasing year by year. There were nearly 300 000 cases of partial tumors, accounting for about 1.6% of all new cases of malignant tumors, ranking 17th in the incidence of malignant tumors [27]. The incidence of brain tumors in China is at the top level in the world, but the huge population base makes China have the highest annual incidence. It is estimated that there are more than 76 000 new cases of brain tumors in China every year, accounting for more than a quarter of the global incidence. Glioma is the most common type of primary brain tumor in adults. The World Health Organization (WHO) classifies glioblastoma multiforme (GBM) based on pathological findings. Grade I–IV [26]. Among them, grades I and II are low-grade gliomas, and grades III and IV are high-grade gliomas. The higher the grade, the higher the malignant degree and the worse the prognosis.

4.6.1 Nanomedicine Therapy Strategy Combined with Chemotherapy

The current standard treatment for brain tumors includes surgical removal, followed by radiation and chemotherapy. Due to the poor targeting of traditional chemotherapy drugs to tumors, short circulation time in the body, low ability to penetrate tissues, and low therapeutic index, the efficacy of the treatment of tumors is limited. The use of nanocarriers can improve tumor targeting, prolong cycle time, and stabilize the tumor.

Now the controlled release of drugs targeting tumor cells can further enhance the specificity and effectiveness of brain tumors so that those drugs that have been

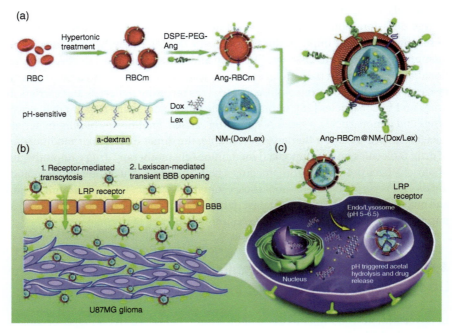

Figure 4.5 Effective and targeted human orthotopic glioblastoma xenograft therapy via a multifunctional biomimetic nanomedicine. (a) Ang-RBCm@NM-(Dox/Lex), (b) mechanisms by which Ang RBCm@NM-(Dox/Lex) crosses the BBB and accumulates in tumor, and (c) active tumor uptake and pH triggered drug release. Source: Reproduced from Ref. [28] with permission from John Wiley & Sons.

abandoned due to systemic toxicity can be effectively used. TMZ is the first line of systemic chemotherapy for glioma.

Anthracyclines are a variety of antitumor drugs, represented by Dox. Since Dox is difficult to pass BBB, glioma is treated by intratumoral injection, resulting in poor drug compliance and greater risk for patients. Therefore, it is very necessary to develop a glioma-targeting drug delivery system for glioma treatment. The low-density lipoprotein receptor-targeting peptide angiopep-2 can be coated with red cell membrane (RBCm) on the surface of pH-sensitive nanoparticles composed of polymers, Dox and lexiscan (Lex). A novel multifunctional bionic nanomedicine Ang-RBCm@NM-(Dox/Lex) has been prepared (see Figure 4.5). Tumor inhibition experiments were conducted in U87 human glioblastoma nude mice, and the results showed that Ang-RBCm@NM-(Dox/Lex) nanomaterials can significantly improve blood circulation time, have excellent BBB penetration ability, and can achieve effective tumor inhibition prolonging survival time. It is suggested that the bionic nanoplatforms can be used as a flexible and powerful GBM treatment system, and can be applied to the treatment of other CNS diseases [28].

4.6.2 Nanomedicine Therapy Strategy Combined with Immunotherapy

Cancer immunotherapy is a promising new cancer treatment method, which is to activate the patient's own immune system to fight the tumor. Cancer immunotherapy mainly includes cancer vaccine therapy, adoptive cell therapy, immune checkpoint inhibitor therapy, and other strategies. Among them, cancer vaccines enhance tumor-specific T cell activity by providing tumor-associated antigens or adjuvants, thereby improving antitumor immune response.

At present, a neoantigen-based vaccine against advanced melanoma and glioma has achieved good results in preliminary clinical trials [29]. However, current delivery methods for neonatal antigens, such as direct injection, usually cause them to precipitate at the injection site, accumulate and generate inflammation, reduce lymphatic drainage, and thus lead to immune tolerance [30]. Therefore, new strategies are needed to improve the resistance of neoantigen and adjuvant molecules to lymphoid tissue delivery efficiency of primary presenting cells (APC) to achieve effective antitumor immunity. Scheetz et al. [31] developed an immunotherapy vector based on synthetic high-density lipoprotein loaded with cytosine guanine (CpG) and tumor-specific neoantigens to target GBM and induce immune-mediated tumor regression. Studies have shown that the combination of the neogenic peptide sHDL/CpG and antiprogrammed death protein ligand 1 (PD-L1) monoclonal antibody induces a specific T cell response, leading to regression of GL261 glioma in situ in 33% of mice, and achieving long-term survival and immune memory.

4.6.3 Combination Treatment Strategy

In recent years, the treatment of cancer has gradually developed from a single-treatment model to a combination treatment model, such as chemical gene-combined therapy, chemical phototherapy combined therapy, and immune radiotherapy-combined therapy [23]. Multiple types of therapeutic drugs can be assembled into a nanostructure through physical adsorption and chemical forces, which can be achieved at lower concentrations. In GBM stem cells, neither miR-21 inhibitor nor TMZ alone can induce apoptosis of GBM cells, while the combination of miR-21 inhibitor and TMZ can significantly increase apoptosis of GBM cells, indicating that combination therapy may be an effective treatment strategy for GBM.

4.6.4 Perspectives of Nanomedicines for Nervous System Diseases

With the improvement of medical levels and the advancement of treatment technology, the treatment of nerves diseases (e.g. Parkinson, Alzheimer, depression) and brain tumors has made great breakthroughs in the past few years. However, the presence of BBB and the high heterogeneity of these nerve or brain diseases have resulted in brain or nerves diseases that are not sensitive to chemotherapy drugs and have poor treatment effects. The emergence of nanodelivery systems provides a new strategy for the treatment of brain or nerve diseases. Different from traditional treatment schemes, nanomaterials achieve effective delivery of therapeutic drugs by precisely adjusting their size, shape, and surface targeting ligands to overcome BBB. However, because the degradation behavior and excretion pathway of nanomaterials in vivo are still unclear, and there are problems such as individual differences and repeatability, nanomaterials are important in the treatment of brain or nerve diseases.

Although nanomedicines still face challenges in clinical translation, the innovative development of multifunctional brain or nerve system-targeting nanomedicines with high BBB penetration and disease selectivity holds significant promise. The functional structure of nanomaterials can be precisely controlled and adjusted, enabling their integration with other therapeutic strategies such as chemotherapy, radiotherapy, photodynamic therapy, and gene therapy for the development of novel drugs for nerve or brain disease treatment. The functional integration of nanomedicine offers a viable approach to achieve multitarget synergistic therapy and enhance the therapeutic efficacy against nerve or brain diseases. Furthermore, for further advancements in nanomedicine study for brain or nerve diseases, attention should also be given to the latest progress in basic research. From a biosafety perspective, it is crucial to employ biodegradable materials that are nonimmunogenic or biotoxic in the research and development of nanomedicine for brain or nerve disease treatment. Additionally, a standardized preparation method should be developed to ensure uniformity, stability, and controllability of obtained nanomaterials, thereby guaranteeing their clinical effectiveness and providing assurance for precise treatment of brain and/or nerve system diseases.

References

1 Jiyu, Z., Genxin, L., Peixing, W. et al. (2007). Research status and prospect of nanomedicine. *Anhui Agronomy Bulletin* 13 (18): 139–142.
2 Xijin, W., Zhang, Y., and Shengdi, C. (2010). Advances in pathogenesis and treatment of Parkinson's disease in ten years. *Chinese Journal of Modern Neurology* 10 (1): https://doi.org/10.3969/j.issn.1672-6731.2010.01.004.
3 Ming, S., Yongzhuo, H., Limei, H. et al. (2012). Multifunctional polymer nano-drug delivery system. *Journal of Controlled Release* 33 (4): 6. doi: CNKI:SUN: YJSK.0.2012-04-007.
4 Yao, K., Suzhen, W., Jiangli, F. et al. (2018). Research progress of inorganic nano drug carriers in tumor diagnosis and treatment. *Journal of Chemical Industry* 69 (1): 13. https://doi.org/10.11949/j.issn.0438-1157.20171179.
5 Zhou, J., Yang, Y., and Zhang, C.Y. (2015). Toward biocompatible semiconductor quantum dots: from biosynthesis and bioconjugation to biomedical application. *Chemical Reviews* 115 (21): 11669–11717. https://doi.org/10.1021/acs.chemrev.5b00049.
6 Wang, S., Riedinger, A., Li, H. et al. (2015). Plasmonic copper sulfide nanocrystals exhibiting near-infrared photothermal and photodynamic therapeutic effects. *ACS Nano* 9 (2): 1788–1800.
7 Melamed, J.R., Riley, R.S., Valcourt, D.M. et al. (2016). Using gold nanoparticles to disrupt the tumor microenvironment: an emerging therapeutic strategy. *ACS Nano* 10 (12): 10631–10635. https://doi.org/10.1021/acsnano.6b07673.
8 Li, J., Liu, J., and Chen, C. (2017). Remote control and modulation of cellular events by plasmonic gold nanoparticles: implications and opportunities for

biomedical applications. *ACS Nano* 11 (3): 2403–2409. https://doi.org/10.1021/acsnano.7b01200.

9 Zhou, W., Gao, X., Liu, D. et al. (2015). Gold nanoparticles for in vitro diagnostics. *Chemical Reviews* 115 (19): 10575–10636. https://doi.org/10.1021/acs.chemrev.5b00100.

10 Na, L., Pengxiang, and Astruc, D. (2014). Anisotropic gold nanoparticles: synthesis, properties, applications, and toxicity. *Angewandte Chemie International Edition* 53 (7): 1756–1789. https://doi.org/10.1002/anie.201300441.

11 Xiaolian, S., Huang, X., Yan, X. et al. (2014). Chelator-free 64Cu-integrated gold nanomaterials for positron emission tomography imaging guided photothermal cancer therapy. *ACS Nano* 8 (8): 8438–8446. https://doi.org/10.1021/nn502950t.

12 Song, J., Yang, X., Jacobson, O. et al. (2015). Sequential drug release and enhanced photothermal and photoacoustic effect of hybrid reduced graphene oxide-loaded ultrasmall gold nanorod vesicles for cancer therapy. *ACS Nano* 9 (9): 9199–9209.

13 Qu, J., Tian, M., Wang, Q. et al. (2016). Photo-thermal properties of MWCNT-H2O nanofluid. *CIESC Journal* 67 (S2): 113–119.

14 Yin, P.T., Shan, S., Chhowalla, M. et al. (2015). Design, synthesis, and characterization of graphene-nanoparticle hybrid materials for bioapplications. *Chemical Reviews* 115 (7): 2483–2531.

15 Geeorgakrlas, V., Tiwariga, J.N., Kemp, K.C. et al. (2016). Noncovalent functionalization of graphene and graphene oxide for energy materials, biosensing, catalytic, and biomedical applications. *Chemical Reviews* 116 (9): 5464–5519.

16 Liang, C., Diao, S., Wang, C. et al. (2014). Tumor metastasis inhibition by imaging-guided photothermal therapy with single-walled carbon nanotubes. *Advanced Materials* 26 (32): 5646–5652.

17 Luo, Y.H., Yue, K., Zhao, L.Y. et al. (2009). Theoretical study on heating effect of Fe3O4 magnetic fluid on tumor tissues in alternating magnetic field. *CIESC Journal* 60 (4): 833–839.

18 Hu, P., Chang, T., Chen, Z.Y. et al. (2017). Surface modification and application in biomedicine and environmental protection of magnetic Fe3O4 nanoparticles. *CIESC Journal* 68 (7): 2641–2652.

19 Laurent, S., Forge, D., Port, M. et al. (2008). Magnetic iron oxide nanoparticles: synthesis, stabilization, vectorization, physicochemical characterizations, and biological applications. *Chemical Reviews* 108 (6): 2064–2110.

20 Lee, N., Yoo, D., Ling, D. et al. (2015). Iron oxide based nanoparticles for multimodal imaging and magneto responsive therapy. *Chemical Reviews* 115 (19): 10637–10689.

21 Li, J.C., Zheng, L.F., Cai, H.D. et al. (2013). Polyethylene imine-mediated synthesis of folic acid-targeted iron oxide nanoparticles for in vivo tumor MR imaging. *Biomaterials* 34 (33): 8382–8392.

22 Zhang, Y., Shen, T.T., Deng, X. et al. (2015). Design of a versatile nanocomposite for 'seeing' drug release and action behavior. *Journal of Materials Chemistry B* 3 (43): 8449–8458.

23 Zhao, M., Van Straten, D., Broekman, M.L.D. et al. (2020). Nanocarrier-based drug combination therapy for glioblastoma. *Theranostics* 10 (3): 1355–1372.

24 Shuang-Shuang, W., Jing-Yue, Z., Han, C. et al. (2018). ROS-induced NO generation for gas therapy and sensitizing photodynamic therapy of tumor. *Biomaterials* 185: S0142961218306355. https://doi.org/10.1016/j.biomaterials.2018.09.004.

25 Kris, M.K. and Petrelli, N.J. (2010). Clinical cancer advances. www.asco.org/www.cancPr.net, November 8, 2010.

26 Tang, W., Fan, W.P., Lau, J. et al. (2019). Emerging blood–brain–barrier-crossing nanotechnology for brain cancer theranostics. *Chemical Society Reviews* 48 (11): 2967–3014.

27 Bray, F., Ferlay, J., Soerjomataram, I. et al. (2018). Global cancer statistics 2018: GLOBOCAN estimates of incidence and mortality worldwide for 36 cancers in 185 countries. *CA: A Cancer Journal for Clinicians* 68 (6): 394–424.

28 Zou, Y., Liu, Y.J., Yang, Z.P. et al. (2018). Effective and targeted human orthotopic glioblastoma xenograft therapy via a multifunctional biomimetic nanomedicine. *Advanced Materials* 30 (51): 1803717. https://doi.org/10.1002/adma.201803717.

29 Taiarol, L., Formicola, B., Magro, R.D. et al. (2020). An update of nanoparticle-based approaches for glioblastoma multiforme immunotherapy. *Nanomedicine* 15 (19): 1861–1871.

30 Janjua, T.I., Rewatkar, P., Ahmed-Cox, A. et al. (2021). Frontiers in the treatment of glioblastoma: past, present and emerging. *Advanced Drug Delivery Reviews* 171: 108–138. https://doi.org/10.1016/j.addr.2021.01.012.

31 Scheetz, L., Kadiyala, P., Sun, X. et al. (2020). Synthetic high-density lipoprotein nanodiscs for personalized immunotherapy against gliomas. *Clinical Cancer Research* 26 (16): 4369–4380.

5

Nanomaterial Translational Nanomedicine for Anti-HIV and Anti-bacterial

Hao Luo[1] and Yujun Song[1,2,3]

[1]*University of Science and Technology Beijing, Center for Modern Physics Technology, School of Mathematics and Physics, 30 Xueyuan Road, Haidian District, Beijing 100083, China*
[2]*Zhengzhou Tianzhao Biomedical Technology Company Ltd., Zhengzhou New Technology Industrial Development Zone, 7B-1209 Dongqing Street, Zhengzhou 451450, China*
[3]*Key Laboratory of Pulsed Power Translational Medicine of Zhejiang Province, Hangzhou Ruidi Biotechnology Company Ltd., Room 803, Bldg. 4, 4959 Yuhangtang Road, Cangqian Street, Hangzhou 310023, China*

Nanomedicine pertains to a new group of drug formulations undergoing rapid developing promoted by cross-innovation of nanotechnology and biomedicine particularly those nanoscale materials preparation methods [1]. The application progresses of nanomedicines, which includes nanoantitumor drugs, nanopeptide–protein drugs, and nanoformulations of nonviral vector gene drugs, are among the most promising directions in nanotechnology and are considered one of the foremost areas for industrialization [2]. The creation of nanomedicines has thus emerged as a critical focus in the current international pharmaceutical industry.

As modernpharmacy is fused with nanotechnology, they produce varieties of nano-synthesis technologies. The core of these technologies are nanoscale materials synthesis and conjugation technology of nanomaterials and functional biomolecules without losing their activity and even achieving synergistic biomedical effects in the disease treatment, which includes direct nanosizing of drugs and nanodrug-carrying systems [2].

The former directly prepares drug nanoparticles using nanoprecipitation techniques or ultrafine powdering techniques such as mechanical ball milling and high-pressure homogenization techniques as well as some advanced synthesis methods (e.g. supersonic spray-drying with a microfluidic nebulator for amorphous nanomedicines [3]; multimode Au@CoFeB-Rg3 nanomedicines by continuous flow microfluidic process [4]; and nanomedicine hydrogel microcapsules by droplet microfluidic process [5]). The latter is accomplished through the use of polymer nanospheres/nanocapsules (collectively known as nanoparticles), solid lipid nanoparticles (SLNs), microemulsions/submicroemulsions, nanoliposomes, magnetic nanoparticles, polymeric micelles, dendrimers, as well as inorganic nanocarriers (e.g. silica nanospheres, carbon nanotubes [CNTs]) to dissolve, disperse, encapsulate, adsorb, couple, etc., and make the drug into a nanodispersion.

Nanomedicine: Fundamentals, Synthesis, and Applications, First Edition. Edited by Yujun Song.
© 2025 WILEY-VCH GmbH. Published 2025 by WILEY-VCH GmbH.

After the drug is nanosized, its physical and chemical properties such as saturated solubility, dissolution rate, crystal form, particle surface hydrophobicity and hydrophilicity, physical field responsiveness (such as light, electric field, and magnetic field responsiveness, pH sensitivity, temperature sensitivity, etc.), and biological properties (such as specific molecule affinity) are changed. This impacts drug absorption, distribution, metabolism, and elimination, which refer to the bio-pharmacological efficacy and pharmacokinetic behavior of the drug, including bio-adhesion, chemical stability in the gastrointestinal tract, oral bioavailability, sustained-release and controlled-release performance, targeting, long-circulation, and abilities related to transdermal/transmucosal/transmembrane/blood–brain barrier (BBB). The aim of changing conventional drugs into nanoscale species is to enhance the effectiveness of drugs, to decrease adverse drug reactions, to boost drug therapy indices, and to promote adherence to drug formulations.

This chapter will summarize those nanotechnologies for the synthesis of nanomedicines for Anti-HIV and Anti-Bacterial, as well as their microstructure and property control.

5.1 Concepts, Anti-HIV Theory, and Types and Features of Current Nanomedicines

Human immunodeficiency virus (HIV) is a viral infection that targets the immune system of the body. Acquired immunodeficiency syndrome (AIDS) is the final stage of the infection. HIV attacks the white blood cells in the body, thus weakening the immune system. This makes infected people more susceptible to diseases such as tuberculosis and certain cancers. HIV is transmitted through the body fluids of an infected person, including blood, breast milk, semen, and vaginal fluids. It can also be transmitted from mother to baby. It is not spread by kissing, hugging, or sharing food. HIV can be treated and prevented with antiretroviral therapy (ART). Untreated HIV usually progresses to AIDS over the years. The World Health Organization (WHO) now defines advanced AIDS as a CD4 cell count of less than 200 cells/mm^3 in adults and adolescents. All HIV-infected children under the age of five are considered to have advanced AIDS.

According to a report published jointly by the WHO and UNAIDS in July 2023, as of the end of 2022, there were an estimated 39 million (33.1–45.7 million) people living with HIV, two-thirds of whom (25.6 million) were in the WHO African region. As of the end of 2022, there were an estimated 39 million (33.1–45.7 million) people living with HIV, two-thirds of whom (25.6 million) were in the WHO African region. In 2022, 630 000 (480 000–880 000) people will die from HIV-related causes and 1.3 million (1.0–1.7 million) people will be living with HIV. In addition, more than 1.5 million people are newly infected with HIV every year. By the end of 2022, 29.8 million of the 39 million people living with HIV were receiving ART, amounting to 76% of all people living with HIV; of these, nearly three-quarters (71%) of those on treatment had suppressed viral loads. Globally, there are 39 million people still living with and infected by AIDS, with 7500 new HIV infections occurring every day. AIDS is far from over for any part of the world, and globally it poses a serious public health challenge.

When humans are infected with HIV, the glycoprotein gp120 on the surface of the virus binds to one of the cell's two transmembrane receptors: the T helper lymphocyte (CD_4^+) and one of the two chemokine receptors, CCR_5 and $CXCR_4$, and enters the cell and replicates in the cell. HIV can infect macrophages (M) and T-helper lymphocytes (T), and AIDS is an advanced stage of HIV infection in which the body's CD_4^+ is about to be depleted to nothing, with symptoms including opportunistic infections such as *Pneumocystis carinii* pneumonia and *Mycobacterium tuberculosis*, dementia, and cancer. T-cellophilic viruses tend to replicate in T cells, while M-type viruses overwhelmingly replicate in the brain. In addition to eroding macrophages (Mac), the lymph nodes, bone marrow, spleen, lungs, and central nervous system are the most important sites attacked by the virus after infection, causing significant neuronal destruction and damage, and the development of HIV-associated dementia, which is almost always the cause of death in untreated HIV-infected patients within 5–10 years. There are 25 different types of ARV medications authorized by pharmacological organizations worldwide. Among these, 10 are HIV protease inhibitors (PIs), 8 are nucleoside and nucleotide reverse transcriptase inhibitors (NRTIs), 4 are non-nucleoside reverse transcriptase inhibitors (NNRTIs), and 1 is an HIV integrase inhibitor (NI). These medications are administered conjointly, known as HAART (highly active ART) or cocktail therapy, to prevent viral replication, retard the progression of the illness, and enrich the patients' standard of living as well as prolong their survival.

5.1.1 Anti-HIV Reverse Transcriptase Inhibitors

HIV reverse transcriptase (RT) is a multifunctional enzymatic protein encoded by the HIV pol gene that plays a key role in the HIV replication cycle. It has three functions, namely, to catalyze negative-stranded DNA synthesis using HIV RNA as a template; to function as RNase H to degrade the RNA template in the RNA–DNA hybrid strand; and to synthesize positive-stranded DNA using viral negative-stranded DNA to generate a double-stranded DNA provirus. RT is an ideal target for drug action and the first in vitro target enzyme model for screening anti-HIV drugs. A total of 11 varieties of anti-HIV RTIs have been approved for marketing, of which 8 are nucleoside HIV RTIs: zidovudine, didanosine, zalcitabine, stavudine, lamivudine, abacavir, tenofovir, and emtricitabine; and 3 non-nucleoside HIV RTIs: nevirapine, delavirdine, and efavirenz. The current varieties of available HIV RTIs studied for their nanoloading and release systems are four drugs including zidovudine, stavudine, desipramine, and zalcitabine. ARV enzyme-inhibiting drugs are mostly pharmaceutically and drug-carrying in the form of polymeric nanoparticles and liposomes synthesized by emulsion polymerization.

5.1.2 Anti-HIV Protease Inhibitors

HIV protease is a dimer composed of two 99 amino acid monomers, which belong to asparagine protease. Gag and pol genes in the HIV genome each encode a polyprotein precursor (p55 and p160), which needs to be hydrolyzed by viral protease and processed into mature structural and functional proteins, which play a key role in the HIV replication cycle. Once HIV protease is inhibited, it can prevent the cleavage of virus precursor peptides into proteases, and prevent the offspring of HIV from maturing and becoming noninfectious virions. HIV protease and HIV RT are both enzymes essential for HIV replication and are important in vitro target enzyme models for screening anti-HIV drugs. There are 10 approved anti-HIV PIs in the market: saquinavir, indinavir, ritonavir, nelfinavir, amprenavir, lopinavir, amprenavir's prodrug fosamprenavir, atazanavir, tipranavir, and darunavir. There are six kinds of HIV PIs: shaquinavir, ritonavir, lopinavir, indinavir, amponavir, and teranavir. Anti-PIs are mostly pharmaceuticals and drug carriers in the form of nanoparticles and self-emulsifying nanoparticles.

In human history, infectious diseases were once the greatest threat to human survival. With the progress of human society and civilization, the discovery and application of antibiotics have gradually brought some classic infectious diseases under control. Smallpox was eradicated globally in 1979, and infectious diseases such as plague and neonatal tetanus, measles, diphtheria, scarlet fever, and poliomyelitis have since been virtually eliminated. However, with the aging of the population and the increasing number of other immunocompromised populations (e.g. AIDS patients, patients undergoing cancer chemotherapy, and organ transplants), infections caused by a number of nonpathogenic, weakly virulent, and conditionally pathogenic organisms have become increasingly common in clinical practice over the past two to three decades. In particular, the wide application, misuse, and abuse of antibiotics have led to microecological imbalance in the organism, the increasing number of drug-resistant bacteria, and the growing severity and complexity of the disease, which have brought many new problems and difficulties to clinical treatment. In recent years, the re-emergence of tuberculosis and other diseases, mostly caused by drug-resistant strains or mutated strains, and the resurgence of tuberculosis, which has put the lives of some three million people around the world at risk of death, have caused widespread global concern. Therefore, bacterial infection, including emerging and reproducing bacterial infection, conditional pathogenic bacterial infection, and drug-resistant bacterial infection, is still one of the main threats to human health in the world, and infectious diseases are still concerned by countries all over the world. The dramatic increase in urban population and mobility brought about by the Industrial Revolution also contributed to the proliferation of diseases. New and more effective antimicrobials need to be made available as quickly as possible for emerging and re-emerging new pathogens, and new preventive measures and new therapeutic drugs are more urgently needed for infections with drug-resistant strains because of the rapid growth of drug resistance.

Infectious diseases have their own specific pathogens. After the pathogens invade the human body, some molecules or organelles on the surface of the pathogen are first used as adhesins to combine with the corresponding receptors on the surface of mucosal epithelial cells. If the pathogen has weak invasiveness to the tissue, it will only grow and multiply locally to produce toxins and cause disease. Some pathogens invade the mucosa to breed and produce toxins, which can destroy the mucosa and submucosa and form lesions or ulcers, but do not enter the bloodstream. Other pathogens such as *Staphylococcus aureus* can produce hyaluronidase and fibrinolytic protease, which are beneficial to enter the subsurface tissue and further spread, repeatedly entering the bloodstream and spreading to the whole body. The target site requirements for the action of antimicrobials, including nanoantimicrobials, are broad, including pathogens at the site of invasion, pathogens in the bloodstream, intracellular pathogens, and up to pathogens localized on the tissues of organs throughout the body [6]. After nanoparticles are injected intravenously, they are rapidly removed from the circulation by the organism as a foreign body. So, what happens to pathogens that colonize the bloodstream? For intracellular bacterial infections, central nervous system, bone marrow, and other parts of the infection, nano-antimicrobial drugs have to overcome certain physiological barriers to enter the infected tissue to play a role. With the continuous in-depth research on nanocarriers, nanoantimicrobial drugs are expected to gradually overcome these therapeutic challenges.

The pharmacological action, efficacy, and mechanism of action of antimicrobial drugs are determined by the chemical structure of the antibiotic itself, which has been studied more clearly, and will not be repeated here, and the characteristics of nanoantimicrobial drug carriers are mainly discussed here. After administration, the drug has to pass through many biological barriers to reach the inflammatory target site of bacterial infection, to achieve distribution in organs and intracellular distribution, which is related to the route of administration. Microorganisms are easy to grow and multiply in cells and cause pathological changes, and the ability of free antibiotics to enter infected cells is relatively weak, or because of their poor stability, diffusion, and retention in cells, the therapeutic effect of intracellular infection is very poor. As a drug carrier with lysosomal properties, nanoparticles can enter cells through endocytosis and fusion because their particle size is similar to that of microorganisms, and can have intracellular targeting effects after carrying antimicrobials, which can significantly improve the efficacy in the treatment of intracellular infection. However, most diseases do not occur in the reticuloendothelial system. So, to reduce or avoid the tendency of nanoparticles to phagocytes during transportation and prolong their circulation time in body, it is necessary to develop other dosage forms to modify the surface of nanoparticles.

As the size of drugs is reduced to nanoscale, their cell uptake rate can be greatly enhanced.

Nanoparticle carriers with small particle size are regarded as foreign substances after entering the blood circulation through intravenous injection and are quickly phagocytosed by monocytes or macrophages in the blood, reaching target sites such as liver, spleen, lung, bone marrow, lymphatic, etc., where the distribution of reticuloendothelial system is concentrated. This phagocytosis by macrophages can help the drug carrier system efficiently deliver drug molecules to these cells, contributing to the killing or clearance of bacteria within the reticuloendothelial system. This effect is one of the passive targets of nanocarriers, which solves the intracellular infection which is difficult to be treated with traditional antibiotics.

The mechanism is to enter the cell through endocytosis, which is the most common form of nonpermeable carriers passing through cells. Endocytosis is categorized into phagocytosis and pinocytosis After entering the circulation, the nanocarrier is adsorbed by plasma opsins, which can bind to the surface receptors of monocytes or macrophages and be recognized and swallowed by these cells. After the ingestion of the carrier drug, the phagocytic vesicles (or phagosomes) fuse with one or more lysosomes to form phagocytic lysosomes. In the phagocytic lysosomes, the drug-loaded particles are hydrolyzed by the acid hydrolase in the lysosome, and the drug is released to produce drug effect. For this reason, the drugs can be specifically concentrated in the target cells, so that the drugs which are not easy to pass through the plasma membrane (plasma membrane) can reach the lysosome, which is very ideal for the treatment of reticuloendothelial system diseases. In addition, phagocytosis and uptake can also be achieved through the recognition of related receptors on macrophages by a variety of physiological ligands.

Phagocytosis is more common than phagocytosis and exists in almost all cells, including phagocytes, without the mediation of serum opsins or any external stimulation. Macromolecular substances are internalized by adsorption to any site on the surface of the cell membrane or specific receptors on the cell surface. Once internalized, phagocytosis vesicles interact with other cellular bodies such as endosomes and lysosomes, which are rich in phagocytosis receptors and ATP energy-supplying proton pumps, which maintain that the pH of the endosomes is between 5.0 and 5.5. This slightly acidic pH environment dissociates the receptor–drug carrier complex and produces drug efficacy (Figure 5.1).

As the size of drugs is reduced to nanoscale, their lymphatic uptake rate can be greatly enhanced, too:

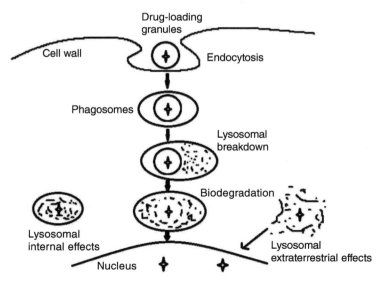

Figure 5.1 Schematic representation of drug-carrying systems through lysosomal involvement in swallowing, release, and treatment.

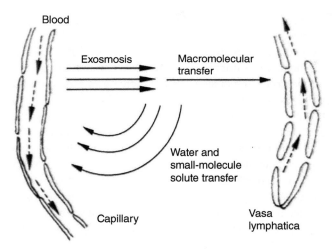

Figure 5.2 Lymphatic uptake mechanisms.

After the drug molecules are permeated from the blood vessels, there are two ways to be reabsorbed into the blood: one is to re-enter the blood directly through the window of the enlarged capillary endothelial cells, and the other is to enter the lymphatic system and return to the blood circulation with the lymph. Subcutaneous, muscular, transdermal, and peritoneal administration is mainly absorbed through the lymphatic system and then enters the blood circulation of the whole body (Figure 5.2).

As the size of drugs is reduced to nanoscale, their distribution in the inflammatory sites can be more uniform by comparing with their conventional counterparts:

The targeting of nanocarriers can be divided into two types: passive targeting and active targeting. Passive targeting includes two types: one is the phagocytosis of nanoparticles by the reticuloendothelial system, and the other is the special permeability of the vascular wall under physiological or pathological conditions, so that the nanocarrier has a certain size and can penetrate the blood vessel into the tissue. In general, nanoparticles cannot exudate from blood vessels after intravenous administration, unless the tissue has intermittent capillary endothelium, which is limited to the liver, spleen, and bone marrow. Even in these organs, the gap between endothelial cells is about 100 nm, meaning that only the smallest particles can penetrate the blood vessels. However, under the pathological condition, the vascular permeability of the inflammatory site is higher than that of the normal, and the carrier may pass through the blood vessel to enter the inflamed or infected site, and the integrity of vascular endothelium is affected, and the endothelial cell window can reach 700 nm.

Through the epithelial barrier: drugs with low molecular weight can be absorbed from the oral cavity, nasal cavity, vagina, and rectum. In general, the transport pathways of drugs across epithelial disorders are passive diffusion, active absorption, facilitated diffusion, and selective and nonselective endocytosis. Oral administration of nanoparticles has attracted more and more attention. Oral administration of nanoparticles can not only prevent the drug from being damaged by gastric acid but also reduce the irritation and toxicity of the drug to the digestive tract. At the same time, when the nanoparticles pass through the gastrointestinal tract, the surface adhesion of nanoparticles on the gastrointestinal tract increases, the residence time is prolonged, and the bioavailability is improved. If the surface of nanoparticles is coupled with bioadhesive materials, it can significantly increase intestinal uptake and achieve local efficacy in the intestinal tract, such as drugs for the treatment of ulcerative colitis. The oral absorption of nanoparticles depends on the particle size, lipophilicity, surface charge, and the composition of polymers. If nanoparticles can be used to make traditional solid dosage forms, such as granules and tablets, while maintaining the advantages of colloidal carriers, that will have a bright future. The oral drug delivery system of nanoparticles has the characteristics of drug protection and slow release. From the therapeutic point of view, long-time and low-dose antibiotics are easy to produce bacterial drug resistance. So, it is not recommended to prepare sustained-release preparations, but it is effective in clinic. The conflict between the design of this kind of sustained-release preparations and the therapeutic point of view, whether maintaining effective bacteriostatic concentration for a long time is easy to produce tolerance and the mechanism of different antibiotic tolerance is worthy of further study.

At present, there are many kinds of antiviral drugs. Since the advent of the first antiviral drug iodoside in 1963, more than 90 antiviral drugs have been used to treat human infectious diseases. These include small molecular drugs, protein drugs that stimulate the immune response, nucleotide drugs, and so on. These drugs usually need to act specifically on related enzymes necessary for virus replication in cells. Therefore, it generally has great side effects on normal cells and tissues. At the same time, because of its special replication mechanism, most viruses mutate rapidly, produce drug resistance, and greatly reduce the therapeutic effect. In recent years, the development and application of nanotechnology have provided a new strategy to solve many problems faced by antiviral drugs while achieving targeted antiviral therapy and reducing side effects. Nanodrugs can also improve the bioavailability of drugs and reduce or delay the occurrence of virus resistance.

Nanodosage forms or nanocarriers can be generally divided into the following seven categories: liposomes, micelles, nanospheres, dendrimers, nanoparticles, nanoemulsion, and polymeric nanocarriers.

Liposomes are usually bilayer hollow structures made from lecithin and ceramide. Liposomes of different sizes (particle size 0.04–10 μm) were obtained according to different preparation methods [7]. According to the number of layers containing lipid bilayers, liposomes can be divided into multilayer, small multilayer, oligo-multilayer, and single-compartment liposomes. Hydrophilic drug molecules can be encapsulated in the liposome cavity, while lipophilic drugs can be inserted into the liposome layer [8]. The entrapment efficiency of the drug mainly depends on the size, surface charge, and membrane fluidity of the liposome, which also affect the stability of the drug-loading system and the drug release effect in cell. Nanoliposome is an ideal gene transport carrier, which has high entrapment efficiency for plasmid DNA and can protect DNA from being degraded by plasma ribozyme [9]. Acyclovir is a synthetic acyclic purine nucleoside analogue, which is mainly used to prevent and treat HSV and VZV infection. However, the oral bioavailability of acyclovir is low and its efficacy is unstable. The bioavailability and antiviral efficacy of acyclovir can be increased by liposomes. The positively charged acyclovir liposome has good corneal penetration and can be used in the treatment of ocular infection.

The micelle is a nanostructure formed when the surfactant aqueous solution reaches a certain concentration (critical micelle concentration), the hydrophobic parts of the excess surfactant molecules attract and associate with each other [10]. The lipophilic ends of amphiphilic molecules gather inside the micelles to avoid contact with polar water molecules. The hydrophilic end is distributed outside the micelle. Therefore, micelles can be used as a good carrier for most insoluble antiviral drugs, which can not only improve the water solubility and bioavailability of drugs but also have organ targeting.

Efavirenz is a non-nucleoside RTI and the first line of clinically effective ART for the treatment of HIV infection (highly active ART [HAART], also known as cocktail therapy). However, efavirenz has poor water solubility and very limited oral bioavailability. In view of this, Chiappetta et al. developed a polyethylene oxide–polypropylene oxide (PEO-PPO) polymer micelle loaded with efavirenz to target the CNS. Following intranasal use in a rat model, a substantial increase in the accumulation of this nanocolloid in the brain was observed. The bioavailability of the drug in the CNS increased fourfold in comparison to the intravenous delivery system, and the relative exposure index rose fivefold. The study highlights the potential of this cost-effective nanocolloid in treating AIDS.

Nanospheres are mostly composed of biodegradable polymers and come in two main forms: monolithic spheres and capsular spheres, which are also known as nanocapsules. Nanospheres can contain many different drugs and are biocompatible. In addition, functionalized nanomicrospheres can be used as adjuvants for delivery of HIV vaccines. Nanoadjuvants not only help the viral vaccine pass through multiple physical barriers in the body and prolong the half-life of the vaccine but also mitigate the toxic side effects associated with the HIV vaccine. Importantly, the nanoadjuvant itself can interfere with the HIV assembly process and inhibit HIV replication.

Liu et al. utilized nanomicrospheres formed from polylactic acid–hydroxyacetic acid copolymer (PLGA) and polyethyleneimine (PEI) to encapsulate an HIV DNA vaccine. These degradable nanomicrospheres improved DNA delivery to APCs compared to conventional methods, and the nanomicrospheres also protected the DNA vaccine from degradation by DNA enzymes, significantly increasing serum antibody levels and cytotoxic T-lymphocyte counts.

Dendrimers are linear polymers with dendritized motifs on each repeating unit, which have unique and perfect dendritic nanostructures that diverge from the central core to form a three-dimensional tree-like structure [11]. Dendrimers themselves have nanorod-shaped structures ranging in size from a few nanometers to tens of nanometers, with intramolecular cavities and multiple binding sites capable of binding different ligands, which makes them highly promising for drug transport. Dendrimers can be used to transport DNA, siRNA, and antiviral drugs [12].

The most widely known nanoproduct for antiviral therapy is VivaGel™, whose active ingredient is the dendritic molecule SPL7013 gel, which has demonstrated its efficacy against HIV and HSV in in vitro experiments and animal models.

Nanoparticles are usually particles of solids, colloids, polymers, etc., with a diameter of 1–1000 nm [13]. Different materials such as proteins, lipids, inorganic materials, etc., can be prepared in the form of nanoparticles. Various types of nanoparticles are used as carriers to transport active antiviral drug molecules to specific sites of infection in the human body through encapsulation, absorption, solubilization, conjugation, etc. Surface modification of nanoparticles is also possible to endow them with active targeting or conditioned stimulus-responsive release. In addition, by adding hydrophilic components such as polyethylene glycol (PEG) chains to the surface of the nanoparticles, unnecessary phagocytosis of the nanoparticles by the human body can be avoided to achieve the "stealth effect," thus maximizing the biocompatibility of the nanoparticles and reducing the cytotoxicity.

SLNs made from lipids that are solid at room temperature can also be used for delivery of antiviral drugs [14]. Due to their stable solid form, such lipid nanoparticles are able to protect drug molecules with unstable chemical properties while enabling slow drug release. Compared with liposomes and conventional emulsions, SLN is able to overcome the stability problems of membranes associated with drug leaching.

In addition, inorganic nanoparticles have shown good antiviral effects. Studies have shown that inorganic nanoparticles have some broad-spectrum antiviral effects in addition to efficient antibacterial properties. Silver nanoparticles can interact with viral surface recognition proteins and adsorb to the surface of the virus in large quantities, thus preventing it from infecting host cells. At the same time, silver ions released from silver nanoparticles can directly bind to phosphorus- and sulfur-containing proteins and nucleic acid molecules, affecting viral protein synthesis and interfering with nucleic acid replication. Meanwhile, silver nanoparticles are considered a potential vaginal antiviral reagent to stop HIV-1 transmission.

Nanoemulsion is a homogeneous, transparent, and stable liquid with uniform appearance formed spontaneously by immiscible oil, water, and surfactant, and this transparent oil–water-miscible liquid is considered to be a colloidal dispersion system formed by the dispersion of emulsion droplets with a particle size of 10–100 nm in another immiscible liquid. The hydrophobic core of the micelle can encapsulate the drug, and such micelles are sometimes referred to as nanoemulsions (with a hydrophobic liquid core, "oil") or nanoparticles (with a solid core) [15]. The hydrophilic shell of micelles can also encapsulate hydrophilic drugs, but in relatively low amounts. The main advantages of nanoemulsions include: the ability to increase the solubility of difficult-to-solve drugs, improve the stability of easily hydrolyzed drugs, and can also be used as a carrier for sustained-release or targeted drug delivery systems. Nanoemulsions are highly stable, with particle size and turbidity that hardly or slowly change over time, unaffected by serum proteins, and have a long lifespan in the circulatory system, as in the case of O/W nanoemulsions in which more than 25% of the oil phase is still in the bloodstream after 24 hours of injection.

There are two basic types of emulsions, namely oil-in-water (O/W) and water-in-oil (W/O). In the former, oil is dispersed in water in the form of small particles, while in the latter, it is the other way around. The presence of domains of different polarities in the nanoemulsions makes it possible to act as a carrier for both hydrophobic and water-soluble drugs, in addition to oil-water-insoluble drugs. In the nanoemulsion, hydrophobic drugs are distributed in the nonpolar microregions or adsorbed in the hydrophobic part of the surfactant-oriented adsorbent layer, while water-soluble drugs are mainly distributed in the water region. For this reason, oil-in-water nanoemulsions can be used as a carrier for hydrophobic drugs, and water-in-oil nanoemulsions can extend the release time of the water-soluble drugs to play the role of slow release. Nanoemulsions can be used for various routes of drug delivery. For example, low viscosity nanoemulsion systems for oral, injectable, nongastrointestinal, pulmonary, and ocular drug delivery, etc., and high viscosity nanoemulsion systems for dermal drug delivery.

Most of the clinical oral preparations of nanoemulsions are used [15]. Nanoemulsions are absorbed through the lymphatic vessels after oral administration, thus overcoming the first-pass effect and the obstacles in passing through the epithelial cell membrane of the gastrointestinal tract. At the same time, due to the low surface tension, they become a hydration layer that is easy to pass through the gastrointestinal wall, so that the drugs can be in direct contact with the gastrointestinal epithelial cells and promote the absorption of the drugs so that the nanoemulsions are the ideal releasers of certain drugs such as steroids, hormones, and antibiotics.

Early nanodrug carriers are mainly liposomes, which have many advantages. However, there are some problems in practical applications, such as lower encapsulation rate, rapid release of water-soluble drugs in the bloodstream, and lower storage stability. Currently, biodegradable polymer nanoparticles are gaining attention as drug carriers in the field of drug delivery and control. The particle size of polymeric nanocarriers ranges from about 10 to 1000 nm. The drug is dissolved, encapsulated, or adsorbed on the nanoparticle matrix to prepare nanospheres and nanocapsules, collectively known as nanoparticles. Nanospheres are essentially a homogeneous system in which the drug can be dispersed uniformly; nanocapsules are a porous system in which the drug is encapsulated in a membrane capsule (contained in the aqueous or oil phase) through a single polymer membrane. Polymer nanoparticles are composed of polymeric materials, including natural polymers (albumin, polysaccharides, etc.), or synthetic polymers such as poly(lactic acid) (PLA), poly(hydroxyglycolic acid) (PGA), and copolymers of the two (PLGA), and poly(hydroxycarboxylic acid) such as poly(hydroxybutyric acid) (PBA). Some of the other polymers used in the production of nanoparticles are polyalkyl acrylic acid, polyalkyl cyanoacrylic acid, polymers of polyalkyl vinyl pyrrolidone, and acrylic acid. Nonbiodegradable ones such as ethyl fibers are less commonly used in nanoparticles.

5.1.3 Characteristics of Polymer Nanocarriers

There should be four main characteristics of polymer nanocarriers for anti-HIV drug transportation: targeting, slow-release and long-lasting time, increased bioavailability, and ability to alter the membrane transport mechanisms for enhanced cell membrane permeability.

The combination of drugs with carriers can avoid molecular degradation and change their distribution in the body so that they can be delivered to certain target organs and target cells in a controllable way. Nanoparticles can be used as carriers of this targeting strategy, including passive targeting and active targeting [16]. After intravenous injection, nanoparticles are generally absorbed by monocyte–macrophage system, mainly distributed in liver, spleen, and lung, a small amount into bone marrow, and have certain targeting of a reticuloendothelial system, such as liver and spleen. Nanoparticles enter the circulation and are adsorbed by plasma conditioners, the adsorbed conditioner particles adhere to the surface of the macrophage (the relevant receptor on the macrophage), and are ingested by the macrophage through intrinsic biochemical effects (endocytosis, fusion, etc.), known as endosomes or phagolysosomes, which fuse with lysosomes in the cell to form a phagolysosome, and in the phagolysosome, the nanoparticles are rapidly digested by the lysosomes, and biodegradation occurs. The released drug must pass through the membrane of the phagolysosome in order to produce an active effect in the target subcellular range, such as bacterial body, at which time the antimicrobial drug can produce a bactericidal or bacteriostatic effect on bacteria, showing that the antimicrobial drug can be transported by the nanocarrier to play the function of its chemical structure. The inrelease of drugs from nanoparticles is achieved by diffusion and matrix biodegradation or bioerosion, where polymers or biomolecules are biodegraded by hydrolysis or enzymatic degradation and the rate of degradation depends on the local water content, pH, and enzyme type and concentration. If nanoparticles are coated or modified by connecting ligands, antibodies, etc., the clearance by liver and spleen macrophages can be reduced or avoided, and these drug-carrying nanoparticles can actively reach the target sites where receptors, antigens, etc., are located, demonstrating active targeting for indistribution.

The drug-carrying nanoparticles that reach the target site have different drug release rates due to different types or ratios of carrier materials, and adjusting the carrier materials or ratios can adjust the rate of drug release so as to prepare drug-carrying nanoparticles with slow-release characteristics.

Due to the adhesion on the surface of the drug-carrying nanoparticles and the small particle size, it is beneficial to both the increase of retention during topical administration and the increase of the contact area and contact time between the drug and the intestinal wall, which increases the bioavailability, while it is beneficial to the oral administration of the drug.

Nanoparticles as drug carriers can change the membrane transport mechanism and increase the permeability of drugs to biological membranes, which is favorable for transdermal absorption and intracellular drug efficacy. Recent studies have revealed that colloidal drug delivery systems such as nanoparticles can help drugs cross the BBB.

5.2 Polymer-based Nanomedicines: Synthesis Methods and Typical Application

In general, polymeric materials are used in three main approaches to enable the emerging science of nanomedicine. First, to overcome the inherent issues of bioavailability of orally dosed poorly water-soluble drugs, solid drug nanoparticles

(SDNs) are often prepared using a variety of techniques such as nanomilling, homogenization, or precipitation. Second, polymers are used as containers for drug molecules either by forming solid polymer matrix nanoparticles to encapsulate drugs (e.g. bio-erodible poly(D,L-lactide-co-glycolide):PLGA, microspheres), or through the self-assembled construction of vehicles such as block copolymer liposomes/vesicles or micelles. Encapsulation is also a key component of implant technologies, including those that erode over time and those that are physically removed. Finally, direct noncovalent or covalent conjugation of drugs to polymers to form prodrugs has been successfully used to enhance circulatory times, reduce toxicity, and deliver drugs through triggered/controlled release.

The preparation methods of polymer nanoparticles include chemical reaction methods (also known as polymerization), physicochemical methods (e.g. emulsification, condensed phase separation, nanoprecipitation, etc.), and physical methods (e.g. mechanical pulverization, vacuum drying, etc.) [17]. There are eight proven methods: the polymerization method, the basic principle of the emulsification method, condensed phase separation methods, nanoprecipitation, supercritical fluid technology, polymer micelles, spray drying method, and dendritic macromolecules.

The polymerization method can be divided into free radical polymerization, ionic (anionic and cationic) polymerization according to the polymerization mechanism, and free radical polymerization can be divided into intrinsic polymerization, solution polymerization, suspension polymerization, and emulsion polymerization according to the way of polymerization reaction implementation. There are also methods of chemical preparation of polymer nanoparticles that are classified into aqueous phase polymerization reactions of monomers or oligomers, polymerization reactions in hydrophilic O/W emulsions, polymerization reactions in ultrafine lipophilic W/O emulsions, interfacial polymerization reactions, and interfacial condensation reactions.

Emulsion polymerization is one of the most commonly used chemical methods for the preparation of polymer nanoparticles. In emulsion polymerization, the monodispersity of the polymer emulsion particle size and its distribution is an important controlling parameter for high-performance latexes and a necessary condition for the preparation of polymer nanoparticles. Designing nanoparticles by polymerization reaction of monomers makes it easier to give new functions to the nanocarriers than physical preparation methods, but the disadvantage is that unreacted monomers or oligomers may remain in the nanoparticles.

Currently, polymer nanoparticles prepared by polymerization include dendritic macromolecules, poly(alkyl ester) of polycyanoacrylate polymer nanoparticles, polyacrylamide polymer nanoparticles, and polyglutaraldehyde nanoparticles.

The basic principle of the emulsification method is to emulsify the less polar solvent phase (oil phase) containing polymer and the more polar solvent phase (water phase) under certain conditions to form an emulsion, and then remove the solvent to obtain nanoparticles. Depending on the method of solvent removal, emulsification can be divided into emulsification-solvent evaporation, emulsification-solvent diffusion, and salt precipitation.

The first two methods often use water-insoluble organic solvents such as dichloromethane and chloroform, which are more toxic. So, the salting-out method, also known as emulsification-solvent diffusion/extraction, was developed [18]. It is characterized by the use of a salting out agent (electrolyte) to separate the organic solvent from the aqueous solution, or by adding a large amount of the polymer's undesirable solvent, which is miscible with the polymer solvent, and as the undesirable solvent continues to be added, the polymer solvent is gradually extracted, and the polymer containing the drug is quickly solidified to form nanoparticles.

Condensed phase separation methods are generally divided into two categories: simple condensed phase separation and compound condensed phase separation [15]. In the simple condensed phase separation method, the polymer is dissolved in the appropriate solvent, and the insoluble drug particles are suspended as the core, or dissolved in water and then added to the polymer solution to form a suspension. By changing the composition of the system, adding appropriate salts (such as sodium sulfate), changing the system temperature, pH or adding a large number of bad solvents of polymer to form two phases of the original suspension, namely condensed phase (also known as dense phase) and continuous phase (also known as dilute phase). With the continuous addition of nonsolvent, the solvent is gradually extracted, and the condensed phase containing drugs is solidified to form nanoparticles (Figure 5.3).

Complex cohesive phase separation is the formation of water-insoluble complex nanoparticles by ionic cross-linking (ionic bonding) of two polyelectrolytes with opposite charges in aqueous solution under certain conditions (e.g. changing pH or temperature). The commonly used polyelectrolyte pairs are gelatin and gum arabic, albumin and gum arabic, alginate and polylysine, alginate and chitosan, and so on.

The condensed phase separation method is simple to operate, the preparation process is easy to control, and it is prepared at room temperature, which is particularly suitable for polymer nanoparticles of heat-sensitive drugs as well as biomacromolecule drugs that are susceptible to denaturation at high temperatures.

Nanoprecipitation was first proposed by Fessi et al. Through the interfacial commotion phenomenon and the conversion of solvent system when the organic phase is miscible with the aqueous phase, the polymer material coats the drug and forms drug-carrying nanoparticles, which migrate to the interface and precipitate continuously with the evaporation of the solvent. The biggest advantage of nanoprecipitation method is that it can avoid the use of chlorinated solvents and surfactants, which reduces the harm to human beings and pollution to the environment.

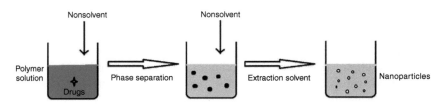

Figure 5.3 Simple condensed phase separation.

The advantage of preparing polymer nanoparticles by supercritical fluid technology is that there is no residual organic solvent in the nanoparticles, the purity of the prepared nanoparticles is high, and the preparation process has no environmental pollution. Supercritical fluid technology includes rapid expansion of the supercritical solution (RESS) method and supercritical anti-solvent (SAS) method.

Spray drying method is commonly used in industry to form powders. This method can be used to coat both fat-soluble and water-soluble drugs in the preparation process. Unlike the spray drying method, the spray freeze drying method is carried out in cryogenic liquid nitrogen, which requires enormous energy, is very costly, and is currently mainly used to encapsulate peptide–protein drugs.

Polymer micelles are formed by the spontaneous assembly of amphiphilic polymers in a selective solvent (which is a good solvent for one chain segment and a bad solvent for the other) through physical interactions such as intermolecular hydrogen bonding, electrostatic interactions, and van der Waals forces.

Conventional polymer micelles are prepared by two types of methods: one is the selective solvent method, the block copolymers or graft copolymers will be dissolved in each component of the common solvent, the formation of polymer solution, and then gradually add one of the components of the precipitant, so that the formation of micelles, and then further through the dialysis or evaporation of the original common solvent to remove the method. The other type is the direct dissolution method in which the block copolymer is dissolved directly in a selective solvent (or a mixture of solvents), and the dissolution and micelle formation are promoted by heating, for example. However, this method requires the polymer to have sufficiently long soluble blocks to ensure micelle formation and stable dispersion.

Currently, the research on polymer micelles is developing rapidly, and new methods of preparing polymer micelles have been developed by physical or chemical means, including the use of special interaction forces between polymer molecules, the temperature dependence of the solvation ability of the block, the adjustment of the pH of the medium, and chemical reactions. These methods of polymer micellization are different from the traditional methods of polymer micelle preparation and have greatly broadened the scope of research and application of polymer micelles (Figure 5.4).

Dendritic macromolecules are macromolecules with a dendritic and highly branched structure obtained by stepwise repetitive reactions of multifunctional group monomers, and their molecular structure consists of three parts: the initial nucleus (ammonia or ethylenediamine, etc.), the branching (number of generations), and the terminal group, and the size of the molecule increases with the number of generations. Dendritic macromolecules have the following characteristics: regular, delicate, and highly symmetric structure; molecular volume, shape, and the type and number of functional groups can be precisely controlled. It has internal cavities to encapsulate drug molecules, and the end groups can be modified to link bioactives such as genes and antibodies. The dense end groups allow each dendrimer to bind more active substances, forming lipophilic, hydrophilic, nonionic, cationic, and anionic surfaces. There are two basic methods, divergence and convergence. The principle is shown in Figure 5.5.

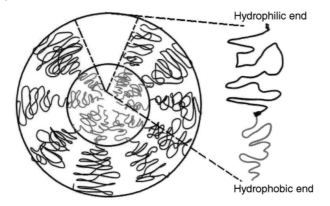

Figure 5.4 Polymer micelles.

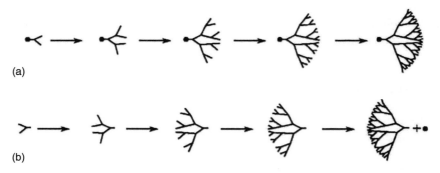

Figure 5.5 (a) Divergence and (b) convergence.

5.3 Inorganic-based Nanomedicines: Synthesis Methods and Typical Application

Inorganic nanomaterials are nanomaterials prepared using inorganic oxides such as silicon oxide, titanium oxide, tin oxide, zinc oxide, and alumina. Such inorganic nanomaterials are now produced in large quantities in industry and are widely used in various consumer and industrial products.

Among them, nanosilicon dioxide has superior properties and is the most widely used class of inorganic nanomaterials. Nanosilica is amorphous flocculent, semitransparent, commonly used particle size of 7–40 nm, with high chemical purity, high temperature resistance, good electrical insulation, and dispersion properties, and easy to prepare. Nanosilicon dioxide is prepared by fumed, sol–gel, reversed-phase microemulsion, precipitation, silica monomer, and wollastonite synthesis methods. The more commonly used methods are gas phase, sol–gel, and precipitation [16]. The fumed-phase method involves the hydrolysis of silane in a hydroxide flame, which produces silica molecules that condense into nanoparticles. The nanosilicon dioxide produced by this method is in the form of spherical particles with high purity. The sol–gel method involves the formation of silica sols by hydrolytic polymerization of ethyl orthosilicate, which has the advantage that it can be carried out under mild reaction conditions, with high silica purity and good dispersion and suspension properties in solution. Precipitation method is a liquid phase chemical synthesis of high-purity nanosilica, using water glass and hydrochloric acid or other acidifying agents as raw materials, adding appropriate amount of surfactant, and controlling a certain synthesis temperature, the synthesis product obtained by centrifugal separation, washing, drying, and high-temperature burning to get the finished product. Nanosilicon dioxide has good hydrophilicity, thermal and chemical stability as well as biocompatibility, and avoids rapid clearance by the reticuloendothelial system. At the same time, the surface of silica nanoparticles has a large number of active groups such as hydroxyl groups, which can be used to bind chemical drugs as well as biomolecules such as enzymes, antibodies, and DNA to silica nanoparticles through physical adsorption and chemical coupling. Controlled release of drugs can also be achieved by preparing silica nanoparticles with different internal structures. Therefore, silica nanoparticles are a new type of promising drug delivery system, especially as gene-drug carriers, which has been a hot research topic in recent years.

In inorganic nanomedicines, apart from metals and their compounds, the most studied are carbon nanomaterials such as fullerenes (Coo) and CNTs [18]. Coo, also known as carbon-caged alkene, spherical carbon, etc., is a three-dodecahedral cage-like structure composed of 60 carbon atoms, containing 12 five-membered rings and 20 six-membered rings. The molecular structure is shaped like a football with high symmetry, belonging to the I_h point group, with a diameter of about 0.71 nm and an inner cavity diameter of about 0.3 nm, which can accommodate a variety of atoms and small molecular groups. The C60 molecule itself has high electronegativity, and excessive electronegativity can be toxic to cells. Through chemical modification, the electronegativity of C60 can be lowered, thus reducing its cytotoxicity and enabling its application in biomedical fields. It can be used as a carrier for plasmid DNA, governing drugs and trace elements to increase the concentration of drugs in the lesion site.

CNTs are seamless nanotubes consisting of single or multiple layers of graphite flakes convoluted [19]. CNTs are classified into two categories: single-walled CNTs (SWCNTs) and multiwalled CNTs (MWCNTs). The number of layers of cylindrical MWC-NTs varies from 2 to 50, which are formed by several to dozens of single-walled tubes in coaxial sleeve configuration, and the spacing of adjacent tubes is 0.34 nm, which is close to the spacing of graphite layers (0.335 nm). When MWCNTs are formed, the interlayers are prone to form trap centers and trap various defects, and thus there are more defects in MWCNTs. The typical diameters and lengths of MWCNTs are 2–50 nm and 0.1–50 μm, respectively, and the typical diameters and lengths of SWCNTs are 0.4–30 nm and 1–50 μm, respectively. Both SWCNTs and MWCNTs have high aspect ratios, generally about 100.

CNTs are mainly prepared by arc discharge, catalytic cracking, laser, plasma injection, ion beam, solar energy, electrolysis, and combustion. Among them, the catalytic cracking method is one of the more mature preparation methods currently applied, and its basic principle is to synthesize CNTs by using the deposition of isotropic carbon atoms cracked from hydrocarbons under the action of iron-, cobalt-, and nickel-based catalysts.

The purification and separation of CNTs are also very important after the preparation of CNTs, and at present, CNTs are mainly purified by oxidation method, which includes two types of oxidation method and strong acid method, due to the structural stability of CNTs and strong resistance to oxidation, so the graphite particles and metal particles in CNTs can be removed by oxidation.

Functionalized and modified CNTs have good monodispersity and biocompatibility, and their low cytotoxicity can be used as nanodrug-carrying systems for the indelivery of chemical drugs as well as biological macromolecules such as proteins and DNA [20].

Balasubramanian et al. modified CNTs to obtain water-soluble CNTs, which were then labeled with fluorescent dyes fluorescein isothiocyanate (FITC) and fluorescent peptides, respectively, and after co-incubation with human-origin or murine-origin fibroblasts for 1 hour at 37 °C, it was observed that fluorescently modified CNTs entered into the cells, mainly in the cytoplasm, and the peptide-modified CNTs even entered into the nucleus. Pantarotto et al. [20] transfected HeLa cells with modified CNTs loaded with plasmid DNA, and the transfection efficiency was more than 10-fold that of the system without CNTs.

5.4 Metallic-based Nanomedicines: Synthesis Methods and Typical Application

Metallic nanomaterials are inorganic nanomaterials prepared from transition metals such as palladium, platinum, gold, silver, copper, and their alloys. Among them, noble metal nanomaterials have good physical properties and are typical systems for studying the light quantum-limited effect, magnetic quantum-limited effect, and the unique properties of nanomaterials, which are widely used in many fields of catalytic chemistry, optics, informatics, electronics, and biology. The control of the size, shape, and structure of metallic nanomaterials, especially gold nanomaterials, as well as the corresponding physical properties, has always been a cutting-edge hot spot in the field of materials science. People can not only prepare spherical nanoparticles with different sizes but also control their morphology and prepare many nanoparticles with different morphologies. The most commonly used preparation methods include chemical reduction, co-precipitation, high-temperature pyrolysis, micelle/microemulsion synthesis, and hydrothermal synthesis.

The most commonly used synthesis method for gold nanoparticles (GNPs) is the chemical reduction method, which is based on the following principle: add an appropriate amount of reducing agent to a certain concentration of chloroauric acid solution to reduce gold ions to gold atoms, and at the same time, add a certain amount of stabilizers to the reaction system to prevent the agglomeration of the nanoparticles. Brust et al. synthesized the surface-functionalized by a one-step method of GNPs by using $NaBH_4$ in the presence of sulfhydryl stabilizers and reducing $AuCl_4^-$ salts to obtain GNPs with sulfhydryl groups on the surface and monodispersity, and the core diameter can be adjusted in the range of 1.5–6 nm by adjusting the amount of sulfhydryl groups. On this basis, Murray et al. developed a method to synthesize multifunctional GNPs by introducing exogenous sulfhydryl groups to replace the sulfhydryl groups on the outer envelope layer of the original GNPs.

Co-precipitation method for the preparation of metal nanoparticles generally involves the use of water-soluble metal cation salt solutions, the addition of alkaline solutions under the protection of inert gases, the formation of hydroxide sols, and then heat treatment to prepare metal oxide nanoparticles. The advantages of the co-precipitation method are its simplicity, ease of implementation, and good reproducibility. The specific saturation magnetization intensity of the prepared metal nanoparticles is generally in the range of 30–50 emu/g". However, the co-precipitation method is difficult to provide good control of the particle size of the metal nanoparticles, which usually have a wide particle size distribution. In order to improve the particle size control of the prepared metal nanoparticles by the co-precipitation method, it is usually necessary to add some stabilizers during the co-precipitation process. The nanoparticle size is regulated by changing the ratio of stabilizers to metal ions.

High-temperature pyrolysis is currently the most commonly used method for the preparation of magnetic nanoparticles, and the prepared magnetic nanoparticles have the advantages of small particle size and narrow particle size distribution. The preparation process is generally as follows: the organometallic precursor is dissolved in a high-boiling organic solvent (phenyl ether, benzyl ether, etc.), and then heated to 200–300 °C under the stabilizing effect of surfactants, to form magnetic nanoparticles. In this method, the particle size, particle size distribution, and morphology of the nanoparticles can be effectively regulated by adjusting the proportion of starting materials, the heating procedure, the reaction time, and other parameters. The prepared magnetic nanoparticles are mostly monodisperse magnetic nanoparticles with good monodispersity, and the particle size is generally 4–20 nm.

LaMer et al. reported a generalized high-temperature pyrolysis method for the preparation of a variety of magnetic nanoparticles such as Cr_2O_3, MnO, Co_3O_4, and NiO. The method employs the high-temperature pyrolysis of finger acid metal salts in nonaqueous solvents, which can effectively regulate the particle size and shape of magnetic nanocrystals. The particle size can be regulated in the range of 3–50 nm. Its reaction rate increases with the shortening of the chain length of the lipoarginine used.

The hydrothermal synthesis method can solve the problem that nanoparticles prepared by high temperature pyrolysis cannot be dispersed in aqueous solution. A simple method is as follows: $FeCl_3$–$6H_2O$ was used as a starting material and 2-pyrrolidone was used as a high-boiling solvent. The materials were subjected to high-temperature pyrolysis at 245 °C, and water-soluble Fe_3O_4 nanoparticles with three particle sizes of 4, 12, and 60 nm obtained by controlling the reflux time [21].

Wang et al. [22] prepared a variety of magnetic nanoparticles and other nanomaterials using a liquid–solid–solution (LSS) phase transfer separation mechanism. The phase transfer process occurs at the interface of these three phases, and various nanoparticles were obtained by the reduction of high-valent metal ions at the

Table 5.1 Comparison of four methods.

Synthesis method	Co-precipitation method	High-temperature pyrolysis	Hydrothermal synthesis	Micelles/Microemulsions
Synthesis condition	Easy	Complex	Complex	Easy
Reaction temperature (°C)	20–90	100–320	20–50	220
Reaction time	Minutes	Days	Hours	Days
Solvents	Water	Organic solvent	Organic solvent	Water–ethanol
Topography control	Bad	Excellent	Nice	Excellent

interface by ethanol. The LSS phase transfer separation technique is applicable to most of the main group of metal ions, and it can be used to prepare a variety of nanoparticles including noble metals (Au, Ag, Pt, Rh, Ir, etc.), quantum dots (ZnSe, CdSe, Ag_2S, PbS, etc.), metal oxides (Fes_3O_4, $BaTiO_3$, $CoFe_2O_4$, TiO_2) and rare earth compounds ($NaYF_4$, YF_4, LaF, YbF_3, etc.), and many other nanoparticles.

It is well known that micelles/microemulsions are widely used as nanoreactors for the preparation of nanomaterials. This technique was used to prepare a reversed-phase micellar/micellar emulsion system (water-in-oil type) containing starting materials, with octane as the oil phase and cetyltrimethylammonium bromide and 1-butanol as surfactant and co-surfactant, respectively. Various magnetic nanoparticles such as metallic cobalt nanoparticles, diamond/platinum alloy nanoparticles, and gold-coated diamond/platinum alloy nanoparticles were prepared in this system. Micellar/microemulsion synthesis techniques are also widely used for the synthesis of spinel-type ferrite nanoparticles, where the particle size is directly related to the water/oil ratio. Table 5.1 is a comparison of the four synthesis methods.

5.5 Multi-functional (Target) Nanomedicines

In addition to the treatment of circulatory or multiorgan diseases (infections, fever, pain, etc.), which require drugs to have a broad-spectrum effect, the treatment of local or single-organ diseases in general (tumors, hepatic or pulmonary disorders, etc.) requires that the drug be able to act selectively on the local or single organ in which these lesions are found. Unfortunately, the vast majority of drugs are widely distributed in the body and are inevitably accompanied by significant adverse reactions when treating local or single-organ diseases. For example, treatment with nonselective cytotoxic antitumor drugs is usually accompanied by severe toxic reactions that harm normal cells and tissues. Therefore, the use of pharmacological means to achieve targeted delivery of drugs to diseased areas is of great significance in improving the therapeutic efficacy of local or single-organ diseases and reducing adverse drug reactions [23].

Targeted agent, also known as targeted drug system (TDS), is a drug delivery system in which the carrier concentrates the drug selectively in the target tissue, target organ, target cell, or intracellular structure through local administration or systemic blood circulation. By making the drug into a targeted drug delivery system that can reach the target area, the efficacy of the drug can be improved, adverse reactions can be reduced, and the safety, efficacy, reliability, and patient compliance of the drug can be improved. Targeted drug delivery system not only requires the drug to selectively reach the target tissues, target organs, target cells, and even intracellular structures at specific sites but also requires a certain concentration of drug retention time for drug efficacy, while the carrier should be free of residual adverse reactions. A successful targeted drug delivery system should have three elements: localized concentration, controlled release, and nontoxic biodegradability. Targeted drug delivery systems are classified into passive targeted drug delivery systems, active targeted drug delivery systems, and physicochemical targeted drug delivery systems, according to the targeting source force.

Passive targeted drug delivery systems, also known as naturally targeted drug delivery systems, are mainly related to the particle size of the drug-carrying particles. Emulsions, liposomes, microspheres, and nanoparticles are common passive targeted drug delivery carriers.

Active targeted drug delivery systems use a modified drug carrier as a "missile" to deliver the drug to the target area in a concentrated manner. For example, after surface modification, drug-loaded particles are not recognized by macrophages, or because they have specific ligands that can bind to the receptors of target cells, or connect monoclonal antibodies to become immune particles, which can avoid the uptake of macrophages, prevent concentration in the liver, change the natural distribution of particles in the body, and reach a specific target site. The drug can also be modified into a prodrug, that is, a pharmacologically inert substance that can be activated at the active site and activated in a specific target area.

Physicochemical targeted drug delivery systems are the application of certain physicochemical methods to deliver drugs at specific sites. For example, magnetic targeting agents are made from magnetic materials and drugs, which are guided by a strong external magnetic field and are located in a specific target area through blood vessels; or heat-sensitive agents are made by using a temperature-sensitive carrier to release drugs in the target area under the local action of thermotherapy; pH-sensitive agents can also be used to prepare pH-sensitive agents to release drugs in a specific pH target area. Embolization agents are used to block the blood supply and nutrition of the target area, which plays the dual role of embolization and targeted chemotherapy.

Drug-loaded particles are absorbed by macrophages of monocyte/macrophage system and transported to liver, spleen, and other organs through normal physiological process. After the particles of the passive targeted drug delivery system are injected intravenously, the distribution in the body first depends on the particle size. Usually, when the particle size is 2.5–10 μm, most of the particles accumulate in macrophages; the particles larger than 7 µm are usually intercepted by the smallest capillary bed of the lung by mechanical filtration and absorbed by mononuclear leukocytes into the lung tissue or alveoli; when smaller than 7 µm, they are generally absorbed by macrophages in the liver and spleen. The nanoparticles of 200–400 nm were quickly cleared by the liver after the liver, while the nanoparticles smaller than 10 nm were slowly accumulated in the bone marrow. In addition to the particle size, the surface properties of the particles also play an important role in the distribution. Therefore, nanodrugs play an important role in passive targeting.

Drug carriers (liposomes, nanoparticles, nanoemulsions, etc.) have a large number of modifiable groups on their surfaces, which provide space for active targeting and have become an important part of drug targeting technology [24]. After the surface of the drug carrier is properly modified, the hydrophobic surface can be replaced by the hydrophilic surface, which can reduce or avoid the phagocytosis of monocyte/macrophage system and prolong its retention time in the circulatory system. It is beneficial to target tissues lacking monocyte/macrophage system other than liver and spleen. Using antibody modification, an immune-targeting drug delivery system targeting to cell surface antigens can be made. Adding a certain antibody to the surface of the drug carrier has the ability to recognize the target cell at the molecular level and can improve the specific targeting of the drug carrier [25]. Binding antibodies or ligands to the end of PEC can not only maintain a long cycle but also maintain the recognition of targets. Drugs and ligands are made into common substances, which can direct drugs to specific target tissues. The common modifications are folic acid, transferrin, various sugar residues, low-density lipoprotein, and so on.

5.6 Future Development

Magnetic nanospheres are mature technology, and magnetically guided drug delivery technology is exactly the development and application of magnetic nanospheres. pH-sensitive nanomaterials, temperature-sensitive nanomaterials, and nanomaterials that are both pH- and temperature-sensitive offer the possibility of physicochemical targeting of drugs. Magnetic targeted drug delivery system includes three main components: magnetic material, drug carrier, and drug. The magnetic nanospheres have strong magnetic permeability and magnetic induction strength, which enables the magnetic carrier to exert strong physical targeting and thermotherapy effects. The nanomagnetic spheres have maximum biocompatibility and minimum antigenicity. The nanomagnetic spheres are nontoxic, and degradable, and the degradation products are nontoxic and are excreted from the body at a certain time. The smaller the particle size, the better, generally the particle size is 10–100 nm. drug carriers include liposomes, nanoparticles, microspheres, and so on. The backbone materials involved are natural poly-amino-acids (albumin, gelatin), polysaccharides (dextran, starch, etc.), lipid-like components (phosphatidylcholine, phosphatidylglycerol, etc.), and other backbone materials (ethyl cellulose, polyalkylcyanoacrylates, etc.) [26]. Magnetic targeted drug delivery system can not only locate the carrier in the target area through the magnetic field, but under the action of the alternating magnetic field, the magnetic material can also absorb the energy of the magnetic field to generate heat so as to achieve the effect of heat therapy at the same time of chemotherapy [27].

The pH value of intracellular matrix is lower than that of extracellular matrix. Based on this principle, pH-sensitive nanocarriers are designed as carriers of genes, nucleic acids, peptides, proteins, and chemical drugs, which can rapidly release drugs in intracellular matrix and improve the curative effect. For example, using pH-sensitive histidine-galactose-cholesterol cationic liposomes as gene carriers, the transfection effect is 100 times higher than that of ordinary galactosylated liposomes. For another example, nystatin was prepared as pH-sensitive liposomes, and the antibacterial effect was significantly increased due to the rapid release of drugs in the acidic environment of macrophage lysosomes. pH-sensitive immune liposomes can target effective carriers of cells, receptor-mediated endocytosis, liposomes enter the acidic environment of endosomes or lysosomes and become unstable and become fusion components, which is an effective plasmid DNA transport system.

The use of nanobiomaterials and nanotechnology in the diagnosis and drug treatment of diseases has shown great vitality and application prospects. Combining the special electrical, optical, and mechanical properties of nanomaterials, their antimicrobial mechanisms and activities have been investigated [28]. Provide the possibility for multifunctional drug therapy systems to meet different biological and therapeutic requirements through nanotechnology. Improving the problems of antimicrobial dose and toxic side effects and overcoming drug resistance through biocompatibility, precise target release, and relatively controllable physicochemical properties provides a very promising research strategy.

How to realize the targeted delivery of drugs to improve their efficacy and safety through the distribution characteristics of nanocarriers, the physical stability of nanocarrier structure, the increase of drug loading capacity, how to regulate the release of drugs, the thermotherapy/photodynamic therapy of the materials, as well as the research on multifunctional system combining the delivery of drugs are all the directions of further enhancing the antidisease ability of nanomedicines.

Most of the anti-HIV and antibacterial nanomedicines are still in the research and development stage. Continuing to explore the unique advantages of nanomedicines and designing nanomedicines with no cytotoxicity, highly efficient targeting of viral infection sites, and controllable antiviral activity will be the focus of future research [29].

References

1 Rosen, H. and Abribat, T. (2005). The rise and rise of drug delivery. *Nature Reviews Drug Discovery* 4 (5): 381–385.
2 Moshfeghi, A.A. and Peyman, G.A. (2005). Microparticulates and nanoparticulates. *Advanced Drug Delivery Reviews* 57 (14): 2047–2052.
3 Amstad, E., Gopinadhan, M., Holtze, C. et al. (2015). Production of amorphous nanoparticles by supersonic spray-drying with a microfluidic nebulator. *Science* 349: 956–960.

4 Zhang, W., Zhao, X., Yuan, Y. et al. (2020). Microfluidic synthesis of multi-mode Au@CoFeB-Rg3 nanomedicines and their cytotoxicity and anti-tumor effects. *Chemistry of Materials* 32: 5044–5056.

5 Liu, R., Wu, Q., Huang, X. et al. (2021). Synthesis of nanomedicine hydrogel microcapsules by droplet microfluidic process and their pH and temperature dependent release. *RSC Advances* 11 (60): 37814–37823.

6 Hans, M.L. and Lowman, A.M. (2002). Biodegradable nanoparticles for drug delivery and targeting. *Current Opinion in Solid State and Materials Science* 6 (4): 319–327.

7 Barratt, G.M. (2000). Therapeutic-applications of colloidal drug carriers. *Pharmaceutical Science & Technology Today* 3 (5): 163–171.

8 Kreuter, J. (1994). *Colloidal Drug Delivery System*, vol. 66, 254–255. Marcel Dekker Inc.

9 Sahoo, S.K. and Labhasetwar, V. (2003). Nanotech approaches to drug delivery and imaging. *Drug Discovery Today* 8 (24): 1112–1120.

10 Discher, D.E. and Kamien, R.D. (2004). Self-assembly: towards precision micelles. *Nature* 69992: 519–520.

11 Gohy, J.F. (2005). Block copolymer micelles. *Advanced Polymer Science* 190: 65–136.

12 Tomalia, D.A., Baker, H., Dewald, J. et al. (1985). A new class of polymers: starburst-dendritic macromolecules. *Polymer Journal* 17 (1): 117–132.

13 Legrand, P., Lesieur, S., Bochot, A. et al. (2007). Influence of polymer behaviour in organic solution on the production of polylactide nanoparticles by nanoprecipitation. *International Journal of Pharmaceutics* 344 (1): 33–43.

14 Zhang, J.Y., Shen, Z.G., Zhong, J. et al. (2006). Preparation of amorphous cefuroxime axetil nanoparticles by controlled nanoprecipitation method without surfactants. *International Journal of Pharmaceutics* 323 (1/2): 153–160.

15 Quintanar-Cuerrero, D., Tamayo-Esquivel, D., Ganem-Quintanar, A. et al. (2005). Adaptation and optimization of the emulsification-diffusion technique to prepare lipidic nanospheres. *European Journal of Pharmaceutical Sciences* 26: 211–218.

16 Barh, C., Bartlett, J., Kong, L. et al. (2004). Silica particles: a novel drug delivery system. *Advanced Materials* 16 (21): 1959–1966.

17 Diederich, F. and T'hilgen, C. (1996). Covalent fullerene chemistry. *Science* 271 (5247): 317–323.

18 Zhang, X.F., Zhang, X.B., Van, T.'G. et al. (1993). Carbon nano-tubes: their formation process and observation by electron microscopy. *Journal of Crystal Growth* 130 (3/4): 368–382.

19 Balasubramanian, K. and Burghard, M. (2005). Chemically functionalized carbon nanotubes. *Small* 1 (2): 180–192.

20 Pantarotto, D., Singh, R., McCarthy, D. et al. (2004). Functionalized carbon nanotubes for plasmid DNA gene delivery. *Angewandte Chemie International Edition* 43 (39): 5242–5246.

21 Li, Z., Sun, Q., and Gao, M. (2005). Preparation of water-soluble magnetite nanocrystals from hydrated ferric salts in 2-pyrrolidone; mechanism leading to Feg04. *Angewandte Chemie International Edition* 44 (1): 123–126.

22 Wang, X., Zhuang, J., Peng, Q. et al. (2005). A general strategy for nanocrystal synthesis. *Nature* 437 (7055): 121–124.

23 Chiappetta, D.A., Hocht, C., Opezzo, J.A.W. et al. (2013). Intranasal administration of antiretrovial-loaded micelles for anatomical targeting to the brain in HIV. *Nanomedicine* 8 (2): 223–237.

24 Liu, Y. and Chen, C.Y. (2016). Role of nanotechnology in HIVIAIDS vaccine development. *Advanced Drug Delivery Reviews* 103: 76–89.

25 Rupp, R., Rosenthal, S.L., and Stanbery, L.R. (2007). VivaGel(SPL7013gel) a candidate dendrimer microbicide for the prevention HIV and HSV infection. *International Journal of Nanomedicine* 2 (4): 561–566.

26 Mehnert, W. and Mäder, K. (2001). Solid lipid nanoparticles. production, characterization and applications. *Advanced Drug Delivery Reviews* 47 (2/3): 165–196.

27 Wang, A.Z., Langer, R., and Farokhzad, O.C. (2012). Nanoparticle delivery of cancer drugs. *Annual Review of Medicine* 63 (1): 185–198.

28 Zhao, J.S. and Castarnova, V. (2011). Toxicology of nanomaterials used in nanomedicine. *Journal of Toxicology and Environmental Health, Part B* 14 (8): 593–632.

29 Ferrari, M. (2005). Cancer nanotechnology opportunities and challenges. *Nature Reviews. Cancer* 5 (3): 161–171.

6

Nanomedicine for Next-generation Dermal Management

Haibin Wu[1], Qian Chen[2], and Shen Hu[3,4]

[1]*Hangzhou Medical College, School of Pharmaceutical Sciences, Yikang Street, Hangzhou 311399, China*
[2]*Zhejiang University, Innovation Center of Yangtze River Delta, National Key Laboratory of Chinese Medicine Modernization, Zhongxing Street, Jiaxing 314100, China*
[3]*Zhejiang University, The Second Affiliated Hospital of Zhejiang University School of Medicine, Department of Obstetrics, Jianghong Street, Hangzhou 310009, China*
[4]*Harvard University, The Harvard T.H. Chan School of Public Health, Department of Epidemiology, Huntington Ave, Boston, MA 02115, USA*

6.1 Introduction

Cutaneous wounds including acute wounds and chronic nonhealing wounds have been considered a silent epidemic, posing a rapidly increasing burden to society worldwide due to productivity loss of patients and drain on healthcare resources [1–5]. Importantly, chronic nonhealing wounds are known to have devastating effects on the quality of patient life and often result in limb amputations, life-long disability, and premature death [1, 5–8]. Besides, the prevalence of chronic nonhealing wounds is expected to increase enormously in the future because of the rapid rise of aging population, diabetes mellitus, and the obesity epidemic, which constitute leading mechanisms underlying impaired wound healing [9–11]. At the fundamental level, normal wound healing process involves a series of well-orchestrated events including hemostasis, inflammation, granulation tissue formation, and wound remodeling [3, 12]. However, in the case of large deep wounds and chronic nonhealing wounds, these sequential phases of the healing process are often disrupted, and disorders can manifest as either underhealing or overhealing [3, 12].

An ideal solution to the underhealing and overhealing issues should have the ability to normalize the disrupted wound healing process and eliminate interference factors. In addition to wound therapy, efficient management of cutaneous wounds, especially chronic nonhealing wounds, also necessitates a comprehensive evaluation of the characteristics of both the wound tissue and the wound microenvironment [13–16]. The cutaneous wound can be readily healed by careful medical practice if the underlying pathology mechanisms and risks are

Nanomedicine: Fundamentals, Synthesis, and Applications, First Edition. Edited by Yujun Song.
© 2025 WILEY-VCH GmbH. Published 2025 by WILEY-VCH GmbH.

detected early; however, if detected at late stages, the treatments are often difficult, costly, and yield modest improvements. Thus, it is necessary not only to develop novel therapeutic strategies but also to design accurate diagnostics to improve the treatment outcome and empower personalized wound management. While some clinical assessments of wound healing are established, such as visual inspection and histological examination, it is usually highly subjective and generally demands invasive procedures, which would probably introduce new wounds and greatly interfere with the assessment processes [16]. Moreover, such invasive procedures could greatly limit the real-time monitoring of wound progression, resulting in difficulties in making timely therapeutic decisions and adjusting the therapeutic strategy in a rational manner. Besides, the mainstays of therapy for these acute wounds and chronic nonhealing wounds have included sutures, staples, surgical debridement, antibiotics, growth factor supplementation, hyperbaric oxygen (HBO) therapy, and skin grafting. However, most of the wound therapeutics, especially those used in chronic nonhealing wounds, achieved only limited success to date, with a huge number of patients having to undergo amputation eventually [1, 5, 6]. Therefore, new and effective strategies for therapeutic and diagnostic management of cutaneous wounds would be valuable.

Owing to their unique physicochemical properties, nano-biomaterials-based solutions may hold the key to cope with various demanding requirements for cutaneous wound management where present conventional treatments and diagnostics are inadequate [17–19]. Furthermore, nano-biomaterials can also be equipped with microspheres, stem cells, scaffolds, or hydrogels for enhanced performance [20–24]. Recently, a variety of nano-biomaterials have been designed and investigated for their potential applications in cutaneous wound assessment and therapy, including but not limited to contrast agents, tissue adhesives, infection control, hemostasis, promoting angiogenesis, engineering the wound microenvironment, and delivering therapeutics [17–19]. The outstanding features enabled by nano-biomaterials make them highly attractive to address the challenges associated with management of cutaneous wounds. Taken together, nano-biomaterials-empowered therapeutic and diagnostic management of cutaneous wounds represents a paradigm shift toward next-generation wound care.

In this review, we will make efforts to conceptualize the design strategies and review key examples of the recent advances in nanomaterial-based approaches to improve management of cutaneous wounds (Figure 6.1). Also, it is intended to bridge the gaps between nano-biomaterials science and regenerative medicine, especially in terms of cutaneous wounds. We first outline the strategies, advances, and advantages of nano-biomaterials-based approaches for cutaneous wound therapy, either through their intrinsic activity or delivery of therapeutic cargoes. Following that, applications of nano-biomaterials for wound assessment are described. Finally, fundamental challenges facing the development of wound therapeutic and diagnostic nano-biomaterials and future opportunities in this emerging field are discussed. It is expected that by gaining a deeper insight into the patho-mechanism underlying dysregulated wound healing and development of

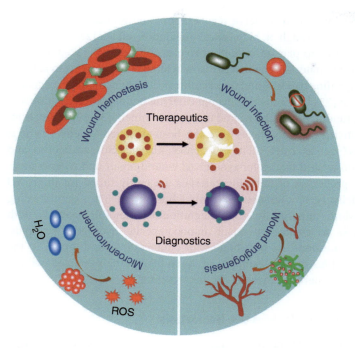

Figure 6.1 Overview of representative nano-biomaterial strategies for cutaneous wound management. Nano-biomaterial properties can be precisely engineered to promote or monitor the wound healing process.

effective and sophisticated nano-biomaterials, we will witness the blooming growth in this highly interdisciplinary field during the next decades.

6.2 Nano-biomaterials-based Therapeutics for Wound Healing

Continuous efforts from the past decade have generated a large number of nano-biomaterials with diversified capabilities for cutaneous wound repair and regeneration. Previously, nanomaterial-based wound therapeutics are mainly focused on infection control and delivering various therapeutic agents into wound beds. In recent years, a variety of innovative nanomaterial-based strategies have been proposed to overcome some of the challenges associated with conventional cutaneous wound therapeutics. These nanomaterial-based strategies not only can exert the therapeutic function of delivered cargoes but also have the intrinsic capability to modulate the hostile wound microenvironment, tune the gene/signal pathway, bond the separate wound edges, or provide biophysical cues, and thus significantly promote the repair and regeneration of cutaneous wound. For example, the aqueous solutions of Stöber silica or iron oxide nanoparticles exhibited robust adhesion properties, which can be used for rapid and strong closure of deep cutaneous wounds [25, 26]. Nanotopographical features in biomaterials are reported to modulate cell behaviors, improve nerve regeneration, and reduce scar formation in the wound tissues through a biophysical regulation effect [27–31]. Copper ions have been reported to play an important role in angiogenesis [32–34]. A few copper-based nano-biomaterials enabled the slow release of copper ions to activate the angiogenesis signaling pathway and thus significantly enhance angiogenesis and wound regeneration [35–37]. Some engineered catalytic nanozymes have attracted great attention for modulation of the wound microenvironment [38, 39]. Apart from the intrinsic bioactivities, nanoparticles have also been exploited to deliver various therapeutic agents (such as drugs, therapeutic gases, nucleic acids, antibiotics, and growth factors) into wound tissues for synergistic wound repair and regeneration. Furthermore, the properties of nano-biomaterials are easily tuned for improved bioactivity and delivery of therapeutic agents, thus increasing wound-healing efficiency [38]. More importantly, the integration of nano-biomaterials with polymeric networks to create synergistic nanocomposite hydrogels, scaffolds, and nanofibers provides endless possibilities for cutaneous wound repair. In the following sections, illustrative examples of nano-biomaterials-enabled wound therapeutics and the underlying mechanism are presented.

6.2.1 Nano-biomaterials for Effective Hemostasis

Blood is responsible for transporting oxygen and nutrients to tissues, removing metabolites, and providing immune surveillance, etc. [40]. Loss of blood from traumatic wounds is one of the greatest causes of morbidities and mortalities [41]. The normal wound-healing sequence including four partly overlapping stages: hemostasis, inflammation, proliferation, and tissue remodeling [3, 12]. The first hemostasis stage of wound repair encompasses a complex network of hemostatic events to prevent blood loss at the wound site. Although some hemostats have been developed and shown improvement in wound hemostasis, they remain largely deficient regarding several desired criteria such as their low portability and applicability in challenging situations [40, 41]. Thus, given the increasing prevalence of cutaneous wounds, intensive interdisciplinary research efforts have been focused on developing hemostatic agents for rapid and efficient hemostasis [40–44]. In this subsection, nano-biomaterials-based hemostatic technologies in terms of materials and mechanisms are provided and discussed.

Self-assembling nanofibers are attractive biomaterials with superior hemostatic activity for wound hemostasis applications [41, 45, 46]. Hammond et al. developed a self-assembling peptide nanofibers-based hemostatic dressing to accelerate hemostasis in cutaneous wounds [41]. In addition, the application of these nanofiber-based hemostatic dressings was found to promote efficient nanofiber clot formation with red blood cells (RBCs) [41]. The intercalated clay nanosheets with a high aspect ratio and large interfacial area that can facilitate chemical bonding or physical adsorption offer unique avenues for multifunctional adhesive materials. Inspired by the adhesion mechanism of mussels, a polydopamine-clay-polyacrylamide (PDA-clay-PAM) nanocomposite hydrogel was developed by Li et al. [47]. Owing to the presence of enough free catechol groups and nanoreinforcement effect, both durable and repeatable adhesiveness as well as excellent mechanical properties were achieved for full-thickness skin wound repair [47].

On the other hand, nanoparticles with tunable biophysicochemical properties offer great opportunities for designing and creation of hemostats [25, 26, 38, 48]. Highly charged nanoparticles, such as synthetic silicate nanoplatelets and polyethylenemine (PEI) nanoparticles, have been proposed to induce blood coagulation either standalone or in combination with other bulk materials by concentrating clotting factors [43, 48, 49]. For example, Yang et al. have reported on the hemostatic capability of a kaolinite nanoclay composite (α-Fe_2O_3-kaolin$_{KAc}$) with high surface areas to render hemostasis in cutaneous wounds [42]. This nanocomposite material acts through efficient absorption of the fluid in blood, activation of blood platelets, and induction of the coagulation cascade by kaolinite and the aggregation of RBCs induced by α-Fe_2O_3 (Figure 6.2a–c) [42]. Positive nanohemostats formed by cholic-acid-mediated self-assembly of PEI were found to be able to effectively induce aggregation and activation of platelets in different bleeding models [48]. Utilizing a similar concept, Olsen's group developed injectable shear-thinning nanocomposite hydrogels composed of synthetic silicate nanoplatelets and gelatin that can significantly increase in vitro blood clotting and in vivo coagulation (Figure 6.2d) [43]. The incorporation of charged silicate nanoplatelets into gelatin matrix significantly improved the physiological hemostatic performance (Figure 6.2e,f) [43].

Recently, Leibler et al. proposed a novel particle nanobridging concept, adhesion by aqueous nanoparticle solutions, for wound closure and hemostasis [25, 26]. Strong, rapid adhesion between biological tissues can be achieved at room temperature by spreading a droplet of aqueous nanoparticle solution on one tissue's surface and then bringing the other surface into contact with it, where the nanoparticles can adsorb onto tissue's surface and act as connectors between them (Figure 6.3a) [25, 26]. Topical application of aqueous solution of Stöber silica or iron oxide nanoparticles yields robust and rapid hemostasis and wound healing even in the presence of blood flow, which is hard to achieve by polymer tissue adhesives (Figure 6.3b) [26]. From a material's perspective, this particle nanobridging concept can also be extended to the preparation of multifunctional nanomaterials by rational choices of sizes, forms, surface chemistry, and assembling. Ling's group developed a highly versatile ROS-scavenging tissue adhesive nanocomposite by immobilizing ultrasmall ROS-scavenging ceria nanocrystals onto the surface of nanobridging silica nanoparticles, which not only possesses strong tissue adhesion strength but also significantly limits the ROS exacerbation-mediated deleterious effects. More importantly, significantly accelerated wound closure and improved quality of the healed skin featured by skin appendage morphogenesis and decreased scar formation were achieved by the ROS-scavenging tissue adhesive nanocomposite due to the desirable synergy (Figure 6.3c,d) [38]. Therefore, equipped with favorable simplicity, rapidity, and robustness nanobridging effect, these multifunctional nanobridging materials could provide useful options for clinical applications and regenerative medicine.

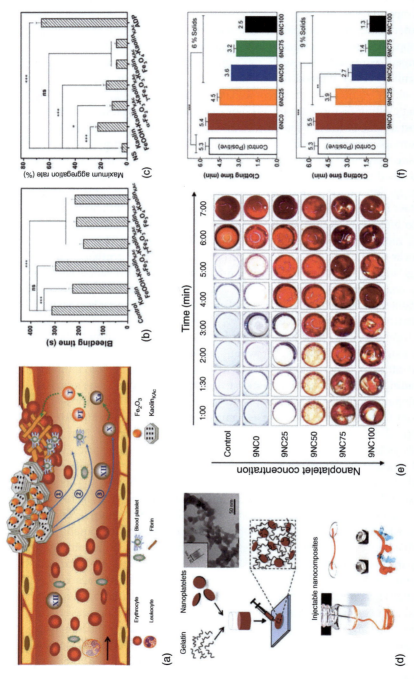

Figure 6.2 (a) Illustration of the presumed action mechanism of the α-Fe$_2$O$_3$-kaolin$_{KAc}$ composite in wound hemostasis. (b) Bleeding time of kaolin, FeOOH-kaolin$_{KAc}$, α-Fe$_2$O$_3$-kaolin$_{KAc}$, γ-Fe$_2$O$_3$-kaolin$_{KAc}$, and Fe$_3$O$_4$-kaolin$_{KAc}$-treated wounds or left untreated using the mouse tail vein incision model. (c) Aggregation of platelet induced by kaolin, FeOOH-kaolin$_{KAc}$, α-Fe$_2$O$_3$-kaolin$_{KAc}$, γ-Fe$_2$O$_3$-kaolin$_{KAc}$, Fe$_3$O$_4$-kaolin$_{KAc}$, normal saline, and ADP represented as the maximal percentage aggregation of sodium citrate-treated whole blood. Source: (a–c) Reproduced with permission from Ref. [42], © 2017, John Wiley & Sons. (d) Schematic illustration of the preparation of the nanocomposite. (e) Clot formation as a function of the incubation time and composition of nanocomposite. (f) Quantitative clot formation times for 6% and 9% nanocomposite hydrogels. Source: (d–f) Reproduced with permission from Ref. [43], © 2014, American Chemical Society.

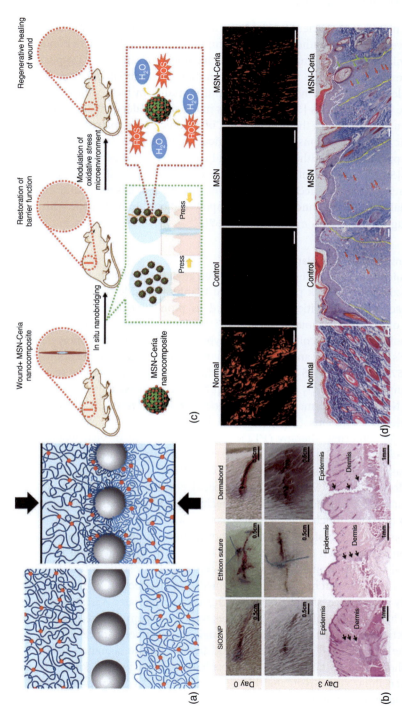

Figure 6.3 (a) Schematic illustration of the nanobridging concept for gluing swollen polymer networks together using particles. Network chains are adsorbed on nanoparticles surface and anchor particles to gel pieces and thus nanoparticles can act as connectors between gel surfaces. Source: (a) Reproduced with permission from Ref. [25], © 2013, Nature Publishing Group. (b) In vivo comparison of wound repair enabled by SiO_2 NP nanobridging, by suturing, and by cyanoacrylate glue of full-thickness dorsal skin injury in an incisional wound model. Source: (b) Reproduced with permission from Ref. [26], © 2014, John Wiley & Sons. (c) Schematic illustration of efficient cutaneous wound repair and tissue regeneration by using MSN-Ceria as a ROS-scavenging tissue nanoadhesive. (d) Representative images of sections stained with Picrosirius red and Masson's trichrome from unwounded normal skin and wounded skin treated as indicated at day 12 post-wounding. Scale bar = 100 μm. Source: (c and d) Reproduced with permission from Ref. [38], © 2018, Elsevier.

6.2.2 Antibacterial Nano-biomaterials

The intact skin provides an efficient barrier toward surrounding invading microorganisms. However, injury or wounding causes disruption in the physical barrier and can introduce microorganisms, thus increasing the likelihood of infections. Despite the intensive research and wide application of antibiotics, bacterial infection is still by far the leading cause of hospitalization and mortality in burn wounds or chronic diabetic wound patients [50, 51]. There is increasing evidence that wound infection plays a crucial role in the development of chronic and impaired wound healing. Chronic wound infections and acute surgical site infections impose a huge burden on the patient and healthcare system and can cause increased morbidity and mortality [52]. Therefore, efficient infection control of wound tissue was considered a prerequisite for successful wound healing. Recent advancements in nano-biomaterials and nanotechnologies have empowered our ability to cope with these demanding wound infection issues [51, 53–57].

6.2.2.1 Inorganic Antibacterial Nano-biomaterials

Inorganic antimicrobial nanomaterials, such as silver, gold, copper oxide, molybdenum disulfide, zinc oxide, black phosphorus, and titanium dioxide nanoparticles with unique bactericidal activities have emerged as promising alternatives for the treatment of cutaneous wound infections [51, 53–55, 57–60]. Zhao et al. developed a polyethylene glycol-modified molybdenum disulfide nanoflowers (PEG-MoS$_2$ NFs) for efficient and synergetic wound antibacterial applications (Figure 6.4a) [57]. The PEG-MoS$_2$ NFs combine the peroxidase-like activity (catalyze the decomposition of H$_2$O$_2$ to generate highly toxic hydroxyl radicals) with near-infrared (NIR) photothermal effect, rendering antibiotic-resistant bacteria more vulnerable to treatments (Figure 6.4b–e) [57]. Based on the exogenous ·OH-enhanced photothermal therapy (PTT), rapid and effective in vivo wound disinfection was achieved by this synergistic antibacterial system PEG-MoS$_2$ NFs [57]. Ultrasmall copper sulfide nanodots (CuS NDs) were also developed to treat multidrug-resistant (MDR) bacteria-infected chronic wounds based on the PTT mechanism [53]. In addition to PTT mechanism, photodynamic therapy (PDT) has also attracted much attention in diagnosis and treatment of bacterial infection and has been successfully applied as a novel therapeutic modality to treat wound infection [58, 61]. Wu et al. fabricated a hybrid hydrogel embedded with two-dimensional (2D) black phosphorus nanosheets via electrostatic interaction between the BPs and chitosan hydrogel for PDT-mediated disinfection [58]. Owing to its capability to produce singlet oxygen (^{1}O$_2$) under simulated visible light irradiation, the hybrid hydrogel also demonstrates robust, efficient, and repeatable antibacterial activities against *Escherichia coli* and *Staphylococcus aureus* [58]. Importantly, the embedded BPs can also synergistically activate signaling pathways of phosphoinositide 3-kinase (PI3K), phosphorylation of protein kinase B (Akt), and extracellular signal-regulated kinases 1 and 2 (ERK1/2) for enhanced proliferation and differentiation to speed up the bacteria-accompanied wound healing [58].

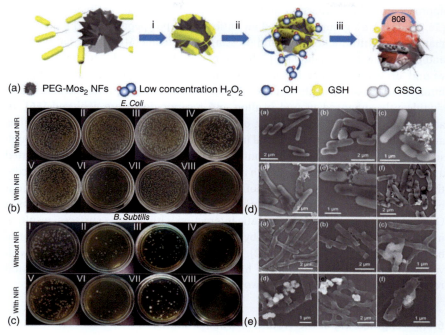

Figure 6.4 (a) Schematic illustration of PEG-MoS$_2$ construction which acts as a combined system for peroxidase catalyst-photothermal synergistic combating of bacteria. (b) Amp^r E. coli and (c) B. subtilis after being treated with (I) PBS, (II) MoS$_2$, (III) H$_2$O$_2$, (IV) MoS$_2$ + H$_2$O$_2$, (V) PBS + NIR, (VI) MoS$_2$ + NIR, (VII) H$_2$O$_2$ + NIR, and (VIII) MoS$_2$ + H$_2$O$_2$ + NIR. Concentration: MoS$_2$ 100 μg ml^{-1}, H$_2$O$_2$ 100 μM. Field emission scanning electron microscopy images (FE-SEM) of (d) Amp^r E. coli and (e) B. subtilis treated with (a) PBS, (b) H$_2$O$_2$ (100 μM), (c) MoS$_2$ (100 μg ml^{-1}), (d) MoS$_2$ + NIR, (e) MoS$_2$ + H$_2$O$_2$, and (f) MoS$_2$ + H$_2$O$_2$ + NIR. Irradiation time: 10 min. Source: Reproduced with permission from Ref. [57], © 2016, American Chemical Society.

6.2.2.2 Organic Antibacterial Nano-biomaterials

In addition, organic materials have also been utilized for constructing antibacterial nano-biomaterials, which possess special characteristics that make them promising antibacterial nano-biomaterials [56, 62–66]. Graphene oxide (GO) is a promising material for the development of antimicrobial nano-biomaterials due to its unique physicochemical properties such as sharp GO edges, destructive extraction of lipid molecules, photothermal ablation, and electron transfer-induced oxidation [66–68]. Wu's group developed a hybrid nanosheet of g-C$_3$N$_4$-Zn^{2+}@GO (SCN-Zn^{2+}@GO) by assembling Zn^{2+} doped sheet-like g-C$_3$N$_4$ with GO via electrostatic bonding and π–π stacking interactions to assist infected-wound healing under short-time exposure to 660 and 808 nm light [69]. Furthermore, the gene expression levels of matrix metalloproteinase-2 (MMP-2), collagen, and interleukin β are modulated by GO and released Zn^{2+} to accelerate the healing speed of infected wounds [69]. Antimicrobial peptides such as poly-lysine could disrupt the cellular membrane structure and affect physiological metabolism of microbes, which is a promising approach to inhibit infection and reduce the likelihood of antibiotic resistance [56, 70, 71]. Ma's group developed an antibacterial polypeptide-based nanofibrous matrix as a multifunctional wound dressing to overcome MDR bacterial infection and promote wound healing [56]. The antibacterial polypeptide-based nanofibrous wound dressing can provide a tensile elastomeric modulus similar to that of human skin tissue and prevent the MDR bacteria-derived wound infection, thus significantly enhanced wound healing and regeneration were achieved [56]. In addition to killing bacteria directly, organic nano-biomaterials target the biofilms and also demonstrate great potential in reversing MDR bacterial infection. Through self-assembling of amphiphilic cationic block copolymers, Chan-Park et al. reported a novel polymeric nanomaterial with a thin polysaccharide shell and a cationic core that can effectively eliminate the biofilms of MDR bacteria [62]. Although without bacteria-killing capacity, the polymeric nanomaterial can diffuse into the biofilms of Gram-positive bacteria via electrostatic interaction with bacteria surfaces, resulting in gradual elimination of biofilms by competitively weakening the attachment of bacteria to surrounding biofilms [62]. Taken together, these nano-biomaterials with various antimicrobial mechanisms have great potential in enhancing repair and regeneration of bacteria-colonized wounds.

6.2.3 Engineering of the Wound Microenvironment

The ongoing, bidirectional interaction among cells and their ever-evolving surrounding microenvironment plays pivotal roles in regulating tissue regenerative responses to injury [72]. Dysregulated wound microenvironments such as overexpression of MMPs, excessive reactive oxygen species (ROS) accumulation, and production of inflammatory mediators have been demonstrated to play essential roles in the pathological cutaneous wound healing process. Current researches have primarily focused on delivering therapeutic growth factors, nucleic acids, or stem cells into the wound bed. However, success with these approaches has been moderate because of the premature degradation of therapeutic biomacromolecules and poor cell survival in the harsh wound microenvironment. Therefore, efficient reprogramming of the harsh wound microenvironments can open a window for improved wound repair and regeneration. Interestingly, several nanoparticles exhibit the capability to modulate the microenvironment of injured tissues by scavenging toxic ROS, downregulating MMPs, and inflammatory mediators' expression. In this subsection, we seek to lay out the therapeutic strategies that target the hostile wound microenvironment and the ways in which nano-biomaterials are being designed and applied to remodel the wound microenvironment. We detail the nano-biomaterials on the cutting edge and how their interactions with the hostile wound microenvironment make them desirable for enhanced cutaneous wound healing. Several novel nano-biomaterials-based strategies have started aiming to modulate the hostile wound microenvironment. The latter, rich in toxic ROS, inflammatory mediators, and MMPs, are responsible for poor efficacy of some promising therapeutics and represent a crucial issue in wound healing abnormalities.

6.2.3.1 Redox Modulation of the Wound Microenvironment

Although the pathological mechanisms involved in the impaired cutaneous wound healing response are still poorly understood, excessive ROS accumulation in the wound microenvironment is reported to play a critical role in impaired cutaneous wound healing. The excessive ROS would cause oxidative damage to essential growth factors, nucleic acids, and cells in the wound bed. Nanozymes are nano-biomaterials with enzyme-mimetic properties, which have been demonstrated to address the drawbacks of natural enzymes and conventional artificial enzymes [73]. As a representative nanozyme, ceria nanocrystals have recently raised great expectations for the treatment of oxidative-stress-associated diseases due to their facile synthesis, good biocompatibility, excellent multiple antioxidant enzyme-mimetic activity, and recyclable catalytic performance [74, 75]. It is noteworthy that Mattson et al. first demonstrated that topical application of water-soluble cerium oxide nanoparticles accelerates the healing of full-thickness dermal wounds by reducing oxidative damage to cellular membranes and proteins [39]. This finding raises the possibility that topical treatment of wounds with antioxidant nano-biomaterials greatly promotes wound-healing process by alleviating oxidative damage responses in the wound microenvironment [39].

To increase the therapeutic efficacy of antioxidant nano-biomaterials, a major focus is to find appropriate combinations to act synergistically to trigger wound healing. A recent study by Ling et al. demonstrated a "seed-and-soil" concept that by simultaneously reshaping the oxidative wound microenvironment into a pro-regenerative one (the "soil") and providing pro-angiogenic miRNA cues (the "seed"), remarkably accelerated wound closure and enhanced quality of the healed wound as featured by highly ordered alignment of collagen fiber, skin appendage morphogenesis, and functional blood vessel growth was achieved [76]. By using a miRNA-impregnated, redox-modulatory ceria nanozyme-reinforced self-protecting hydrogel (PCN-miR/Col), they found that not only the hostile oxidative wound microenvironment was reshaped into a friendly one but also the integrity of encapsulated pro-angiogenic miRNA in the oxidative microenvironment was largely preserved (Figure 6.5) [76]. Altogether, these studies suggest that acting alone and in combination, antioxidant nano-biomaterials can protect either the endogenous stem cells, growth factors, and nucleic acids or the externally administered ones in the harsh oxidative wound microenvironment.

6.2.3.2 Regulation of Microenvironmental MMP Activity

MMPs are present in both acute and chronic cutaneous wounds. The proteolytic microenvironment induced by MMP overexpression has also been suggested to be responsible for a wound's chronicity, which leads to significant extracellular matrix (ECM) destruction, impaired epithelial closure of the wound, and increased susceptibility of open wounds to infection. Therefore, selective inhibition of MMPs in the wound microenvironment is considered a promising way in achieving improved wound healing. For example, Yoo et al. fabricated an MMP-responsive nanofibrous matrix to control the release of MMP-2 small interfering RNA (siRNA) in response to the high concentration of MMPs in diabetic ulcers [77]. The siRNA-incorporated

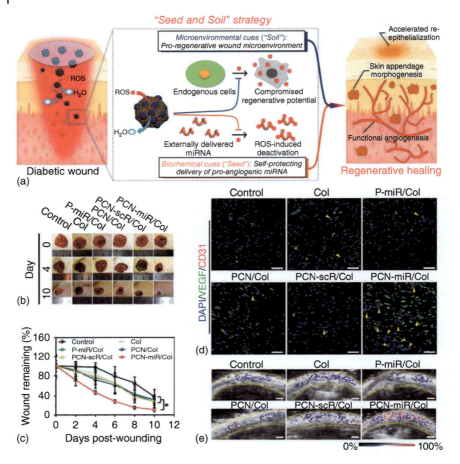

Figure 6.5 (a) Schematic illustration of the simultaneous self-protecting delivery of proangiogenic miRNA cues and creation of pro-regenerative wound microenvironment to drive highly efficient functional angiogenesis and regenerative diabetic wound healing by PCN-miR/Col nanocomposite hydrogel. Wound images at days 0, 4, and 10 after the indicated treatment (b) and quantification of wound closure, (c) as a percentage of the initial wound area, $n = 5$. (d) Representative confocal images of VEGF and CD31 double-stained sections at day 28. Scale bars, 50 μm. (e) Photoacoustic images of oxygenated hemoglobin from each group at day 28. Scale bars, 1 mm. Source: (b–e) Reproduced with permission from Ref. [76], © 2019, American Chemical Society.

nanofibrous meshes dramatically increased the neo-collagen accumulation at the wound sites and the wound recovery rates of diabetic ulcers with reduced side effects [77]. Similarly, a 4PEG-siRNA/LPEI cluster with higher resistance against RNase attack and enhanced transfection efficiency of MMP-2 siRNA in response to MMPs was developed [78]. Diabetic ulcers treated with the 4PEG-siRNA/LPEI clusters displayed an accelerated wound closure rate with a lower expression level of MMP-2 [78]. In addition to MMP-2, overexpression of matrix metalloproteinase-9 (MMP-9) is also involved in refractory wound healing process [79, 80]. Hammond et al. report the development of a nanometer-scale coating in a commercially available nylon bandage using the layer-by-layer (LbL) technology to sustain the delivery of MMP-9 siRNA for up to two weeks in the wound bed (Figure 6.6) [79]. This MMP-9 siRNA-incorporated nano-coating successfully reduces the MMP-9 level within the wound tissues, significantly accelerating the wound closure rate and improving the quality of healed wound [79].

6.2.3.3 Targeting the Pro-inflammatory Mediators

Excessive inflammatory responses play crucial roles in impaired wound healing by following instructions from the upregulated pro-inflammatory mediators in wound microenvironment. In addition, inflammation in wound tissue has also been shown to result in increased scarring. Furthermore, persistent inflammation in the impaired nonhealing wound greatly predisposes wound tissue to tumor progression. Nano-biomaterials-based therapeutics aimed to target wound microenvironmental hyper-inflammation may promote impaired wound healing by abolishing these deleterious effectors. There has been considerable progress made in developing novel nano-biomaterials for promoting wound healing processes. For example, Wang et al. demonstrated a novel pH-controllable H_2S-releasing nanofibrous coating to inhibit excessive inflammation responses and accelerate wound regeneration [81]. Schilling et al. presented a curcumin/gelatin-blended nanofibrous mats by electrospinning process to overcome the shortcomings of curcumin including hydrophobicity, instability, poor absorption, and rapid systemic elimination for enhanced wound healing [82]. Topical application of curcumin/gelatin-blended nanofibrous mats leads to sustained release of bioactive curcumin, which in turn greatly decreases the expression of pro-inflammatory mediator monocyte chemoattractant protein-1 (MCP-1) [82]. Owing to the lasting inhibition of the inflammatory response during the wound healing process, the wound microenvironment turns out to be friendly for wound healing process with efficient re-epithelialization, granulation tissue formation, as well as collagen deposition achieved at the late stage of wound healing [82].

6.2.4 Nano-biomaterials-enabled Biophysical Regulation of Wound Healing

In addition to the well-documented effects of aforementioned microenvironmental biochemical factors (ROS, MMPs, pro-inflammatory mediators), the influence of biophysical cues such as topography, stiffness, elasticity, porosity, nano-scale patterning, adhesiveness, and electrical stimulation on cutaneous wound healing has also been increasingly recognized [83]. Cells in the wound tissue may sense and respond to these biophysical cues through different signaling mechanisms, which are responsible for regulating cell fate [31]. Nano-biomaterials equipped with particular biophysical cues have emerged as an attractive approach in modulation cutaneous wound healing process, which greatly influences cellular proliferation, migration, organization, and fate by stimulating changes in cellular form and composition [84, 85]. Hence, in the following sections, we will elaborate on the biophysical regulation of wound healing enabled by nano-biomaterials.

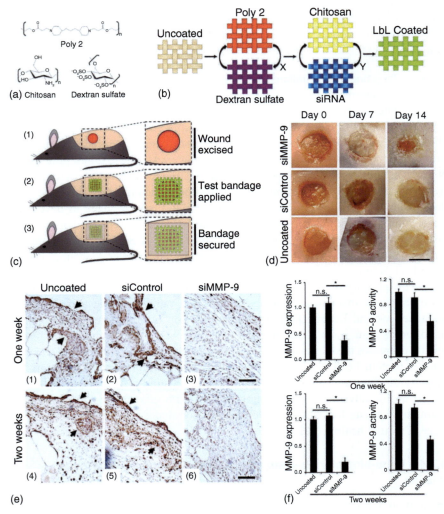

Figure 6.6 (a) Chemical structures of polymers utilized in this study. (b) Schematic illustration of the fabrication process and hierarchical structure of LbL films with a nanocoating. (c) Schematic illustration of how bandages are applied to full-thickness excisional wounds. (d) Digital images of wounds following surgery (0 d), and after 7 or 14 d of various treatments. Scale bars, 5 mm. (e) Immunohistochemical (IHC) staining of MMP-9 within the wound treated with uncoated ((1) and (4)), siControl ((2) and (5)), and siMMP-9 ((3) and (6)) bandage. Scale bars 50 μm. (f) Quantitative MMP-9 activity and real-time PCR analysis of MMP-9 expression relative to the housekeeping gene β-actin after one week or 2 weeks of treatment. Source: Reproduced with permission from Ref. [79], © 2015, John Wiley & Sons.

6.2.4.1 Surface Nanotopographical Features

Human tissues including skin tissue are complex ensembles of multiple cell types organized in well-defined 3D structures of ECM, which is hierarchical from nano- to microscale [86]. Inspired by these natural design strategies, nanotopography-guided approaches for tissue repair and regeneration have been intensively investigated over the last several decades [27, 86–89]. Particularly, the importance of nanotopography in the guidance of skin cell migration, cell division, and organization of ECM is becoming increasingly recognized and has tremendous potential for cutaneous wound repair and regeneration [90, 91]. For example, Suh et al. reported the fabrication of an ECM scaffold with uniformly spaced nanogrooved surfaces for cutaneous wound healing [90]. Benefited from the nanotopographical cues similar to the natural organization of collagen fibers, the migration speed, cell division, and ECM production were greatly improved without supplementation of growth factors [90]. In addition to skin cell migration and division, the nanotopographical cues also demonstrated potential to promote neurite outgrowth, a highly desired property since neuropathy plays a crucial role in the development and progression of chronic diabetic wounds [91, 92]. Li et al. developed aligned nanofibrous scaffolds by immobilizing ECM protein and growth factor onto nanofibers, which significantly induced neurite outgrowth and enhanced skin cell migration during wound healing in comparison to randomly oriented nanofibers.

In addition to regulating the somatic cell's behaviors, the cell–nanoarchitecture interactions can also drive specific stem cell differentiation for tissue regeneration [93]. Yim et al. developed nanogratings with 250 nm line width on polydimethylsiloxane (PDMS) and found that nanotopography-induced human mesenchymal stem cell (hMSC) displayed an upregulation of neurogenic and myogenic differentiation markers (Figure 6.7a) [93]. The in vitro investigations clearly revealed the superiority of the nanopatterned substrates compared to the unpatterned ones (Figure 6.7b,c). As shown in Figure 6.7d, further mechanistic study revealed that this regulation was dependent upon the integrin-activated focal adhesion kinase (FAK) pathway. Furthermore, Fu et al. demonstrated that nanotopography could provide potent regulatory signals to regulate human embryonic stem cell (hESC) behaviors, including cell morphology, adhesion, proliferation, clonal expansion, and self-renewal [89]. The nanotopography-induced stem cell modulation may open a new avenue for the design and fabrication of stem cell transplantation technologies with nanotopographical features as cutaneous wound therapeutics.

6.2.4.2 Mechanical Cues

It has been widely accepted that mechanical parameters, such as stiffness, elasticity, and porosity are essential design variables in the construction of tissue engineering biomaterials [94, 95]. These mechanical cues are converted to various biochemical responses that may affect cell fate and behaviors via multiple intracellular pathways [96]. Besides, the mechanical match between the host and implanted foreign therapeutics is one of the perquisites for successful biomaterials-assisted tissue engineering. Changes in the mechanical properties of dermis during skin aging or

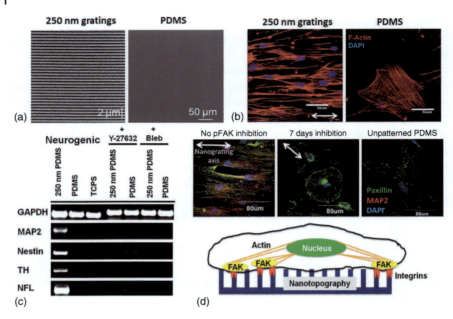

Figure 6.7 (a) SEM images of 250 nm gratings and unpatterned polydimethylsiloxane (PDMS) substrates and (b) fluorescently labeled hMSCs on these substrates for F-actin (red) and DAPI (blue). (c) RT-PCR analysis of neuronal differentiation markers microtubule-associated protein 2 (MAP2), nestin, tyroxine hydroxylase (TH), and neurofilament-light (NFL) in hMSCs cultured on nanogratings with indicated inhibitors. (d) The nanotopography-induced cellular differentiation was mediated by FAK phosphorylation pathway. Source: Reproduced with permission from Ref. [93], © 2013, American Chemical Society.

tissue remodeling process greatly affect the activity of resident skin cells [97]. Thus, nano-biomaterials with dynamically adjustable mechanical properties present an exciting opportunity to drive the wound repair and regeneration process. To enhance the efficacy in wound repair and regeneration, various mechanical-tunable nano-biomaterials were fabricated and investigated [98, 99]. Nanoclay has become an attractive class of nanomaterials owing to its high drug-loading capacity, aqueous stability, and enhanced biointeractions [47]. Hassan et al. found that mechanical properties of nanocomposite hydrogel wound dressing were significantly improved by adding nanoclay into a polymeric matrix to meet the essential requirements for wound dressing [100].

By taking advantage of mechanical cues, the stiffness of nano-biomaterials is suggested to regulate cellular behavior including plasticity of macrophages during the wound healing process [101]. For example, human macrophages are demonstrated to sensitively respond to stiffer 3D nano-fibrillar matrices in terms of polarization toward anti-inflammatory wound healing phenotypes as determined by protein and gene expression of inflammatory cytokines (Figure 6.8) [101]. In addition to stiffness, the porous nature of nano-biomaterials has also been explored for wound healing applications [102, 103]. Faster wound regeneration and lower immune

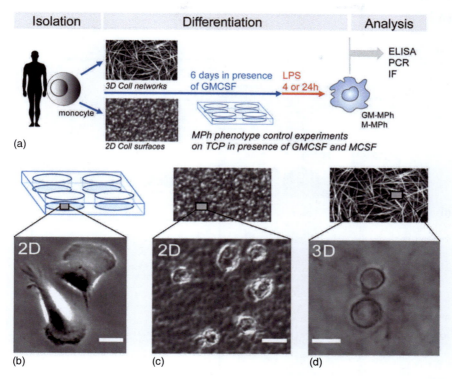

Figure 6.8 (a) Experimental setup to investigate the generation of an inflammatory or a wound-healing MPh phenotype in the presence of various cell culture substrates. Representative microscopy images showing the strong dependence of cell morphology on cell culture substrates being (b) tissue culture plastics (TCPs), (c) dense 2D Coll matrices, and (d) 3D Coll networks (Scale bar: 20 μm). Source: Reproduced with permission from Ref. [101], © 2017, John Wiley & Sons.

response were achieved by microporous hydrogel than nonporous hydrogels [103]. In addition, Ma et al. developed a macroporous nanofibrous scaffold with multilevel porous structures for cutaneous wound healing [102]. The well-interconnected multilevel pore structures were critical for cellular migration and complex cellular network formation within the scaffold [102, 103]. Taken together, these findings suggest a strong influence of porosity and stiffness of nano-biomaterials on cell fate and behaviors during wound healing process and emphasize the need for construction of nano-biomaterials mechanically mimicking the in vivo context.

6.2.4.3 Bio-electrical Stimulation

Various studies have demonstrated that endogenous electrical field (EF) actively participates in cutaneous wound healing by stimulating cell migration, cell proliferation, and cell differentiation, and improving their regenerative potentials [83, 104–107]. Therefore, along with the booming development of nano-biomaterials, bioelectric regulation is regarded as a promising and practical adjunctive treatment for cutaneous wound healing. Kim et al. developed a

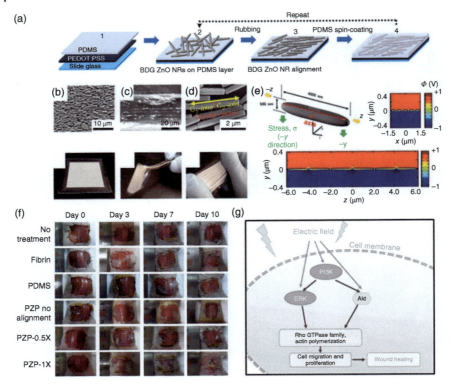

Figure 6.9 (a) The fabrication process of bidirectionally grown zinc oxide nanorod (BDG ZnO NR)-based piezoelectric patch (PZP). SEM image of (b) one-directionally aligned BDG ZnO NRs on the patch after rubbing, (c) cross-sectional view of a nine-layered PZP, and (d) BDG ZnO NRs showing a crystal center (red) and two different growth directions (yellow). (e) Schematic illustration demonstrating BDG ZnO NR and the calculated piezoelectric potential (Φ) corresponding to the BDG ZnO NR under external stress (σ) exerted in $-y$ direction (green arrow). (f) Representative skin wound images at 0, 3, 7, and 10 d post-treatment. (g) Schematic illustration demonstrating the in vivo EF-induced intracellular signaling pathways for cell migration and proliferation during wound healing. Source: Reproduced with permission from Ref. [108], © 2016, John Wiley & Sons.

one-directionally aligned zinc oxide nanorod-based piezoelectric dermal patch to promote wound healing (Figure 6.9a–e) [108]. After being applied on the wound bed of animal models, the piezoelectric dermal patch generated an EF upon animal motion; and induced piezoelectric potentials at the wound site significantly promoted the wound healing process via enhanced cellular metabolism, migration, and protein expression (Figure 6.9f) [108]. As shown in Figure 6.9g, mechanistic study revealed that this piezoelectric dermal patch-mediated bio-electrical stimulation of wound healing was dependent on the phosphorylation of protein kinase B (Akt), phosphoinositide 3-kinase (PI3K), and extracellular signal-regulated kinases 1 and 2 (ERK1/2) pathway [108]. Wang et al. report an efficient self-powered electric dressing for enhanced skin wound healing by generating an alternating discrete electric field via a wearable nanogenerator that can convert mechanical energy from skin movements into electricity [107]. The accelerated skin wound healing could be attributed to electric field-facilitated cellular migration, proliferation, and differentiation [107]. Similarly, Ma et al. reported an injectable carbon nanotube (CNT)-based conductive cryogel to transmit endogenous electrical signals in wounded tissue for improved wound healing [44].

6.2.5 Angiogenic Nano-biomaterials

Wounded tissues need to reestablish vasculature networks to satisfy their need for oxygen and nutrients as well as accomplish other functions through the growth of new blood capillaries from pre-existing capillaries in angiogenesis process [109, 110]. Angiogenic programming in wounded tissue is a multidimensional process regulated by a variety of skin cells and immune cells as well as their bioactive products such as cytokines, growth factors, and secreted microvesicles [110–113]. While the etiology of chronic cutaneous wounds is multifaceted, the progression to a nonhealing phenotype is closely related to poor vascular networks [114–116]. Despite significant advances that have been made in improving wound angiogenesis, current therapeutic options remain limited and ineffective and failed to translate into meaningful clinical improvement [117–120]. Nano-biomaterials that exhibit intrinsic angiogenic properties or are employed as delivery vehicles for angiogenic agents are receiving increasing interest in addressing the defect angiogenesis issues during wound healing process [32, 35, 36, 121–125]. Herein, this subsection discusses recent progress of angiogenic nano-biomaterials that may offer promise to improved wound healing outcomes.

6.2.5.1 Nano-biomaterials as Delivery Vehicles for Angiogenic Therapeutics

Clinical translation of angiogenic growth factor- and nucleic acid-based therapies has been greatly hampered by delivery barriers, including susceptibility to premature degradation, membrane impermeability, and the inability to achieve sustained delivery [122, 125–127]. Numerous nanotechnological strategies have been designed to overcome these issues and enhance the pharmaceutical properties of growth factor- and nucleic acid-based angiogenic therapies for enhanced wound angiogenesis [125, 126]. Duvall et al. developed a biodegradable tissue regenerative nanocomposite scaffold for temporally controlled delivery of angiogenic siRNA [126]. By silencing prolyl hydroxylase domain protein 2 (PHD2) through externally delivered siRNA, the enhanced expression of pro-angiogenic genes controlled by proangiogenic transcription factor hypoxia-inducible factor-1α (HIF-1α) and scaffold vascularization in vivo were achieved [126]. Similarly, a vascular endothelial growth factor (VEGF) pDNA/PEI polyplexes-loaded porous hyaluronic acid hydrogel was demonstrated to significantly promote angiogenesis and accelerate wound closure [128]. In addition to these nucleic acid-based therapeutics, various angiogenic growth factors incorporated in nano-biomaterials also demonstrate efficient angiogenesis response by overcoming issues such as rapid leakage and short half-life of growth factors, which are valuable for wound repair and regeneration [123, 125].

6.2.5.2 Nano-biomaterials as Intrinsic Angiogenic Agents

In addition to being exploited as delivery vehicles, nano-biomaterials with intrinsic angiogenic properties also attracted intense interest in designing wound therapeutics [35, 36, 121]. Copper is proposed to play an important role in stimulating angiogenesis processes such as stabilization of pro-angiogenic HIF-1α, induction of VEGF, and upregulation of the activity of copper-dependent enzymes [32–36]. Ameer et al. developed a folic acid-modified copper metal-organic framework (MOF) nanoparticle (F-HKUST-1) for sustained release of proangiogenic Cu^{2+}, which resulted in prominent angiogenesis and improved wound healing response [35]. The functionalization of folic acid into HKUST-1 not only enabled the slow release of copper ions but also reduced the cytotoxicity in comparison with applications of copper salts or oxides [35]. Owing to their intrinsic NIR region absorption and excellent heat generation ability, copper nanoparticles could also be exploited as photothermal agents [36, 129]. Based on its multifunctionality, Wu et al. fabricated a Cu_2S nanoparticles-embedded micropatterned nanocomposite membrane with bifunctional properties for skin tumor therapy and simultaneous wound angiogenesis as well as wound regeneration [36]. With uniformly incorporated Cu_2S nanoparticles, the membranes demonstrated excellent killing efficiency of skin tumor cells under NIR irradiation and significantly stimulated angiogenesis and wound regeneration in vivo [36].

6.3 Nano-biomaterials for Imaging and Monitoring of Cutaneous Wounds

In addition to wound therapeutics, diagnostic and theranostic modalities are increasingly recognized as complement components for both acute and chronic wound management to reduce hospitalization time, prevent amputations, and improve the understanding of pathological processes underlying impaired healing. Cutaneous wound healing involves a complex network of biochemical events and some important parameters such as wound infection, oxygen level, and enzyme expression have traditionally been monitored with visual inspection or histological examination. However, the interpretation of visual inspection is inherently subjective and often yields inconsistent information. Although histological examinations can overcome some of these limitations, the procedure is usually invasive, introduces a new wound that interferes with the wound repair, and cannot provide continuous evaluation of the wound healing process. Thanks to the development of precise medicine and ever-increasing affordability and power of nanoscience and nanotechnologies, various nano-biomaterials-based imaging and diagnostic modalities have been devised to detect specific biological parameters and markers in the wounded tissue. By leveraging their inherent advantages such as unique magnetic and optical properties, facile surface functionalization, these nano-biomaterials-based imaging and diagnostic modalities not only can overcome challenges faced by traditional visual inspection or histological examinations but may also potentially redefine the management of cutaneous wounds [130–135].

This subsection discusses the state-of-the-art in wound assessment and monitoring of markers and parameters that are associated with wound repair and infection by utilizing nano-biomaterials-based sensors or theranostics.

6.3.1 Wound Infection Monitoring

Despite extensive research and various available wound therapeutics, cutaneous wound management is still extremely difficult and time-consuming owing to its susceptibility to infection. For example, chronic diabetic foot ulcer is highly susceptible to bacterial infection and can result in long-term nonhealing phenotype, which is a major cause of mortality and limb amputation in diabetic patients. Therefore, there is an urgent need for early detection and monitoring of wound infection with high specificity and efficiency. Timely treatment based on early detection can efficiently prevent wound deterioration and decrease spreading of the infections and inhibit evolution of antimicrobial-resistant strains. Accordingly, great efforts have been made to detect the wound infection status by using various nanoprobes or theranostics [131–135].

6.3.1.1 Nanoprobes for Wound Infection Monitoring

Bacterial biofilms have been implicated in both impaired wound healing and occurrence of chronic inflammation [136]. Rotello et al. developed a pH-responsive nanoparticle embedded with transition metal catalysts for rapid and effective imaging of biofilms through bioorthogonal activation of imaging agents inside the acidic microenvironment of biofilms [135]. Furthermore, nanoparticle-based multimodal imaging has also been intensively investigated for infection monitoring [137, 138]. By utilizing upconversion nanoparticles (UCNPs) and gold yolk-shell nanoparticles as the building blocks, Kuang et al. fabricated a heterodimer of UCNP and gold yolk-shell nanoparticles for the quantification of polymyxin-B-resistant *E. coli*, which can produce both the circular dichroism and upconversion luminescence signals for drug-resistant bacteria surveillance [138]. For example, Liu et al. presented an optical- and nuclear-dual-modality probe based on self-assembly of vancomycin on Gram-positive bacteria for imaging bacterial infection, which can form nanoaggregates, and thus significantly increase fluorescence signal on the surface of MRSA [137].

6.3.1.2 Theranostic Nano-biomaterials for Wound Infection Control

In addition to these nanoprobes for imaging of wound infection, various theranostic nanoprobes for simultaneously imaging and therapy of wound infection have also been introduced [131–134]. For example, Chen et al. present an activatable theranostic nanoprobe for highly sensitive NIR fluorescence (NIRF) imaging and PTT of methicillin-resistant *Staphylococcus aureus* (MRSA)-infected wounds [131]. The theranostic nanoprobe was fabricated by coating silica nanoparticles with vancomycin-modified polyelectrolyte-cypate complexes (SiO_2-Cy-Van), which is activated by a bacteria-responsive dissociation mechanism and resulting in MRSA-activated NIRF imaging and photothermal bactericidal effect (Figure 6.10)

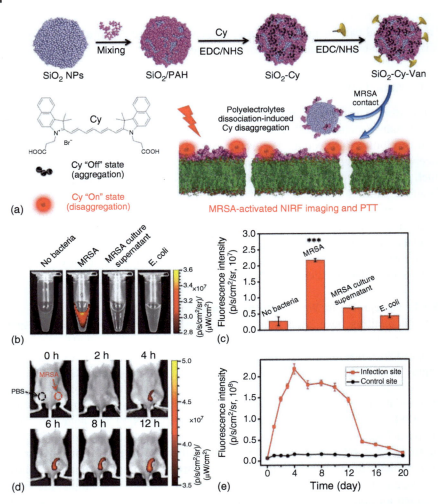

Figure 6.10 (a) Preparation and schematic illustration of the SiO_2-Cy-Van nanoprobes for MRSA-activated NIRF imaging and PTT. (b) Evaluation of fluorescence of SiO_2-Cy-Van nanoprobes-incubated bacteria for 4 h. MRSA and *E. coli* cultures with a concentration of 10^7 CFU were utilized, and controls included culture supernatant of MRSA and the nanoprobes only. (c) Quantification of fluorescence signal intensities of the images shown in (b). *** indicates $P < 0.001$. (d) NIRF images of MRSA-infected mice treated with SiO_2-Cy-Van nanoprobes at different time points post-injection. (e) Quantification of fluorescence intensities of the infection site and control site (PBS) of MRSA (10^7 CFU)-infected mice at different time points post-injection of SiO_2-Cy-Van nanoprobes. Source: Reproduced with permission from Ref. [131], © 2017, American Chemical Society.

[131]. Remarkably, the SiO_2-Cy-Van nanoprobes can enable highly sensitive NIRF monitoring and efficient PTT of MRSA infections in wounded tissues [131]. The unique optical characteristic of aggregation-induced emission luminogens (AIEgens) enables their extensive bioimaging and therapeutic applications [139]. Tang et al. reported a theranostic AIEgen, triphenylethylene-naphthalimide triazole

(TriPE-NT), which is capable of simultaneously imaging and killing Gram-positive and Gram-negative bacteria by photodynamic mechanism [133]. In vivo experiment in infected wound model indicated that TriPE-NT not only can inhibit wound infection but also can be used as a potential fluorescent agent for monitoring the wound bacterial infections [133]. Furthermore, Fan et al. demonstrated an intelligent theranostic nanocomposite wound dressing which can simultaneously sense the pathogenic bacteria and release an encapsulated antimicrobials agent, which would prevent overuse of antimicrobials, decrease the occurrence of antibiotic resistance, as well as improve the stability of encapsulated antimicrobials [132]. Taken together, these theranostic strategies that are based on bacterial-responsive functional nano-biomaterials offer great opportunities to monitor and combat bacterial-infected wounds [131–134].

6.3.2 Imaging of Wound Parameters and Markers

Basic wound parameters, such as pO_2, pH, MMP, and ROS, are known to modify gene expression profiles, modulate cell proliferation/migration, and influence activity of enzymes and growth factors, which is essential for wound repair and regeneration [13, 140–145]. Therefore, monitoring these physicochemical and physiological parameters is of great importance to provide information on wound healing status or underlying pathological mechanisms of impaired healing. Highly sensitive monitoring of specific parameters from the wound tissue can be achieved by carefully designed probes that can produce electrical, magnetic, optical, or photoacoustic signals. Owing to their favorable magnetic, acoustic, and optical properties, and facile functionalization, a growing number of intelligent nanoprobes capable of reporting specific wound analytes have been extensively developed and validated [130].

6.3.2.1 Imaging of Physiological and Pathological Wound Parameters

ROS level of wound tissue is higher than that of normal skin owing to the enhanced infiltration of immune neutrophils and macrophage cells, which can produce large amounts of ROS [146]. Sustained overproduction of ROS in wound microenvironment may cause oxidative tissue damage and thus greatly impair wound healing. Dong et al. reported a redox-sensitive surface-enhanced Raman scattering (SERS) nanoprobe by attaching redox-sensitive anthraquinone onto the surface of gold nanoshell for in situ and noninvasively imaging of spatiotemporal redox evolution during the wound healing process [147]. Additionally, it is known that pH of the wound fluid is of great importance for assessing wound condition [143, 145]. Wang et al. reported a ratiometric pH-activated fluorescent probes based on hairpin-contained DNA i-motif nanostructure for high-resolution, sensitive, reversible measurement of small pH variations [148]. Furthermore, Wu et al. fabricated a dual luminescent porous silicon nanoprobe for wound ROS and pH monitoring, which can undergo fluorescent color evolution from red to blue upon the stimulation of elevated ROS and pH in the wound tissue due to a fluorescence resonance energy transfer (FRET) mechanism [130]. Collectively, these nanoprobes capable of detecting wound status pave a new path for precise wound management, which would greatly promote the understanding and treatment of cutaneous wounds.

6.3.2.2 Imaging of Stem Cell-based Wound Therapy

Stem cell-based regenerative medicines have shown great potential for cutaneous wound repair and regeneration [149, 150]. Tracking and monitoring the in vivo fate, cell viability and regenerative capabilities of transplanted stem cells is vital for improving the safety, accelerating the clinical application, and enhancing therapeutic efficacy of stem cell-based wound therapy [151, 152]. Benefiting from their desirable spatial-temporal resolution and lower autofluorescence, Ag_2S quantum dots (QDs) were employed for high-efficiency labeling of MSCs [151, 153]. Furthermore, fluorescence imaging in the second NIR (NIR-II) window was performed by Wang et al. to visualize the dynamic wound-homing behavior of Ag_2S QDs labeled MSCs [151]. Li et al. introduced a conjugated polymer (CP) nanodot-based noninvasive fluorescent tracker for tracking of MSCs to reveal their in vivo migration behaviors and mechanisms in promoting skin regeneration [154]. Owing to their unique features such as high brightness, good fluorescence stability, and slight interference with stem cell properties, the CP nanodots display significantly enhanced tracking time and sensitiveness without compromising the bioactivities of MSCs in comparison with the commercial QD-based cell labeling kits [154]. This facile reliable stem cell labeling approach is highly desirable in providing valuable insights into in vivo fate and regenerative mechanisms of therapeutic stem cells, which would significantly facilitate the clinical translation of stem cell-based wound therapeutics [151–154].

6.3.2.3 Imaging of Wound Scarring Markers

After discussing the development of nanoprobes capable of monitoring physiological and pathological wound parameters that lead to impaired wound healing, we next focus on the assessment of abnormal scarring resulting from wound overhealing using novel nanoprobes [155–158]. Gene expression at the transcriptional level offers valuable information for monitoring disease progression. Pathological overexpression of connective tissue growth factor (CTGF) mRNA has been shown to be a relevant biomarker for early diagnosis of hypertrophic and keloid scars [159]. Xu et al. demonstrated that gold nanoparticles functionalized with spherical nucleic acids targeting intracellular CTGF mRNA, known as NanoFlares, could serve as a visual indicator of hypertrophic scars and keloids in the skin of live animals and in ex vivo human skin (Figure 6.11) [156]. Xu et al. further constructed nucleic-acid-based dual functional probes to measure the intracellular scar-related CTGF mRNA expression and inhibit its expression with enhanced sensitivity and specificity [157]. In conclusion, these noninvasive nanoprobes pave an innovative way for biopsy-free scar diagnosis, which may eventually facilitate scar wound management and inform therapeutic decisions on the basis of the mRNA-expression profiles.

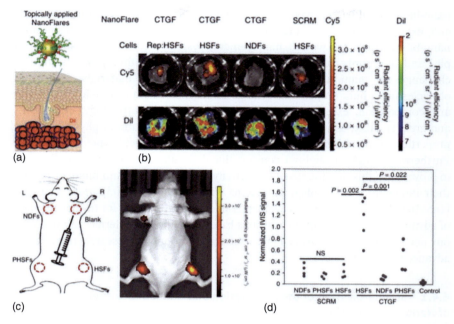

Figure 6.11 (a) Schematic illustration of the topical application of NanoFlares to ex vivo human skin that has been injected through the dermal side with DiI-labeled cells.
(b) Representative in vivo imaging system (IVIS) images of Cy5- and DiI-labeled cells.
(c) Schematic and representative IVIS image of the subcutaneous administration of NDFs (SCRM-Cy5), PHSFs (SCRM-Cy5), HSFs (Cy5), and empty gel (blank). The dotted areas mark the approximate regions of injected cells. (d) Dot plots of the intensities of IVIS signals generated from NDFs, PHSFs, HSFs, and negative controls ($n \geq 4$). Source: Reproduced with permission from Ref. [156], © 2018, Springer Nature.

6.4 Conclusion and Future Outlook

In this review, various nano-biomaterials featured with distinct merits have been summarized, either in imaging or in the treatment of acute and chronic cutaneous wounds, or both [39, 43, 48, 55, 66, 76, 79, 89, 93, 129, 130, 135, 155]. Nano-biomaterials play increasingly important roles in the management of cutaneous wounds, from therapeutic delivery platforms to microenvironment modulators, and they offer powerful tools for wound diagnostics [17, 20, 35, 130, 160–164]. We investigate nano-biomaterial features that are crucial for the induction of successful wound healing and highlight applications of nano-biomaterials for imaging and monitoring of the cutaneous wound. The full potential of nano-biomaterial-based wound solutions has not been realized yet; however, a solid ground for future advancements and clinical translations has been established.

Although nano-biomaterial-based wound therapeutics and diagnostics demonstrate some encouraging results, the development of nano-biomaterials-based wound solutions is still in its infancy stage and there is a lot of potential waiting to be explored in the future. For performance- and cost-competitive nano-biomaterial-based wound management solutions, necessary advances and extensive efforts are urgently needed. As we continue to create novel nano-biomaterials for wound management, systems become increasingly complex and the number of variables increases dramatically, thus posing significant challenges in manipulating these multiple component materials. The future obstacles lie in bulk preparation of nano-biomaterials with well-defined structure and composition, excellent biocompatibility, uniform morphology, high purity, and great stability, as well as comprehensive characterization of physicochemical properties of nano-biomaterials and their interactions with the human body. Furthermore, high-throughput computational simulations and machine learning to provide insights into the structure–activity relationship of nano-biomaterials and their interaction with biological systems will become more important than ever [165–167]. Herein, we outline several possible future directions for the development of advanced cutaneous wound solutions based on nano-biomaterials and hope that our perspectives may provide some useful inspirations for researchers to further advance this rising field.

6.4.1 Standardization of the Preparation and Functionalization of Nano-biomaterials

The most challenging part of transferring an innovative scientific concept from the laboratory to the patient's bedside is to design standard procedures for scale-up production, as well as optimizing the product quality deviation [155]. The further development and clinical translation of nano-biomaterial-based wound solutions are severely limited by the lack of well-defined, quality-controlled, and standardized nano-biomaterial [168, 169]. Therefore, in addition to developing new and sophisticated synthesis techniques, efforts are urgently needed to address the nonuniformity issues in the physical properties of nano-biomaterials, which often lead to nonrepeatable results. Standardized preparation protocols should be capable of producing nano-biomaterials with high purity, fixed physicochemical properties, and good uniformity to enhance reproducibility, quality, and safety, which is crucial not only in experimental models but ultimately in the clinic. A direct translation of the standards of traditional wound therapeutics to nano-biomaterial-based wound solutions would be challenging given the distinct features of nano-biomaterials, but similar standards should be soon established. In general, straightforward and rudimentary methods are encouraged to be used in the production of nano-biomaterials, which can minimize the by-products and facilitate future cost-effective commercialization.

6.4.2 Bio-safety of Nano-biomaterials

The clinical application and performance of biomaterials used in regenerative medicine are usually hampered by their toxicity and immunogenicity [117, 170]. Therefore, biocompatibility should be factored into the design of nano-biomaterials that are used in cutaneous wound management, as nano-biomaterials become sophisticated and integrate complex components. The biocompatibility of a particular nano-biomaterial greatly depends on the composition, biodegradable ability, physicochemical properties, and route of administration. Depending on the level of interaction with human bodies, various levels of biocompatibility evaluations are required for nano-biomaterials used in cutaneous wounds to guarantee their body acceptability [171]. For long-term implantation, the toxicity, immunogenicity, and clearance of nano-biomaterials should be systemically investigated at all levels before their in vivo application [172]. For occasional internal use, contact reactions and immediate immune responses need to be carefully considered. Although topical application allows sufficient therapeutics and diagnostics in the target wound area, the mismatch between the chemical/physical properties of implanted nano-biomaterials and human tissues can cause incompatible microenvironment and poor integration. The biocompatibility of engineered nano-biomaterials can be improved through modulation of nano-biomaterial size, charge, and surface chemistry to reduce inflammation and chemical/physical mismatch between the nano-biomaterials and wound tissue.

6.4.3 Computational Simulation and Machine Learning

Next-generation nanomaterials are envisioned to mimic living tissues in terms of mechanical, chemical, and biological properties and thus could serve as key players in tissue repair and regeneration [173]. Advances in big data and machine learning techniques hold great promise for the acceleration of the development process of nano-biomaterial-based wound therapeutics and diagnostics [162]. One of the directions in the future is to build advanced simulation and computational models to assess how the composition, size, shape, synthesis method, and surface modification of nano-biomaterials can affect wound management performance [174, 175]. It is suggested that machine learning algorithms are well suited for extracting knowledge from a large amount of published data and studies, which could be used to refine the approaches for novel nanomaterials discovery [176]. For example, machine learning models have been successfully applied to predict and optimize the chemical/physical properties of MOFs [177]. Therefore, the application of machine learning techniques also offers great promise for predicting, validating, and optimizing the wound management performance of nano-biomaterials. Although researchers have only just begun to exploit the potential of these machine learning techniques, they may have considerable implications for developing future wound therapeutic and diagnostic nanomaterials.

Acknowledgments

We gratefully acknowledge the support from the National Natural Science Foundation of China (32201161 to H.W.), Zhejiang Provincial Natural Science Foundation of China (LQ21H300004 to H.W., GF22H168862 to S.H.), Hangzhou Medical College Qiuzhen Talent Project (00004F1RCYJ2209 to H.W.), Basic Scientific

Research Funds of Department of Education of Zhejiang Province (KYZD202205 to H.W.), Health Bureau of Zhejiang Province/General Program (2021438429 to S.H.), and Zhejiang Traditional Chinese Medicine Science and Technology Program (2023ZF122 to S.H.).

References

1 Posnett, J. and Franks, P. (2008). The burden of chronic wounds in the UK. *Diabetic Medicine* 14 (5): S7–S85.
2 Sen, C.K., Gordillo, G.M., Roy, S. et al. (2009). Human skin wounds: a major and snowballing threat to public health and the economy. *Wound Repair and Regeneration* 17 (6): 763–771.
3 Eming, S.A., Martin, P., and Tomic-Canic, M. (2014). Wound repair and regeneration: mechanisms, signaling, and translation. *Science Translational Medicine* 6 (265): 265sr6.
4 Sun, B.K., Siprashvili, Z., and Khavari, P.A. (2014). Advances in skin grafting and treatment of cutaneous wounds. *Science* 346 (6212): 941–945.
5 Frykberg, R.G. and Banks, J. (2015). Challenges in the treatment of chronic wounds. *Advances in Wound Care* 4 (9): 560–582.
6 Han, G. and Ceilley, R. (2017). Chronic wound healing: a review of current management and treatments. *Advances in Therapy* 34 (3): 599–610.
7 Powers, J.G., Higham, C., Broussard, K., and Phillips, T.J. (2016). Wound healing and treating wounds: chronic wound care and management. *Journal of the American Academy of Dermatology* 74 (4): 607–625.
8 Gould, L., Abadir, P., Brem, H. et al. (2015). Chronic wound repair and healing in older adults: current status and future research. *Wound Repair and Regeneration* 23 (1): 1–13.
9 Heyer, K., Herberger, K., Protz, K. et al. (2016). Epidemiology of chronic wounds in Germany: analysis of statutory health insurance data. *Wound Repair and Regeneration* 24 (2): 434–442.
10 Tsourdi, E., Barthel, A., Rietzsch, H. et al. (2013). Current aspects in the pathophysiology and treatment of chronic wounds in diabetes mellitus. *BioMed Research International* 2013: 6.
11 Guo, S. and DiPietro, L.A. (2010). Factors affecting wound healing. *Journal of Dental Research* 89 (3): 219–229.
12 Gurtner, G.C., Werner, S., Barrandon, Y., and Longaker, M.T. (2008). Wound repair and regeneration. *Nature* 453: 314.
13 McLister, A., McHugh, J., Cundell, J., and Davis, J. (2016). New developments in smart bandage technologies for wound diagnostics. *Advanced Materials* 28 (27): 5732–5737.
14 Dargaville, T.R., Farrugia, B.L., Broadbent, J.A. et al. (2013). Sensors and imaging for wound healing: a review. *Biosensors and Bioelectronics* 41: 30–42.

15 Hattori, Y., Falgout, L., Lee, W. et al. (2014). Multifunctional skin-like electronics for quantitative, clinical monitoring of cutaneous wound healing. *Advanced Healthcare Materials* 3 (10): 1597–1607.

16 Paul, D.W., Ghassemi, P., Ramella-Roman, J.C. et al. (2015). Noninvasive imaging technologies for cutaneous wound assessment: a review. *Wound Repair and Regeneration* 23 (2): 149–162.

17 Hamdan, S., Pastar, I., Drakulich, S. et al. (2017). Nanotechnology-driven therapeutic interventions in wound healing: potential uses and applications. *ACS Central Science* 3 (3): 163–175.

18 Ashtikar, M. and Wacker, M.G. (2018). Nanopharmaceuticals for wound healing – lost in translation? *Advanced Drug Delivery Reviews* 129: 194–218.

19 Kalashnikova, I., Das, S., and Seal, S. (2015). Nanomaterials for wound healing: scope and advancement. *Nanomedicine* 10 (16): 2593–2612.

20 Castaño, O., Pérez-Amodio, S., Navarro-Requena, C. et al. (2018). Instructive microenvironments in skin wound healing: biomaterials as signal releasing platforms. *Advanced Drug Delivery Reviews* 129: 95–117.

21 Xavier, J.R., Thakur, T., Desai, P. et al. (2015). Bioactive nanoengineered hydrogels for bone tissue engineering: a growth-factor-free approach. *ACS Nano* 9 (3): 3109–3118.

22 Thoniyot, P., Tan, M.J., Karim, A.A. et al. (2015). Nanoparticle–hydrogel composites: concept, design, and applications of these promising, multi-functional materials. *Advanced Science* 2 (1–2): 1400010.

23 Pina, S., Oliveira, J.M., and Reis, R.L. (2015). Natural-based nanocomposites for bone tissue engineering and regenerative medicine: a review. *Advanced Materials* 27 (7): 1143–1169.

24 Zhang, Y.S. and Khademhosseini, A. (2017). Advances in engineering hydrogels. *Science* 356 (6337): eaaf3627.

25 Rose, S., Prevoteau, A., Elzière, P. et al. (2013). Nanoparticle solutions as adhesives for gels and biological tissues. *Nature* 505: 382.

26 Meddahi-Pelle, A., Legrand, A., Marcellan, A. et al. (2014). Organ repair, hemostasis, and in vivo bonding of medical devices by aqueous solutions of nanoparticles. *Angewandte Chemie. International Edition* 53 (25): 6369–6373.

27 Tonazzini, I., Jacchetti, E., Meucci, S. et al. (2015). Schwann cell contact guidance versus boundary interaction in functional wound healing along nano and microstructured membranes. *Advanced Healthcare Materials* 4 (12): 1849–1860.

28 Weng, W., He, S., Song, H. et al. (2018). Aligned carbon nanotubes reduce hypertrophic scar via regulating cell behavior. *ACS Nano* 12 (8): 7601–7612.

29 Keung, A.J., Kumar, S., and Schaffer, D.V. (2010). Presentation counts: microenvironmental regulation of stem cells by biophysical and material cues. *Annual Review of Cell and Developmental Biology* 26 (1): 533–556.

30 Downing, T.L., Soto, J., Morez, C. et al. (2013). Biophysical regulation of epigenetic state and cell reprogramming. *Nature Materials* 12: 1154.

31 Connelly, J.T., Gautrot, J.E., Trappmann, B. et al. (2010). Actin and serum response factor transduce physical cues from the microenvironment to regulate epidermal stem cell fate decisions. *Nature Cell Biology* 12: 711.

32 Zhao, S., Li, L., Wang, H. et al. (2015). Wound dressings composed of copper-doped borate bioactive glass microfibers stimulate angiogenesis and heal full-thickness skin defects in a rodent model. *Biomaterials* 53: 379–391.

33 Harris, E.D. (2004). A requirement for copper in angiogenesis. *Nutrition Reviews* 62 (2): 60–64.

34 Gérard, C., Bordeleau, L.-J., Barralet, J., and Doillon, C.J. (2010). The stimulation of angiogenesis and collagen deposition by copper. *Biomaterials* 31 (5): 824–831.

35 Xiao, J.S., Zhu, Y.X., Huddleston, S. et al. (2018). Copper metal-organic framework nanoparticles stabilized with folic acid improve wound healing in diabetes. *ACS Nano* 12 (2): 1023–1032.

36 Wang, X.C., Lv, F., Li, T. et al. (2017). Electrospun micropatterned nanocomposites incorporated with Cu_2S nanoflowers for skin tumor therapy and wound healing. *ACS Nano* 11 (11): 11337–11349.

37 Xiao, J.S., Chen, S.Y., Yi, J. et al. (2017). A cooperative copper metal-organic framework-hydrogel system improves wound healing in diabetes. *Advanced Functional Materials* 27 (1).

38 Wu, H.B., Li, F.Y., Wang, S.F. et al. (2018). Ceria nanocrystals decorated mesoporous silica nanoparticle based ROS-scavenging tissue adhesive for highly efficient regenerative wound healing. *Biomaterials* 151: 66–77.

39 Chigurupati, S., Mughal, M.R., Okun, E. et al. (2013). Effects of cerium oxide nanoparticles on the growth of keratinocytes, fibroblasts and vascular endothelial cells in cutaneous wound healing. *Biomaterials* 34 (9): 2194–2201.

40 Hickman, D.A., Pawlowski, C.L., Sekhon, U.D.S. et al. (2018). Biomaterials and advanced technologies for hemostatic management of bleeding. *Advanced Materials* 30 (4).

41 Hsu, B.B., Conway, W., Tschabrunn, C.M. et al. (2015). Clotting mimicry from robust hemostatic bandages based on self-assembling peptides. *ACS Nano* 9 (9): 9394–9406.

42 Long, M., Zhang, Y., Huang, P. et al. (2018). Emerging nanoclay composite for effective hemostasis. *Advanced Functional Materials* 28 (10).

43 Gaharwar, A.K., Avery, R.K., Assmann, A. et al. (2014). Shear-thinning nanocomposite hydrogels for the treatment of hemorrhage. *ACS Nano* 8 (10): 9833–9842.

44 Zhao, X., Guo, B.L., Wu, H. et al. (2018). Injectable antibacterial conductive nanocomposite cryogels with rapid shape recovery for noncompressible hemorrhage and wound healing. *Nature Communications* 9.

45 Luo, Z., Wang, S., and Zhang, S. (2011). Fabrication of self-assembling d-form peptide nanofiber scaffold d-EAK16 for rapid hemostasis. *Biomaterials* 32 (8): 2013–2020.

46 Morgan, C.E., Dombrowski, A.W., Rubert Pérez, C.M. et al. (2016). Tissue-factor targeted peptide amphiphile nanofibers as an injectable therapy to control hemorrhage. *ACS Nano* 10 (1): 899–909.

47 Han, L., Lu, X., Liu, K.Z. et al. (2017). Mussel-inspired adhesive and tough hydrogel based on nanoclay confined dopamine polymerization. *ACS Nano* 11 (3): 2561–2574.

48 Cheng, J., Feng, S., Han, S. et al. (2016). Facile assembly of cost-effective and locally applicable or injectable nanohemostats for hemorrhage control. *ACS Nano* 10 (11): 9957–9973.

49 Baker, S.E., Sawvel, A.M., Zheng, N., and Stucky, G.D. (2007). Controlling bioprocesses with inorganic surfaces: layered clay hemostatic agents. *Chemistry of Materials* 19 (18): 4390–4392.

50 Xiong, M.-H., Bao, Y., Yang, X.-Z. et al. (2014). Delivery of antibiotics with polymeric particles. *Advanced Drug Delivery Reviews* 78: 63–76.

51 Ray, P.C., Khan, S.A., Singh, A.K. et al. (2012). Nanomaterials for targeted detection and photothermal killing of bacteria. *Chemical Society Reviews* 41 (8): 3193–3209.

52 Percival, S.L., McCarty, S.M., and Lipsky, B. (2014). Biofilms and wounds: an overview of the evidence. *Advances in Wound Care* 4 (7): 373–381.

53 Qiao, Y., Ping, Y., Zhang, H. et al. (2019). Laser-activatable CuS nanodots to treat multidrug-resistant bacteria and release copper ion to accelerate healing of infected chronic nonhealing wounds. *ACS Applied Materials & Interfaces* 11 (4): 3809–3822.

54 Chen, W.Y., Chang, H.Y., Lu, J.K. et al. (2015). Self-assembly of antimicrobial peptides on gold nanodots: against multidrug-resistant bacteria and wound-healing application. *Advanced Functional Materials* 25 (46): 7189–7199.

55 Mao, C.Y., Xiang, Y.M., Liu, X.M. et al. (2017). Photo-inspired antibacterial activity and wound healing acceleration by hydrogel embedded with Ag/Ag@AgCl/ZnO nanostructures. *ACS Nano* 11 (9): 9010–9021.

56 Xi, Y.W., Ge, J., Guo, Y. et al. (2018). Biomimetic elastomeric polypeptide-based nanofibrous matrix for overcoming multidrug-resistant bacteria and enhancing full-thickness wound healing/skin regeneration. *ACS Nano* 12 (11): 10772–10784.

57 Yin, W.Y., Yu, J., Lv, F.T. et al. (2016). Functionalized nano-MoS_2 with peroxidase catalytic and near-infrared photothermal activities for safe and synergetic wound antibacterial applications. *ACS Nano* 10 (12): 11000–11011.

58 Mao, C.Y., Xiang, Y.M., Liu, X.M. et al. (2018). Repeatable photodynamic therapy with triggered signaling pathways of fibroblast cell proliferation and differentiation to promote bacteria-accompanied wound healing. *ACS Nano* 12 (2): 1747–1759.

59 Yang, X.L., Yang, J.C., Wang, L. et al. (2017). Pharmaceutical intermediate-modified gold nanoparticles: against multidrug-resistant bacteria and wound-healing application via an electrospun scaffold. *ACS Nano* 11 (6): 5737–5745.

60 Chen, J., Andler, S.M., Goddard, J.M. et al. (2017). Integrating recognition elements with nanomaterials for bacteria sensing. *Chemical Society Reviews* 46 (5): 1272–1283.

61 Mao, D., Hu, F., Kenry et al. (2018). Metal–organic-framework-assisted in vivo bacterial metabolic labeling and precise antibacterial therapy. *Advanced Materials* 30 (18): 1706831.

62 Li, J.H., Zhang, K.X., Ruan, L. et al. (2018). Block copolymer nanoparticles remove biofilms of drug-resistant gram-positive bacteria by nanoscale bacterial debridement. *Nano Letters* 18 (7): 4180–4187.

63 Li, Z., Feng, X., Gao, S. et al. (2019). Porous organic polymer-coated band-aids for phototherapy of bacteria-induced wound infection. *ACS Applied Biomaterials* 2 (2): 613–618.

64 Hsiao, C.W., Chen, H.L., Liao, Z.X. et al. (2015). Effective photothermal killing of pathogenic bacteria by using spatially tunable colloidal gels with nano-localized heating sources. *Advanced Functional Materials* 25 (5): 721–728.

65 Landis, R.F., Gupta, A., Lee, Y.W. et al. (2017). Cross-linked polymer-stabilized nanocomposites for the treatment of bacteria biofilms. *ACS Nano* 11 (1): 946–952.

66 Perreault, F., de Faria, A.F., Nejati, S., and Elimelech, M. (2015). Antimicrobial properties of graphene oxide nanosheets: why size matters. *ACS Nano* 9 (7): 7226–7236.

67 Lu, X., Feng, X., Werber, J.R. et al. (2017). Enhanced antibacterial activity through the controlled alignment of graphene oxide nanosheets. *Environmental Sciences* 114 (46): E9793–E9801.

68 Ji, H., Sun, H., and Qu, X. (2016). Antibacterial applications of graphene-based nanomaterials: Recent achievements and challenges. *Advanced Drug Delivery Reviews* 105: 176–189.

69 Li, Y., Liu, X.M., Tan, L. et al. (2018). Rapid sterilization and accelerated wound healing using Zn^{2+} and graphene oxide modified g-C_3N_4 under dual light irradiation. *Advanced Functional Materials* 28 (30): 1800299.

70 Zasloff, M. (2002). Antimicrobial peptides of multicellular organisms. *Nature* 415: 389.

71 Brogden, K.A. (2005). Antimicrobial peptides: pore formers or metabolic inhibitors in bacteria? *Nature Reviews Microbiology* 3: 238.

72 Schultz, G.S., Davidson, J.M., Kirsner, R.S. et al. (2011). Dynamic reciprocity in the wound microenvironment. *Wound Repair and Regeneration* 19 (2): 134–148.

73 Wei, H. and Wang, E. (2013). Nanomaterials with enzyme-like characteristics (nanozymes): next-generation artificial enzymes. *Chemical Society Reviews* 42 (14): 6060–6093.

74 Kim, C.K., Kim, T., Choi, I.Y. et al. (2012). Ceria nanoparticles that can protect against ischemic stroke. *Angewandte Chemie* 51 (44): 11039.

75 You, L., Wang, J., Liu, T. et al. (2018). Targeted brain delivery of rabies virus glycoprotein 29-modified deferoxamine-loaded nanoparticles reverses functional deficits in parkinsonian mice. *ACS Nano* 12 (5): 4123–4139.

76 Wu, H., Li, F., Shao, W. et al. (2019). Promoting angiogenesis in oxidative diabetic wound microenvironment using a nanozyme-reinforced self-protecting hydrogel. *ACS Central Science* 5 (3): 477–485.

77 Kim, H.S. and Yoo, H.S. (2012). Matrix metalloproteinase-inspired suicidal treatments of diabetic ulcers with siRNA-decorated nanofibrous meshes. *Gene Therapy* 20: 378.

78 Kim, H.S., Son, Y.J., and Yoo, H.S. (2016). Clustering siRNA conjugates for MMP-responsive therapeutics in chronic wounds of diabetic animals. *Nanoscale* 8 (27): 13236–13244.

79 Castleberry, S.A., Almquist, B.D., Li, W. et al. (2016). Self-assembled wound dressings silence MMP-9 and improve diabetic wound healing in vivo. *Advanced Materials* 28 (9): 1809–1817.

80 Li, N., Luo, H.-C., Ren, M. et al. (2017). Efficiency and safety of β-CD-$(D_3)_7$ as siRNA carrier for decreasing matrix metalloproteinase-9 expression and improving wound healing in diabetic rats. *ACS Applied Materials & Interfaces* 9 (20): 17417–17426.

81 Wu, J., Li, Y., He, C. et al. (2016). Novel H_2S releasing nanofibrous coating for in vivo dermal wound regeneration. *ACS Applied Materials & Interfaces* 8 (41): 27474–27481.

82 Dai, X., Liu, J., Zheng, H. et al. (2017). Nano-formulated curcumin accelerates acute wound healing through Dkk-1-mediated fibroblast mobilization and MCP-1-mediated anti-inflammation. *NPG Asia Materials* 9: e368.

83 Kloth, L.C. (2014). Electrical stimulation technologies for wound healing. *Advances in Wound Care* 3 (2): 81–90.

84 Yanez-Soto, B., Liliensiek, S.J., Gasiorowski, J.Z. et al. (2013). The influence of substrate topography on the migration of corneal epithelial wound borders. *Biomaterials* 34 (37): 9244–9251.

85 Sundelacruz, S., Li, C., Choi, Y.J. et al. (2013). Bioelectric modulation of wound healing in a 3D in vitro model of tissue-engineered bone. *Biomaterials* 34 (28): 6695–6705.

86 Kim, H.N., Jiao, A., Hwang, N.S. et al. (2013). Nanotopography-guided tissue engineering and regenerative medicine. *Advanced Drug Delivery Reviews* 65 (4): 536–558.

87 Xie, J.W., MacEwan, M.R., Ray, W.Z. et al. (2010). Radially aligned, electrospun nanofibers as dural substitutes for wound closure and tissue regeneration applications. *ACS Nano* 4 (9): 5027–5036.

88 Xi, Y., Dong, H., Sun, K. et al. (2013). Scab-inspired cytophilic membrane of anisotropic nanofibers for rapid wound healing. *ACS Applied Materials & Interfaces* 5 (11): 4821–4826.

89 Chen, W., Villa-Diaz, L.G., Sun, Y. et al. (2012). Nanotopography influences adhesion, spreading, and self-renewal of human embryonic stem cells. *ACS Nano* 6 (5): 4094–4103.

90 Kim, H.N., Hong, Y., Kim, M.S. et al. (2012). Effect of orientation and density of nanotopography in dermal wound healing. *Biomaterials* 33 (34): 8782–8792.

91 Patel, S., Kurpinski, K., Quigley, R. et al. (2007). Bioactive nanofibers: synergistic effects of nanotopography and chemical signaling on cell guidance. *Nano Letters* 7 (7): 2122–2128.

92 Falanga, V. (2004). The chronic wound: impaired healing and solutions in the context of wound bed preparation. *Blood Cells, Molecules, and Diseases* 32 (1): 88–94.

93 Teo, B.K.K., Wong, S.T., Lim, C.K. et al. (2013). Nanotopography modulates mechanotransduction of stem cells and induces differentiation through focal adhesion kinase. *ACS Nano* 7 (6): 4785–4798.

94 Chen, S., Shi, J., Xu, X. et al. (2016). Study of stiffness effects of poly(amidoamine)–poly(n-isopropyl acrylamide) hydrogel on wound healing. *Colloids and Surfaces B: Biointerfaces* 140: 574–582.

95 Leijten, J. and Khademhosseini, A. (2016). From nano to macro: multiscale materials for improved stem cell culturing and analysis. *Cell Stem Cell* 18 (1): 20–24.

96 Ma, Y., Lin, M., Huang, G. et al. (2018). 3D spatiotemporal mechanical microenvironment: a hydrogel-based platform for guiding stem cell fate. *Advanced Materials* 30 (49): 1705911.

97 Achterberg, V.F., Buscemi, L., Diekmann, H. et al. (2014). The nano-scale mechanical properties of the extracellular matrix regulate dermal fibroblast function. *Journal of Investigative Dermatology* 134 (7): 1862–1872.

98 Koosha, M., Mirzadeh, H., Shokrgozar, M.A., and Farokhi, M. (2015). Nanoclay-reinforced electrospun chitosan/PVA nanocomposite nanofibers for biomedical applications. *RSC Advances* 5 (14): 10479–10487.

99 Sandri, G., Aguzzi, C., Rossi, S. et al. (2017). Halloysite and chitosan oligosaccharide nanocomposite for wound healing. *Acta Biomaterialia* 57: 216–224.

100 Kokabi, M., Sirousazar, M., and Hassan, Z.M. (2007). PVA–clay nanocomposite hydrogels for wound dressing. *European Polymer Journal* 43 (3): 773–781.

101 Friedemann, M., Kalbitzer, L., Franz, S. et al. (2017). Instructing human macrophage polarization by stiffness and glycosaminoglycan functionalization in 3D collagen. *Networks* 6 (7): 1600967.

102 Wei, G., Jin, Q., Giannobile, W.V., and Ma, P.X. (2006). Nano-fibrous scaffold for controlled delivery of recombinant human PDGF-BB. *Journal of Controlled Release* 112 (1): 103–110.

103 Griffin, D.R., Weaver, W.M., Scumpia, P.O. et al. (2015). Accelerated wound healing by injectable microporous gel scaffolds assembled from annealed building blocks. *Nature Materials* 14: 737.

104 Song, B., Gu, Y., Pu, J. et al. (2007). Application of direct current electric fields to cells and tissues in vitro and modulation of wound electric field in vivo. *Nature Protocols* 2: 1479.

105 Reid, B., Song, B., McCaig, C.D., and Zhao, M. (2005). Wound healing in rat cornea: the role of electric currents. *FASEB Journal: Official Publication of the Federation of American Societies for Experimental Biology* 19 (3): 379–386.

106 Zhao, M., Song, B., Pu, J. et al. (2006). Electrical signals control wound healing through phosphatidylinositol-3-OH kinase-γ and PTEN. *Nature* 442: 457.

107 Long, Y., Wei, H., Li, J. et al. (2018). Effective wound healing enabled by discrete alternative electric fields from wearable nanogenerators. *ACS Nano* 12 (12): 12533–12540.

108 Bhang, S.H., Jang, W.S., Han, J. et al. Zinc oxide nanorod-based piezoelectric dermal patch for wound healing. *Advanced Functional Materials* 2017, 27 (1): 1603497.

109 Schugart, R.C., Friedman, A., Zhao, R., and Sen, C.K. (2008). Wound angiogenesis as a function of tissue oxygen tension: a mathematical model. *Proceedings of the National Academy of Sciences of the United States of America* 105 (7): 2628–2633.

110 Icli, B., Nabzdyk, C.S., Lujan-Hernandez, J. et al. (2016). Regulation of impaired angiogenesis in diabetic dermal wound healing by microRNA-26a. *Journal of Molecular and Cellular Cardiology* 91: 151–159.

111 Potente, M. and Carmeliet, P. (2017). The link between angiogenesis and endothelial metabolism. *Annual Review of Physiology* 79 (1): 43–66.

112 Kreuger, J. and Phillipson, M. (2015). Targeting vascular and leukocyte communication in angiogenesis, inflammation and fibrosis. *Nature Reviews Drug Discovery* 15: 125.

113 Barrientos, S., Stojadinovic, O., Golinko, M.S. et al. (2008). Perspective article: growth factors and cytokines in wound healing. *Wound Repair and Regeneration* 16 (5): 585–601.

114 Okonkwo, U.A. and DiPietro, L.A. (2017). Diabetes and wound. *Angiogenesis* 18 (7): 1419.

115 Kolluru, G.K., Bir, S.C., and Kevil, C.G. (2012). Endothelial dysfunction and diabetes: effects on angiogenesis, vascular remodeling, and wound healing. *International Journal of Vascular Medicine* 2012: 30.

116 Cho, C.-H., Sung, H.-K., Kim, K.-T. et al. (2006). COMP-angiopoietin-1 promotes wound healing through enhanced angiogenesis, lymphangiogenesis, and blood flow in a diabetic mouse model. *Proceedings of the National Academy of Sciences of the United States of America* 103 (13): 4946–4951.

117 Briquez, P.S., Clegg, L.E., Martino, M.M. et al. (2016). Design principles for therapeutic angiogenic materials. *Nature Reviews Materials* 1 (1): 15006.

118 Li, S., Nih, L.R., Bachman, H. et al. (2017). Hydrogels with precisely controlled integrin activation dictate vascular patterning and permeability. *Nature Materials* 16 (9): 953–961.

119 Wu, Y., Chen, L., Scott, P.G., and Tredget, E.E. (2007). Mesenchymal stem cells enhance wound healing through differentiation and angiogenesis. *Stem Cells* 25 (10): 2648–2659.

120 Wietecha, M.S. and DiPietro, L.A. (2012). Therapeutic approaches to the regulation of wound angiogenesis. *Advances in Wound Care* 2 (3): 81–86.

121 Gao, W.D., Jin, W.W., Li, Y.N. et al. (2017). A highly bioactive bone extracellular matrix-biomimetic nanofibrous system with rapid angiogenesis promotes diabetic wound healing. *Journal of Materials Chemistry B* 5 (35): 7285–7296.

122 Lino, M.M., Simões, S., Vilaça, A. et al. (2018). Modulation of angiogenic activity by light-activatable miRNA-loaded nanocarriers. *ACS Nano* 12 (6): 5207–5220.

123 Xie, Z., Paras, C.B., Weng, H. et al. (2013). Dual growth factor releasing multi-functional nanofibers for wound healing. *Acta Biomaterialia* 9 (12): 9351–9359.

124 Park, H.-J., Lee, J., Kim, M.-J. et al. (2012). Sonic hedgehog intradermal gene therapy using a biodegradable poly(β-amino esters) nanoparticle to enhance wound healing. *Biomaterials* 33 (35): 9148–9156.

125 Losi, P., Briganti, E., Errico, C. et al. (2013). Fibrin-based scaffold incorporating VEGF- and bFGF-loaded nanoparticles stimulates wound healing in diabetic mice. *Acta Biomaterialia* 9 (8): 7814–7821.

126 Nelson, C.E., Kim, A.J., Adolph, E.J. et al. (2014). Tunable delivery of siRNA from a biodegradable scaffold to promote angiogenesis in vivo. *Advanced Materials* 26 (4): 607–614.

127 Raftery, R.M., Walsh, D.P., Castaño, I.M. et al. (2016). Delivering nucleic-acid based nanomedicines on biomaterial scaffolds for orthopedic tissue repair: challenges, progress and future perspectives. *Advanced Materials* 28 (27): 5447–5469.

128 Tokatlian, T., Cam, C., and Segura, T. (2015). Porous hyaluronic acid hydrogels for localized nonviral DNA delivery in a diabetic wound healing model. *Advanced Healthcare Materials* 4 (7): 1084–1091.

129 Ji, M., Xu, M., Zhang, W. et al. (2016). Structurally well-defined Au@Cu_{2-x}S core–shell nanocrystals for improved cancer treatment based on enhanced photothermal efficiency. *Advanced Materials* 28 (16): 3094–3101.

130 Chen, X.S., Wo, F.J., Jin, Y. et al. (2017). Drug-porous silicon dual luminescent system for monitoring and inhibition of wound infection. *ACS Nano* 11 (8): 7938–7949.

131 Zhao, Z., Yan, R., Yi, X. et al. (2017). Bacteria-activated theranostic nanoprobes against methicillin-resistant *Staphylococcus aureus* infection. *ACS Nano* 11 (5): 4428–4438.

132 Zhou, J., Yao, D., Qian, Z. et al. (2018). Bacteria-responsive intelligent wound dressing: simultaneous in situ detection and inhibition of bacterial infection for accelerated wound healing. *Biomaterials* 161: 11–23.

133 Li, Y., Zhao, Z., Zhang, J.J. et al. (2018). A bifunctional aggregation-induced emission luminogen for monitoring and killing of multidrug-resistant bacteria. *Advanced Functional Materials* 28 (42): 1804632.

134 Zhai, X., Song, B., Chu, B. et al. (2018). Highly fluorescent, photostable, and biocompatible silicon theranostic nanoprobes against *Staphylococcus aureus* infections. *Nano Research* 11 (12): 6417–6427.

135 Gupta, A., Das, R., Yesilbag Tonga, G. et al. (2018). Charge-switchable nanozymes for bioorthogonal imaging of biofilm-associated infections. *ACS Nano* 12 (1): 89–94.

136 Percival, S.L., Hill, K.E., Williams, D.W. et al. (2012). A review of the scientific evidence for biofilms in wounds. *Wound Repair and Regeneration* 20 (5): 647–657.

137 Yang, C., Ren, C., Zhou, J. et al. (2017). Dual fluorescent- and isotopic-labelled self-assembling vancomycin for in vivo imaging of bacterial infections. *Angewandte Chemie International Edition* 56 (9): 2356–2360.

138 Sun, M., Qu, A., Hao, C. et al. (2018). Chiral upconversion heterodimers for quantitative analysis and bioimaging of antibiotic-resistant bacteria in vivo. *Advanced Materials* 30 (50): 1804241.

139 Liang, J., Tang, B.Z., and Liu, B. (2015). Specific light-up bioprobes based on AIEgen conjugates. *Chemical Society Reviews* 44 (10): 2798–2811.

140 Meier, R.J., Schreml, S., Wang, X.D. et al. (2011). Simultaneous photographing of oxygen and pH in vivo using sensor films. *Angewandte Chemie. International Edition* 50 (46): 10893–10896.

141 Krismastuti, F.S.H., Cavallaro, A., Prieto-Simon, B., and Voelcker, N.H. (2016). Toward multiplexing detection of wound healing biomarkers on porous silicon resonant microcavities. *Advanced Science* 3 (6): 1500383.

142 Morton, L.M. and Phillips, T.J. (2016). Wound healing and treating wounds: differential diagnosis and evaluation of chronic wounds. *Journal of the American Academy of Dermatology* 74 (4): 589–605.

143 Schreml, S., Meier, R.J., Weiß, K.T. et al. (2012). A sprayable luminescent pH sensor and its use for wound imaging in vivo. *Experimental Dermatology* 21 (12): 951–953.

144 DeRosa, C.A., Seaman, S.A., Mathew, A.S. et al. (2016). Oxygen sensing difluoroboron β-diketonate polylactide materials with tunable dynamic ranges for wound imaging. *ACS Sensors* 1 (11): 1366–1373.

145 Tamayol, A., Akbari, M., Zilberman, Y. et al. (2016). Flexible pH-sensing hydrogel fibers for epidermal applications. *Advanced Healthcare Materials* 5 (6): 711–719.

146 Niethammer, P., Grabher, C., Look, A.T., and Mitchison, T.J. (2009). A tissue-scale gradient of hydrogen peroxide mediates rapid wound detection in zebrafish. *Nature* 459: 996.

147 Sun, J., Han, S., Wang, Y. et al. (2018). Detection of redox state evolution during wound healing process based on a redox-sensitive wound dressing. *Analytical Chemistry* 90 (11): 6660–6665.

148 Ma, W., Yan, L.A., He, X. et al. (2018). Hairpin-contained i-motif based fluorescent ratiometric probe for high-resolution and sensitive response of small pH variations. *Analytical Chemistry* 90 (3): 1889–1896.

149 Sasaki, M., Abe, R., Fujita, Y. et al. (2008). Mesenchymal stem cells are recruited into wounded skin and contribute to wound repair by transdifferentiation into multiple skin cell type. *The Journal of Immunology* 180 (4): 2581–2587.

150 Branski, L.K., Gauglitz, G.G., Herndon, D.N., and Jeschke, M.G. (2009). A review of gene and stem cell therapy in cutaneous wound healing. *Burns* 35 (2): 171–180.

151 Chen, G., Tian, F., Li, C. et al. (2015). In vivo real-time visualization of mesenchymal stem cells tropism for cutaneous regeneration using NIR-II fluorescence imaging. *Biomaterials* 53: 265–273.

152 Chen, G., Zhang, Y., Li, C. et al. (2018). Recent advances in tracking the transplanted stem cells using near-infrared fluorescent nanoprobes: turning from the first to the second near-infrared window. *Advanced Healthcare Materials* 7 (20): 1800497.

153 Chen, G., Tian, F., Zhang, Y. et al. (2014). Tracking of transplanted human mesenchymal stem cells in living mice using near-infrared Ag_2S quantum dots. *Advanced Functional Materials* 24 (17): 2481–2488.

154 Jin, G.R., Mao, D., Cai, P.Q. et al. (2015). Conjugated polymer nanodots as ultrastable long-term trackers to understand mesenchymal stem cell therapy in skin regeneration. *Advanced Functional Materials* 25 (27): 4263–4273.

155 Yeo, D.C., Chew, S.W.T., and Xu, C. (2019). Polymeric biomaterials for management of pathological scarring. *ACS Applied Polymer Materials* 1: 612–624.

156 Yeo, D.C., Wiraja, C., Paller, A.S. et al. (2018). Abnormal scar identification with spherical-nucleic-acid technology. *Nature Biomedical Engineering* 2 (4): 227–238.

157 Zheng, M., Wiraja, C., Yeo, D.C. et al. (2018). Oligonucleotide molecular sprinkler for intracellular detection and spontaneous regulation of mRNA for theranostics of scar fibroblasts. *Small* 14 (49): 1802546.

158 Fang, R.H. and Zhang, L. (2018). Optical detection of abnormal skin scarring. *Nature Biomedical Engineering* 2 (4): 201–202.

159 Castleberry, S.A., Golberg, A., Sharkh, M.A. et al. (2016). Nanolayered siRNA delivery platforms for local silencing of CTGF reduce cutaneous scar contraction in third-degree burns. *Biomaterials* 95: 22–34.

160 Lu, Y., Aimetti, A.A., Langer, R., and Gu, Z. (2016). Bioresponsive materials. *Nature Reviews Materials* 2: 16075.

161 Moroni, L., Burdick, J.A., Highley, C. et al. (2018). Biofabrication strategies for 3D in vitro models and regenerative medicine. *Nature Reviews Materials* 3 (5): 21–37.

162 Sadtler, K., Singh, A., Wolf, M.T. et al. (2016). Design, clinical translation and immunological response of biomaterials in regenerative medicine. *Nature Reviews Materials* 1: 16040.

163 Mehlenbacher, R.D., Kolbl, R., Lay, A., and Dionne, J.A. (2017). Nanomaterials for in vivo imaging of mechanical forces and electrical fields. *Nature Reviews Materials* 3: 17080.

164 Park, S.-M., Aalipour, A., Vermesh, O. et al. (2017). Towards clinically translatable in vivo nanodiagnostics. *Nature Reviews Materials* 2: 17014.

165 Ling, S., Kaplan, D.L., and Buehler, M.J. (2018). Nanofibrils in nature and materials engineering. *Nature Reviews Materials* 3: 18016.

166 Ekins, S., Puhl, A.C., Zorn, K.M. et al. (2019). Exploiting machine learning for end-to-end drug discovery and development. *Nature Materials* 18 (5): 435–441.

167 Butler, K.T., Davies, D.W., Cartwright, H. et al. (2018). Machine learning for molecular and materials science. *Nature* 559 (7715): 547–555.

168 Shi, J., Kantoff, P.W., Wooster, R., and Farokhzad, O.C. (2016). Cancer nanomedicine: progress, challenges and opportunities. *Nature Reviews Cancer* 17: 20.

169 Lim, E.-K., Kim, T., Paik, S. et al. (2015). Nanomaterials for theranostics: recent advances and future challenges. *Chemical Reviews* 115 (1): 327–394.

170 Stabler, C.L., Li, Y., Stewart, J.M., and Keselowsky, B.G. (2019). Engineering immunomodulatory biomaterials for type 1 diabetes. *Nature Reviews Materials* 4 (6): 429–450.

171 Cianchetti, M., Laschi, C., Menciassi, A., and Dario, P. (2018). Biomedical applications of soft robotics. *Nature Reviews Materials* 3 (6): 143–153.

172 Dellacherie, M.O., Seo, B.R., and Mooney, D.J. (2019). Macroscale biomaterials strategies for local immunomodulation. *Nature Reviews Materials* 4 (6): 379–397.

173 Ruskowitz, E.R. and DeForest, C.A. (2018). Photoresponsive biomaterials for targeted drug delivery and 4D cell culture. *Nature Reviews Materials* 3: 17087.

174 Doan, M. and Carpenter, A.E. (2019). Leveraging machine vision in cell-based diagnostics to do more with less. *Nature Materials* 18 (5): 414–418.

175 Kalinin, S.V., Sumpter, B.G., and Archibald, R.K. (2015). Big–deep–smart data in imaging for guiding materials design. *Nature Materials* 14: 973.

176 Copp, S.M., Bogdanov, P., Debord, M. et al. (2014). Base motif recognition and design of DNA templates for fluorescent silver clusters by machine learning. *Advanced Materials* 26 (33): 5839–5845.

177 Moosavi, S.M., Chidambaram, A., Talirz, L. et al. (2019). Capturing chemical intuition in synthesis of metal-organic frameworks. *Nature Communications* 10 (1): 539.

7

Nanomedicine for Targeting Delivery of Gene and Other DNA/RNA Therapies Based Viruses Engineering

Xiangrong Song, Mengran Guo, Zhongshan He, Xing Duan, and Wen Xiao

Sichuan University, State Key Laboratory of Biotherapy and Cancer Center, West China Hospital, Department of Critical Care Medicine, No. 37 Guoxue Lane, Wuhou District, Chengdu 610041, China

Gene-based therapies have revolutionized the field of medicine because of their ability to treat a wide range of diseases, including cancer, genetically related diseases, and viral infectious diseases. Drugs cannot always cure a malfunctioning human body. Sometimes, the only way to fix what ails a person is to tinker with their genes: the blueprint for how biological systems are built and how they function [1]. Gene therapy is a biological therapy that introduces foreign normal genes into target cells through gene transfer technology to correct or compensate for diseases caused by gene defects and abnormalities, and finally achieve the purpose of treatment. It involves the use of genetic material, such as DNA or RNA [2], to treat or cure genetic disorders, acquired diseases, and certain types of cancer. For example, some researchers are using gene-editing technologies such as CRISPR to precisely change DNA sequences [3, 4]. siRNA can induce specific silencing of target genes via RNA interference (RNAi) [5]. mRNA can express any gene of interest instantaneously [2, 6]. However, the delivery of these therapies remains a significant challenge due to their poor bioavailability, instability, and potential off-target effects. To address these issues, targeted delivery of these therapies has emerged as a promising approach, which can improve their precision and efficacy while minimizing side effects.

Viral vectors and non-viral vectors are the two main vector types used in the field of gene-based drugs delivery. Despite the wide application of virus vectors in both ex vivo and in vivo gene therapy delivery, their fundamental drawbacks, such as the risk of carcinogenesis, limited insertion size, immune response, and difficulty in mass, severely constrain their further applications [7, 8]. With the further development of non-viral vectors, lipid- or polymer-based nanocarriers have shown tremendous potential in gene drug delivery [3]. Importantly, nanoparticles can be easily functionalized with specific receptor ligands for more targeted therapeutic delivery. In this section, we mainly summarize the possibilities and expectations of using nanotechnology for targeted delivery of siRNA/mRNA/p-DNA/genome editing to tumors, brain, spleen, lymph node (LN), lung and liver, and discuss the importance of targeted delivery of gene-based nanomedicines for successful treatment outcomes.

Nanomedicine: Fundamentals, Synthesis, and Applications, First Edition. Edited by Yujun Song.
© 2025 WILEY-VCH GmbH. Published 2025 by WILEY-VCH GmbH.

7.1 Targeting Delivery of Gene Therapies to the Tumors

Over the past four decades, cancer has become a global threat to life. According to data from the WHO, cancer has claimed over 10 million lives in 2020, surpassing the number of deaths from the COVID-19 pandemic [9]. Since the 1990s, gene therapy has been proposed to promote tumor eradication by interfering with cell signaling pathways associated with cell growth and invasion, thereby enhancing the therapeutic efficacy of cytotoxic drugs, and reversing chemoresistance [9–11]. At present, gene therapy has become one of the most promising tumor therapies in clinical application [12]. According to the U.S. National Medical Library's clinicaltrials.gov database, as of 2022, there have been nearly 3000 clinical trials involving various tumor gene therapies. Well-designed gene therapy relies heavily on effective gene vectors, and targeted delivery at tumor sites has significant benefits for improving the accuracy and efficiency of cancer gene therapy. Nanocarrier modifications for targeted delivery to tumors are mainly divided into two types: passive targeting and active targeting (Figure 7.1).

7.1.1 Passive Targeting

Passive targeting mainly utilizes the physiological structural characteristics of the tumor microenvironment to allow the drug to generate natural distribution differences in the body, thereby accumulating more at the tumor site. One of the most

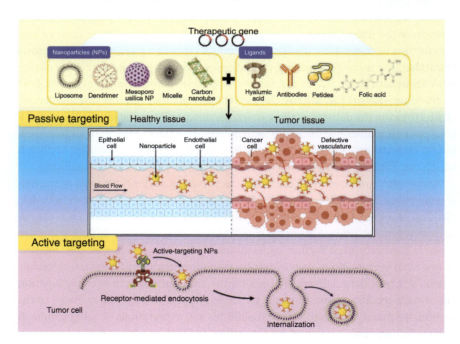

Figure 7.1 Nanocarriers targeting delivery to tumors are mainly divided into two categories: passive targeting and active targeting.

widely known passive targeting is the enhanced permeability and retention (EPR) effect is the most widely known passive targeting approach, which is based on the differences in microvascular structure between solid tumors and normal tissues. This difference results in macromolecules or particles with a diameter of around 100 nm being more prone to accumulate inside the tumor tissue and achieve targeted effects. However, the classic "passive targeting" based on the EPR effect has encountered constraints in clinical practice due to the heterogeneity of tumors and individual differences [12]. In addition, drug release at tumor could also be achieved by utilizing the unique pH [13], enzyme environment [14, 15], and intracellular reductive environment [16] of the tumor site, so as to achieve the purpose of targeted drug delivery.

7.1.2 Active Targeting

Active targeting delivery of tumors refers to the selective recognition and effective absorption of nanocarriers by target cancer cells. The main method is to chemically or physically link probe molecules such as antibodies, peptides, glycans, and nucleic acid aptamers that can specifically bind to target molecules on the surface of nanocarriers, so as to achieve active targeting. Tumor targeting ligands, such as epidermal growth factor receptor (EGFP) [17], folate (FA) [18], arginine–glycine–aspartic acid peptide (iRGD) [19], and so on.

7.2 Targeting Delivery of Gene-based Nanovaccines to the Spleen

Spleen, the body's largest immune organ, is considered as the occurrence center of cellular and humoral immunity. The spleen is rich in a large number of important immune cells and immune factors involved in immune response, which play a very important role in the field of immunotherapy [20, 21]. Splenic delivery is expected to stimulate effective immune activation (Figure 7.2). Therefore, the spleen is an ideal target organ for immunotherapy drugs, such as CAR-T and vaccine [22–24]. It is reported that large (>200 nm), rigid nanoparticles are more easily absorbed by

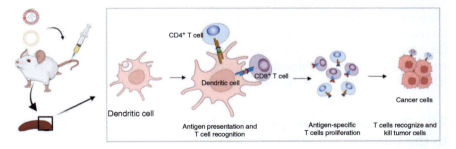

Figure 7.2 Targeting delivery of gene based nanovaccines to dendritic cells in the spleen can induce a strong immune response for cancer therapy.

the spleen [25]. In addition, some studies have shown that the regulation of lipid composition changes the tissue tropism of nanoparticles. Recently, mRNA vaccines, represented by mRNA-LNP vaccine, have attracted much attention. Cheng et al. developed selective organ targeting (SORT) strategies [26]. They found that the addition of SORT molecule, 1,2-dioleyl-*sn*-glycerol-3-phosphate (18 Pa) to LNP enhanced mRNA delivery to the spleen.

7.3 Targeting Delivery of Gene-based Nanovaccines to the LNs

The lymph nodes (LNs) are immune organs composed of immune cells and are another important site for activating the immune system [27]. Therefore, targeting LNs is an effective strategy to enhance the immune response [28], commonly used in vaccine design [29, 30]. Immune cells, substances, and vaccines usually enter the lymph node through the lymphatic system. In this respect, subcutaneous [31], intradermal [32], or intramuscular [33] administration is mainly related to lymphatic transport. It has been reported that 20–50 nm nanoparticles with anionic properties are ideal materials for delivery to LNs. In addition, antigen-presenting cells (APCs/such as dendritic cells, DCs) engulf pathogens, tumor cells, and vaccines in peripheral tissues and transfer them to LNs [34]. DC-mediated lymphatic transport can improve the efficiency of nanoparticle delivery to the lymph nodes. DC-targeted DNA/mRNA vaccines can better initiate an anti-tumor immune response [35, 36].

7.4 Targeting Delivery of Gene-based Nanomedicines to the Liver

The liver is an important organ of human body, is the most important metabolic center. Many diseases can affect the liver, including very serious diseases such as tyrosinemia [37] and autoimmune hepatitis [38]. In addition, the liver can be used as a protein factory to treat a variety of rare diseases through protein replacement therapy [39]. Thus, liver is also a good target for gene therapy, and optimized nanomedicine delivery is important. Nanoparticles with high charge and/or large size have been reported to be rapidly absorbed by Kupfer cells when administered intravenously. Fortunately, LNP delivery to the liver is very effective [40, 41]. LNP was easy to bind with ApoE after intravenous administration (i.m.), which changed the composition of LNP surface. Due to the specific binding of ApoE to the receptors on hepatocytes [42], LNP mainly accumulated in the liver (Figure 7.3). Patisiran (ONPATTRO™) is the first LNPs delivery RNA drug approved by the US FDA [43].

7.5 Targeting Delivery of Gene-based Nanomedicines to the Lung

The lungs are the main gateway for the body to the external environment and are easily exposed to various harmful substances in the air, which may lead to acute and

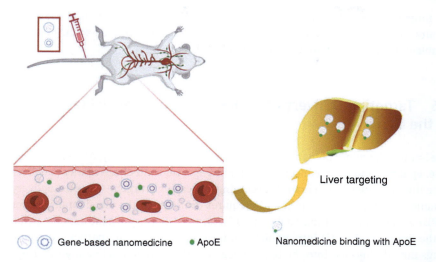

Figure 7.3 LNP binding with ApoE tends to be accumulated in the liver after intravenous administration.

chronic respiratory diseases [44]. Gene therapy provides a new innovative approach for the treatment of refractory lung diseases [45]. For treatments that must reach the lungs, three routes are prioritized: intravenous, intranasal, and inhalation. Local administration through intranasal and inhalation routes can achieve excellent gene-based nanomedicine delivery to the lung [46], especially with the assistance of functionalized lipid nanoparticles (Figure 7.4). For intravenous administration, a selective organ targeting (SORT) strategy developed by Cheng et al. involves adding a SORT molecule, 1,2-dithiolane-3-pentamethylamine (DOTAP), to LNPs, which can specifically target the lungs [26]. In lung-active targeting design, GALA peptide has high affinity for the pulmonary endothelium, and is therefore used as a ligand

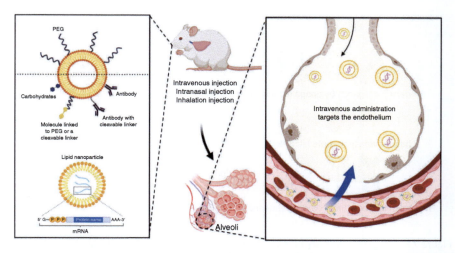

Figure 7.4 Targeting delivery of gene based nanomedicines to the lung can be achieved by functionalized lipid nanoparticles.

for targeting the pulmonary endothelium [47]. Hagino et al. reported that they obtained one of the highest lung-selective nanomedicines reported so far by using double-envelope LNPs with GALA peptides [48].

7.6 Targeting Delivery of Gene-based Nanomedicines to the Brain

The brain is the center of many incurable diseases, such as epilepsy, Alzheimer's disease, epilepsy, and brain cancer [49, 50]. Finding a safe and effective way to deliver gene therapy drugs to the brain is becoming increasingly urgent. Targeted delivery of nanomedicine to the brain mainly requires overcoming the blood–brain barrier (BBB). There are several approaches used for targeted delivery of gene nanomedicine to the brain [51–53]: (i) intracranial administration, using tiny particles carrying drugs directly injected into the brain, bypassing the BBB and delivering the drugs directly to the target and (ii) surface modification of nanomedicines to overcome the BBB, mainly focused on the following areas: permeability regulation of the BBB; screening of targeting ligands using phage display technology; and hitchhiking on endogenous transport pathways (Figure 7.5).

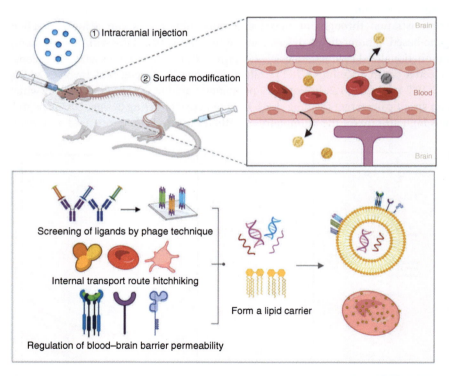

Figure 7.5 Targeting delivery of gene based nanomedicines to the brain can be achieved by intracranial injection and surface modification of lipid nanoparticles.

References

1 Brody, H. (2018). Gene therapy. *Nature* 564: S5.
2 Qin, S., Tang, X., Chen, Y. et al. (2022). mRNA-based therapeutics: powerful and versatile tools to combat diseases. *Signal Transduction and Targeted Therapy* 7: 166.
3 Xu, C.F., Chen, G.J., Luo, Y.L. et al. (2021). Rational designs of in vivo CRISPR-Cas delivery systems. *Advanced Drug Delivery Reviews* 168: 3–29.
4 Zheng, Q., Qin, F., Luo, R. et al. (2021). mRNA-loaded lipid-like nanoparticles for liver base editing via the optimization of central composite design. *Advanced Functional Materials* 31: 2011068.
5 Morales-Becerril, A., Aranda-Lara, L., Isaac-Olive, K. et al. (2022). Nanocarriers for delivery of siRNA as gene silencing mediator. *EXCLI Journal* 21: 1028–1052.
6 Frye, M., Harada, B.T., Behm, M., and He, C. (2018). RNA modifications modulate gene expression during development. *Science* 361: 1346–1349.
7 Li, L., Hu, S., and Chen, X. (2018). Non-viral delivery systems for CRISPR/Cas9-based genome editing: challenges and opportunities. *Biomaterials* 171: 207–218.
8 Hammond, S.M., Aartsma-Rus, A., Alves, S. et al. (2021). Delivery of oligonucleotide-based therapeutics: challenges and opportunities. *EMBO Molecular Medicine* 13: e13243.
9 Nakamura, T., Sato, Y., Yamada, Y. et al. (2022). Extrahepatic targeting of lipid nanoparticles in vivo with intracellular targeting for future nanomedicines. *Advanced Drug Delivery Reviews* 188: 114417.
10 Younis, M.A., Khalil, I.A., Elewa, Y.H.A. et al. (2021). Ultra-small lipid nanoparticles encapsulating sorafenib and midkine-siRNA selectively-eradicate sorafenib-resistant hepatocellular carcinoma in vivo. *Journal of Controlled Release* 331: 335–349.
11 Xu, B., Jin, Q., Zeng, J. et al. (2016). Combined tumor- and neovascular-"dual targeting" gene/chemo-therapy suppresses tumor growth and angiogenesis. *ACS Applied Materials & Interfaces* 8: 25753–25769.
12 Younis, M.A., Khalil, I.A., and Harashima, H. (2020). Gene therapy for hepatocellular carcinoma: highlighting the journey from theory to clinical applications. *Advanced Therapeutics* 3: 2000087.
13 Guan, X.W., Guo, Z., Lin, L. et al. (2016). Ultrasensitive pH triggered charge/size dual-rebound gene delivery system. *Nano Letters* 16: 6823–6831.
14 Xu, B., Xia, S., Wang, F. et al. (2016). Polymeric nanomedicine for combined gene/chemotherapy elicits enhanced tumor suppression. *Molecular Pharmaceutics* 13: 663–676.
15 Yu, T., Xia, S., Wang, F. et al. (2016). Pigment epithelial-derived factor gene loaded novel COOH-PEG-PLGA-COOH nanoparticles promoted tumor suppression by systemic administration. *International Journal of Nanomedicine* 11: 743–759.
16 Xu, F., Zhong, H., Chang, Y. et al. (2018). Targeting death receptors for drug-resistant cancer therapy: codelivery of pTRAIL and monensin using

dual-targeting and stimuli-responsive self-assembling nanocomposites. *Biomaterials* 158: 56–73.

17 Lian, F., Ye, Q., Feng, B. et al. (2020). rAAV9-UPII-TK-EGFP can precisely transduce a suicide gene and inhibit the growth of bladder tumors. *Cancer Biology & Therapy* 21: 1171–1178.

18 He, Z.Y., Wei, X.W., Luo, M. et al. (2013). Folate-linked lipoplexes for short hairpin RNA targeting claudin-3 delivery in ovarian cancer xenografts. *Journal of Controlled Release* 172: 679–689.

19 Xiao, W., Zhang, W., Huang, H. et al. (2020). Cancer targeted gene therapy for inhibition of melanoma lung metastasis with eIF3i shRNA loaded liposomes. *Molecular Pharmaceutics* 17: 229–238.

20 Jindal, A.B. (2016). Nanocarriers for spleen targeting: anatomo-physiological considerations, formulation strategies and therapeutic potential. *Drug Delivery and Translational Research* 6: 473–485.

21 Jiang, Y., Hardie, J., Liu, Y. et al. (2018). Nanocapsule-mediated cytosolic siRNA delivery for anti-inflammatory treatment. *Journal of Controlled Release* 283: 235–240.

22 Kurosaki, T., Kodama, Y., Muro, T. et al. (2013). Secure splenic delivery of plasmid DNA and its application to DNA vaccine. *Biological & Pharmaceutical Bulletin* 36: 1800–1806.

23 Kimura, S., Khalil, I.A., Elewa, Y.H.A., and Harashima, H. (2019). Spleen selective enhancement of transfection activities of plasmid DNA driven by octaarginine and an ionizable lipid and its implications for cancer immunization. *Journal of Controlled Release* 313: 70–79.

24 Kimura, S., Khalil, I.A., Elewa, Y.H.A., and Harashima, H. (2021). Novel lipid combination for delivery of plasmid DNA to immune cells in the spleen. *Journal of Controlled Release* 330: 753–764.

25 Hoshyar, N., Gray, S., Han, H., and Bao, G. (2016). The effect of nanoparticle size on in vivo pharmacokinetics and cellular interaction. *Nanomedicine (London, England)* 11: 673–692.

26 Cheng, Q., Wei, T., Farbiak, L. et al. (2020). Selective organ targeting (SORT) nanoparticles for tissue-specific mRNA delivery and CRISPR-Cas gene editing. *Nature Nanotechnology* 15: 313–320.

27 Najibi, A.J. and Mooney, D.J. (2020). Cell and tissue engineering in lymph nodes for cancer immunotherapy. *Advanced Drug Delivery Reviews* 161–162: 42–62.

28 Yu, X., Dai, Y., Zhao, Y. et al. (2020). Melittin-lipid nanoparticles target to lymph nodes and elicit a systemic anti-tumor immune response. *Nature Communications* 11: 1110.

29 Chen, J., Ye, Z., Huang, C. et al. (2022). Lipid nanoparticle-mediated lymph node-targeting delivery of mRNA cancer vaccine elicits robust CD8(+) T cell response. *Proceedings of the National Academy of Sciences of the United States of America* 119: e2207841119.

30 He, R., Zang, J., Zhao, Y. et al. (2022). Nanotechnology-based approaches to promote lymph node targeted delivery of cancer vaccines. *ACS Biomaterials Science & Engineering* 8: 406–423.

31 Zhong, X., Zhang, Y., Tan, L. et al. (2019). An aluminum adjuvant-integrated nano-MOF as antigen delivery system to induce strong humoral and cellular immune responses. *Journal of Controlled Release* 300: 81–92.

32 Smedley, J.V., Bochart, R.M., Fischer, M. et al. (2022). Optimization and use of near infrared imaging to guide lymph node collection in rhesus macaques (*Macaca mulatta*). *Journal of Medical Primatology* 51: 270–277.

33 Huang, C.H., Huang, C.Y., and Huang, M.H. (2019). Impact of antigen–adjuvant associations on antigen uptake and antigen-specific humoral immunity in mice following intramuscular injection. *Biomedicine & Pharmacotherapy* 118: 109373.

34 Herve, P.L., Plaquet, C., Assoun, N. et al. (2021). Pre-existing humoral immunity enhances epicutaneously-administered allergen capture by skin DC and their migration to local lymph nodes. *Frontiers in Immunology* 12: 609029.

35 Kranz, L.M., Diken, M., Haas, H. et al. (2016). Systemic RNA delivery to dendritic cells exploits antiviral defence for cancer immunotherapy. *Nature* 534: 396–401.

36 Wang, F., Xiao, W., Elbahnasawy, M.A. et al. (2018). Optimization of the linker length of mannose-cholesterol conjugates for enhanced mRNA delivery to dendritic cells by liposomes. *Frontiers in Pharmacology* 9: 980.

37 van Ginkel, W.G., Rodenburg, I.L., Harding, C.O. et al. (2019). Long-term outcomes and practical considerations in the pharmacological management of tyrosinemia type 1. *Paediatric Drugs* 21: 413–426.

38 Czaja, A.J. (2016). Diagnosis and management of autoimmune hepatitis: current status and future directions. *Gut Liver* 10: 177–203.

39 Quiviger, M., Giannakopoulos, A., Verhenne, S. et al. (2018). Improved molecular platform for the gene therapy of rare diseases by liver protein secretion. *European Journal of Medical Genetics* 61: 723–728.

40 Zhang, H., You, X., Wang, X. et al. (2021). Delivery of mRNA vaccine with a lipid-like material potentiates antitumor efficacy through Toll-like receptor 4 signaling. *Proceedings of the National Academy of Sciences of the United States of America* 118: e2005191118.

41 Chen, K.P., Fan, N., Huang, H. et al. (2022). mRNA vaccines against SARS-CoV-2 variants delivered by lipid nanoparticles based on novel ionizable lipids. *Advanced Functional Materials* 32: 2204692.

42 Yan, X., Kuipers, F., Havekes, L.M. et al. (2005). The role of apolipoprotein E in the elimination of liposomes from blood by hepatocytes in the mouse. *Biochemical and Biophysical Research Communications* 328: 57–62.

43 Akinc, A., Maier, M.A., Manoharan, M. et al. (2019). The Onpattro story and the clinical translation of nanomedicines containing nucleic acid-based drugs. *Nature Nanotechnology* 14: 1084–1087.

44 Dobrowolski, C., Paunovska, K., Hatit, M.Z.C. et al. (2021). Therapeutic RNA delivery for COVID and other diseases. *Advanced Healthcare Materials* 10: e2002022.

45 Fan, N., Chen, K., Zhu, R. et al. (2022). Manganese-coordinated mRNA vaccines with enhanced mRNA expression and immunogenicity induce robust immune responses against SARS-CoV-2 variants. *Science Advances* 8: eabq3500.

46 Kim, D.I., Song, M.K., and Lee, K. (2019). Comparison of asthma phenotypes in OVA-induced mice challenged via inhaled and intranasal routes. *BMC Pulmonary Medicine* 19: 241.

47 Kusumoto, K., Akita, H., Ishitsuka, T. et al. (2013). Lipid envelope-type nanoparticle incorporating a multifunctional peptide for systemic siRNA delivery to the pulmonary endothelium. *ACS Nano* 7: 7534–7541.

48 Hagino, Y., Khalil, I.A., Kimura, S. et al. (2021). GALA-modified lipid nanoparticles for the targeted delivery of plasmid DNA to the lungs. *Molecular Pharmaceutics* 18: 878–888.

49 Ouyang, Q., Meng, Y., Zhou, W. et al. (2022). New advances in brain-targeting nano-drug delivery systems for Alzheimer's disease. *Journal of Drug Targeting* 30: 61–81.

50 Godbout, K. and Tremblay, J.P. (2022). Delivery of RNAs to specific organs by lipid nanoparticles for gene therapy. *Pharmaceutics* 14: 2129.

51 Alotaibi, B.S., Buabeid, M., Ibrahim, N.A. et al. (2021). Potential of nanocarrier-based drug delivery systems for brain targeting: a current review of literature. *International Journal of Nanomedicine* 16: 7517–7533.

52 Liu, P. and Jiang, C. (2022). Brain-targeting drug delivery systems. *Wiley Interdisciplinary Reviews. Nanomedicine and Nanobiotechnology* 14: e1818.

53 Rawal, S.U., Patel, B.M., and Patel, M.M. (2022). New drug delivery systems developed for brain targeting. *Drugs* 82: 749–792.

8

Nanomedicine for Bio-imaging and Disease Diagnosis

Ziqi Wang[1] and Yujun Song[1,2,3]

[1]*University of Science and Technology Beijing, Center for Modern Physics Technology, School of Mathematics and Physics, 30 Xueyuan Road, Haidian District, Beijing 100083, China*
[2]*Zhengzhou Tianzhao Biomedical Technology Company Ltd., Zhengzhou New Technology Industrial Development Zone, 7B-1209 Dongqing Street, Zhengzhou 451450, China*
[3]*Key Laboratory of Pulsed Power Translational Medicine of Zhejiang Province, Hangzhou Ruidi Biotechnology Company Ltd., Room 803, Bldg. 4, 4959 Yuhangtang Road, Cangqian Street, Hangzhou 310023, China*

8.1 Concepts, Types, and Features of Current Nanoprobes

It is well known that early and accurate diagnosis of major diseases, especially tumors, is the key to improving the cure rate and the quality of life of patients. Therefore, the development of highly efficient, precise, and sharp diagnostic technology is one of the main goals of modern medicine. The rapid development of imaging, especially molecular imaging, plays an ever increased role in the early diagnosis of diseases, active drug screening, and even real-time evaluation of therapeutic effects. Molecular imaging is an emerging interdisciplinary discipline that describes and measures biological processes in vivo, at the cellular and single molecular levels. Molecular imaging technology combines molecular probing technology with modern analytical imaging instruments, integrates biology, chemistry, optics, data processing, nanotechnology, image processing, and other disciplines and technologies, and analyzes molecules closely related to physiological and pathological activities in vivo. In particular, real-time and noninvasive imaging of genes/molecules and their transmission pathways plays a key role in the occurrence and development of some diseases, which is currently aiming at observing the physiological and pathological changes in the body at the cellular or single molecular level by imaging methods before the disease exhibits some distinct clinical symptoms. As an important part of molecular medicine, molecular imaging technology is playing an increasingly important role in modern diagnostics, enabling early diagnosis of diseases at the single molecular level for the personalized molecular therapy.

In recent years, molecular imaging has been well applied in the clinical diagnosis and the study of the fundamental biomedical mechanism, which not only

Nanomedicine: Fundamentals, Synthesis, and Applications, First Edition. Edited by Yujun Song.
© 2025 WILEY-VCH GmbH. Published 2025 by WILEY-VCH GmbH.

promotes the early diagnosis and treatment of diseases, also introduces a new concept for clinical diagnosis. The core of molecular imaging is the perfect fusion of fast, sensitive, and high-resolution imaging equipment with high-specificity and high-affinity image nanoprobes. The imaging probe refers to the complex that can bind specifically to the target and produce imaging signals (such as light, magnetic, electrical). The ideal molecular imaging probe should meet the following conditions [1]: (i) it has high specific binding and affinity to the target; (ii) it has good permeability and can quickly cross biological barriers, such as blood vessels and cell membranes; (iii) it does not cause obvious immune response of the body. It remains relatively stable in vivo, has a proper clearance period in the blood circulation, and can be fully bound to the target biomolecules without obvious side effects; and (iv) it can be coupled with image signal molecules to amplify the signal that needs to be detected to a certain extent to facilitate imaging and generate effective detection information. At present, the commonly used molecular imaging probes include various conventional nonspecific contrast agents, molecular probes with specific molecular ligands, and nanoprobes developed with nanotechnology in recent years.

The types and characteristics of nanoprobes can be divided according to different fields and principles, including optical nanoprobes, magnetic nanoprobes, CT nanoprobes, nuclide imaging nanoprobes, photoacoustic imaging nanoprobes, and multifunctional influence nanoprobes. This chapter will give the detailed classification, synthesis, characterization and application of these ultrasensitive nanoprobes for disease diagnosis and biomedical mechanism study.

8.1.1 Optical Nanoprobe

Optical molecular imaging has the advantages of noninvasiveness, nonradiation, high safety, high sensitivity, and high resolution. The resolution of NIR fluorescence imaging is 1–2 mm, and it can penetrate 8 cm-thick tissue. The fluorescence imaging signal is strong, the background is low, and the bright signal can be directly emitted. In recent years, optical probes based on precious metal nanomaterials (gold, silver, and other nanoparticles), semiconductor nanomaterials (CdSe quantum dots), inorganic nonmetallic nanomaterials (carbon quantum dots, graphene quantum dots) have developed rapidly. In particular, the application of near-infrared probes in in vivo tracing and the translational medicine endows optical molecular imaging with great potential.

For instance, the nanomedicine (Au@CoFeB–Rg3) prepared by Zhang et al. exhibits multimodal imaging capabilities, and demonstrates remarkable efficacy against tumors [2]. Moreover, the magnetoplasma nanocore–shell structure exhibits exceptional optical properties within the visible wavelength range. This is illustrated in Figure 8.1. Similar magnetic-precious metal core–shell structures were also studied by Ma et al. [3]. The absorption peaks of these nanoparticles in high-frequency bands are broader compared to those of pure noble metal particles. At present, varieties of semiconductor nanomaterials or quantum dots have been rapidly developed for biomedical imaging application. One of these studies conducted by

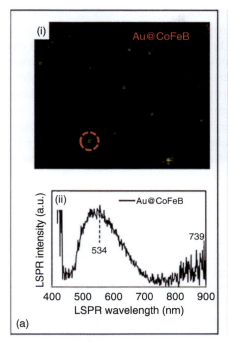
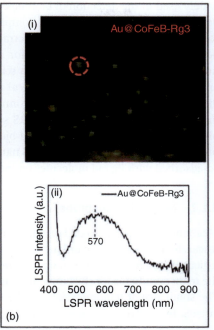

Figure 8.1 LSPR real color images of (a(i) Au@CoFeB NPs and b(i) Au@CoFeB–Rg3 NMs) and their LSPR spectra (a(ii) Au@CoFeB and b(ii) Au@CoFeB–Rg3) measured by dark-field microscopy and spectroscopy. Source: From Ref. [2], 2020, American Chemical Society.

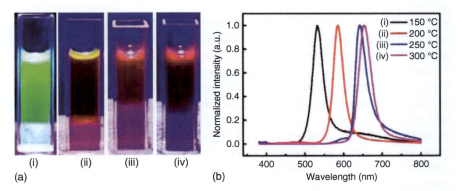

Figure 8.2 Solution color under ultraviolet illumination of CdSe nanocrystal solution (a) and the normalized photoluminescence spectra of CdSe nanocrystals and (b) synthesized at different reaction temperatures of (i) 150, (ii) 200, (iii) 250, and (iv) 300 °C. Source: From Ref. [4], 2017, American Chemical Society.

Wang et al. exclusively synthesized highly crystalline CdSe quantum dots, which exhibit exceptional light absorption and photoluminescence properties within the UV-visible spectrum, as shown in Figure 8.2 [4].

Raman nanoprobe is also an optical nanoprobe, which is a highly sensitive detection technique based on surface-enhanced Raman. At present, it has been widely used in the detection of biomolecules, such as glucose sensing, nucleic acid detection, and protein detection. Compared with fluorescence, surface-enhanced Raman spectroscopy has narrow spectral lines and is not photobleached, so it can be used as a molecular fingerprint for multichannel detection. Recently, Raman imaging technology based on Raman scattering nanoprobes has been successfully applied to the detection of physiological activities of living cells and tumor imaging diagnosis. Due to the complex environment in vivo, SERS probes applied in vivo imaging required not only high SERS sensitivity but also suitable size and morphology as well as good stability and biocompatibility. Porter's research group [5] prepared the first generation of SERS nanoprobes by reporting that molecules were co-adsorbed with ligand molecules on metal nanoparticles.

8.1.2 Magnetic Nanoprobe

Magnetic nanobiomaterials are widely used in clinical diagnosis and biomedical research because of their unique magnetic properties. In addition to utilizing biological effects related to the chemical composition and structure of nanomaterials, the main paradigms for the biomedical applications of magnetic nanomaterials are: The small size of nanomaterials can be used to deliver them to specific biological targets. By applying a safe external magnetic field stimulation without tissue penetration depth limitation, magnetic nanomaterials can produce magnetic, thermal, force, and other physical effects and act on biological targets, and then trigger a variety of biological effects to achieve diagnostic or therapeutic functions.

Magnetic properties are the key to the biomedical applications of magnetic nanomaterials. Material design can effectively regulate the magnetic properties of nanomaterials, so as to improve the application performance. In general, its magnetic performance parameters mainly include: saturation magnetization, magnetic anisotropy, magnetic remanence, and coercivity. Magnetization is a physical quantity that describes the magnetic strength of a macroscopic magnetic medium. For the ferromagnetic ordered nanoferric oxide materials, the maximum magnetization that can be achieved with the continuous enhancement of the external magnetic field is saturation magnetization, which is related to the size, composition, and morphology of the nanomaterial. For example, saturation magnetization increases proportionally with the increase of particle size within the threshold, tends to be constant beyond the threshold, and approaches the magnetization of bulk phase materials. Magnetic anisotropy refers to the difficulty of magnetization in different directions, including magnetic crystal anisotropy, surface anisotropy, and exchange anisotropy. The magnetocrystalline anisotropy is affected by the cell structure. For example, when Fe^{2+} is completely replaced by Co^{2+} in iron oxide nanoparticles, the magnetic anisotropy increases 20 times compared with the original particles. Surface anisotropy and exchange anisotropy are affected by particle morphology. Generally, nanoparticles with larger surface area contribute more significantly to surface anisotropy, and core–shell structure also affects exchange anisotropy. Coercivity can reflect the ability of material to resist demagnetization, which is significantly affected by particle size change. It can be seen that ferromagnetic

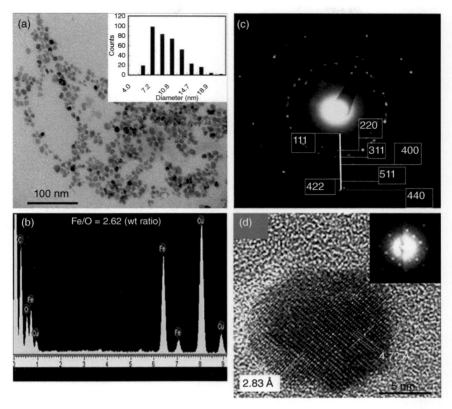

Figure 8.3 (a) Typical TEM image of the Fe_3O_4 NPs; the *inset* shows a histogram of the size distribution of Fe_3O_4 NPs (average diameter: 9.8 ± 3.0 nm), (b) EDS of Fe_3O_4 NPs, (c) electron–diffraction pattern of an ensemble of Fe_3O_4 NPs, and (d) HRTEM image of a typical single crystal of Fe_3O_4 NPs with distinct crystalline lattices. Source: From Ref. [6], 2011, European Journal of Inorganic Chemistry.

nanomaterials exhibit size-dependent magnetic properties. When the size is less than a certain critical value and the magnetic crystal anisotropy energy (KV) is less than the heat energy at room temperature (KT), the coercivity and remanence tend to zero, showing superparamagnetic properties.

Magnetic nanoprobes have also been extensively studied recently, such as Song et al. successfully synthesized Fe_3O_4 nanoparticles with a size of 10 nm using a modified iron salt coprecipitation process, as shown in Figure 8.3 [6]. These nanoparticles exhibited excellent dispersion stability in water and demonstrated remarkable superparamagnetization, displaying a saturation magnetization of 48.0 emu/g under a low magnetic field strength of 0.5 T. Studies have demonstrated minimal toxicity to the examined organs, and following over 48 hours of recycling, they ultimately migrate to the spleen with no significant retention in any associated organs.

The surface modification of Fe and CoFe nanoparticles was expanded by Wang et al. through the doping of zinc oxide and alumina, resulting in the formation of

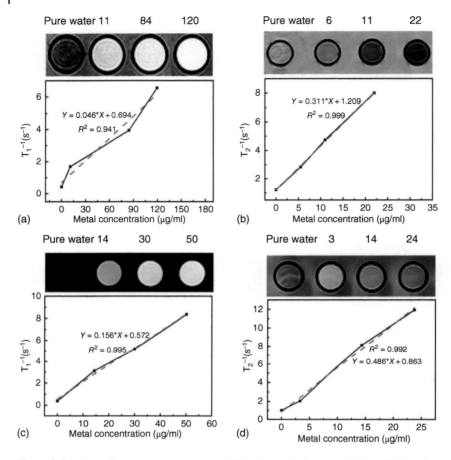

Figure 8.4 Magnetic resonance imaging (MRI) effects of ultra-small $FeZn_x@Zn(1-y)$-$FeyO-(OH)z$ and $(CoFe)(1-x)Alx@Al(1-x)(CoFe)xO-(OH)z$ hybrid nanoparticle solutions. Source: From Ref. [7], 2015, Royal Society of Chemistry.

well-dispersed nanohybrids. The enhanced T_1-weighted spin-echo imaging and T_2-weighted spin-echo imaging exhibited by the agent demonstrate its potential as an exceptional negative MRI contrast agent, as shown in Figure 8.4 [7].

8.1.3 Photoacoustic Imaging Nanoprobe

Photoacoustic imaging is a new and rapidly developed nondestructive biophoton imaging method based on the optical absorption differences in biological tissues and using ultrasound as the medium. It combines the advantages of high contrast characteristic of pure optical imaging and high penetration characteristic of pure ultrasonic imaging, and uses ultrasonic detector to detect photo-sound waves instead of photon detection in optical imaging. In principle, it avoids the influence of optical scattering, and can provide high-contrast and high-resolution tissue images, providing an important means for studying the structural morphology, physiological characteristics, metabolic function, and pathological characteristics of biological tissues, and has a wide application prospect in biomedical clinical diagnosis as well as in the field of tissue structure and function imaging. However, due to the strong scattering and attenuation of incident light energy in biological tissues, the sensitivity of photoacoustic images can be reduced. Therefore, the detection sensitivity of photoacoustic images can be well enhanced by using photoacoustic imaging nanoprobes, and structural imaging, which mainly focuses on morphological observation, can be extended to functional molecular imaging.

8.1.4 CT Nanoprobe

The concept of CT imaging is an imaging technology in which X-ray beams are irradiated from multiple directions along a selected body layer, the X-ray dose is measured, the absorption coefficient of each unit volume of the tissue is calculated after digitization, and the image is reconstructed. The results of CT imaging are mainly compared in two ways: (i) natural contrast and (ii) artificial contrast. (i) Natural contrast refers to the density of human tissue structure, which can be divided into high-density bone tissue, calcification; medium density of cartilage, muscle, parenchymal organs, nerves, connective tissues, and body fluids; low density of fat and gas. The formation of natural contrast depends on the difference in density and thickness of human tissue structure. (ii) Artificial contrast means that many tissues in the human body are not significantly compared due to factors such as small density difference, overlap or thickness, and it is generally necessary to apply artificial comparison methods to display the anatomical structure. Positive contrast media such as barium and iodine can be used for artificial contrast. Negative contrast media can be used simultaneously for enhanced contrast imaging, such as double aero-barium angiography of the digestive tract.

At present, the commonly used commercial CT contrast agents are organic molecules containing iodine. Iodine has a high X-ray absorption coefficient. However, the iodine content is quickly cleared by the kidney, which makes the imaging time short and has renal toxicity. Moreover, X-rays can induce iodine-containing substances to ionize iodine ions, resulting in stronger toxicity. With the development of nanomaterials and biotechnology, nano-CT probes based on gold nanoparticles have been vigorously developed. Gold has a higher atomic number and X-ray absorption coefficient than iodine. The contrast effect of gold was about 2.7 times higher than that of iodine per unit mass [8]. Due to the excellent surface properties, mature preparation route and good biocompatibility of gold nanoparticles, CT contrast agents based on gold nanoparticles are expected to improve the above-mentioned shortcomings of traditional CT contrast agents.

8.1.5 Nuclide Imaging Nanoprobe

The basic principle of nuclear medicine imaging is to directly image cell metabolism and function on the basis of in vivo by injecting radionuclide markers, mainly including positron emission tomography (PET) and single photon emission computed tomography (SPECT) imaging. PET mainly uses positron nuclides such as ^{11}C, ^{13}N, ^{15}O, and ^{18}F, while SPECT uses single photon nuclides such as ^{99}Tc^m, ^{111}In, and ^{201}Tl. PET is significantly better than SPECT in terms of spatial resolution, detection sensitivity, and quantitative accuracy. The spatial resolution of PET is less than 100 μm, and the spatial resolution can reach about 1 mm^3 in experimental animal models, and its sensitivity is 100–1000 times that of SPECT.

The most commonly used PET contrast agent is ^{18}F-fluorodeoxyglucose (FDG). In terms of tumor imaging, FDG is widely used in tumor imaging because most malignant tumor cells have high glucose metabolism. However, the essence of FDG is a glucose metabolism-specific imaging agent rather than a tumor-specific imaging agent, and it cannot accurately show the range and contour of tumor diffusion. Nanoprobes possess imaging signals, targeting functions, and tunable pharmacokinetic properties by modifying isotopes, biological ligands, and PEG. Compared with the original single contrast agent signal molecule, the multifunctional nanoprobe makes the image clearer and image-based diagnosis more accurate.

8.1.6 Multifunctional/Multimodal Nanoprobes

Optical imaging has the advantages of adjustable emission spectrum and short imaging time, but it has some disadvantages such as poor tissue penetration and spontaneous fluorescence interference. MRI has strong tissue penetration, but the imaging sensitivity is low. CT has high resolution for dense tissue, and it is often difficult to diagnose the lesions of hollow organs. PET provides detailed functional and metabolic molecular information of lesions, and has the advantages of sensitivity, accuracy, and specificity, but there are still some false positive and false negative phenomena. Table 8.1 [9] is a comparison of the resolution, imaging depth, imaging time, and other parameters of various biological imaging technologies. As can be seen from the table, each imaging technique and mode has its own advantages and disadvantages. Therefore, with the improvement of disease diagnosis requirements, a single imaging mode cannot fully meet the needs of diagnosis, and it is urgent to develop dual-mode or even multimode imaging technology. PET/CT dual-mode imaging is a very successful example. One of the prerequisites for the development of multimode imaging is to develop multimode multifunction image probe.

8.2 Synthesis Methods

The optical, electrical, and chemical properties of nanoprobes used in medical imaging and diagnostics are affected by small changes in the material itself, such as elemental composition, size, shape, internal structure, surface chemistry, and assembly structure. For example, the strength of the optical signal from surface-enhanced Raman scattering depends mainly on the design of the Raman-enhanced substrate probe material [10]. Compared with the spherical, rod-shaped or star-shaped nanoprobes of precious metals, the electromagnetic field intensity is significantly enhanced, and the signal can be amplified by 1–2

Table 8.1 Comparison of biological imaging techniques.

Technique	Resolution[a]	Depth	Time[b]	Quantitative[c]	Multi-channel	Imaging agents	Target	Cost[a],[d]	Main small-animal use	Clinical use
MRI	10–100 μm	No limit	Minutes to hours	Yes	No	Paramagnetic chelates, magnetic particles	Anatomical, physiological, molecular	$$$	Versatile imaging modality with high soft-tissue contrast	Yes
CT	50 μm	No limit	Minutes	Yes	No	Iodinated molecules	Anatomical, physiological	$$	Imaging lungs and bone	Yes
Ultrasound	50 μm	cm	Seconds to minutes	Yes	No	Microbubbles	Anatomical, physiological	$$	Vascular and interventional imaging||	Yes
PET	1–2 mm	No limit	Minutes to hours	Yes	No	^{18}F-, ^{64}Cu-, or ^{11}C-labeled compounds	Physiological, molecular	$$$	Versatile imaging modality with many tracers	Yes
SPECT	1–2 mm	No limit	Minutes to hours	Yes	No	^{99m}Tc- or ^{111}In-labeled compounds	Physiological, molecular	$$	Imaging labelled antibodies, proteins, and peptides	Yes
Fluorescence reflectance imaging	2–3 mm	<1 cm	Seconds to minutes	No	Yes	Photoproteins, fluorochromes	Physiological, molecular	$	Rapid screening of molecular events in surface-based disease	Yes
FMT	1 mm	<10 cm	Minutes to hours	Yes	Yes	Near-infrared fluorochromes	Physiological, molecular	$$	Quantitative imaging of fluorochrome reporters	In development
Biolumine-scence imaging	Several mm	cm	Minutes	No	Yes	Luciferins	Molecular	$$	Gene expression, cell, and bacterium tracking	No
Intravital microscopy	1 μm	<400–800 μm	Seconds to hours	No	Yes	Photoproteins, fluorochromes	Anatomical, physiological, molecular	$$	All of the above at higher resolutions but limited depths and coverage	In development[e]

a) For high-resolution, small-animal imaging systems. (Clinical imaging systems differ.)
b) Time for image acquisition.
c) Quantitative here means inherently quantitative. All approaches allow relative quantification.
d) Cost is based on purchase price of imaging systems in the United States: $, <US$100,000; $$, US$100,000–300,000; $$$, >US$300,000.
e) For microendoscopy and skin imaging.

orders of magnitude. Compared with the spherical, rod-shaped or star-shaped nanoprobes of precious metals, the electromagnetic field intensity is significantly enhanced, and the signal can be amplified by 1–2 orders of magnitude. Even with the same morphology, the monodispersion of the probe particle size can make the excitation wavelength of the light source and the absorption wavelength of the probe material reach the maximum resonance coupling, and again improve the optical detection signal by two orders of magnitude. In addition, the particle spacing of the nanoprobe also has a great influence on its sensitivity. When the distance between adjacent nanoparticles is less than 20 nm, the optical hot spot effect will be generated between the nanoprobes, which again improves the sensitivity of biological detection by 1–2 orders of magnitude. In view of the great influence of the nanoprobe material itself, its precise synthesis and controllable self-assembly are an effective way to improve the performance of nanobiologic probes in the field of medical imaging and diagnosis. There are a variety of methods to prepare nanoprobe materials, and different methods can obtain nanomaterials with different morphologies and properties. The following will introduce the synthesis method of nanoprobe materials by taking magnetic nanoparticles and precious metal nanoparticles as examples.

8.2.1 Preparation and Synthesis of Noble Metal Nanoprobes

The preparation of noble metal nanoparticles can be divided into two categories: physical methods and chemical methods. The metal is made into nanoscale small particles by physical means, which is called physical method, including gas phase method, metal vapor solvation method, mechanical grinding method, etc. Compared with physical methods, noble metal nanoparticles prepared by chemical methods have the advantages of uniform particle size and uniform dispersion, and chemical methods can synthesize noble metal nanoparticles with different morphologies by controlling reaction time and stabilizing ligands. Therefore, rapid development has been achieved in recent years. At present, there are several common chemical methods for the synthesis of precious metal nanoparticles.

8.2.1.1 Chemical Synthesis Method

The precious metal nanoparticles were obtained by dissolving the precious metal salts in the solution and then reducing the precious metal elements in the solution with a reducing agent. The reducing agents used are sodium borohydride, ascorbic acid, glucose, and hydrazine hydrate. In addition, in order to improve the dispersion and stability of precious metal nanoparticles, a dispersant is often added to the reaction solution because the dispersant can effectively avoid agglomeration and crosslinking caused by collisions between particles. At present, widely used dispersants include: cetyltrimethyl ammonium bromide, cetyltrimethyl ammonium chloride, n-dodecyl mercaptan, dihexadecylpyridine dithiophosphate (PyDDP), aniline, thioglycolic acid, and formaldehyde sulfonate naphthalene sodium salt. By changing the temperature, pH value, the initial amount of reducing agent, dispersant, noble metal compound, and the order of adding different substances,

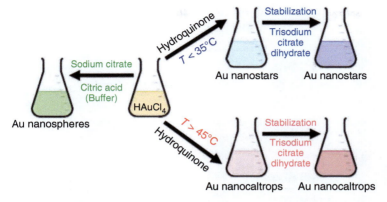

Figure 8.5 Chemical reduction method. Source: Richard et al. [11], American Chemical Society.

the controlled synthesis of noble metal nanoparticles with different morphology and particle size can be achieved by chemical reduction method (Figure 8.5).

The core–shell structure of nanohybrids is also a hot topic recently. Wang et al. have developed a novel approach utilizing nuclear alloying and shell gradient doping for the synthesis of nanohybrids, which is achieved through competitive reactions or sequential reduction nucleation and co-precipitation reactions of mixed metal salts during microfluidic and batch cooling processes [12]. By employing this approach, alloys of CoM, FeM, AuM, and AgM (where M = Zn or Al) can serve as cores while transition metal gradient-doped ZnO or Al_2O_3 can act as shells for the synthesis of diverse gradient core–shell nanohybrids. These nanohybrids exhibit distinctive magnetic and optical properties. Moreover, the ultra-small (2.3 ± 0.7 nm) CoFe–Wo$_x$ nanohybrid with tiny CoFe nuclei and mixed metal oxide shells ($CoWO_4$-Fe_2WO_6) were synthesized by Wu et al. through precise microfluidic control, as shown in Figure 8.6 [13].

8.2.1.2 Photochemical Method

Photochemical method refers to the method of using light to synthesize nanoparticles. The basic principle is: when the reaction liquid is irradiated with light of specific wavelength, photoelectrons and reducing free radicals will be generated in the reaction liquid, and then the metal cations will be reduced by these free radicals to form nanoparticles. Compared with other methods, photochemical method has the advantages of safety, simplicity, cheapness, and less harm to the material itself. In recent years, researchers have used this method to obtain precious metal nanoparticles with different shapes, colors, high dispersion, and good stability.

8.2.1.3 Template-based Method

Template-based method is one of the most commonly used methods in the synthesis of nanoparticles, which can effectively control the growth of nanoparticles. So, it has advantage that other methods cannot be achieved. Common templates include vesicles, micelles, microemulsions, membranes, liquid crystals, nanosmall

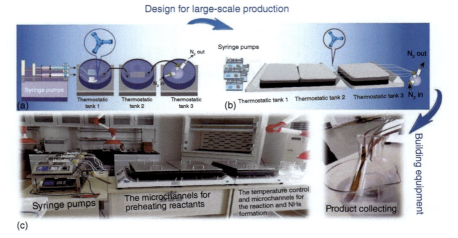

Figure 8.6 Schematic presentation of the SPMP (a), design of the PMS (b), and material objects of the PMS equipment (c). Source: From Ref. [13], 2023, Chemical Engineering Journal.

molecules, self-assembled polymers, and biomacromolecules. These templates have a wide variety of morphologies and have a limited domain effect on nanoparticles so that nanoparticles with different structures can be synthesized. In addition, the method is simple to operate and easy to synthesize. So, it is widely liked by people.

8.2.1.4 Electrochemical Process

Electrochemical method is a simple, fast, and environmentally friendly synthesis method. The equipment required is relatively simple, and the synthesis conditions are very mild. High-purity precious metal nanoparticles can be prepared by this method, especially the metal nanoparticles with high electronegativity, which is difficult to achieve by other methods. Alaa et al. [14] first modified the polyelectrolyte onto the surface of cowpea Mosaic virus by electrochemical method and then synthesized a gold nanoparticle with a very narrow particle size distribution using the modified virus as a template.

8.2.2 Preparation and Synthesis Method of Magnetic Nanoprobe

Magnetic nanoparticles can be synthesized by physical and chemical methods, including molecular beam epitaxy, high-pressure nanocrystalline, vapor deposition, steam rapid cooling, and pulverization. Magnetic nanoparticles can be synthesized by physical and chemical methods, including molecular beam epitaxy, high-pressure nanocrystalline, vapor deposition, steam rapid cooling, and pulverization.

8.2.2.1 Coprecipitation Method

The coprecipitation method is usually carried out under N_2 or Ar_2 protection, and the alkali is added to the salt solution of Fe^{2+}/Fe^{3+} to obtain the corresponding magnetic nanoparticles. In this process, the synthesis of magnetic nanoparticles with different compositions, sizes, and morphology can be achieved by changing the ionic strength, pH value, temperature, salt type, and proportion of the solution. Magnetic nanoparticles synthesized by coprecipitation method have wide particle size distribution, uneven morphology, poor dispersion, and easy aggregation, which cannot meet the needs of practical applications. So, some organic molecules are often added as dispersants in the synthesis process.

8.2.2.2 Thermal Decomposition Method

Thermal decomposition method refers to the method of thermal decomposition of metal-organic compounds in high boiling point organic solvents to obtain magnetic nanoparticles. Acetylacetone compounds, metal salts of carbonyl group and metal copper-iron reagents are commonly used in this process. In order to make the magnetic nanoparticles have good stability and dispersion, different surfactants are often added to the magnetic nanoparticles synthesized by thermal decomposition method as dispersing agents. When magnetic nanoparticles are synthesized by thermal decomposition method, the size and morphology of magnetic nanoparticles are determined by the types and concentration ratios of metallo-organic compounds, solvents, and surfactants in the reaction system. In addition, reaction time and temperature also affect the formation and growth of magnetic nanoparticles to a great extent.

8.2.2.3 Hydrothermal Synthesis Method

Hydrothermal synthesis is a method of synthesizing nanoparticles in a hydrothermal reactor under high temperature and high pressure, using aqueous solution or organic solvent as reaction medium magnetic nanoparticles with different sizes and morphologies can be synthesized by changing the reaction time, temperature, and reaction medium.

8.3 Typical Application Examples in the Disease Diagnosis and Study of Biological Events

The essence of molecular imaging is to visualize the interaction between molecular probes and target molecules through a series of advanced imaging technologies. The basic principle is as follows: Molecular probe → intracellular introduction of living tissue → Interaction of labeled molecular probe with target molecule → Signaling → Signal detection by imaging equipment → Computer processing imaging → Display of living tissue molecular image, functional metabolism image, gene transformation image. Among them, the study of molecular imaging probes is the most important and the prerequisite for molecular imaging research.

8.3.1 Application of Optical Nanoprobes

At present, semiconductor nanomaterials and inorganic nonmetallic nanomaterials based on quantum dots are the research hotspots of optical nanoprobes. Quantum dots have unique optical properties. Its fluorescence emission wavelength is adjustable, the fluorescence emission range covers the band of 300–2400 nm, and it can achieve single excitation and multiple emission. It has good photochemical stability and long fluorescence life. At the same time, quantum dots have a small size, long circulation time in vivo, and a good passive targeting effect on tumors. These superior properties make quantum dots the first to be used as nanofluorescent probes for fluorescence imaging in vivo. Carbon quantum dots and graphene quantum dots are a kind of ultra-small carbon nanoparticles. Their excellent optical properties, low toxicity, good biocompatibility, and low preparation cost make them the research hotspot of nanoprobes. They emit near-infrared fluorescence for imaging in vivo.

Nie and others were the first to use quantum dots to locate and image tumors in vivo [15]. They coated the quantum dots with layers of polymer nanoparticles and polyethylene glycol and attached them to prostate-specific monoclonal antibodies. This quantum dot probe connected with specific tumor-targeting ligands was then injected into tumor-bearing nude mice, and it was found that the quantum dots could gather around the tumor tissue so that real-time information about tumor size and location in the animal could be obtained through fluorescence imaging. This work has greatly promoted the application of quantum dots in the field of fluorescence imaging in vivo (Figures 8.7 and 8.8).

A research group [16] conjugated photosensitizer onto carbon quantum dots to prepare nanoprobes that have the dual effect of fluorescence imaging and photodynamic therapy. The experimental results in mice show that the probe has good efficacy in tumor-targeting imaging and photodynamic therapy (Figure 8.9).

The biggest problem in surface Raman enhancement with nanoprobes is that the probes themselves are easily agglomerated and their spectra change due to the absorption of external impurities. Many researchers have solved these problems by wrapping and surface-modifying nanoparticles in various ways. The microbead modification method of polydivinyl benzene can preserve the particles well, but because of the need for secondary modification, the particles are too large to be used for cell detection. The probes prepared by silicon shell encapsulation have been widely used because of their small particle size and easy biomodification, but the probes prepared by this method have serious nonspecific adsorption. Qian and

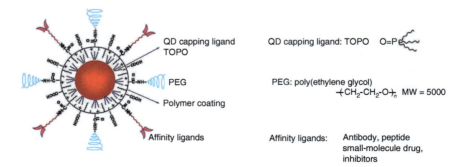

Figure 8.7 Structure of a multifunctional QD probe. Source: From Ref. [15], 2004, Springer Nature.

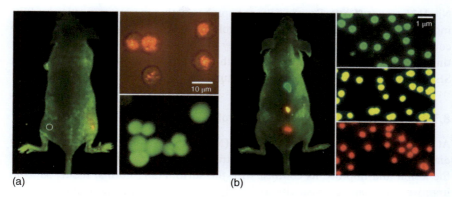

Figure 8.8 Sensitivity and multicolor capability of QD imaging in live animals. (a and b) Sensitivity and spectral comparison between QD-tagged and GFP-transfected cancer cells (a) and simultaneous in vivo imaging of multicolor QD-encoded microbeads (b). Source: From Ref. [15], 2004, Nature Biotechnology.

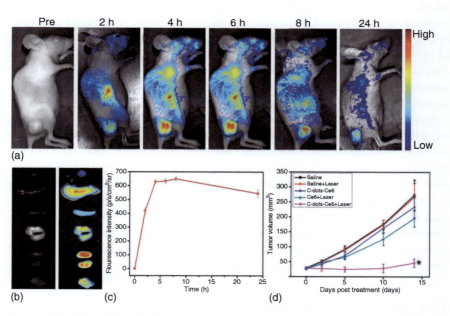

Figure 8.9 (a) Real-time in vivo NIR fluorescence images after intravenous injection of C-dots-Ce6 in nude mice at different time points; (b) ex vivo images of mice tissues (from top to bottom: heart, liver, spleen, lung, kidneys, tumor); (c) the average fluorescence intensities from the tumor area at 24 hours post-injection ($n = 5$); and (d) MGC803 tumor growth curves after various treatments ($n = 5$). ($^*P < 0.05$ for other groups versus C-dots-Ce6 + laser group). Source: From Ref. [16], 2012, John Wiley & Sons.

Nie [17] prepared a series of SERS nanoprobes with superior properties by PEG modification. Using this PEG-modified gold nanoparticle-based SERS nanoprobe coupled with tumor-targeting ligand, they used Raman spectroscopy for the first time to image nude mice transplanted with head and neck tumors, and found that SERS nanoprobes could efficiently target tumors.

8.3.2 Application of Magnetic Nanoprobes

At present, magnetic resonance imaging (MRI) is widely used in clinical imaging diagnosis and disease monitoring. Contrast agents can enhance the signal of the lesion and improve its contrast with the surrounding normal tissue, so as to improve the imaging sensitivity and early detection ability. The contrast agents usually enhance the contrast between normal and diseased tissue by shortening the longitudinal relaxation time of hydrogen protons (T_1 contrast agent) or transverse relaxation time (T_2 contrast agent). T_1 contrast agent can make the target area brighter and is a positive contrast agent, while T_2 contrast agent can darken the imaging of diseased tissue and is a negative contrast agent. The relaxation enhancement efficiency of T_1 and T_2 contrast agents was measured by the relaxation rates r_1 and r_2, respectively. In general, to obtain a large r_2 value, a high magnetization of the nanoparticles is required. At the same time, the widely used gadolinium chelates (such as Gd-DTPA) as contrast agents for magnetic resonance limit their application in molecular MRI due to their shortcomings such as too fast metabolism in vivo, nonspecific distribution, difficult modification, and nontargeting.

Superparamagnetic iron oxide nanoparticles (SPIONs) and gadolinium-containing nanomaterials are new directions in magnetic resonance molecular imaging. For example, Fe_3O_4 nanorings have a large magnetic resonance r_2 value due to their high magnetization and high magnetic field inhomogeneity, which is four times stronger than traditional superparamagnetic ferric oxide. In addition, iron oxide is biodegradable and can be metabolized by cells into the normal plasma iron pool, binding with red blood cell hemoglobin or used in other metabolic processes. Therefore, magnetic iron oxide nanoparticles have become the preferred materials for constructing new MRI contrast agents due to their excellent in vivo safety, tumor tissue specificity, and high magnetic sensitivity.

8.3.3 Application of Photoacoustic Imaging Nanoprobes

There are also many innovative research results in the field of photoacoustic bimodal probe imaging. Researchers have developed a new kind of photoacoustic probe based on new nanomaterial–gold nanocages [18]. They first used this hollow gold nanocage to photoacoustic image the lymphatic system, and the results showed that the probe could image the lymph nodes very well. They also modified the surface of the gold nanocages to target melanoma ligand peptides. The cell experiments showed that gold nanocages targeted by ligand peptides were three times more able to enter target cells than gold nanocages modified with PEG only. The mouse tumor transplantation experiments show that ordinary gold nanocages

Figure 8.10 TEM images of Au nanocages (AuNCs) with average edge lengths of (a) 50 nm and (b) 30 nm. (c) UV–vis extinction spectra of aqueous suspensions of AuNCs with edge lengths of 50 nm (solid line) and 30 nm (dashed line). Source: From Ref. [18], 2011, American Chemical Society.

can also enhance the photoacoustic imaging of tumors because of their passive targeting ability, while gold nanocages modified with targeted peptides can clearly display tumors with a size of 5 mm around numerous blood vessels (Figure 8.10).

Due to the strong near-infrared absorption capacity of gold and carbon materials, compared with carbon and gold, the toxicity is lower, the biological affinity is better, and the surface functionalization is easy. So, if you wrap gold in carbon, you get the best of both worlds. Below is a gold-coated carbon nanotube (CNT@gold) used as a photoacoustic imaging probe to achieve better imaging results [19]. At the same time, modification of lymphatic-targeting ligands on the gold-@carbon nanotubes enables the probe to target lymph nodes. The results of the study showed that the lymphatic vessel imaging signal of the probe was increased by 64 times compared with the conventional probe (Figure 8.11).

8.3.4 Application of CT Nanoprobe

The emergence of nanotechnology provides a new solution to solve the insufficiency of CT contrast agents in clinical. Studies have shown that when the size of particles is greater than 8 nm, they will not be filtered by the glomerulus and excreted by the renal tubules, but enter the liver parenchymal cells and spleen according to different sizes, and are taken up by the reticuloendothelial system and collected in the liver and spleen, and finally be further decomposed. After surface modification, it can further delay the phagocytosis by macrophages, so the half-life in vivo can be greatly increased. At the same time, the surface of the particles can be modified with the introduction of targeted ligands, so that the particles have specific recognition ability. These advantages make researchers begin to try to use nanoparticles as CT contrast agents.

The University of Connecticut and Nanoprobes, Inc. [20] were the first to use gold nanoparticles for CT imaging in 2006. They used 1.9 nm gold nanoparticles to X-ray mice transplanted with subcutaneous breast tumors. The results showed that compared with iodide contrast imaging, the tumor site can be more clearly distinguished, and even blood vessels with a diameter of only 100 μm can be seen. And the tumor imaging time is very long. This is due to the high atomic number of gold compared to iodine (Au: 79; I: 53) and X-ray absorption coefficient (Au: 5.16 cm^2 g^{-1} at 100 keV;

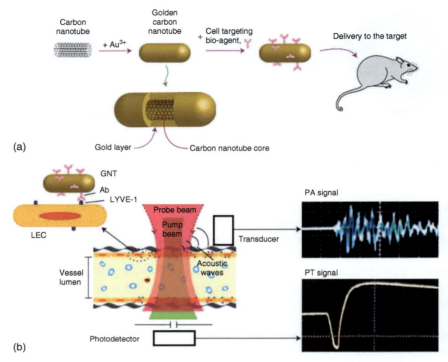

Figure 8.11 Schematics of GNT-assisted PA/PT molecular diagnostics and therapeutics. Source: From Ref. [19], 2009, Springer Nature.

I: 1.94 cm^2 g^{-1}), which can significantly inhibit the propagation of X-rays. Then the use of gold nanoparticles as CT contrast agents began to be paid attention to.

8.4 Future Development

Molecular imaging technology can detect disease changes at the cellular and molecular levels, thus making it possible to study the disease development process of the overall microenvironment in the living body, which will have a revolutionary impact on modern and future medical models. As the basis of the development of molecular imaging, probe determines the direction of the development of medical imaging technology Nanoprobes overcome the defects of traditional imaging probes, solve the problems that molecular probes cannot solve to a certain extent, and have made remarkable achievements in the field of imaging research. At present, more and more nanoprobes are moving from basic research to clinical application.

At present, with the application of nanoprobes in MRI imaging, many magnetic nanoprobes have entered the clinic. In terms of CT and PET imaging, nanoprobes prepared from various precious metals and isotopic materials are mainly in the basic research stage. Optical correlation imaging technologies such as optical and photoacoustic imaging have the advantages of nondestructive and high resolution, and will provide technical support for the development of a new generation of biomedical imaging. Fluorescent, Raman, photoacoustic, and other nanoprobes will also develop in the direction of targeting, high sensitivity, multimode, and nontoxic with the progress of modern science and technology.

Existing molecular imaging techniques have advantages and disadvantages in terms of temporal and spatial resolution, penetration depth, energy extension, probe availability, detection limits, etc., and the combination of multiple imaging techniques will provide more comprehensive information. The design of a new multifunctional imaging probe is a concrete step in the future development of molecular imaging. Future nanoprobes should be both "smart" and "capable," capable of imaging well under a variety of imaging technology conditions, and have therapeutic effects.

References

1 Weissleder, R. and Mahmood, U. (2001). Molecular imaging. *Radiology* 219: 316–333.
2 Zhang, W., Zhao, X., Yuan, Y. et al. (2020). Microfluidic synthesis of multi-mode Au@CoFeB–Rg3 nanomedicines and their cytotoxicity and anti-tumor effects. *Chemistry of Materials* 32: 5044–5056.
3 Ma, J., Wang, J., Zhang, G. et al. (2019). Magnetic and optical properties of Ag-CoFe nanohybrids prepared by a sequenced microfluidic process. *ChemistrySelect* 4: 14157–14161.
4 Wang, J., Zhao, H., Zhu, Y., and Song, Y. (2017). Shape-controlled synthesis of CdSe nanocrystals via a programmed microfluidic process. *Journal of Physical Chemistry C* 7 (121): 3567–3572.
5 Park, H.Y., Lipert, R.J., and Porter, M.D. (2004). Single-particle Raman measurements of gold nanoparticles used in surface-enhanced Raman scattering (SERS)-based sandwich immunoassays. *Nanosensing: Materials and Devices* 5593: 464–477.
6 Song, Y., Wang, R., Rong, R. et al. (2011). Synthesis of well-dispersed aqueous-phase magnetite nanoparticles and their metabolism as an MRI contrast agent for the reticuloendothelial system. *European Journal of Inorganic Chemistry* 3301–3313.
7 Wang, J., Zhao, K., Shen, X. et al. (2015). Microfluidic synthesis of ultra-small magnetic nanohybrids for enhanced magnetic resonance imaging. *Journal of Materials Chemistry C* 3: 12418–12429.
8 Hubbell, J.H. (1990). X-ray cross-sections and crossroads – a citation-classic commentary on photon mass attenuation and energy-absorption coefficients. *Current Contents. Engineering, Technology & Applied Sciences* 51: 22.
9 Weissleder, R. and Pittet, M.J. (2008). Imaging in the era of molecular oncology. *Nature* 452: 580–589.
10 Lee, H.K., Lee, Y.H., Koh, C.S.L. et al. (2019). Designing surface-enhanced Raman scattering (SERS) platforms beyond hotspot engineering: emerging

opportunities in analyte manipulations and hybrid materials. *Chemical Society Reviews* 48 (3): 731–756.

11 Richard, E., Darienzo, Wang, J. et al. (2019). Surface-enhanced Raman spectroscopy characterization of breast cell phenotypes: effect of nanoparticle geometry. *ACS Applied Nano Materials* 2 (11): 6960–6970.

12 Wang, R., Yang, W., Song, Y. et al. (2015). A general strategy for nanohybrids synthesis via coupled competitive reactions controlled in a hybrid process. *Scientific Reports* 5: 9189.

13 Qiong, W., Liu, R., Miao, F. et al. (2023). Multimodal ultra-small CoFe–WO$_x$ nanohybrids synthesized by a pilot microfluidic system. *Chemical Engineering Journal* 452 (2): 139355.

14 Alaa, A.A., George, P.L., and David, J.E. (2011). CPMV-polyelectrolyte-templated gold nanoparticles. *Biomacromolecules* 12: 2723–2728.

15 Gao, X.H., Cui, Y.Y., Levenson, R.M. et al. (2004). In vivo cancer targeting and imaging with semiconductor quantum dots. *Nature Biotechnology* 22: 969–976.

16 Huang, P., Lin, J., Wang, X. et al. (2012). Light-triggered theranostics based on photosensitizer-conjugated carbon dots for simultaneous enhanced-fluorescence imaging and photodynamic therapy. *Advanced Materials* 24: 5104–5110.

17 Qian, X.M. and Nie, S.M. (2008). Single-molecule and single-nanoparticle SERS: from fundamental mechanisms to biomedical applications. *Chemical Society Reviews* 37: 912–920.

18 Cai, X., Li, W., Kim, C.H. et al. (2011). In vivo quantitative evaluation of the transport kinetics of gold nanocages in a lymphatic system by noninvasive photoacoustic tomography. *ACS Nano* 5: 9658–9667.

19 Kim, J.W., Galanzha, E.I., Shashkov, E.V. et al. (2009). Golden carbon nanotubes as multimodal photoacoustic and photothermal high-contrast molecular agents. *Nature Nanotechnology* 4: 688–694.

20 Hainfeld, J.F., Slatkin, D.N., Focella, T.M. et al. (2006). Gold nanoparticles: a new X-ray contrast agent. *The British Journal of Radiology* 79: 248–253.

9

Magnetic Nanoparticles and Their Applications

Xiuyu Wang and Yuting Tang

College of Energy Engineering, Zhejiang University, Lingyin Street, Xihu District, Hangzhou 310027, China

9.1 Introduction

Magnetic nanoparticles (MNPs) are nanoparticles with at least one dimension between approximately 1 and 100 nm and showing a response to an applied magnetic field. MNPs, as a remarkably versatile class of nanomaterials, have garnered extensive attention and interest in the field of biomedical applications. Since the synthesis of monodisperse Fe_3O_4 nanoparticles with size below 20 nm in the laboratory of Sun Shouheng in 2002 [1], the medical applications of MNPs have flourished, a trend that is prominently reflected in the rapid growth of related research articles on PubMed. As of 2022, there has been an exponential increase, with more than 1330 articles focusing on the application of MNPs in biomedical sciences (Figure 9.1).

The magnetic responsiveness and small size endow MNPs with enormous potential in biomedical applications, such as magnetic resonance imaging (MRI), magnetic hyperthermia, targeted drug delivery, neuromodulation, cell permeability, and tissue cryopreservation, etc. Across diverse research domains, the application of MNPs is progressively revealing its significance in enhancing diagnostic precision, therapeutic efficacy, and innovative medical solutions. For instance, MNPs enable precise tumor localization through magnetic guidance, facilitating efficient surgical resection or treatment. Moreover, their application in MRI is poised to elevate imaging contrast, furnishing clinicians with comprehensive insights into pathological conditions. Additionally, their distinct magnetocaloric properties present novel prospects for magnetic hyperthermia therapy in the field of cancer.

The performance of MNPs in different biomedical applications depends on specific magnetic properties. For MRI, high saturation magnetization (M_s) value of MNPs leads to short T_2 relaxation time and then strong contrast imaging. For magnetic hyperthermia therapy, high anisotropy (K) value of MNPs leads to strong heating capacity, and then high tumor treatment efficiency. Moreover, a suitable Curie temperature (T_c) value, to avoid overheating problem, is also considered for hyperthermia therapy. For targeted drug delivery, high M_s value leads to stronger

Nanomedicine: Fundamentals, Synthesis, and Applications, First Edition. Edited by Yujun Song.
© 2025 WILEY-VCH GmbH. Published 2025 by WILEY-VCH GmbH.

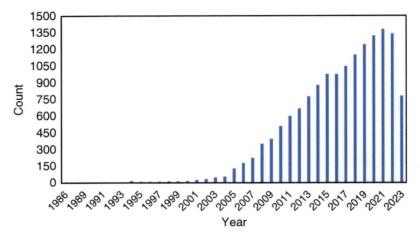

Figure 9.1 Statistics on the articles focusing on the application of magnetic nanoparticles in biomedical science.

magnetic force that enhances their ability to navigate through complex biological environments and target tumor sites. Therefore, tailoring magnetic properties of MNPs for specific biological applications is imperative.

This chapter will provide an overview of recent developments related to the optimization of magnetic properties of MNPs and their biomedical applications, mainly including MRI, magnetic hyperthermia, targeted drug delivery, and neuromodulation, etc. The chapter will also conclude the challenges currently faced and propose potential strategies for addressing them, aiming to provide readers with a comprehensive understanding of this burgeoning field.

9.2 Classification of Magnetic Nanoparticles

The material magnetism originates from the magnetic moment of the atom, which is mainly dominated by microscopic magnetic structures, including the spin and orbital angular momentum of the electron in the atom. According to the macroscopic magnetic behavior when exposed to external magnetic field, magnetic materials could be classified into magnetically soft materials and magnetically hard materials. Magnetically soft materials can be magnetized but do not tend to stay magnetized, with high magnetic permeability and low coercivity. In contrast, magnetically hard materials have high coercivity that makes them difficult to demagnetize (see Figure 9.2). From the microscopic magnetic structure, materials are classified into ferromagnetism (FM), ferrimagnetism (FiM), antiferromagnetism (AFM), paramagnetism (PM), and diamagnetism (DM). Magnetic materials usually refer to FM, FiM, and AFM (see Figure 9.3). Their magnetic properties arise in the presence of unpaired valence electrons located on the metal atoms or metal ions, and they can spontaneously magnetize to form a secondary magnetic structure. Ferromagnetic materials display inherent magnetization in the absence of an

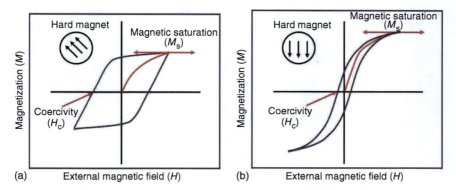

Figure 9.2 (a) Hard magnetic nanoparticles and (b) soft magnetic nanoparticles.

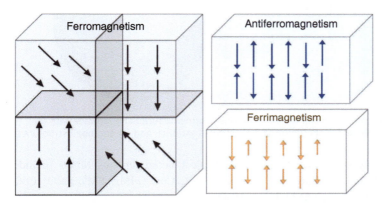

Figure 9.3 The magnetic moment structure of magnetic nanoparticles.

external magnetic field, where magnetic moments in the same magnetic domains are aligned in the same direction. They have a significant, observable magnetic permeability (μ) with a range from 10^1 to 10^6, which strongly induces magnetization in the presence of external magnetic field. FM is usually exhibited in alloys or compounds of iron, cobalt, nickel, and certain rare-earth metals. In contrast, ferrimagnetic materials, where some magnetic moments point in the opposite direction but with different magnitudes, exhibit lower μ value than FM materials, with $1 < \mu < 10^3$. Iron oxide, Fe_3O_4 is the typical ferrimagnetic material. Some ferrites (spinel type, magneto-plumbite type, and garnet type) are also described as ferrimagnetic materials. In antiferrimagnetic materials, such as hematite, chromium, iron manganese, and nickel oxide, the magnetic moments of atoms have a regular aligning pattern with neighboring spins pointing in opposite directions but a vanishing total magnetization. Based on the structure and composition, MNPs can be also classified into single phase and heterogeneous phases (e.g. FM/FiM, inverted FiM/FM, FM/AFM, inverted AFM/FM, FiM/AFM, and FiM/FiM). Heterogeneous MNPs describe a system in which distinct magnetic phases are constructed by one or multiple reaction interfaces into core–shell, dumbbell, and

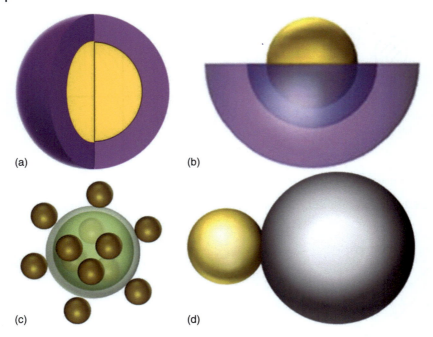

Figure 9.4 Schematic representation of various heterostructure nanoparticles with (a) core–shell, (b) core–shell–shell multilayers, (c) core–satellite, and (d) dumbbell structures.

core–satellite structures (see Figure 9.4). Typical heterogeneous MNPs include Ni/NiO (FM/AFM), Co/CoO (FM/AFM), γ-Fe_2O_3/CoO (FiM/AFM), Fe_3O_4/CoO (FiM/AFM), FeO/Fe_3O_4 (AFM/FiM), MnO/Mn_3O_4 (AFM/FiM), $CoFe_2O_4$/Fe_3O_4 (FM/FiM), $CoFe_2O_4$/FeCo (FM/FM), and Fe_3O_4/$CoFe_2O_4$ (FiM/FM) MNPs. The magnetic interaction between different magnetic phases could lead to superior magnetic properties and novel magnetic effects, which would hold more effective applications in biomedical field.

9.2.1 Magnetic Regulation

In past decades, MNPs as zero-dimension magnetic materials within nanoscale have attracted considerable attention due to their novel magnetic properties and abundant magnetic effects, such as tunable saturation magnetization, controlled coercivity, exchange bias effects, magnetic proximity effect, and giant magnetoresistance. The precise regulation and design of magnetic properties of MNPs are of great importance in facilitating their performance in biomedical applications. Tailoring strategies commonly include changing size, morphology, and composition (doping element) (Table 9.1).

9.2.1.1 Size

As the size of nanomagnetic materials decreases, the specific surface area of the materials increases, resulting in significant changes in basic magnetic properties.

Table 9.1 Tunable magnetic properties important for biomedical applications [2].

Tunable properties	Application
Saturation magnetization (M_s)	Biosensing, drug delivery, magnetic resonance imaging (MRI)
Coercivity (H_c)	Biosensing, hyperthermia
Blocking temperature (T_B)	Biosensing, drug delivery, hyperthermia
Néel and Brownian relaxation time of nanoparticles (t_N and t_B)	Biosensing, hyperthermia

These basic magnetic properties include saturation magnetization M_s, remnant magnetization M_r, blocking temperature T_B, and coercivity H_c.

The reduction in size of MNPs leads to a drop in saturation magnetic M_s, which can be attributed to two factors: decreased magnetic moments and surface canting spins. Size-dependent magnetic moments can be described by Eq. (9.1):

$$m = M\rho V \tag{9.1}$$

where m refers to magnetic moments, M refers to the particle's mass magnetization, ρ refers to the density of the particle, and V refers to the particle's volume.

Due to the large proportion of surface canting spins, the reduced ΔM_s can be calculated by Eq. (9.2):

$$\Delta M_s = M_{sb}[(r-d)/r]^3 \tag{9.2}$$

As stated above, M_{sb} refers to M_s value of the bulk, r refers to the radius of MNPs, and d refers to the thickness of the surface of disordered spins. These disordered spins originate from the lower coordination number and the broken magnetic exchange bonds.

In contrast, coercivity exhibits a nonmonotonic trend with the size due to the evolution of domain structure. In larger multidomain particles, coercivity typically decreases with increasing size. This behavior can be attributed to the dynamics of domain wall formation and motion. When the size is reduced to the critical single-domain diameter (D_{sd}), a shift in the coercivity occurs, where the magnetization reversal process transforms from being dominated by domain wall movement to being driven primarily by the anisotropy energy. Below the critical size D_{sd}, the anisotropy energy, E_a, increases with increasing size, and thus the energy required to overcoming the anisotropy energy for magnetization reversal also increases. Further reducing the particle size, below the superparamagnetic diameter D_{sp}, the thermal energy surpasses the anisotropy energy, leading to the free fluctuation of the magnetic moments. Therefore, the MNPs exhibit zero coercivity.

In addition, size-dependent magnetic properties of MNPs also include blocking temperature (T_B), which refers to the temperature at which thermal energy equals the anisotropy energy (the onset of superparamagnetism). As particle size decreases, the T_B value decreases (see Figure 9.5).

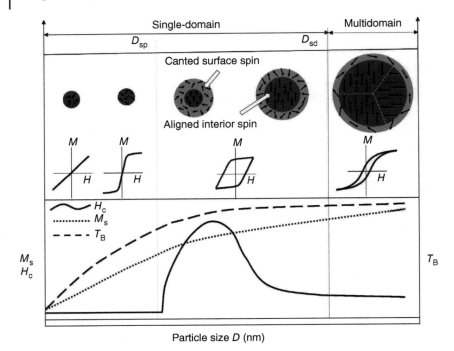

Figure 9.5 The curve of magnetic properties with nanoparticle diameter.

Guaedia et al. have reported an increasing trend for the M_s value of the iron oxide nanoparticles observed in the size range of 4 to 45 nm (the largest size, 45 nm, exhibited the highest M_s value, 92 emu g^{-1} at 5 K, close to the expected value for bulk magnetite) [3]. According to the report by Xuemin He and co-workers, a similar trend was observed in the case of Ni nanoparticles [4]. The saturation magnetization M_s is strongly size-dependent, with a maximum value of 52.01 and 82.31 emu g^{-1} at room temperature and 5 K, respectively. They also observed a change in the Curie temperature T_c, 593, 612, 622, 626, and 627 K for the 24, 50, 96, 165, and 200 nm Ni nanoparticles, respectively. In addition to size-dependent magnetic properties, the performance of MNPs in biomedical applications, such as cellular uptake, biodistribution, and pharmacokinetics, also depends on the particle size. Very small-sized nanoparticles with a hydrodynamic diameter less than 5.5 nm are excreted renally. Medium-sized nanoparticles exhibit a broader distribution within the body, including the bone marrow, heart, kidney, and stomach. On the other hand, larger particles with a hydrodynamic diameter exceeding 100 nm are readily taken up by phagocytes and therefore tend to quickly accumulate in fenestrated tissues such as the liver and spleen [5]. Therefore, size can serve as a design parameter that can be readily manipulated to tune the magnetic properties of M_s, H_c, and T_B, and the performance of MNPs in specific biomedical applications.

9.2.1.2 Shape

The shape of MNPs is also a key factor of direction-dependent magnetic properties, including saturation magnetization, coercivity, remanent magnetization, and anisotropy. For instance, spherical MNPs can be magnetized homogeneously in any

Figure 9.6 (a) Schematic of demagnetization field lines of a magnetized ellipsoid and (b) a prolate spheroid, *a, b, c* refers to axis lines.

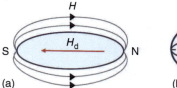
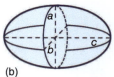

direction, while magnetization of elongated MNPs will be easier along their long axis than the short axis. This anisotropic behavior can affect coercivity and remanence. The shape can also affect the packing behavior of magnetic moments, thus changing total magnetic magnetization. For nonspherical MNPs, the demagnetizing field (H_d), which is opposite to the external field and is proportional to magnetization, is a critical concept when considering the magnetic behavior of MNPs (Figure 9.6). The value of demagnetizing field (H_d) can be estimated by Eq. (9.3) [6].

$$H_d = -N_d M \tag{9.3}$$

where N_d is the demagnetizing factor, which is shape- and direction-dependent. For example, $N_d = 1/3$ for a spherical particle in all directions, while for a prolate spheroid, it is determined by the aspect ratio.

Kovalenko and co-workers successfully synthesized spherical, cubic iron oxide, Fe_3O_4 with controlled shape by careful adjustment of the stabilizer composition (fatty acid salts) and the reaction temperature profile. Cubic Fe_3O_4 MNPs exhibit a higher saturation magnetization than the spherical particles with almost the same volume, while they both have the same value of blocking temperature with independence of shape [7]. Zhao and co-workers successfully synthesized 30 nm octapod-shaped iron oxide nanoparticles by introducing chloride anions [8]. The octapod-shaped iron oxide can induce stronger inhomogeneity of the local magnetic fields, which can enhance the transverse relaxation process. Mitra and co-workers found that very small octahedral Fe_3O_4 MNPs exhibit a distinctive Verwey transition around 120 K, which is not present in spherical similarly sized (4–13 nm) nanoparticles. This unique behavior can be attributed to better surface coordination, reduced surface spin disorder, higher saturation magnetization, and improved stoichiometry [9]. The interesting phenomenon has been further demonstrated by Mohapatra and co-workers [10]. Mohapatra synthesized Fe_3O_4 MNPs with different shapes including spheres, octahedrons, cubes, rods, wires, and multipods, by adjusting reaction conditions (the ratio of precursor to surfactant content and heating rate, etc.). In addition, they also found the enhancement of magnetic properties for anisotropic-shaped NPs (the wire, rod, and octahedron), and higher performance of hyperthermia applications.

9.2.1.3 Composition

Changing composition serves as a powerful strategy for achieving versatile MNPs with tunable magnetic properties. The approach involves changing or doping elements (with different unpaired electrons or magnetic moments) and changing cation distribution sites (the octahedral sites and the tetrahedral sites) in spinel or inverse spinel crystal structures (Figure 9.7). Ferrites with a spinel structure, such as MFe_2O_4

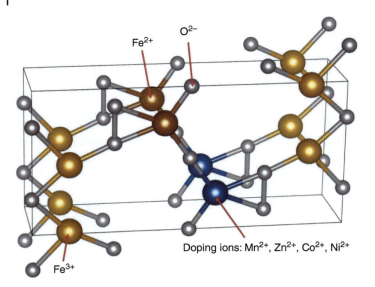

Figure 9.7 The lattice structure of Fe_3O_4 with doping ions (such as Mn^{2+}, Zn^{2+}, Co^{2+}, Ni^{2+}).

with M = Fe, Ni, Zn, Co, and Mn, can provide high flexibility in terms of chemical composition and the distribution of cations in octahedral and tetrahedral sites. The overall magnetic properties of ferrites are strongly dependent on the Curie temperature, magnetic anisotropy, and magnetic moment of the substitution metal M [11]. Therefore, ferrites are an advantageous class of materials to explore the dependence of magnetic properties on composition. According to the report by Deng et al., the M_s values at 300 K for 200 nm MFe_2O_4 microspheres exhibited significant differences. They observed the highest magnetization for Fe_3O_4, with M_s values of 81.9 emu g^{-1}, while the lowest value was 53.2 emu g^{-1} for $MnFe_2O_4$ [12]. Interestingly, Lee et al. found the opposite result by comparing $MnFe_2O_4$, $FeFe_2O_4$, $CoFe_2O_4$, and $NiFe_2O_4$ MNPs of the same 12-nm size. $MnFe_2O_4$ showed the highest magnetization with 110 emu g^{-1} [13]. They attributed this result to the special crystal structure of $MnFe_2O_4$ MNPs, which had a mixed spinel structure (Mn^{+2} and Fe^{+3} occupying both Oh and Td sites), and the rest had an inverse spinel structure (Mn^{+2} and Fe^{+3} occupying Oh sites but only Fe^{+3} occupying the Td sites).

In addition, some other ions like Al^{3+} and rare-earth ions like Sm^{3+} have been used to change the cation ion concentration in crystal sub-lattices to regulate the magnetic moment. According to a current report by Shao and co-workers, doping Ni–Mg–Co ferrite with Al^{3+} and rare-earth ions has a certain effect on its size and magnetic properties. They found that the average crystallite size decreased from 56.7 to 36.6 nm after doping with rare-earth ions while doping with Al^{3+} caused the average crystallite size to increase to 90.7 nm. The remanent magnetization (M_r), square degree (M_r/M_s), coercivity rectangle degree (S^*), and anisotropy constant (K) all increase after doping with rare-earth ions, but they decrease after doping with Al^{3+} [14].

9.2.2 Exchange–Coupling Interaction

Exchange–coupling describes a magnetic interaction across the interface between different magnetic phases. Compared to single-phase MNPs, exchange–coupling MNPs not only break the limitation of theoretical value of the bulk but also provide high free degree of manipulating magnetism, which can be theoretically achieved by controlling the physical and chemical properties of magnetic phases and their interface.

Core–shell MNPs, where both the core and the shell exhibit magnetic properties (FM, FiM, and AFM), have long been studied because the spherical heterointerface maximizes the exchange–coupling interaction (see Figure 9.8). Currently, numerous efforts have been dedicated to enhancing magnetic properties via constructing exchange–coupling MNPs with core–shell structures. For instance, Kevin Sartori and co-workers successfully synthesized $Fe_{3-x}O_4/CoFe_2O_4$ core–shell MNPs with average size of 7.2 and 12.3 nm by the thermal decomposition method [15]. According to their report, $Fe_{3-x}O_4/CoFe_2O_4$ MNPs exhibited larger T_B (159 K for the 7.2 nm particles, 284 K for the 12.3 nm particles) value than $Fe_{3-x}O_4$ core (62 K for the 7 nm, 142 K for 11 nm), high H_c (>21 000 Oe), and larger M_s value of 67 emu g^{-1} (M_s value of $Fe_{3-x}O_4$ with a similar size of 7 nm was only 55 emu g^{-1}). Such results have also been observed in $FePt/Fe_3O_4$ core–shell MNPs with higher saturation magnetization and improved thermal stability [16]. These authors give a rational explanation that the enhancement of magnetic properties is led by the effective magnetic coupling

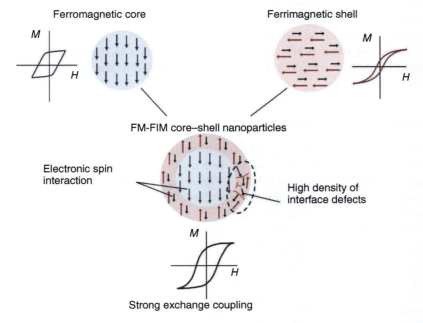

Figure 9.8 Schematic illustration of exchange–coupling between ferromagnet core and ferrimagnetic shell.

between two phases, which is dependent on interfacial characteristics like vacancies and misalignment.

In addition to superior magnetic properties, core–shell MNPs also exhibit a high degree of freedom to tailor the magnetic properties. Most researchers control the physicochemical properties of the core and the shell to regulate the saturation magnetization M_s, coercivity H_c, blocking temperature T_B, and anisotropy K, etc. For example, De and co-workers observed that the exchange bias and coercivity of Co/Co_3O_4 MNPs, exhibit a nonlinear trend with increasing Co_3O_4 shell thickness [17].

9.3 Biomedical Applications

MNPs with novel magnetic properties can produce abundant physical effects at the nanoscale when exposed to an external magnetic field. These physical effects have the potential to act on biological targets at the micro-nanoscale and amplify resulting biological effects, thereby providing valuable insights into the underlying molecular mechanisms in complex biological systems. For example, the localized magnetic field induced by MNPs can influence the relaxation process of hydrogen protons excited by radiofrequency pulses, significantly reducing relaxation times and thereby enhancing the magnetic resonance signal in the region. Heating capability of MNPs in the presence of alternating magnetic field (with high frequency of 100 kHz to 1 MHz), which can convert the magnetic field energy to heat through hysteresis losses, Brownian and Néel spin relaxations, has been used to kill cancer cells with minimum harm to the surrounding healthy tissues. The following section will introduce the contribution of magnetic properties of MNPs to their performance in monitoring MRI, target drug delivery, magnetic hyperthermia, and advanced neuromodulation fields.

9.3.1 Magnetic Resonance Imaging

MRI, an innovative medical imaging technique, relies on the differential magnetic resonance signals generated by various tissues within a biological organism when exposed to an external magnetic field (see Figure 9.9). The strength of these resonance signals is determined by the relaxation times of protons in water molecules within the tissues. Local magnetic fields produced by unpaired electron spins in certain constituents can either shorten or lengthen the relaxation times of protons, which are categorized as longitudinal relaxation time T_1 and transverse relaxation time T_2, thereby enhancing the strength of magnetic resonance signals in the adjacent region and improving image contrast.

MNPs with unpaired electrons, which can induce local magnetic field to change the relaxation times of protons, have been developed as an effective contrast agent to enhance the imaging quality and sensitivity in MRI. At present, nanoiron oxide magnetic liquid has been utilized to the clinical application stage as MRI contrast agent. Larger-sized iron oxide MNPs have higher relaxivity values r_2 in the MRI.

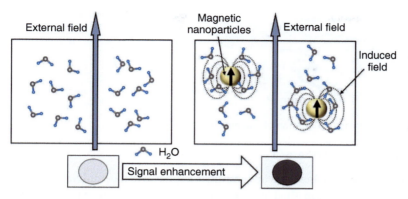

Figure 9.9 Schematic illustration of magnetic resonance imaging.

For example, Cheon and co-workers studied iron oxide of 4, 6, 9, and 12 nm, which showed size-dependent M_s of 25, 43, 80, and 102 emu g^{-1}, respectively, and their relaxivity values r_2 were positively correlated with the diameter of the iron oxide NPs [18]. Brollo and co-workers obtained magnetic iron oxide nanoparticles with sizes between 8 and 15 nm and finally colloids suitable for their use as contrast agents on MRI. The relaxivity values r_2 normalized to the square of the saturation magnetization were shown to be constant and independent of the particle size, which means that the saturation magnetization is the main parameter controlling the efficiency of these MNPs as MRI-contrast agents.

With the advancement of nanotechnology, researchers have achieved a range of performance-controllable MNPs through various synthesis methods. MNPs like alloys (such as FeCo, FePt) and doped MNPs (such as MeFe$_2$O$_4$, Me = Mn, Zn, Co) have attracted much focus due to their tunable magnetic properties. According to the report by Mikhaylov and co-workers, cobalt ferrite (CoFe$_2$O$_4$), MNPs possess outstanding properties for contrast MRI (r_1 = 22.1 s^{-1} mM^{-1} and r_2 = 499 s^{-1} mM^{-1}) that enabled high-resolution T_1- and T_2-weighted MRI-based signal detection [19]. For ZnFe$_2$O$_4$ nanoparticles, their saturation magnetization and crystal structures can be easily tuned by adjusting their Zn-doping contents, thereby precisely tuning their contrast properties in MRI [20].

In addition, it has been demonstrated that core–shell MNPs exhibit improved performance as T_2 contrast agents in MRI due to their high saturation magnetization induced by exchange–coupling interaction. For example, Fe/MFe$_2$O$_4$ (M = Fe, Mn, Co) core–shell NPs have demonstrated higher transverse relaxation (r_2) than their single-core counterpart within similar size [21]. The monometallic Fe core leads to high overall magnetization and the superparamagnetic ferrite shell can protect the core from oxidation. For Zn$_{0.5}$Mn$_{0.5}$Fe$_2$O$_4$/Fe$_3$O$_4$ core–shell nanoparticles, the coating of the Zn$_{0.5}$Mn$_{0.5}$Fe$_2$O$_4$ nanoparticle by a magnetite or maghemite shell can minimize the effect of the magnetic dead layer at the core surface, improving the magnetic properties and offering thus outstanding relaxation r_2 values higher than 300 mM^{-1} s^{-1} at H 1.5 T, thereby a vast potential as MRI contrast agents [22].

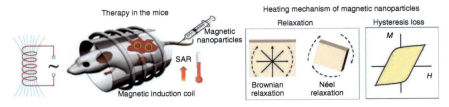

Figure 9.10 Schematic illustration of magnetic hyperthermia.

9.3.2 Magnetic Hyperthermia

Magnetic hyperthermia therapy is an innovative approach to the minimally invasive treatment of cancer. This method involves using local hyperthermia with MNPs that are delivered only to the cancer cells or tissues and heated externally by an alternating magnetic field, which makes it a noninvasive method, minimizes the side effects, and allows for targeting deep cancer cells (see Figure 9.10). Heat capacity depends on the heating mechanisms of MNPs, including hysteresis loss, Brownian relaxation, and Néel relaxation. The physical properties of the MNPs are crucial for the heating mechanism. In previous research, it has been found that Fe_3O_4 and γ-Fe_2O_3 NPs with different sizes have different heat generation mechanisms under the external alternating field. For larger MNPs (>40 nm), hysteresis loss is the main mechanism of heating, while the relaxation losses (Brownian and Néel relaxations) for superparamagnetic nanoparticles are about 10 nm.

For tumor hyperthermia, MNPs are required to exhibit a large saturation magnetization (M_s) to generate large amount of heat in the tumor cells under an alternating magnetic field, and the large M_s values enable precise control over the movement of the MNPs in the blood [23]. MNPs should also be superparamagnetic to achieve good colloidal stability. When the MNP size reduces below critical diameter, MNPs exhibit superparamagnetic state, and the dipole–dipole interaction energy is small, minimizing the particle aggregation in the existence of applied magnetic field. In addition, Curie temperature (T_c) of MNPs is used to precisely control the temperature of tumor site. When exposed to an alternating magnetic field, MNPs initially experience induction heating. As their temperature rises and is higher than T_c value, MNPs lose their magnetism, resulting in a sharp decline in their heating capabilities. When the temperature decreases below the Curie temperature, MNPs regain magnetic properties and restore their heating ability. Eventually, this process raises the temperature of the targeted area to a level near the Curie point (T_c) of MNPs. This achieves the automatic temperature control and heating required for magnetic hyperthermia therapy, making it especially significant for the treatment of deep-seated tumors. Therefore, it is important to design magnetic hyperthermia with high saturation magnetization, small size, and suitable Curie temperature.

The core–shell nanoparticles due to their superior magnetic properties and high degrees of modulation have recently been explored for magnetic hyperthermia. Jalili and co-workers successfully synthesized hard/soft ($CoFe_2O_4$/Fe_3O_4) and soft/hard (Fe_3O_4/$CoFe_2O_4$) nanocomposites (core/coating) using a facile and eco-friendly co-precipitation method [24]. The $CoFe_2O_4$/Fe_3O_4 MNPs, with good exchange–coupling interaction between the hard and soft magnetic phases, presented a larger saturation magnetization than the $CoFe_2O_4$ NPs, which is effective for their use in magnetic hyperthermia. Seung-Hyun Noh and co-workers synthesized core–shell cube MNPs comprised of $Zn_{0.4}Fe_{2.6}O_4$ core (50 nm in edge) and $CoFe_2O_4$ shell (5 nm in thickness) with a total size of 60 nm by the seed-mediated growth method [25]. They found that the core–shell structure brings about a 14-fold increase in the coercivity with an exceptional energy conversion of magnetic field into thermal energy of $10\,600\,W\,g^{-1}$, the largest reported to date. Such capability of the CS-cube is highly effective for drug-resistant cancer cell treatment.

9.3.3 Targeted Drug Delivery

MNPs loaded with therapeutic drugs or target antibodies can be systematically administered to the tumor site in the human body guided by an external magnetic field, and release drug components, enhancing the drug's concentration specifically only at the tumor sites (see Figure 9.11). Compared to traditional drug delivery system, MNPs as targeted carriers yield improved treatment efficacy and the reduction of drug doses with serious side effects. However, MNPs are prone to be cleaned by the reticuloendothelial system (RES) due to their poor biocompatibility, reducing their half-life in the bloodstream. To make MNPs loaded with therapeutic agents safely move in the bloodstream, it is therefore essential to reduce particle size and modify surface chemistry by coating the MNPs with neutral and hydrophilic substances such as PEG, polysaccharides and dysopsonins. In addition, they should also have a high saturation magnetization and a large susceptibility, which directly influence the magnetic force.

The use of MNPs for drug delivery has been evolving since 1963, and MNPs consisting of Fe_3O_4 or maghemite cores coated with biocompatible and functional polymers have become a major research focus for targeted drug delivery [26]. According to the report by Pourjavadi and co-workers, Fe_3O_4@SiO_2@PNG-Hy MNPs were fabricated as carriers for delivery of doxorubicin (DOX). They found

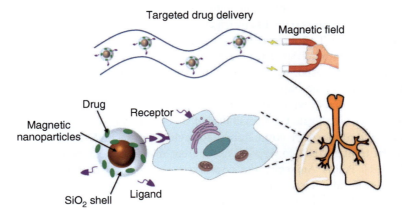

Figure 9.11 Schematic illustration of targeted drug delivery.

that the DOX release from DOX-loaded Fe_3O_4@SiO_2@PNG-Hy MNPs exhibited a good pH and temperature-dependent release behavior [27]. Mansouri designed a pH-responsive sustained release system with entrapment efficiency of DOX of $47.59 \pm 3.98\%$. In their report, the nucleolin-targeted AS1411 aptamer is conjugated to guanidinium groups of epibromohydrin-functionalized TiO_2@γ-Fe_2O_3 nanoparticles (AS1411@GMBS@EG@TiO_2@Fe_2O_3, NP–Apt) to increase drug delivery in targeted tumor tissues [28].

Core–shell MNPs with magnetic exchange-coupling have been explored as a multifunctional platform that integrates diagnosis and therapy of cancer. According to the report by Yumeng Liu and co-workers, PEGylated FePt–Fe_2O_3 core–shell nanoparticles conjugated with folic acid (FA) not only showed a high DOX (an antitumor drug)-loading efficiency of 38% but also exhibited a reduction of the T_2-weighted signal intensity of the tumor, demonstrating their potential to serve as a nanoplatform for imaging-guided drug delivery [29].

However, there are also challenges to consider, such as the need for precise control of the magnetic field and ensuring that the nanoparticles effectively release the drug at the tumor site. The field of nanomedicine is actively researching and developing these techniques to improve their effectiveness in cancer treatment and other targeted therapies.

9.3.4 Neuromodulation

MNPs have emerged as a promising tool for noninvasive neuromodulation through several modulation mechanisms including magnetothermal – conversion of the magnetic field into heat; magnetomechanical – conversion of the magnetic field into force or torque; and magnetoelectric – conversion of the magnetic field into electric potential (see Figure 9.12). This innovative approach enables deep tissue penetration and freely behaving animal studies.

Neuromodulation via the magnetothermal mechanism of MNPs can be achieved through two primary avenues. The first involves the direct activation of temperature-gated cation channels, specifically TRPV1, TRPA1, and TMEM16A ion channels. The second method entails the indirect stimulation of neurons expressing G-protein-coupled receptors (GPCRs) by triggering the release of thermal-sensitive designer drugs. Jacob T. Robinson and co-workers combined 15 nm Co-doped iron oxide MNPs with a rate-sensitive thermoreceptor (TRPA1-A), achieving sub-second behavioral responses in Drosophila melanogaster. Heat generated by MNPs when applied to alternative magnetic field, activated the dTRPA1-A protein channel and led to a rapid wing extension response with a latency of 510 ± 186 ms [30]. They further investigated the feasibility of this magnetothermal stimulation mechanism in multichannel stimulation, achieving sub-second, multichannel stimulation by tuning MNPs to respond to different magnetic field strengths and frequencies [31]. Polina Anikeeva and co-workers developed a thermally sensitive liposome-embedded target drugs and MNPs, which release small molecules (DREADD) from thermally sensitive lipid vesicles with a 20 seconds latency in the presence of alternative magnetic field, achieving the enhanced swimming behavior of mice [32]. The use of MNPs greatly improves the practicality of chemically manipulated neurons.

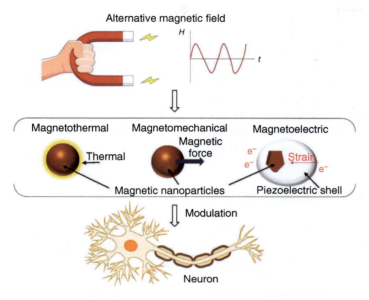

Figure 9.12 Schematic illustration of the application of MNPs in neuromodulation.

When exposed to gradient or rotating magnetic field, MNPs can generate mechanical forces or torques at nanoscale, which act on mechanically sensitive ion channels or receptors in proximity to ion channels, activating ion channels to regulate cellular functions and fate. Mechanical-sensitive ion channels primarily include transient receptor potential vanilloid 4 (TRPV4) and two-pore domain potassium channels (TREK-1). These mechanical forces and torques depend on the intrinsic properties of MNPs (such as size and saturation magnetization) as well as the external magnetic field (including strength, gradient, and direction) [33]. Jinwoo Cheon and co-workers connected cube-shaped MNPs with cellular membranes of an inner ear hair cell, yielding pico-Newtons of mechanical force on the cells and tens of nanometers of displacements of the stereocilia with sub-millisecond temporal resolution [34]. They recently designed a magnetic toolkit, which is composed of assembled 25 nm inverse-spinel iron oxide with stronger FM, activating mechanosensitive ion channel Piezo1 and achieving consistent and reproducible neuromodulation in freely moving mice [35]. Andy Tay and his colleagues have demonstrated that starch-coated MNPs can stretch lipid membranes and then increase the probability of opening N-type calcium ion channels and calcium influx, achieving the treatment of fragile X syndrome (FXS) disease [36].

Neuromodulation via magnetostrictive effect of MNPs has also attracted much attention. These MNPs are composed of a ferromagnetic core and a piezoelectric shell, which are defined as magnetoelectric nanoparticles (MENPs). MENPs can generate local electric field in targeted tissues when wirelessly controlled by an external magnetic field, activating the central nervous system with spatial resolution of a single nerve and temporal resolution of less than 10 ms. It has been demonstrated that such nanoparticles can be utilized to modulate brain activity, recovering damaged signals in the Thalamic region and restoring periodicity in other regions in Parkinson's disease model [37]. In recent research, Nguyen and co-workers employed MENPs with a core–shell structure to stimulate brain activity of mice, achieving fast neuronal responses in cortical slices ex vivo and a noninvasive and contactless deep brain stimulation [38].

These researches have confirmed the effectiveness of utilizing MNPs for stimulating the nervous system. Apart from the above-discussed biomedical application, MNPs have also been applied in tissue cryopreservation, regenerative medicine, and endothelial cellular penetration. This opens up exciting possibilities for the development of biomedical research and health sciences.

9.4 Conclusion and Outlook

It has been shown that MNPs have a great potential for cancer diagnosis and therapy, like MRI, hyperthermia, target drug delivery, and neuromodulation. The exploration of such materials with enhanced or adjustable magnetism will facilitate the development of these emerging applications, and offer innovative solutions for diagnosing and treating an array of diseases and disorders. In addition, it is crucial to delve deeper into the underlying mechanisms and effects of MNPs on cellular and molecular processes in biology and medicine. This will bring the extension of MNP applications to other biomedical domains, such as tissue engineering, gene therapy, immunotherapy, and biosensing, opening up new avenues for groundbreaking research.

References

1 Sun, S.H. and Zeng, H. (2002). Size-controlled synthesis of magnetite nanoparticles. *Journal of the American Chemical Society* 124 (28): 8204–8205.
2 Kolhatkar, A.G., Jamison, A.C., Litvinov, D. et al. (2013). Tuning the magnetic properties of nanoparticles. *International Journal of Molecular Sciences* 14 (8): 15977–16009.
3 Guardia, P., Labarta, A., and Batlle, X. (2011). Tuning the size, the shape, and the magnetic properties of iron oxide nanoparticles. *The Journal of Physical Chemistry C* 115 (2): 390–396.
4 He, X. and Shi, H. (2012). Size and shape effects on magnetic properties of Ni nanoparticles. *Particuology* 10: 497–502.
5 Ling, D.S., Lee, N., and Hyeon, T. (2015). Chemical synthesis and assembly of uniformly sized iron oxide nanoparticles for medical applications. *Accounts of Chemical Research* 48 (5): 1276–1285.
6 Ma, Z., Mohapatra, J., Wei, K. et al. (2023). Magnetic nanoparticles: synthesis, anisotropy, and applications. *Chemical Reviews* 123 (7): 3904–3943.
7 Kovalenko, M.V., Bodnarchuk, M.I., Lechner, R.T. et al. (2007). Fatty acid salts as stabilizers in size- and shape-controlled nanocrystal synthesis: the case of inverse spinel iron oxide. *Journal of the American Chemical Society* 129 (20): 6352–6353.

8 Zhao, Z., Zhou, Z., and Bao, J. (2013). Octapod iron oxide nanoparticles as high-performance T_2 contrast agents for magnetic resonance imaging. *Nature Communications* 4: 2266.

9 Mitra, A., Mohapatra, J., Meena, S.S. et al. (2014). Verwey transition in ultrasmall-sized octahedral Fe_3O_4 nanoparticles. *The Journal of Physical Chemistry C* 118 (33): 19356.

10 Mohapatra, J., Xing, M., Beatty, J. et al. (2020). Enhancing the magnetic and inductive heating properties of Fe_3O_4 nanoparticles via morphology control. *Nanotechnology* 31: 275706.

11 Umut, E. (2019). Magnetic properties of manganese ferrite $MnFe_2O_4$ nanoparticles synthesized by co-precipitation method. *Hittite Journal of Science and Engineering* 6 (4): 243–249.

12 Deng, H., Li, X., Peng, Q. et al. (2005). Monodisperse magnetic single-crystal ferrite microspheres. *Angewandte Chemie, International Edition* 44: 2782.

13 Lee, J.H., Huh, Y.M., and Yw, J. (2007). Artificially engineered magnetic nanoparticles for ultra-sensitive molecular imaging. *Nature Medicine* 13: 95–99.

14 Shao, L., Sun, A., and Zhang, Y. (2021). Comparative study on the structure and magnetic properties of Ni–Mg–Co ferrite doped with Al and rare earth elements. *Journal of Materials Science: Materials in Electronics* 32: 5339.

15 Sartori, K., Cotin, G., Bouillet, C. et al. (2019). Strong interfacial coupling through exchange interactions in soft/hard core–shell nanoparticles as a function of cationic distribution. *Nanoscale* 11: 12946–12958.

16 Chai, Y., Feng, F., Li, Q. et al. (2019). One-pot synthesis of high-quality bimagnetic core/shell nanocrystals with diverse exchange coupling. *Journal of the American Chemical Society* 141 (8): 3366–3370.

17 De, D., Iglesias, O., Majumdar, S., and Giri, S. (2016). Probing core and shell contributions to exchange bias in Co/Co_3O_4 nanoparticles of controlled size. *Physical Review B* 94: 184410.

18 Jun, Y.W., Huh, Y.-M., Choi, J.-S. et al. (2005). Nanoscale size effect of magnetic nanocrystals and their utilization for cancer diagnosis via magnetic resonance imaging. *Journal of the American Chemical Society* 127: 5732.

19 Mikhaylov, G., Mikac, U., Butinar, M. et al. (2022). Theranostic applications of an ultra-sensitive T_1 and T_2 magnetic resonance contrast agent based on cobalt ferrite spinel nanoparticles. *Cancers* 14 (16): 4026.

20 Ma, Y., Xia, J., Yao, C. et al. (2019). Precisely tuning the contrast properties of $Zn_xFe_{3-x}O_4$ nanoparticles in magnetic resonance imaging by controlling their doping content and size. *Chemistry of Materials* 31 (18): 7255–7264.

21 Yoon, T., Lee, H., Shao, H., and Weissleder, R. (2011). Highly magnetic core–shell nanoparticles with a unique magnetization mechanism. *Angewandte Chemie, International Edition* 50: 4663–4666.

22 Cardona, F.A., Urquiza, S.A., Presa, P. et al. (2016). Enhanced magnetic properties and MRI performance of bi-magnetic core–shell nanoparticles. *RSC Advances* 6: 77558.

23 Obaidat, I.M., Narayanaswamy, V., Alaabed, S. et al. (2019). Principles of magnetic hyperthermia: a focus on using multifunctional hybrid magnetic nanoparticles. *Magnetochemistry* 5 (4): 67.

24 Jalili, H., Aslibeiki, B., Hajalilou, A. et al. (2022). Bimagnetic hard/soft and soft/hard ferrite nanocomposites: structural, magnetic and hyperthermia properties. *Ceramics International* 48: 4886.

25 Noh, S., Na, W., Jang, J.-t. et al. (2012). Nanoscale magnetism control via surface and exchange anisotropy for optimized ferrimagnetic hysteresis. *Nano Letters* 12 (7): 3716–3721.

26 Hu, F.X., Neoh, K.G., and Kang, E.T. (2006). Synthesis and in vitro anti-cancer evaluation of tamoxifen-loaded magnetite/PLLA composite nanoparticles. *Biomaterials* 27: 5725–5733.

27 Pourjavadi, A., Kohestanian, M., and Streb, C. (2020). pH and thermal dual-responsive poly(NIPAM-co-GMA)-coated magnetic nanoparticles via surface-initiated RAFT polymerization for controlled drug delivery. *Materials Science and Engineering: C* 108: 110418.

28 Mansouri, N., Jalal, R., Akhlaghinia, B. et al. (2020). Design and synthesis of aptamer AS1411-conjugated EG@TiO_2@Fe_2O_3 nanoparticles as a drug delivery platform for tumor-targeted therapy. *New Journal of Chemistry* 44: 15871–15886.

29 Liu, Y., Yang, K., Cheng, L. et al. (2013). PEGylated FePt@Fe_2O_3 core–shell magnetic nanoparticles: potential theranostic applications and in vivo toxicity studies. *Nanomedicine: Nanotechnology, Biology and Medicine* 9: 1077–1088.

30 Sebesta, C., Torres Hinojosa, D., and Wang, B. (2022). Subsecond multichannel magnetic control of select neural circuits in freely moving flies. *Nature Materials* 21: 951–958.

31 Rao, S., Chen, R., and LaRocca, A.A. (2019). Remotely controlled chemomagnetic modulation of targeted neural circuits. *Nature Nanotechnology* 14: 967–973.

32 Wu, C., Shen, Y., Chen, M. et al. (2018). Recent advances in magnetic-nanomaterial-based mechanotransduction for cell fate regulation. *Advanced Materials* 30: 1705673.

33 Lee, J.-H., Kim, J., Levy, M. et al. (2014). Magnetic nanoparticles for ultrafast mechanical control of inner ear hair cells. *ACS Nano* 8 (7): 6590–6598.

34 Lee, J., Shin, M., Lim, Y. et al. (2021). Non-contact long-range magnetic stimulation of mechanosensitive ion channels in freely moving animals. *Nature Materials* 20: 1029–1036.

35 Tay, A. and Carlo, D.D. (2017). Magnetic nanoparticle-based mechanical stimulation for restoration of mechano-sensitive ion channel equilibrium in neural networks. *Nano Letters* 17: 886–892.

36 Yue, K., Guduru, R., Hong, J. et al. (2012). Magneto-electric nanoparticles for non-invasive brain stimulation. *PLoS One* 7 (9): e44040.

37 Pandey, P., Ghimire, G., Garcia, J. et al. (2021). Single-entity approach to investigate surface charge enhancement in magnetoelectric nanoparticles induced by AC magnetic field stimulation. *ACS Sensors* 6: 340–347.

38 Nguyen, T., Gao, J., and Wang, P. (2021). In vivo wireless brain stimulation via non-invasive and targeted delivery of magnetoelectric nanoparticles. *Neurotherapeutics* 18: 2091–2106.

10

Nanomedicine-mediated Immunotherapy

Wei Hou[1] and Yujun Song[1,2,3]

[1] University of Science and Technology Beijing, Center for Modern Physics Technology, School of Mathematics and Physics, 30 Xueyuan Road, Haidian District, Beijing 100083, China
[2] Zhengzhou Tianzhao Biomedical Technology Company Ltd., Zhengzhou New Technology Industrial Development Zone, 7B-1209 Dongqing Street, Zhengzhou 451450, China
[3] Key Laboratory of Pulsed Power Translational Medicine of Zhejiang Province, Hangzhou Ruidi Biotechnology Company Ltd., Room 803, Bldg. 4, 4959 Yuhangtang Road, Cangqian Street, Hangzhou 310023, China

10.1 Immune System and Immune Response

The immune system is an important system that performs immune responses and functions in the body, playing an important role in defending against foreign substances and maintaining the stability of the physiological environment. The immune system consists of immune organs (including central immune organs: thymus and bone marrow, as well as other peripheral immune organs and tissues: spleen, lymph nodes, and mucosal immune system), immune cells (including lymphocytes, monocytes, macrophages, and dendritic cells [DCs]), and immune molecules (including cytokines, chemokines and their receptors, complement and its regulatory molecules, adhesion molecules, and antibodies) (Figure 10.1) [1]. Through the recognition, response, regulation, and memory abilities of the immune system, it plays three important functions: immune defense, immune surveillance, and maintaining internal environmental stability. When pathogens such as bacteria, viruses, and parasites invade the body, the immune system plays its role in fighting infections to protect the body from pathogen attacks. This is known as immune defense [2]. The immune system can effectively recognize and eliminate cancerous or decaying cells in the body promptly, known as immune surveillance [3]. In addition, the immune system and the neuroendocrine system balance each other through autoimmune tolerance and immune regulation to jointly maintain the stability of the internal environment of the body.

The immune response initiated and participated by the immune system can be divided into innate immunity (also known as nonspecific immune response) and adaptive immunity (also known as specific immune response). Skin and various mucosal layers serve as protective barriers against pathogen invasion

Nanomedicine: Fundamentals, Synthesis, and Applications, First Edition. Edited by Yujun Song.
© 2025 WILEY-VCH GmbH. Published 2025 by WILEY-VCH GmbH.

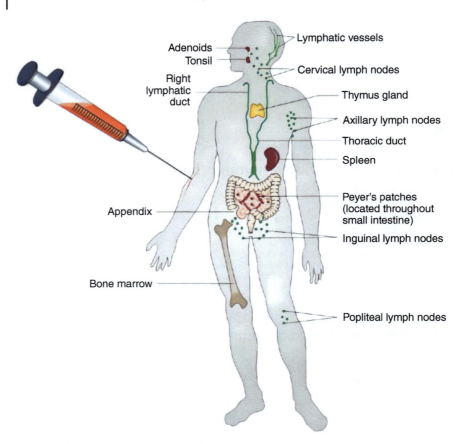

Figure 10.1 Immune organizations of human beings. Source: Reproduced with permission of Ref. [1], © 2017, Springer Nature.

and are the first line of defense against infection in the innate immune system [4]. Secondly, innate immunity can also induce existing antibacterial components such as complements and cytokines to clear pathogens and pathological self-components. Subsequently, natural killer cells (NK), macrophages, and DCs are activated, initiating an inflammatory response by recognizing pathogen-associated molecular patterns (PAMPs) to clear pathogens. The adaptive immune response consists of antigen recognition, activation, and effector functions of lymphocytes. According to the different mediators mediated effector responses, it can be divided into T cell-mediated cellular immune response and B cell-mediated humoral immune response. Cellular immunity refers to the proliferation and specific cytotoxic effect of T lymphocytes in the body stimulated by antigens presented by antigen-presenting cells, thereby killing pathogens. Humoral immunity refers to the activation and differentiation of B lymphocytes in the body after receiving the first signal of antigen stimulation and the second signal of helper T cells (Th) assistance, resulting in the production of plasma cells that produce antibodies, mainly IgG, to neutralize pathogenic bacteria, toxins, and other antigenic foreign substances.

At this point, cells and molecules involved in immune response work together to maintain the stability of the internal environment of the immune system.

10.2 Mechanism of Tumor Immunotherapy

The mechanism of immunotherapy for tumors is shown in Figure 10.2 [5]. In the adaptive immune process, antigens transmit "maturation" signals to DCs and promote DC maturation. At the same time, antigens can be internalized and processed by mature DCs, further inducing DC maturation and antigen presentation. Based on MHC I or II, DCs can present antigens to T cells. In this process, immature DCs present antigens to regulatory T cells (Treg) to generate immune therapy tolerance [6]. APC activation of T cells mainly relies on the following three pathways: the binding of MHC I or II complexes to receptors on the surface of T cells, the expression of co-stimulatory molecules on the surface of cells, and the secretion of cytokines (such as IL-12 and IFN-y) [7]. In this process, multiple types of T cells participate in the formation and control of immune responses, including helper T cells (CD4$^+$ T cells), cytotoxic T cells (CTLs, CD8$^+$ T cells), memory T cells, and regulatory T cells (Tregs) [8]. The helper T cells help stimulate the killing and clearance of tumor cells by CTLs and other immune cells (e.g. NK cells, macrophages, or granulocytes). In addition, after exposure to antigen, memory T cells are formed and circulate in the body, which will remain for a long time to respond quickly to eliminate foreign antigens. Tumor-specific T cells must enter the tumor bed to exert their functions, but the immunosuppressive microenvironment limits the infiltration of T cells into the tumor interior. The immunosuppressive microenvironment of tumors mainly includes the following aspects [9]: (i) Tumors inhibit the activity of effector T cells by triggering "false" immune responses or local accumulation or expansion of Treg. For example, immature DCs can

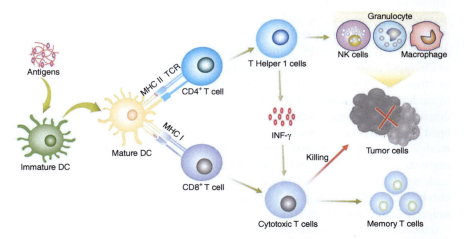

Figure 10.2 Mechanisms of tumor immunotherapy. Source: Reproduced with permission of Ref. [5], © 2021, John Wiley & Sons.

present antigens to Tregs, inducing an immunosuppressive response. (ii) Tumor cells downregulate the expression of MHC class I molecules and tumor antigens to reduce immunogenicity. (iii) Tumor cells can also produce multiple immune checkpoints (e.g. PD-L1 or PD-L2), which bind to receptors (PD-1) on the surface of T cells, resulting in the inactivation of effector T cells. The expression of this inhibitory ligand may be related to carcinogenic mutations such as PTNE loss seen in many cancers. (iv) Immunosuppressive molecules released by tumors, such as indoleamine 2,3-dioxygenase (IDO), can decompose tryptophan into kynurenine and limit the function of T cells.

10.3 Mechanism of Nanomedicine-enhancing Immunotherapy

Cancer is a major challenge facing global health today. Although the clinical application of surgical, chemotherapy, and radiotherapy treatments has become increasingly sophisticated, problems such as incomplete therapeutic efficacy, tumor metastasis and recurrence, and significant adverse reactions have yet to be resolved. In 1893, William Bradley Coley, known as the "father of immunotherapy," used inactivated bacteria to treat tumors, inspiring countless scientific researchers to further study immunotherapy. For example, in 1986, cytokine interferon-α (IFN-α) was approved by the US Food and Drug Administration (FDA) as the first immunotherapy drug for hairy cell leukemia to be launched [10]. In 1992, based on the results of multi-institutional studies, recombinant interleukin-2 (IL-2), as an immunotherapy agent, was approved by the FDA for the treatment of patients with metastatic renal cancer, and subsequently approved for the treatment of metastatic melanoma [11, 12]. In 2011, CTLA-4 was approved for the treatment of advanced melanoma. The emergence of new therapies such as the monoclonal antibody ipilimumab and chimeric antigen receptor (CAR) T cell therapy is a milestone in cancer immunotherapy and was ranked as 1 of the top 10 breakthroughs of 2013 evaluated by the journal of "Science" [13]. James Allison and Tasuku Honjo discovered CTLA-4 and PD-1 immune checkpoints, making great contributions to the rapid development of immunotherapy, and thus won the 2018 Nobel Prize in Physiology or Medicine [14]. Cancer immunotherapy has been widely explored in clinical practice due to its ability to induce the immune system to function and demonstrate enhanced efficacy. As shown in Figure 10.3 [15], since Dr. Coley developed the "Coley toxin" treatment method in 1893, many tumor immunotherapy methods have been developed. Especially after confirming the existence of immune negative regulation mechanisms generated by "immune checkpoint" molecules such as CTLA-4 and PD-L1 in the tumor microenvironment, multiple immune checkpoint-related drugs have been approved by the FDA for the treatment of multiple malignant tumors [16].

Immunotherapy automatically rejects tumor cells by regulating the immune system, but due to the unique adaptive mutations of tumors, cancer cells are seen as "self" and develop tolerance to immunotherapy. Tumors have multiple

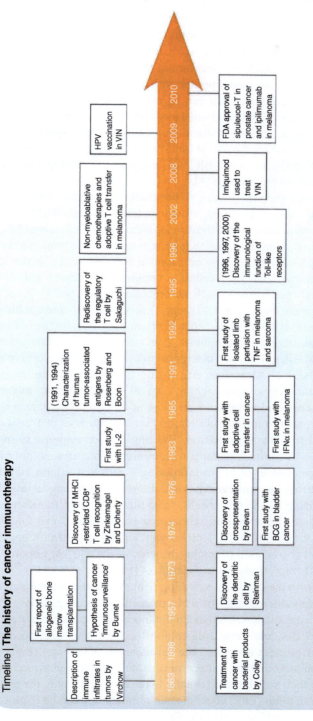

Figure 10.3 Development history of tumor immunotherapy. Source: Reproduced with permission of Ref. [15], © 2011, Springer Nature.

tolerance mechanisms to immunotherapy, such as regulatory immune cells, immunosuppressive cytokines and chemokines, and immune checkpoints. Combining other treatment methods with immunotherapy can effectively regulate the anti-tumor immune effect to eliminate residual tumor cells. Tumor immunotherapy activates the immune system's clearance of tumors by regulating the immune system and the immune microenvironment of tumors. At the same time, tumor cells are transformed into forms that can be recognized by the immune system, ultimately being recognized and eliminated by the immune system. The anti-tumor immune response triggered by immunotherapy can promote systemic immune monitoring, and eliminate local residual tumor cells and metastatic tumor cells. In addition, immunotherapy may establish long-term immune memory and immune protection against tumor recurrence [17].

However, for tumor immunotherapy to be widely applied in clinical practice, many challenges still need to be overcome. One of the main obstacles is the limited response rate to immunotherapy and the development of drug resistance of patients. Research has shown that the main reason for tumor resistance to immunotherapies, such as immune checkpoint therapy or adoptive therapy (ACT), is the lack of recognition of tumor antigens by T cells. On the one hand, cancer cells alter the antigen presentation mechanism (such as proteasome subunits or transporters related to antigen processing), alter B2M (beta-2-microglobulin) or major histocompatibility complexes (MHC) to avoid displaying tumor antigens on the cell surface. On the other hand, immune checkpoints, T cell depletion, immunosuppressive cell populations (MDSCs, Treg, M2 macrophages), and the release of tryptophan metabolites, TGF, CSF-1, adenosine, etc., in the tumor microenvironment also contribute to resistance to immunotherapy. Clinical data shows that only a small portion of patients respond to immune checkpoint therapy. The characteristic of nonimmunogenic patient tumors (cold tumors) is a low number of T cells or low expression of PD-L1 in the tumor tissue, resulting in poor response to immune checkpoint therapy. In contrast, patients with immunogenic tumors (hot tumors) that contain a large number of tumor-infiltrating T cells and exhibit high PD-L1 expression benefit from immune checkpoint blockers and have long-lasting clinical responses. By inducing ICD in tumor cells, "cold tumors" can be transformed into "hot tumors" [18]. At the same time, enhancing endogenous T cell function, in vitro expansion of tumor-infiltrating lymphocytes, and antigen-specific engineered T cells (CAR-T) can also be used to combat drug resistance and low responsiveness in tumor immunotherapy, as shown in Figure 10.4 [19].

In recent years, with the continuous development of nanotechnology, nanomaterials have been widely studied in the field of nanomedicine as drug carriers. Various types of nanomedicine carriers, such as lipid nanocarriers, polymer nanocarriers, inorganic nanoparticles, and some drug conjugates, have been developed for the delivery of tumor treatment drugs and the development of vaccines (Figure 10.5) [20]. In terms of immunotherapy, nanocarriers exhibit excellent pharmacokinetics in the body due to their suitable size (usually 1–100 nm), and due to the enhanced permeability and retention (EPR) effect of solid tumors, they can be retained for a long time in tumors, providing a platform for many drugs with poor water-soluble

10.3 Mechanism of Nanomedicine-enhancing Immunotherapy | 251

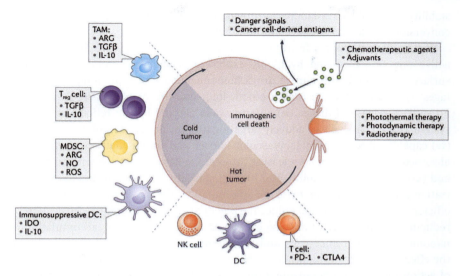

Figure 10.4 Transforming "cold tumors" into "hot tumors" to overcome immune therapy tolerance. Source: Reproduced with permission of Ref. [19], © 2019, Springer Nature.

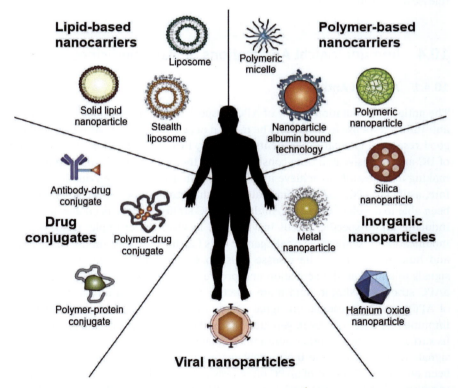

Figure 10.5 Scheme illustrations of currently nano drug carriers such as drug conjugates, lipid-based nanocarriers, polymer-based nanocarriers, inorganic nanoparticles, and virtual nanoparticles. Source: Reproduced with permission of Ref. [20], © 2011, Springer Nature.

stability and low bioavailability to provide possibilities for promoting the clinical conversion of safe and effective immunotherapy.

Nanomedicine benefits from passive targeting mediated by the EPR effect and active targeting mediated by highly expressed specific receptors on the tumor surface. Compared with simple drugs, nanomedicines have the following advantages: (i) stable presence in the blood, long half-life of blood circulation, and no sudden release of drugs; (ii) able to evade capture by the reticuloendothelial system; (iii) by passively or actively targeting tumors, drug accumulation increases; (iv) capable of penetrating the vascular barrier and tumor matrix; (v) effectively absorbed by tumor cells and releasing drugs at the desired location within the cells; and (vi) good biocompatibility and biodegradability. Immune drug nanomaterialization can significantly increase the accumulation of drugs in tumors, improve the efficiency of drug action, and reduce the potential systemic toxicity of drugs, thereby reducing the incidence of adverse events. At the same time, the accumulation of nanomedicines in tumor cells and tumor microenvironments can also enhance the effectiveness of immunotherapy. Nanomaterials also provide new mechanisms of action for immunotherapeutic drugs, including the ability to display ligands to immune cells, regulate intracellular drug penetration, and control the time of drug release or activation [21].

10.4 Immunological Applications of Nanomedicines

10.4.1 Artificial Antibody

The antigen presentation ability of APCs (especially DC) determines the degree of anti-tumor response in the body. The method of activating DC in vitro has achieved good results in clinical experiments. However, in vitro cultivation and induction of DC are expensive and time-consuming, and there is a lack of unified standards, making it impossible to achieve industrialized pharmaceutical production. Therefore, artificial APCs (aAPCs) have attracted the attention of researchers and have been able to develop. Currently, multiple aAPC strategies have been developed for inducing tumor-specific T cells in vitro and in vivo. For example, using genetic modification of xenogeneic or allogeneic cells (such as fly cells, mouse fibroblasts, and human red and white disease cells) to express pMHC and co-stimulatory signals to stimulate the activation and proliferation of corresponding T cells. The aAPC strategy is simpler and more effective than in vitro culture and induction of APCs, and its presentation signal is easy to determine, thus better controlling immune response. However, genetically modified aAPC has immunogenicity, and its surface may contain other nontumor antigens, stimulation signals, or inhibitory signals [22]. To overcome these problems, noncellular structures of aAPCs have been proposed. This type of aAPC often uses nanoparticles or liposomes as carriers to activate T cells through surface modification of activated antibodies and pMHC. For example, in spherical polystyrene beads (4–5 μm), surface modification with CD3 and CD28 antibodies can selectively stimulate $CD4^+T$ cell activation but not

CD8$^+$T cell activation. Subsequent experiments have shown that modifying the melanoma antigen MHC complex on the surface of polystyrene nanoparticles can activate specific T cell activation [22]. However, polystyrene nanoparticles cannot be degraded and can easily cause toxicity or thrombosis in the body, so they can only be used to activate T cells in vitro. The expanded T cells obtained through this method must be removed from polystyrene nanoparticles before being injected back into the patient's body, which increases the difficulty of their preparation. Therefore, Levine et al. developed magnetic aAPC, hoping to simplify the operation by using a magnetic field to remove aAPC after activating T cells. The above proposed aAPC only provides antigen signals and co-stimulation signals required for T cell activation, lacking cytokine signals. However, CD8$^+$T cells require continuous acquisition of cytokine signals during the activation process [23]. Therefore, designing aAPCs that can provide three types of signals is of great significance in clinical conversion. To achieve this idea, relevant researchers have proposed using a combination of degradable nanoparticles and cytokines as carriers. For example, nanoparticles composed of biodegradable poly(lactic-*co*-glycolic acid) (PLGA) and cytokines [24]. These biodegradable and sustained-release cytokine nanoparticles can effectively promote the secretion of IFN-γ by mouse and human T cells. In addition, these aAPCs that can provide cytokine signals significantly improve the activation and proliferation of CD4$^+$T cells and CD8$^+$T cells. However, this degradable aAPC cannot function for a long time, and the noncytokine components released after degradation also have a certain impact on the activation of T cells. To avoid these issues, it is possible to consider converting cytokine signals into surface receptor signals similar to CD25 for presentation. In summary, to better simulate natural APCs, it is necessary to design and investigate the ability of aAPCs of different materials, sizes, surface ligands, and shapes to activate T cells. Designing a novel and optimized aAPC can improve the clinical efficacy of tumor immunotherapy and achieve industrial manufacturing, with unlimited potential in tumor immunotherapy.

10.4.2 Reprogrammed Immunity

In 1909, Paul Ehrlich proposed a hypothesis that the human body continuously produces vegetative cells, and the immune system continuously eliminates these vegetative cells [25]. On this basis, Burnet and Thomas proposed the "tumor immune monitoring" hypothesis, which states that the immune system can recognize and eliminate malignant tumors, thereby inhibiting their occurrence and development [26]. Subsequently, researchers observed phenomena such as normal mice clearing adoptive tumors from homologous nude mice and spontaneous regression of melanoma in patients with autoimmune diseases in clinical practice [27]. These phenomena demonstrate the theory of "tumor immune monitoring" and also indicate that tumor cells express specific molecules, namely tumor antigens, that are different from normal cells. These tumor antigens enable tumor cells to be "monitored" by immune cells, which in turn can be recognized and cleared. However, some phenomena contradict the hypothesis of "tumor immune monitoring." For example, Stutman et al. conducted tumor studies on nude mice and wild-type mice, indicating that there was no difference in the occurrence probability and latency of both induced and nonviral spontaneous tumors between the two models [28]. This raises doubts about the "tumor immune monitoring" hypothesis, and Prehn even proposed an opposite hypothesis, namely the "tumor immune stimulation" hypothesis, which suggests that the immune system promotes tumor growth.

In the 1990s, researchers conducted mouse model experiments and found that the immune system not only played an anti-tumor role but also participated in the remodeling of tumor immunogenicity and assisted tumor cells in escaping immune system surveillance. In 2002, Dunn and Schreider proposed the "tumor immune editing" hypothesis, which believed that the immune system played a dual role in the occurrence and development of tumors, encompassing both immune surveillance and immune escape. This further improved the "tumor immune surveillance" hypothesis [29]. The hypothesis of "tumor immune monitoring" divides immunity into three stages during the occurrence and development of tumors [30], as shown in Figure 10.6. The first stage is the immune clearance stage. During this stage, tumor cells proliferate rapidly and are in the stage of tumor neogenesis. They possess strong immunogenicity and are easily recognized and cleared by the immune system. The entire immune system of the body, including nonspecific innate immunity and specific acquired immunity, is involved in the

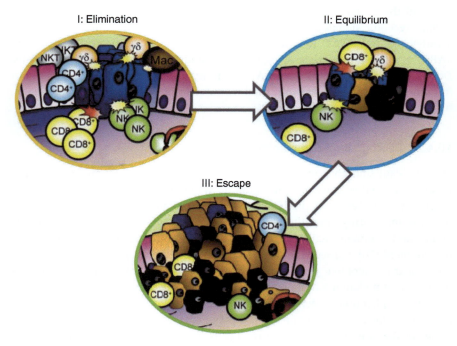

Figure 10.6 The three Es of cancer immunoediting are elimination, equilibrium, and escape.

recognition and clearance process of tumor cells. For example, intrinsic effector cells recruited to the tumor site (natural killer T cells, NKT; natural killer cells, NK; some T cells, γδ-T) Identify tumor cells first and release perforin and interferon-γ (IFN-γ). Alternatively, tumor cells can be killed through the Fas/FsaL pathway [31]. The tumor antigens released by tumor cell death are captured and ingested by DCs. Stimulated mature DCs migrate to lymph nodes and present captured antigen information to helper T cells ($CD4^+T$ cells) or cytotoxic T cells ($CD8^+T$ cells), promoting their proliferation and differentiation. Finally, $CD4^+T$ and $CD8^+T$ cells with tumor specificity will infiltrate the tumor site and clear tumor cells expressing tumor antigens. The effective activation of CD8+T cells is considered the most critical step in tumor immunotherapy. After all tumor cells are cleared, the "tumor immune editing" process ends. However, in most cases, tumor cells will undergo genetic mutations, resulting in the inability of tumor cells to be eliminated. At this point, tumor cells and the immune system will enter the second stage, which is the immune balance stage. At this stage, tumor cells and the immune system are in a relatively balanced state, and the immune system exerts a strong selective pressure on tumor cells, which can suppress but cannot eliminate rapidly mutated tumor cells. Although many immune escape variants of tumor cells are eliminated, tumor cells remain in a state of genetic mutation, and new variants continue to emerge. When the genetic mutations in a tumor accumulate to a certain extent, it will ultimately produce tumor cells that can resist, evade, or even suppress the anti-tumor immune response. Subsequently, the balance between tumor cells and the immune system will be disrupted, and their interaction mode will enter the third stage, which is the immune escape stage. At this stage, tumor cells not only reduce their antigen expression levels and alter their apoptotic signaling pathways, but also release molecules with immunosuppressive functions to suppress immune cells or induce the generation of negative regulatory T lymphocytes to suppress immune cells, forming a tumor immunosuppressive microenvironment. Once the anti-tumor mechanism of the immune system is suppressed, tumor cells will enter a rapid growth phase and easily metastasize to other parts.

10.4.3 Nanomedicines as Agonists

Nanoparticles can improve the safety and effectiveness of drugs while altering the pharmacokinetics of immunomodulatory drugs. Tumor targeting based on nanomedicine is usually achieved through two main mechanisms, namely passive targeting and active targeting. Passive targeting relies on enhanced permeation and retention effects (EPR), while active targeting relies on nanoparticles modified by targeted ligands such as antibodies or peptides, which can specifically recognize receptors overexpressed at tumor sites. Due to the complex matrix signals within tumors, reshaping the immune microenvironment of tumors is more important than directly treating tumor cells with cytotoxic drugs. The use of nanoparticles loaded with immune stimulators such as TLR agonists, STING agonists, and other hazard sensor ligands can reduce systemic toxicity after administration. The nanoparticles loaded with STING agonists target tumors through the EPR effect. After the drug is

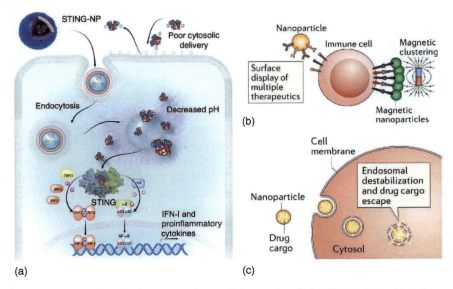

Figure 10.7 (a) STING-NPs enhance intracellular uptake of cGAMP and, in response to decreased pH within endosomal compartments, disassemble and promote endosomal escape of cGAMP to the cytosol. IKK, IκB kinase; IκB, an inhibitor of κB; IRF3, IFN regulatory factor 3; TBK1, TANK-binding kinase 1. Source: Reproduced with permission of Ref. [32], © 2019, Springer Nature. (b) Magnetic nanoparticles mimic antigen presentation on the surface of immune cells. (c) Schematic of nanoparticle entry into cells. Source: Reproduced with permission of Ref. [33], © 2020, Springer Nature.

internalized by cells, it interacts with STING receptors in the cytoplasm to increase drug efficacy (as shown in Figure 10.7a). Under systemic administration conditions, STING agonists have lower efficacy, but intravenous administration can be achieved using nanoparticles to enhance immunotherapy [32].

The use of surface critical immune regulatory receptor interactions between cells to generate co-stimulatory signals. Adjusting the size of nano drugs to the physiological scale of cells and modifying receptor mimics on the surface can simulate the transmission of such co-stimulating signals. For example, the agonist anti-CD137 antibody bound to the surface of liposomes can activate CD8$^+$T cells, with an effect approximately 10 times that of soluble anti-CD137 antibodies at the same concentration [34]. Bind multiple ligands onto the surface of magnetic nanoparticles and use magnetic force to aggregate the nanoparticles to enhance the stimulation of immune cells, as shown in Figure 10.7b. Nanoparticles can destabilize the endocytic membrane and promote drug entry into the cytoplasm, as shown in Figure 10.7c [33].

To fight against the pneumonia epidemic caused by the novel coronavirus, Liu et al. synthesized a cationic liposome. The activator of TLR-4, the activator of amphiphilic monoacyl lipid A (MPLA), and the activator of TLR-9, CpG were encapsulated at the same time, and the S1 subunit of SARS-CoV-2 was assembled on the surface of the liposome to prepare a safer and more effective nanovaccine. The nanovaccine can effectively trigger the humoral immune response in mice and

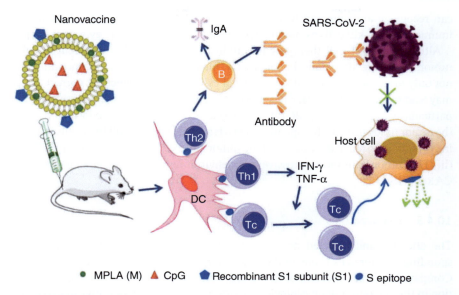

Figure 10.8 Schematic of liposomal subunit nanovaccine for anti-SARS-CoV-2. Source: Liu et al. [35]/American Chemical Society.

produce antibodies that effectively inhibit novel coronavirus. In addition, compared to free S1, nanovaccines can trigger strong cellular immunity by activating $CD4^+$ and $CD8^+$T cells (Figure 10.8) [35].

10.4.4 Nanomedicine-combined CAR-T Therapy

In the late 1980s, scientists began conducting cancer treatment experiments using CAR-T cells and genetically engineered T cells to express antibodies that recognize antigens [36, 37]. In 2011, Carl and colleagues reported successful cases of chronic lymphocytic leukemia cured based on anti-CD19 CAR-T 19 [38]. In 2017, Kymarinh was approved by the FDA for the use of CAR-T in the treatment of refractory acute lymphoblastic leukemia, marking a milestone in immunotherapy and indicating the commercial clinical application of CAR-T. Specifically, the first step in engineered T cell therapy is to isolate T cells from the patient's blood, and then activate, amplify, and genetically modify them to express CAR for antigen recognition. At present, the most common way to genetically modify T cells is to use engineered viral vectors, such as Y retroviruses or lentiviruses, to carry CAR builders [39]. In short, T cells are activated and then transduced by viral particles, which release their contents into T cells, causing the CAR builder to be inserted into the genome of the transduced T cells. Afterward, engineered T cells were re-injected into the same patient, during which CAR-T cells could recognize target antigens on tumor cells and cause tumor cell death [40]. CAR is a transmembrane receptor that includes both extracellular and intracellular parts: the extracellular part contains a specific antibody scFv extracellular domain, while the intracellular part contains CD3; and other synergistic stimulation signal regions (such as CD28, 4-1BB, and OX-40). CAR can recognize antigens and further activate CAR-T cells, induce T cell-mediated immune effects, kill cells expressing specific antigens, and secrete cytokines [41].

Although CAR-T cell therapy has certain advantages in the treatment of solid tumors, there are also some challenges. (i) Due to the expression of tumor antigens not only on tumor cells but also on normal cells, failure to select the correct target may lead to off-target effects, such as B cell aplastic anemia [42]. Target selection is particularly important in CAR-T design. (ii) CAR-T cells need to be located deep in the tumor tissue to exert therapeutic effects. However, it is difficult for T cells injected intravenously to reach the interior of the tumor and exert effective anti-tumor effects. (iii) The tumor immunosuppressive microenvironment limits the effective effect of CAR-T cells [43].

10.4.5 Nanovaccines

The effective and targeted delivery of immune-stimulating molecules to corresponding effector cells is one of the important keys to cancer immunotherapy [44]. Compared to direct injection of free drugs, nanocarriers can prevent drug degradation in the surrounding physiological environment, improve their pharmacokinetic behavior, and promote their delivery to target cells. For example, Kwong et al. reported a method of simultaneous loading α-Liposome nanocarriers for CD40 and CpG (two highly effective anti-tumor agents). These ligands are coupled to the surface of liposomes to maintain high levels in the tumor and surrounding tissues after intratumoral injection and can limit their entry into the systemic circulation or reach the distal lymphatic organs while allowing them to be presented to APC at the tumor and tumor-draining lymph nodes, thereby blocking systemic toxicity while maintaining the anti-tumor effects of two different immune stimulators (Figure 10.9a) [45]. Li et al. designed a DNA tetrahedral structure for multivalent loading and delivery of CpG. By loading DNA nanostructured complexes, CpG can resist nuclease degradation and maintain basic integrity in fetal bovine serum and cells. This complex can also effectively and noninvasively enter RAW264.7 cells without the assistance of transfectants. After cell uptake, the CpG sequence is recognized by Toll-like receptor 9 (TLR9), which activates downstream pathways to induce immune stimulation effects, leading to high levels of secretion of various pro-inflammatory cytokines (such as TNF-α, IL-6, and IL-12) (Figure 10.9b) [46].

In addition, using nanocarriers to deliver tumor-related antigens to antigen-presenting cells to exert the role of tumor vaccines and enhance immune response mechanisms is also a promising strategy for improving immunotherapy [47]. For example, Lai et al. co-encapsulated CpG ODN and melanoma-specific TRP2 (180–188) peptide in a mannose-modified liposome, specifically targeting DCs and demonstrating synergistic effects (Figure 10.10a) [48]. Compared to the administration of free drugs, these DC-targeted liposomes enhanced the anti-tumor response and prolonged the survival time of melanoma-bearing mice due to the increase of effector T cells. Zhou et al. developed a method based on COF@ICG @OVA's treatment platform [49]. By using COF as a carrier, the stability of OVA antigen and the photostability of ICG are greatly improved. CO@ICG @After being engulfed

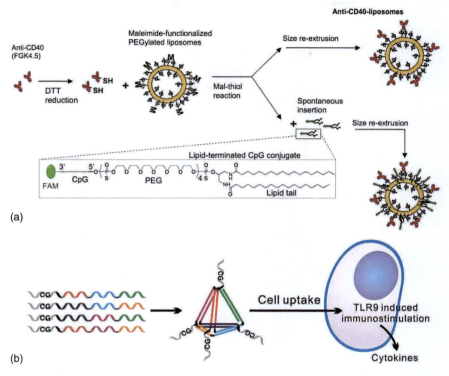

Figure 10.9 (a) Schematic illustration of the synthesis of PEGylated liposomes bearing lipid-anchored anti-CD40 and CpG oligonucleotides. Source: Reproduced from [45] with permission from © Elsevier 2011. (b) Schematic showing of the assembly of CpG-bearing DNA tetrahedron and its immunostimulatory effect. Source: Reproduced with permission of Ref. [46], © 2011, American Chemical Society.

by DC, OVA nanoparticles can effectively promote antigen presentation, combined with photothermal therapy and photodynamic therapy to ablate primary tumors, and further combined with PD-L1 checkpoint blockade therapy to induce systemic immunity, effectively inhibiting tumor metastasis (Figure 10.10b,c) [49].

Overall, vaccines come in two forms: preventive and therapeutic [50]. Human papillomavirus (HPV), as a preventive vaccine, can prevent cancer from viral sources. IL-2, as an immune stimulator, was co-administered with a short peptide (antigen) derived from glycoprotein 100 (gp100), showing an enhanced tumor response. Compared with IL-2 alone, the survival period of advanced melanoma patients was prolonged [51]. Cancer vaccines promote the maturation of DCs, activate tumor-specific T cells [52–54] after antigen presentation, and induce tumor-specific immune responses. CTLs directly eliminate specific tumor cells and assist T cells in secreting cytokines to enhance the function of CTLs [55]. The specific antigen cancer vaccines extracted from different types of tumor cells can stimulate the immune system to fight against different tumor cells, thereby expanding the applicability of tumor vaccines [56]. Overall, the use of tumor vaccines

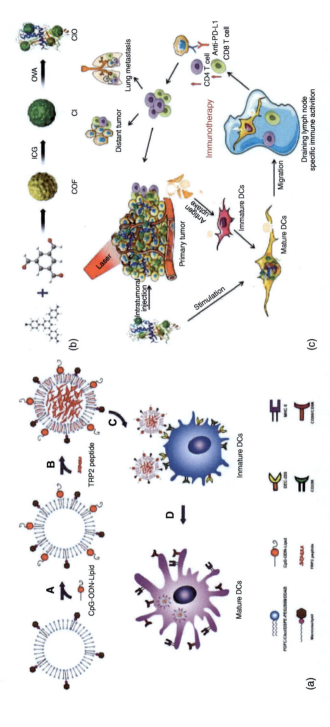

Figure 10.10 (a) Schematic description of the preparation of M/CpG-ODN-TRP2-Lipo and ability of DC targeting and activation. Source: Lai et al. [48]/ Ivyspring International Publisher. (b) Synthesis process of COF@ICG@OVA. (c) Schematic illustration of the mechanism of CIO nanoparticle-enhanced phototherapy combined with immunotherapy. Source: Reproduced with permission of Ref. [49], © 2020, The Royal Society of Chemistry.

can co-deliver tumor-specific antigens and immune adjuvants to DCs, triggering the maturation and antigen presentation processes of DCs, and subsequently triggering a strong antigen-specific T cell response, effectively killing tumor cells. In addition, autologous tumor cells (ATCs) obtained from patients can also be used as tumor vaccines to achieve personalized immunotherapy. Patient-specific antigens released from ATC can induce patient-specific anti-tumor immunity [57]. Vaccines derived from ATC lysates, especially when combined with other treatment methods, can be effectively used to treat tumors and prevent tumor recurrence, such as ovarian cancer [58]. However, the immune response mediated by tumor vaccines is often negatively regulated in the immunosuppressive microenvironment. Specifically, certain autologous vaccines may trigger the secretion of INF-γ, induce the expression of PD-L1 in tumors, and thus cause adaptive immune resistance. In addition, the preparation of ATC vaccines requires sufficient tumor samples, and the types or stages of certain tumors limit the scope of application of such vaccines. Overall, the emergence of cancer vaccines based on antigens and ATC lysates will drive the development of cancer treatment and bring hope to patients. However, the clinical application of tumor vaccines still faces many challenges, including breaking through many treatment limitations, ensuring the most suitable clinical environment, shortening turnover time, and expanding production scale.

10.4.6 Nanomedicines Affect Cytokines

Cytokines are secreted by activated immune cells (such as T lymphocytes, macrophages, B lymphocytes, monocytes, and NK cells) and some stromal cells (such as epidermal cells, fibroblasts, vascular endothelial cells, and keratinocytes). These molecules have a wide range of biological activities and a variety of biological functions, including stimulating or inhibiting immune functions, regulating a variety of cellular physiological functions, etc. At present, interleukin, interferon, and granulocyte-macrophage colony-stimulating factor (GM-CSF) are the most widely used cytokines in tumor immunotherapy. For example, under certain conditions, both IL-2 and IL-15 can be used to treat metastatic cancer by stimulating anti-tumor immune cells (including NK cells and CTLs) or by controlling the proliferation and differentiation of CTLs. IL-12 can stimulate T cell proliferation and differentiation, and stimulate the secretion of cytokines (such as IFN-γ, TNF-α, and GM-CSF) while inducing CTLs and NK cytotoxicity. IFN-γ can activate macrophages, DSs, lymphocytes or mature NK cells to enhance the anti-tumor ability of the body. In addition, IFN-γ can promote the expression of MHC II molecules on the surface of macrophages to increase antigen presentation, and the highly expressed Fc receptor can increase the phagocytosis of macrophages. In addition, GM-CSF improves the immune response by promoting the differentiation of DSs and increasing tumor-specific antigen presentation. Sockolosky et al. verified that the engineered IL-12 receptor could amplify the immunotherapeutic effect of effector T cells while avoid the toxicity of IL-2.

The emergence of nanomedicine provides a novel possibility with decreased toxicity and increased targeting for treating tumors, strengthening immune cells, stimulating abscopal effects, and changing the tumor microenvironment (TME) and tumor immune microenvironment (TIME). Metal and/or metal oxide nanoparticles (NPs) have attracted increased research interest due to their therapeutic functions, in which iron-based NPs (e.g. Fe_3O_4 nanoparticles) demonstrate significant advantages as multifunctional carriers for chemotherapeutic drugs delivery by their unique physiochemical properties (e.g. excellent biocompatibility and magnetic properties), auto-liver targeting function characteristics and lesion targeted abilities guided by magnetic fields. Our group synthesized and conjugated $Fe@Fe_3O_4$ nanoparticles with ginsenoside Rg3 (NpRg3) to develop a novel type of nanomedicine [59, 60]. Nanoparticle conjugation of ginsenoside Rg3 significantly inhibited HCC formation and development. It restricts abnormal division of tumor cells and reduces the number of mitochondria inside the cells.

Tumor metastasis to the lung is the common death cause for HCC patients. Tumor angiogenesis and circulating tumor cells were two main factors affecting HCC metastasis. Ginsenoside Rg3 could suppress tumor growth and tumor angiogenesis by inhibiting the mobilization of endothelial progenitor cells from the bone marrow microenvironment to the peripheral circulation and by modulating VEGF-dependent tumor angiogenesis. [59, 60] After NpRg3 leaves the liver, the long-term circulation of iron-based NpRg3 might have therapeutic effects on circulating tumor cells by activating immune cells (e.g. T or B cells) in the blood. [59, 60]. Based on these results, NpRg3 application can not only inhibit primary HCC tumors in the liver but also eliminate HCC metastasis to the lung.

Gut microbial alterations contribute to the onset and progression of alcoholic liver diseases, nonalcoholic fatty liver disease, and liver cirrhosis. Importantly, gut microbial alterations promote HCC development by the microbiota-liver axis, mainly because microbial products activate the immune system to drive the proinflammatory response. Notably, TLR-4-mediated inflammatory signals induced by the gram-negative bacterial component lipopolysaccharide (LPS) are important for promoting liver fibrosis and HCC development. To directly measure liver or tumor inflammation in tumor-bearing mice, the liver tissues were collected and smashed. Luminex multiplex technology was used to detect cytokines and chemokines in liver tissues. A total of 29 inflammatory factors were detectable, and the inflammatory markers with significant differences were presented. Compared with DEN-30W group, the levels of inflammatory factors killing tumor cells or inhibiting tumor progression, including tumor necrosis factor (TNF-α), interferon-γ (IFN-γ), interleukin-1β (IL-1β), IL-12 (p70), IL-17, IL-15, and granulocyte CSF (G-CSF), were significantly reduced in the DEN-42W group. NpRg3 administration significantly elevated these cytokines compared to DEN-42W group. In contrast, the levels of chemokines promoting tumor progressions, such as keratinocytes, chemokines (KC), and LPS-induced CXC chemokines (LIX), were obviously increased in the DEN-42W group versus DEN-30W group but decreased after NpRg3 administration. This study demonstrates that NpRg3 enhances the levels of inflammatory factors killing tumor cells or inhibiting tumor progression, but decreases the levels of chemokines promoting tumor progression, thereby significantly inhibiting HCC development.

10.5 Summary and Outlook

Nanomedicine-mediated immunotherapy represents a promising frontier in the field of medical treatment, where the convergence of nanotechnology and immunotherapy holds significant potential for groundbreaking advances. In recent years, substantial progress has been made, laying a solid foundation for the development of innovative therapeutic approaches. The integration of nanotechnology enables the precise design and fabrication of nanoparticles, which can serve as carriers for drugs, antigens, and gene-editing tools, among others, to enhance the responsiveness of the immune system. Furthermore, nanotechnology facilitates improved delivery efficiency within the body, minimizing toxic side effects and allowing targeted interventions in the tumor microenvironment.

The future of nanomedicine-mediated immunotherapy holds several exciting prospects as below.

Precision therapy. In-depth understanding of tumor heterogeneity in TIME will allow for more precise treatment strategies, and customizing therapies for different subtypes and individual characteristics.

Labeling and diagnostics. Leveraging the biological imaging properties of nanoparticles in the monitor of tumor cells (e.g., CTC cells), TME and TIME will enable early tumor diagnosis and real-time monitoring, providing dynamic data support for adjusting treatment plans.

Optimization of combination therapies. Nanomedicine is expected to facilitate more effective combinations of immunotherapy with other treatment modalities, maximizing treatment efficacy while minimizing toxicity and side effects.

Gene editing and personalized treatment. Advancements in nanomedicine will accelerate the application of gene-editing tools, offering more possibilities for personalized treatment and genetic repairs or modification that can regulate the TME or TIME intrinsically.

Clinical application expansion. As studies deepen into the immunological effects of nanomedicines , their clinical applications are likely to be broadened, encompassing various cancer types and extending into other areas of chronic or intractable diseases, providing patients with an array of treatment options based on immunity regulation.

The ongoing development of nanomedicine-mediated immunotherapy not only accelerates progress in cancer treatment but also promises to bring about further innovations in future medical practices. By integrating nanotechnology with immunology, much more precise, smart, and personalized approaches for the treatmentof intractable or chronic diseases are anticipated, ultimately providing patients with improved therapeutic outcomes and survival prospects.

References

1 Brodin, P. and Davis, M.M. (2017). Human immune system variation. *Nature Reviews Immunology* 17 (1): 21–29.
2 Paces, J., Strizova, Z., Daniel, S. et al. (2020). COVID-19 and the immune system. *Physiological Research* 69 (3): 379.
3 Simon, A.K., Hollander, G.A., and McMichael, A. (1821). Evolution of the immune system in humans from infancy to old age. *Proceedings of the Royal Society B: Biological Sciences* 2015 (282): 20143085.
4 Iwasaki, A. and Medzhitov, R. (2015). Control of adaptive immunity by the innate immune system. *Nature Immunology* 16 (4): 343–353.
5 Liang, J.L., Luo, G.F., Chen, W.H. et al. (2021). Recent advances in engineered materials for immunotherapy-involved combination cancer therapy. *Advanced Materials* 33 (31): 2007630.
6 Darrasse-Jèze, G., Deroubaix, S., Mouquet, H. et al. (2009). Feedback control of regulatory T cell homeostasis by dendritic cells in vivo. *Journal of Experimental Medicine* 206 (9): 1853–1862.
7 Gilboa, E. (2004). The promise of cancer vaccines. *Nature Reviews Cancer* 4 (5): 401–411.
8 DiToro, D., Harbour, S.N., Bando, J.K. et al. (2020). Insulin-like growth factors are key regulators of T helper 17 regulatory T cell balance in autoimmunity. *Immunity* 52 (4): 650–667.e10.
9 Mellman, I., Coukos, G., and Dranoff, G. (2011). Cancer immunotherapy comes of age. *Nature* 480 (7378): 480–489.
10 Ahmed, S. and Rai, K.R. (2003). Interferon in the treatment of hairy-cell leukemia. *Best Practice & Research Clinical Hematology* 16 (1): 69–81.
11 Nam, J., Son, S., Park, K.S. et al. (2019). Cancer nanomedicine for combination cancer immunotherapy. *Nature Reviews Materials* 4 (6): 398–414.
12 Hodi, F.S., O'Day, S.J., McDermott, D.F. et al. (2010). Improved survival with ipilimumab in patients with metastatic melanoma. *New England Journal of Medicine* 363 (8): 711–723.
13 Couzin-Frankel, J. (2013). Breakthrough of the year 2013. Cancer immunotherapy. *Science* 342 (6165): 1432–1433.
14 Mackall, C.L. (2018). Engineering a designer immunotherapy. *Science* 359 (6379): 990–999.
15 Lesterhuis, W.J., Haanen, J.B.A.G., and Punt, C.J.A. (2011). Cancer immunotherapy – revisited. *Nature Reviews Drug Discovery* 10 (8): 591–600.
16 Sharma, J., Rocha, R.C., Phipps, M.L. et al. (2012). A DNA-templated fluorescent silver nanocluster with enhanced stability. *Nanoscale* 4 (14): 4107–4110.
17 Sharma, P., Hu-Lieskovan, S., Wargo, J.A. et al. (2017). Primary, adaptive, and acquired resistance to cancer immunotherapy. *Cell* 168 (4): 707–723.
18 Corrales, L., Glickman, L.H., McWhirter, S.M. et al. (2015). Direct activation of STING in the tumor microenvironment leads to potent and systemic tumor regression and immunity. *Cell Reports* 11 (7): 1018–1030.

19 Riley, R.S., June, C.H., Langer, R. et al. (2019). Delivery technologies for cancer immunotherapy. *Nature Reviews Drug Discovery* 18 (3): 175–196.

20 Wicki, A., Witzigmann, D., Balasubramanian, V. et al. (2015). Nanomedicine in cancer therapy: challenges, opportunities, and clinical applications. *Journal of Controlled Release* 200: 138–157.

21 Shi, Y. and Lammers, T. (2019). Combining nanomedicine and immunotherapy. *Accounts of Chemical Research* 52 (6): 1543–1554.

22 Butler, M.O., Lee, J.S., Ansen, S. et al. (2007). Long-lived antitumor CD8+ lymphocytes for adoptive therapy generated using an artificial antigen-presenting cell. *Clinical Cancer Research* 13 (6): 1857–1867.

23 Wuest, S.C., Edwan, J.H., Martin, J.F. et al. (2011). A role for interleukin-2 trans-presentation in dendritic cell-mediated T cell activation in humans, as revealed by daclizumab therapy. *Nature Medicine* 17 (5): 604–609.

24 Sunshine, J.C. and Green, J.J. (2013). Nanoengineering approaches to the design of artificial antigen-presenting cells. *Nanomedicine* 8 (7): 1173–1189.

25 Ehrlich, P. (1909). Über den jetzigen Stand der Chemotherapie. *Berichte der Deutschen Chemischen Gesellschaft* 42 (1): 17–47.

26 Burnet, M. (1957). Cancer – a biological approach: III. Viruses associated with neoplastic conditions. IV. Practical applications. *British Medical Journal* 1 (5023): 841.

27 Leduc, E.H., Coons, A.H., and Connolly, J.M. (1955). Studies on antibody production: II. The primary and secondary responses in the popliteal lymph node of the rabbit. *The Journal of Experimental Medicine* 102 (1): 61–72.

28 van den Broek, M.E., Kägi, D., Ossendorp, F. et al. (1996). Decreased tumor surveillance in perforin-deficient mice. *The Journal of Experimental Medicine* 184 (5): 1781–1790.

29 Dunn, G.P., Bruce, A.T., Ikeda, H. et al. (2002). Cancer immunoediting: from immunosurveillance to tumor escape. *Nature Immunology* 3 (11): 991–998.

30 Dunn, G.P., Old, L.J., and Schreiber, R.D. (2004). The three Es of cancer immunoediting. *Annual Review of Immunology* 22: 329–360.

31 Kim, R. (2007). Cancer immunoediting: from immune surveillance to immune escape. *Cancer Immunotherapy* 9–27.

32 Shae, D., Becker, K.W., Christov, P. et al. (2019). Endosomolytic polymersomes increase the activity of cyclic dinucleotide STING agonists to enhance cancer immunotherapy. *Nature Nanotechnology* 14 (3): 269–278.

33 Irvine, D.J. and Dane, E.L. (2020). Enhancing cancer immunotherapy with nanomedicine. *Nature Reviews Immunology* 20 (5): 321–334.

34 Kwong, B., Gai, S.A., Elkhader, J. et al. (2013). Localized immunotherapy via liposome-anchored Anti-CD137+ IL-2 prevents lethal toxicity and elicits local and systemic antitumor immunity. *Cancer Research* 73 (5): 1547–1558.

35 Liu, L., Liu, Z., Chen, H. et al. (2020). Subunit nanovaccine with potent cellular and mucosal immunity for COVID-19. *ACS Applied Biomaterials* 3 (9): 5633–5638.

36 Gill, S., Maus, M.V., and Porter, D.L. (2016). Chimeric antigen receptor T cell therapy: 25 years in the making. *Blood Reviews* 30 (3): 157–167.

37 Gross, G., Waks, T., and Eshhar, Z. (1989). Expression of immunoglobulin-T-cell receptor chimeric molecules as functional receptors with antibody-type specificity. *Proceedings of the National Academy of Sciences* 86 (24): 10024–10028.

38 Porter, D.L., Levine, B.L., Kalos, M. et al. (2011). Chimeric antigen receptor-modified T cells in chronic lymphoid leukemia. *New England Journal of Medicine* 365 (8): 725–733.

39 Fesnak, A.D., June, C.H., and Levine, B.L. (2016). Engineered T cells: the promise and challenges of cancer immunotherapy. *Nature Reviews Cancer* 16 (9): 566–581.

40 Lim, W.A. and June, C.H. (2017). The principles of engineering immune cells to treat cancer. *Cell* 168 (4): 724–740.

41 Srivastava, S. and Riddell, S.R. (2015). Engineering CAR-T cells: design concepts. *Trends in Immunology* 36 (8): 494–502.

42 Huang, R., Li, X., He, Y. et al. (2020). Recent advances in CAR-T cell engineering. *Journal of Hematology & Oncology* 13 (1): 1–19.

43 Hill, J.A., Giralt, S., Torgerson, T.R. et al. (2019). CAR-T–and a side order of IgG, to go? Immunoglobulin replacement in patients receiving CAR-T cell therapy. *Blood Reviews* 38: 100596.

44 Zhang, Y., Li, N., Suh, H. et al. (2018). Nanoparticle anchoring targets immune agonists to tumors enabling anti-cancer immunity without systemic toxicity. *Nature Communications* 9 (1): 6.

45 Kwong, B., Liu, H., and Irvine, D.J. (2011). Induction of potent anti-tumor responses while eliminating systemic side effects via liposome-anchored combinatorial immunotherapy. *Biomaterials* 32 (22): 5134–5147.

46 Li, J., Pei, H., Zhu, B. et al. (2011). Self-assembled multivalent DNA nanostructures for noninvasive intracellular delivery of immunostimulatory CpG oligonucleotides. *ACS Nano* 5 (11): 8783–8789.

47 Iwama, T., Uchida, T., Sawada, Y. et al. (2016). Vaccination with liposome-coupled glypican-3-derived epitope peptide stimulates cytotoxic T lymphocytes and inhibits GPC3-expressing tumor growth in mice. *Biochemical and Biophysical Research Communications* 469 (1): 138–143.

48 Lai, C., Duan, S., Ye, F. et al. (2018). The enhanced antitumor-specific immune response with mannose-and CpG-ODN-coated liposomes delivering TRP2 peptide. *Theranostics* 8 (6): 1723.

49 Zhou, Y., Liu, S., Hu, C. et al. (2020). A covalent organic framework as a nanocarrier for synergistic phototherapy and immunotherapy. *Journal of Materials Chemistry B* 8 (25): 5451–5459.

50 Palucka, K., Banchereau, J., and Mellman, I. (2010). Designing vaccines based on biology of human dendritic cell subsets. *Immunity* 33 (4): 464–478.

51 Schwartzentruber, D.J., Lawson, D.H., Richards, J.M. et al. (2011). gp100 peptide vaccine and interleukin-2 in patients with advanced melanoma. *New England Journal of Medicine* 364 (22): 2119–2127.

52 Zhou, J., Kroll, A.V., Holay, M. et al. (2020). Biomimetic nanotechnology toward personalized vaccines. *Advanced Materials* 32 (13): 1901255.

53 Liu, S., Jiang, Q., Zhao, X. et al. (2021). A DNA nanodevice-based vaccine for cancer immunotherapy. *Nature Materials* 20 (3): 421–430.
54 Li, W.H. and Li, Y.M. (2020). Chemical strategies to boost cancer vaccines. *Chemical Reviews* 120 (20): 11420–11478.
55 Chen, Y., De Koker, S., and De Geest, B.G. (2020). Engineering strategies for lymph node targeted immune activation. *Accounts of Chemical Research* 53 (10): 2055–2067.
56 Zheng, D.W., Gao, F., Cheng, Q. et al. (2020). A vaccine-based nanosystem for initiating innate immunity and improving tumor immunotherapy. *Nature Communications* 11 (1): 1985.
57 Sahin, U. and Türeci, Ö. (2018). Personalized vaccines for cancer immunotherapy. *Science* 359 (6382): 1355–1360.
58 Cully, M. (2018). Personalized cancer vaccines hit the spot. *Nature Reviews Drug Discovery* 17 (6): 393–393.
59 Zhao, X., Jicheng, W., Guo, D. et al. (2022). Dynamic ginsenoside-sheltered nanocatalysts for safe ferroptosis-apoptosis combined therapy. *Acta Biomaterialia* 151 (1): 549–560.
60 Ren, Z., Chen, X., Hong, L. et al. (2019). Nanoparticle conjugation of ginsenoside Rg3 inhibits hepatocellular carcinoma development and metastasis. *Small* 1905233.

11

Nanomedicine-mediated Ultrasound Therapy

Qingwei Liao, Yaoyao Liao, and Wei Si

Key Laboratory of Sensors, Beijing Information Science and Technology University, Yayuncun street, Beijing 100192, China

11.1 Concept, Therapy Theory, and Devices

11.1.1 The Concept of Ultrasound Therapy and Its Therapy Theory

Ultrasound is a mechanical wave with a frequency exceeding the hearing limit of the human ear (>20 kHz), and its average propagation rate is 340 m s^{-1} in air, 1540 m s^{-1} in human soft tissues and fluids, and 3860 m s^{-1} in the skull. Ultrasound has been widely used in the field of medicine because of its characteristics of nonradiation, noninjury, and high spatial and temporal sensitivity. There are two main types of ultrasounds used in medicine – one is ultrasound diagnosis and the other is ultrasound therapy. Diagnostic ultrasound refers to the use of ultrasound to diagnose diseases, including ultrasound imaging, ultrasound bone examination, and brightness scan. It is based on the principle that differences in the propagation laws, tissue properties, and tissue geometry of ultrasound in biological tissues make differences in the propagation laws and fluctuation phenomena such as transmission, reflection, scattering, and interference of ultrasound waves, which, in turn, change the amplitude, frequency, and phase of the signal at the receiving end. And then these parameters are measured, identified and processed by the outside world to judge the disease of the organism [1]. Ultrasound therapy refers to the use of ultrasound waves absorbed by the body's biological mechanism to treat diseases, usually used in the frequency range of 1–3 mHz [2]. Focused ultrasound (FUS) is a high-precision, high-sensitivity ultrasound technology in the body on a wide range of target areas without damaging the surrounding normal T-cells. In medicine, two intensities of ultrasound are commonly used to enable the treatment of patients: high-intensity focused ultrasound (HIFU) and low-intensity pulsed ultrasound (LIPUS).

HIFU is an ultrasound with an intensity greater than 5 W cm^{-2}, and its working principle is: using the characteristics of ultrasound's visibility, tissue penetration, and focusing, the low-energy ultrasound in the body, through a specific ultrasound-focusing transducer, unit transducer, or multitransducer projections,

Nanomedicine: Fundamentals, Synthesis, and Applications, First Edition. Edited by Yujun Song.
© 2025 WILEY-VCH GmbH. Published 2025 by WILEY-VCH GmbH.

is focused in a very small spatial focal domain (also known as the focal spot), and the diseased tissues located in the focal domain, such as tumors, hyperplasia, etc., is irradiated with a short burst of radiation so that the diseased tissue in the focal domain is quickly heated to over 70 °C and coagulative necrosis occurs. As ultrasound is a mechanical wave, it does not cause harm to the surrounding normal tissues, thus enabling the noninvasive removal of tumors [3].

LIPUS is ultrasound with an intensity of less than $3\,W\,cm^{-2}$. Compared with HIFU, the thermal effect of LIPUS is much smaller, mainly due to mechanical effects. At the same time, LIPUS does not require precise positioning of the target point and can achieve repeated treatment and accurate diagnosis of the target area, which has high clinical application value. In addition, LIPUS can also realise the fast and easy targeted treatment of microbubbles and related molecules.

The application of ultrasound in the medical field is based on its biological properties, i.e. thermal, mechanical, cavitation, thixotropic, and acoustic chemical effects, as follows.

11.1.1.1 Thermal Effect

After the tissue is subjected to ultrasound, the ultrasonic energy in the ultrasound will be absorbed and converted into heat, thus warming the tissue locally. At different temperatures and different exposure times, the size, shape, and response of the target area will change differently. The tendency of the temperature to rise also depends on the intensity and duration of the sound waves. The thermal effect of ultrasound at a reasonable level can increase the metabolism of the new city, accelerate blood circulation, and improve the enzyme activity of the body. However, when ultrasound reaches a certain level, it will make the temperature in the tissues rise sharply, and the cells in the tissues will be greatly damaged or even die. The results of clinical trials show that: ultrasound has the greatest effect on bone and connective tissue, and the smallest effect on fat and blood; when HIFU acts on sensitive areas such as the rectum, external sphincter, neurovascular bundles, etc., it does not have a great effect on the neighboring areas due to the temperature difference between the target area and the peripheral areas [2].

11.1.1.2 Mechanical Effects

The vibration of ultrasound causes cell flow, rotation, friction, extrusion, and other phenomena, which affect the permeability of cells and promote cell diffusion. The mechanical effect of HIFU is to use the negative pressure of ultrasound waves to cause pressure on human tissues, causing them to produce bubbles, which continue to expand and eventually reach resonance, resulting in the shattering of bubbles. The enormous pressure generated by the explosion also causes damage, even destruction, to the cells in the body. In the tissues, the combined effect of bubbles, acoustic flow, pressure, and cavities creates a mechanical effect. The medical field applies this effect to ultrasonic lithotripsy and soft tissue removal [4].

11.1.1.3 Cavitation Effect

Bubbles in biological tissues are numerous and they are in regular motion. This equilibrium is apparently disturbed by the radiation of ultrasound, the addition of which

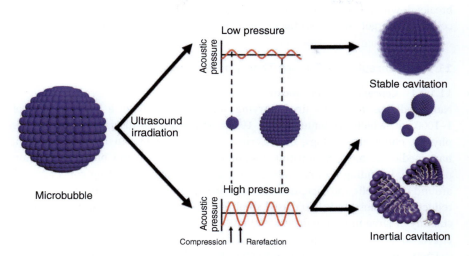

Figure 11.1 A diagram of the microbubble cavitation phenomenon caused by changes in incident sound pressure. Source: Reproduced with permission from Ref. [6], © 2017, Springer Nature.

causes the bubbles to vibrate due to mechanical action. Cavitation consists of two types: inertial cavitation and stable cavitation. Inertial cavitation means that at lower wave amplitudes, the bubble is in a steady state and continues to oscillate linearly, increasing in size until it reaches the size of the resonance. This is followed by stable cavitation where the wave intensity is higher and becomes nonlinear. So, the oscillations are no longer stable, causing the bubble to grow and break up rapidly [5]. Since the oscillations are faster than the speed of sound, this process produces a large release of energy leading to enormous pressures. The compression of the gas due to excessive pressure and low density creates a shock wave that causes injury and destroys the surrounding cells. The specifics of the two types of cavitation are shown in Figure 11.1 [6].

11.1.1.4 Thixotropic Effect

Ultrasound irradiation causes a decrease in viscosity, thinning of plasma, and precipitation of blood in living organisms. It has been found that when the intensity of ultrasound is small, its effect is reversible, that is, when the ultrasound stops radiating, its effect will gradually recover over time. Once the radiation is too strong, it will cause irreversible damage to the human body [7].

11.1.1.5 Acoustic Impulse and Acoustic Chemical Effects

In two media with a large difference in impedance, ultrasound will cause a sharp increase in radiation pressure and a tearing force on the tissue, resulting in the formation of an acoustic flow. The higher the intensity of the acoustic impulse, the more likely it is to cause damage to the surrounding tissue, which is known as the acoustic impulse effect. At the moment of cell membrane collapse, in extreme physical environments such as high temperatures and pressures, the cell membrane breaks down and reacts with other substances in the tissue, thus damaging other cells in the target area. This effect is also known as the acoustic chemical effect.

11.1.2 Concepts and Therapy Theory of Nanomedicines

11.1.2.1 Concept of Nanomedicine

Nanomedicine refers to drug-carrying nanoparticles with a particle size of 1–1000 nm, which originated in the 1990s, and is a cross-discipline in designing physics, chemistry, biology, engineering, and other fields. In medical research, nanomedicines have demonstrated significant superiority and broad application prospects, including (i) improved drug release; (ii) increased drug concentration around target cells and increased efficacy; (iii) improved drug delivery kinetics; (iv) reduced drug dosage; (v) increased solubility of hydrophobic drugs; (vi) enhanced stability of large molecule drugs; (vii) low therapeutic cost; and (viii) reduced side effects on healthy tissues [8].

There are organic materials, inorganic materials, and organic–inorganic hybrid nanocarrier materials for nanomedicines. Organic materials are small molecules, dendritic macromolecules, polymers, biomacromolecules, etc., and inorganic materials are metal/alloy nanomaterials, semiconductor nanomaterials, rare earth-doped nanomaterials, carbon nanomaterials, etc., as shown in Figure 11.2 [9, 10]. Among these materials, the most commonly used are polymer micelles, polymer–drug couplings, dendrimers, and liposomes.

11.1.2.2 Therapy Theory of Nanomedicine

The use of nanomedicines for cancer therapy is based on their enhanced permeability and retention (EPR) effect in solid tumors to improve the delivery of chemotherapeutic drugs. The effect was proposed by Professor Maeda from Japan in 1986 [11]. The EPR effect shows that tumor blood vessels grow faster and in greater numbers compared to normal blood vessels, which creates a large number of vascular voids, allowing enough space for nanomedicines to enter the tumor site. In addition, since the tumor loses the function of lymph, there is no lymphatic circulation. So, macromolecule-like substances and lipid particles will gather in the tumor site and

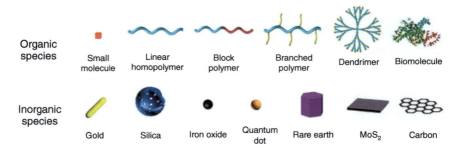

Figure 11.2 Classification of nanomedicines. Source: Reproduced with permission from Ref. [9], © 2018, American Chemical Society.

sludge in the tumor. To address the problems of high toxicity, poor water solubility, and poor targeting of traditional chemotherapeutic drugs, it may be worthwhile to use nanomedicine as a carrier and with the help of the EPR effect, it is expected to achieve highly efficient and highly targeted treatment of tumors. Due to its low retention in normal tissues and high enrichment and retention in tumor tissues, it has a better therapeutic effect and fewer toxic side effects [8].

11.1.3 Equipment for Nanomedicine-mediated Ultrasound Therapy

Devices for nanomedicine-mediated ultrasound therapy are also a major research hotspot. In the process of synergistic action of nanomedicines and ultrasound, nanomedicines and ultrasound devices play an extremely important role.

Among them, ultrasound devices often appear in the form of ultrasound transducers to focus the beam on a point to achieve the intensity required to produce various effects of ultrasound. Its beam can be divided into geometric focusing and electronic focusing (as shown in Figure 11.3a). Geometric focusing uses the concave spherical or cylindrical surface of the HIFU transducer to bring the ultrasound to the focal point. Electronic focusing uses a phased array transducer consisting of multiple piezoelectric elements, each of which has a specific excitation signal, and changing the phase of this excitation signal enables the focal point to move [12]. The most commonly used transducers are shown in Figure 11.3b, which are concave focusing transducers with fixed aperture and focal length, phased array

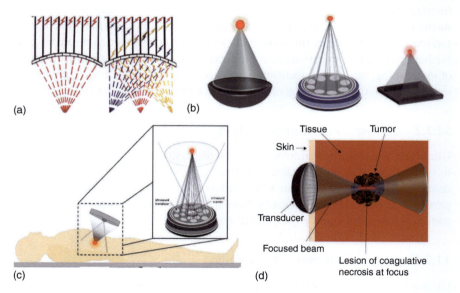

Figure 11.3 (a) Focused classification of ultrasound transducers. (b) Commonly used types of ultrasound transducers. Source: Reproduced with permission from Ref. [9], © 2018, American Chemical Society. (c) General schematic of treatment using ultrasound transducers. (d) Local schematic of tumor treatment using nanomedicines and ultrasound transducers. Source: Reproduced with permission from Ref. [9], © 2020, American Chemical Society.

transducers, and flat transducers. Figure 11.3c shows a general schematic of ex vivo focusing using an ultrasound transducer, and Figure 11.3d shows a local schematic of tumor treatment using nanomedicines and an ultrasound transducer [13].

11.2 Nanomedicine Synthesis with Ultrasound Field Response

Using a series of actions of ultrasound as an auxiliary tool for nanomedicine synthesis can accelerate the reaction rate, promote the generation of new phases, and produce a shearing effect on the agglomerates. The preparation of nanoparticles by ultrasonic method can significantly reduce the reaction time and simplify the reaction conditions compared with other traditional preparation methods, and the obtained nanoparticles are well dispersed, with small particle size and uniform distribution. There are several methods to synthesize nanodrugs using ultrasonic field response as follows.

11.2.1 Ultrasound Chemical Precipitation

The ultrasonic chemical precipitation method is a simple, low-cost process that results in a well-dispersed drug with a small particle size. This is because the intervention of the ultrasonic field provides energy for the system to overcome the interfacial barriers between the particles so that the generation rate of precipitation nuclei increases by several orders of magnitude; ultrasound-generated bubbles are located on the surface of the solid particles, which will impede the nuclei from further growth; the cavitation effect produces a shockwave and micro-jet particles of its crushing, mixing, emulsification, effectively organized the nuclei growth and agglomeration, the distribution of the grains is more homogeneous [14].

11.2.2 Ultrasonic Atomization Pyrolysis

Ultrasonic atomization technology uses the dispersive effect of ultrasound energy to obtain homogeneous ultrafine powders by pyrolysis of micron-sized droplets containing the mother liquor in a high-temperature reactor. In this technique, the particle size of the nanomedicine is affected by the reaction temperature, solution concentration, ultrasound power, and ultrasound time.

11.2.3 Ultrasonic Electrochemical Method

Ultrasonic electrochemical technology combines acoustics and electrochemistry to modulate electrochemical reactions using ultrasonic energy acceleration. The ultrasonic electrochemical method still oscillates the metal deposited on the cathode surface by electrolysis through the high pressure or jet produced by the vibration and cavitation of ultrasonic waves, causing it to be rapidly dislodged from the cathode surface and suspended in the electrolyte in the form of fine particles, preventing its growth. The greatest advantage of this method is that it enables the nontoxic and nonpolluting preparation of nanoparticles [15].

11.2.4 Ultrasonic Reduction Method

Ultrasonic reduction is a method that uses ultrasound to create a reducing agent in an aqueous or alcoholic solution, thereby reducing the corresponding metal salt to become a nanomaterial.

11.3 Application Examples

The application of nanomedicine-mediated ultrasound therapy is mainly based on the use of the nonthermal mechanism of ultrasound, i.e. the use of various forms of energy including cavitation, acoustic flow, and radiative forces. The ways in which ultrasound and nanomedicines can be used synergistically can be divided into three main categories: first, ultrasound can help nanomedicines penetrate tissues, skin, etc. The skin, as a major barrier to the organism, largely prevents the drugs from entering the body, making it impossible for the drugs to fully utilize their therapeutic effects [16]. Second, ultrasound can chemically activate the nanomedicine, which is called "son dynamic therapy." The so-called sonodynamic therapy refers to the chemical reaction of nanomedicines under the action of ultrasound to achieve the therapeutic purpose. Third, ultrasound can have a direct effect on cell membranes, altering the permeability or uptake of drugs into cells and tissues, a process that is often carried out by microbubbles [17]. The combination of ultrasound and microbubbles for treatment is an ultrasonic perforation process that can significantly enhance drug delivery to the target area [18]. This ultrasonic perforation process, also known as the acoustic pore effect, is also based on the cavitation effect of ultrasound, which refers to the appearance of a pore in the cell membrane when the bubble ruptures, through which macromolecules can enter the cell, while the microbubbles rupture by irradiation, releasing the drug they encapsulate, providing specific and highly concentrated drug delivery to the target tissue.

Delalande et al. [19] developed microbubbles of smaller size, specificity, and targeting, which are capable of delivering nucleic acids, based on an ultrasound perforation method. The overall process of this technique is shown in Figure 11.4, where targeted microbubbles containing nanodrugs are first injected into the body and after they bind specifically to cells, ultrasound irradiation of the target region activates them to deliver genes. Zhao et al. [20] found that the use of ultrasound in conjunction with Dox liposomes not only inhibited tumor growth but also rendered the tumor more sensitive to the drug. Hynynen et al. improved the performance of adriamycin liposomes using MRI-guided FUS. Hynynen et al. [21] used MRI-guided FUS to improve the accumulation of Adriamycin liposomes in the brain and found that the delivery of this nanomedicine was increased by 3.5-fold, prolonging the survival time of mice by 24%.

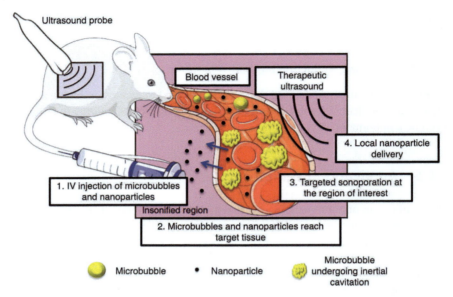

Figure 11.4 Principle of ultrasound-mediated delivery of nanomedicines in mice. Source: Reproduced with permission from Ref. [19], © 2012, Elsevier.

References

1 Kaneko, O.F. and Willmann, J.K. (2012). Ultrasound for molecular imaging and therapy in cancer. *Quantitative Imaging in Medicine and Surgery* 2 (2): 87.
2 Lopez, W., Nguyen, N., Cao, J. et al. (2021). Ultrasound therapy, chemotherapy and their combination for prostate cancer. *Technology in Cancer Research & Treatment* 20: 15330338211011965.
3 Canavese, G., Ancona, A., Racca, L. et al. (2018). Nanoparticle-assisted ultrasound: a special focus on sonodynamic therapy against cancer. *Chemical Engineering Journal* 340: 155–172.
4 Chaussy, C.G. and Thüroff, S. (2017). High-intensity focused ultrasound for the treatment of prostate cancer: a review. *Journal of Endourology* 31 (S1): S-30–S-37.
5 Elkhodiry, M.A., Momah, C.C., Suwaidi, S.R. et al. (2016). Synergistic nanomedicine: passive, active, and ultrasound-triggered drug delivery in cancer treatment. *Journal of Nanoscience and Nanotechnology* 16 (1): 1–18.
6 Lee, H., Kim, H., Han, H. et al. (2017). Microbubbles used for contrast enhanced ultrasound and theragnosis: a review of principles to applications. *Biomedical Engineering Letters* 7: 59–69.
7 Chen, Q., Sun, T., and Jiang, C. (2021). Recent advancements in nanomedicine for 'cold' tumor immunotherapy. *Nano-micro Letters* 13: 1–47.
8 Yang, P., Zhi, X., Xu, Y. et al. (2023). Nanomedicines enhance minimally invasive therapy of pancreatic cancer. *Nano Today* 51: 101891.

9 Zhao, N., Yan, L., Zhao, X. et al. (2018). Versatile types of organic/inorganic nanohybrids: from strategic design to biomedical applications. *Chemical Reviews* 119 (3): 1666–1762.

10 Blanco, E., Kessinger, C.W., Sumer, B.D. et al. (2009). Multifunctional micellar nanomedicine for cancer therapy. *Experimental Biology and Medicine* 234 (2): 123–131.

11 Matsumura, Y. and Maeda, H. (1986). A new concept for macromolecular therapeutics in cancer chemotherapy: mechanism of tumoritropic accumulation of proteins and the antitumor agent smancs. *Cancer Research* 46 (12_Part_1): 6387–6392.

12 Elhelf, I.A.S., Albahar, H., Shah, U. et al. (2018). High intensity focused ultrasound: the fundamentals, clinical applications and research trends. *Diagnostic and Interventional Imaging* 99 (6): 349–359.

13 Izadifar, Z., Izadifar, Z., Chapman, D. et al. (2020). An introduction to high intensity focused ultrasound: systematic review on principles, devices, and clinical applications. *Journal of Clinical Medicine* 9 (2): 460.

14 Bhanvase, B.A. and Sonawane, S.H. (2015). Effect of type and loading of surfactant on ultrasound-assisted synthesis of $CaZn_2(PO_4)_2$ nanoparticles by chemical precipitation. *Chemical Engineering and Processing: Process Intensification* 95: 347–352.

15 Karatutlu A, Barhoum A, Sapelkin A. Liquid-phase synthesis of nanoparticles and nanostructured materials. *Emerging Applications of Nanoparticles and Architecture Nanostructures. Elsevier*, 2018: 1–28.

16 Tachibana, K. and Tachibana, S. (2001). The use of ultrasound for drug delivery. *Echocardiography* 18 (4): 323–328.

17 Tachibana, K. and Tachibana, S. (1998). Application of ultrasound energy as a new drug delivery system. *Nihon rinsho. Japanese Journal of Clinical Medicine* 56 (3): 584–588.

18 Chowdhury, S.M., Lee, T., and Willmann, J.K. (2017). Ultrasound-guided drug delivery in cancer. *Ultrasonography* 36 (3): 171.

19 Delalande, A., Kotopoulis, S., Postema, M. et al. (2013). Sonoporation: mechanistic insights and ongoing challenges for gene transfer. *Gene* 525 (2): 191–199.

20 Zhao, Y.Z., Dai, D.D., Lu, C.T. et al. (2012). Using acoustic cavitation to enhance chemotherapy of DOX liposomes: experiment in vitro and in vivo. *Drug Development and Industrial Pharmacy* 38 (9): 1090–1098.

21 Treat, L.H., McDannold, N., Zhang, Y. et al. (2012). Improved anti-tumor effect of liposomal doxorubicin after targeted blood–brain barrier disruption by MRI-guided focused ultrasound in rat glioma. *Ultrasound in Medicine & Biology* 38 (10): 1716–1725.

12

Nanomedicine-mediated Photodynamic and/or Photothermal Therapy

Shangqing Jing[1] and Yujun Song[1,2,3]

[1] University of Science and Technology Beijing, Center for Modern Physics Technology, School of Mathematics and Physics, 30 Xueyuan Road, Haidian District, Beijing 100083, China
[2] Zhengzhou Tianzhao Biomedical Technology Company Ltd., Zhengzhou New Technology Industrial Development Zone, 7B-1209 Dongqing Street, Zhengzhou 451450, China
[3] Key Laboratory of Pulsed Power Translational Medicine of Zhejiang Province, Hangzhou Ruidi Biotechnology Company Ltd., Room 803, Bldg. 4, 4959 Yuhangtang Road, Cangqian Street, Hangzhou 310023, China

Cancer is the second most prevalent disease that seriously affects human health worldwide. According to the latest global cancer statistics released by the International Agency for Research on Cancer in 2020, there are 19.3 million new cancer cases and nearly 10 million cancer deaths worldwide. In 2040, global cancer cases are expected to reach 28.4 million, a 47% increase from 2020 [1]. Therefore, effective cancer treatments and prevention measures are crucial to global cancer control. However, traditional cancer treatment, such as surgery, chemotherapy, and radiotherapy, could cause severe complications, damage normal tissues or cells, and tend to metastasize and recurrence. Consequently, more safe and effective treatments have been waiting to be discovered and applied in clinics. With the development of optics, thermal and nanotechnology in modern medicine, photodynamic and photothermal therapy (PTT) combined with nanomedicine serve as a new alternative showing great potential in clinical cancer therapy, attracting extensive attention.

Light plays a critical role in disease diagnosis (photodiagnosis) and therapy (phototherapeutic). Phototherapies, a relatively recent cancer therapeutic strategy, can be broadly classified into photodynamic therapy (PDT) and PTT. PDT induces apoptosis by reactive oxygen species (ROS) created by the interaction between light and photosensitizer (PS). PTT is based on photothermal agent that absorbs light energy and converts it into heat energy, then leading to tissue ablation [2].

The selection of PSs/photothermal agents has been the major research on photodynamic and PTT. The first generation of PSs is hematoporphyrin and photofrin. In the 1970s, hematoporphyrin as the first PS was used in clinical settings at the Roswell Park Cancer Institute in the United States [3]. In 1983, photoprotein was developed and was the first to receive regulatory approval for the treatment of various cancers in more than 40 countries around the world, including the United

Shangqing Jing and Yujun Song contributed equally to this chapter.

Nanomedicine: Fundamentals, Synthesis, and Applications, First Edition. Edited by Yujun Song.
© 2025 WILEY-VCH GmbH. Published 2025 by WILEY-VCH GmbH.

States (by the Food and Drug Administration [FDA] in 1995). However, these PSs also have notable drawbacks. First, weak light absorption of hematoporphyrin and photofrin in near-infrared wavelengths results in low penetration of light in tumor tissue and prevents PSs from having an effect on disease treatment. Furthermore, due to the side effects of these PSs, patients should not be exposed to sunlight for at least four weeks after treatment, which poses limitations on the clinical application [4].

To overcome the limitations of the first generation of PSs, the second generation of PSs based on porphyrin and nonporphyrin molecules was developed. Porphyrinoid-based PSs include porphyrins, chlorins, pheophorbides, texaphyrins, porphycenes, phthalocyanines (Pc), naphthalocyanines (Nc), and so on. Compared to porphyrinoid-based PSs, nonporphyrinoid-based PSs have higher water solubility. Some nonporphyrinoid-based PSs such as acridine orange, rose Bengal (RB), and methylene blue (MB) were used during the dipyrromethene boron difluoride (BOD-IPY) derivates, ruthenium(II) complexes, tetraphenylethene derivates, fullerene, and so on. However, most of them possess large planar structures which lead to weak fluorescence and poor ROS generation efficiencies [5]. Particularly, low selective accumulation in targeted tissue similar to first generation made it not an ideal PS.

In general, both the first- and second-generation PSs are not an efficient alternative for PDT. The new-generation PSs, nanomedicine, are showing the potential to overcome the limitations of the traditional PSs.

Nanomedicine is a significant application of nanotechnology in medical field, having enabled nanoparticle (NP) drug-delivery vehicles a promising prospect. The abnormal branching and leaky walls of tumor vasculature lead to the presence of large pores, allowing NPs to be delivered to the tumor tissue through the passive targeting of the enhanced permeability and retention (EPR) effect. This is an effective strategy that leverages the unique physical and physiological features of tumor blood vessels to achieve targeted drug delivery in cancer treatment [6]. Also, with rational design, specific ligands of NPs could actively target and increase the aggregation in tumor cells. The passive and active delivery limit the NPs accumulating in normal tissue and reduce the side effects of drugs on system. They also impair lymphatics of solid tumors and prolong the retention time of carried drugs. To further improve treatment efficacy and overcome the limitations of single nanomedicine therapies, nanomedicine-mediated combination therapies are rapidly advancing. These innovative treatments include chemodynamic therapy, focused ultrasound therapy, pulsed electric field therapy, and phototherapies. Particularly, nanomedicine-mediated photodynamics and PTT showed exciting therapeutic effect.

In this chapter, mechanisms connected with PDT/PTT and some types of PSs and photothermal agents were illustrated, particularly on some nanomaterials that can be both used in PDT and PTT will be discussed in detail. Furthermore, we will give a brief introduction to challenges that photodynamic and photothermal therapy will face, and will prospect regarding the future development of nanomedicine-mediated photodynamics and PTT.

12.1 Photodynamic and Photothermal Mechanism for Anti-cancer

Light has been employed in disease treatment for almost thousands of years. As early as ancient Egypt, ancient India and ancient China, light was used to treat psoriasis, osteomalacia, vitiligo, skin diseases, etc. [7, 8]. In the early twentieth century, researchers made a groundbreaking discovery regarding the combination of light and specific chemicals, leading to cellular death [9]. In 1907, Von Tappeiner and Jodlbauer pioneered the use of topical eosin and white light to treat skin tumors, introducing the concept of "photodynamic action" to explain the oxygen requirement in photosensitization reactions [10, 11]. From that time, photodynamic therapy (PDT) starts to bloom.

The mechanism of photodynamic and PTT is essentially a molecular photophysical and photochemical process, requiring the accumulation of light-absorbing molecules at relevant positions. Under a specific laser irradiation, molecules in PSs absorb photon energy and transition from the ground state to excited state, and then relax and release energy. The radiation relaxation leads to fluorescence used for imaging to display lesions. The vibrational relaxation and intersystem crossing-induced triplet states are the generation of heat and ROS, respectively, thus exerting photodynamic and photothermal effects to induce apoptosis of tumor cells.

Electrons in the ground state absorb photon energy and transitions to various vibrational energy levels of excited state under illumination. Due to the instability of excited state, the electrons in lowest vibrational level release energy and return to the ground state mainly through three relaxation pathways: fluorescence emission, vibrational relaxation, and intersystem crossing relaxation (Figure 12.1). Molecules in the lowest vibrational energy level of the excited state fall to different

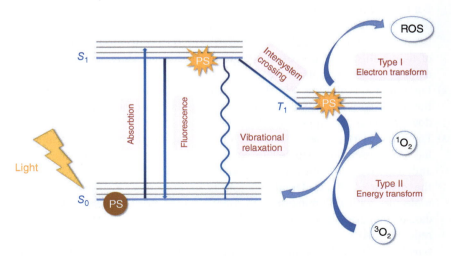

Figure 12.1 The mechanism of photodynamic and photothermal therapy.

vibrational energy levels in the ground state and emit corresponding photons, that is fluorescence. Fluorescence as an imaging tool has been applied in investigating both microscopic and macroscopic chemical and biological phenomena.

Distinguishing from the radiative mechanism of fluorescence emission, vibrational relaxation is a nonradiative mechanism by which relaxed molecules transfer energy to surrounding molecules through collisions between molecules and the surrounding environment. The vibrational relaxation mechanism is used to induce the photothermal effect and raise the temperature of the targeted tissue in PTT. (The details about PTT will be described in the following sections.)

The other nonradiative relaxation pathway is intersystem crossing relaxation. Through intersystem crossing, an excited molecule can convert a spin singlet to a spin triplet state. Once a molecule populates to a triplet state, it can generate ROS in two types of photochemical reactions (type I and type II). For type I, the interaction between triplet state PS molecule and nearby substrate (e.g. lipids, proteins, nucleic acids), caused by electron transfer, leads to generating radicals or radical ions which continue to react with oxygen to form other ROS, such as superoxide anion radicals, hydroxyl radical (OH·), and hydrogen peroxide (H_2O_2). OH· is extremely reactive and can cause oxidative damage to lipids, proteins, DNA, and amino acids. For type II, due to oxygen having a triplet spin ($^3O_2$) in the ground state, it may interact with oxygen to form toxic ROS, as well as be called singlet oxygen ($^1O_2$) in the way of energy transfer. $^1O_2$ is able to react with cellular components including unsaturated fatty acids, proteins, nucleic acids, and mitochondria to induce cell death. This process underlies the principles of PDT [12–14].

As a result of the short lifetime and limited diffusion distance of ROS, oxidative damage may only occur at the location of PSs. Different locations of PSs in tumor tissue will cause ROS to induce different biological responses. There are three main effects: the direct cytotoxic effect on the cancer cells, the damage to tumor vasculatures, and the stimulation of immune response.

(1) For tumor cell: The generation of ROS will trigger different mechanisms of cell death, including apoptosis, autophagy, and necrosis, which are decided by the treatment conditions. High damage such as high PS concentration or high dose of light will cause necrosis; moderate damage will induce apoptosis; minor damage induces autophagy [13].

(2) For tumor vasculatures: Nutrients provided by tumor vasculatures promote tumor cell growth. Different targeted PSs could break vasculatures and shut down the nutrient supply after PDT.

(3) For immune response: PDT is not just recognized as a topical treatment, it has been extensively studied to confirm that PDT can also activate innate and adaptive immunity, which will conduce to the long-term control of the disease. Cell accidental necrosis is accompanied by rapid release of molecules with inflammatory and immunomodulatory properties, known as danger-/damage-associated molecular patterns (DAMPs). DAMPs include calreticulin (CALR) and heat shock proteins (HSP), adenosine triphosphate (ATP), type I interferon (IFN), high-mobility group box 1 (HMGβ1), and annexin A1 (ANXA1), need to recognize by pattern-recognition receptors (PRRs) (Toll-like

receptors, NOD-like receptors, and RIG-1-like receptors) in immune cells [15]. Dendritic cells (DCs) are recruited to tumor cells by ATP and engulf tumor antigens with stimulation by CALR/HSP. After that, DCs migrate to the lymph nodes and present tumor antigens to naive T cells. Activated T cells (cytotoxic CD8+ T cells) migrate through the body and kill the tumor cells, both at the primary site of irradiation and nonirradiated cancer lesions (metastasis) [14–18]. This type of cell death is known as immunogenic cell death (ICD).

12.2 Nanomaterial-based PSs (Nano-PSs) for PDT

Nanomaterial-based PSs with their unique physiochemical properties exhibit great potential in PDT. Importantly, strong absorption in the NIR region and high ROS production efficiency are required for PS. Moreover, it should be safe, biodegradable, and biocompatible. There are four types of nanomaterials based PSs: organic-based, metal-based, carbon-based, and semiconductor-based nanomaterials.Organic PSs with their low molar extinction coefficients tend to suffer from photobleaching and enzymatic degradation [19]. On the contrary, metal nanostructures such as gold, silver, and platinum possess 5–6 orders of molar extinction coefficients, better photostability, and enhanced resistance to enzymatic degradation [20].

12.2.1 Metal-based Nanomaterials

12.2.1.1 Au NPs

It has been found that metal NPs (e.g. Au, Ag, Pt, Fe, Cu, etc.) have the ability to produce ROS; particularly, some metals, such as Au and Ag are able to form $^1O_2$ upon photo-irradiation at the localized surface plasmonic resonance (LSPR) absorption bands [21]. Huang et al. [22] first proposed that Au NRs can sensitize formation of $^1O_2$ and be used in PDT under NIR light irradiation (915 nm, < 130 mW cm^{-2}). The PDT effect mediated by Au NRs is at least 10-fold than the doxorubicin (DOX) (a conventional anti-cancer drug) as well as the PTT effect (780 nm, 130 mW cm^{-2}), in killing B16F0 melanoma tumor in mice.

The biological effect of Au NPs depends on their particle size, shape, surface modification, and concentrations. For example, Au NPs with smaller size and negatively charged coatings were more toxic than those with larger size and positively charged coatings. Besides, compared to Au nanospheres which were nontoxic to HaCaT cells at 100 mg ml^{-1}, Au nanorods at only a dose of 25 mg ml^{-1} significantly decreased the viability of HaCaT cells and caused the production of ROS and upregulation of several genes involved in cellular stress and toxicity.

Furthermore, compared to the irradiation of continuous wave (CW) laser source, irradiation with pulsed laser sources has a better effect on the generation of $^1O_2$ and ROS. Pasparakis [23] synthesized citrate-stabilized GNPs, followed by using CW laser irradiation (50 mW cm^{-2}, 532 nm) for 10 minutes and nanosecond Nd:YAG laser source (7 ns pulse, 532 nm), while total irradiation dose was same as CW laser. The calculated results show that the efficiency of $^1O_2$ production for CW and pulsed

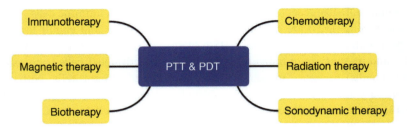

Figure 12.2 Combination of different therapy methods with PDT and PTT.

source was 0.03 and 0.07, respectively, which is about 4–5 orders of magnitude higher than common organic dyes by, thus leading to particle fragmentation.

To sum up, the generation of $^1O_2$ for Au NPs in PDT depends on various complex factors, including their sizes, shapes, surface modification, concentrations, their intrinsic catalytic performance, and even laser source.

12.2.1.2 Ag NPs

Ag NPs have also been employed in PDT for the properties of producing $^1O_2$. It was reported that metal NPs show strong morphology-dependent sensitization of $^1O_2$ [22]. In the work of Hwang group, $^1O_2$ can be produced by Ag decahedrons and Ag triangular plates instead of Ag nanocubes under light irradiation. For instance, when Ag decahedron-containing solution was excited by 885 nm light, a strong $^1O_2$ phosphorescence emission at ~1263 nm was observed (Figure 12.2). Ag nanocubes and decahedral NPs were incubated, respectively, with HeLa cells for four hours. The result indicated that Ag decahedral NPs induced ~60% cell death, which was fourfold higher than that of Ag nanocubes only ~13% $^1O_2$-mediated cellular death.

In addition to forming $^1O_2$, Ag NPs can also induce cell death by generating hydroxyl radicals via Fenton-like reactions. Hydroxyl radicals enable to be produced by hydrogen peroxide in two ways: UV irradiation and Fenton or Fenton-like reactions involving transition metal ions such as Fe^{2+} [24]. In reaction (3), Ag NPs can also serve as "Fenton NPs" resulting in the formation of hydroxyl radicals with the presence of H_2O_2. The production of hydroxyl radicals will be accelerated by increasing the Ag NPs or hydrogen peroxide concentration, or under acidic conditions [25].

$$H_2O_2 + hv \rightarrow 2^{\bullet}OH \qquad (12.1)$$

$$H_2O_2 + M^{n+} \rightarrow M^{n+1} + {}^{\bullet}OH + OH^- \qquad (12.2)$$

$$Ag + H_2O_2 + H^+ = Ag^+ + {}^{\bullet}OH + H_2O \qquad (12.3)$$

12.2.1.3 Metal Clusters

Metal nanoclusters (NCs) are a type of ultrasmall nanomaterial (<2 nm) possessing distinct physicochemical and biomedical characteristics. When the size is less than 2 nm, NCs lose the plasma properties of metal NPs and have more active electron transition and strong light absorption, resulting in enhanced fluorescence emission. At sizes below 2 nm, NCs deviate from the plasma properties exhibited by metal

NPs, displaying enhanced fluorescence emission attributable to heightened electron transition activity and intense light absorption. Furthermore, metal NCs feature facile surface modification, exceptional photostability, and favorable biocompatibility. Coupling strategies can effectively enhance metal targeting and stability in complex biological environments.

Metal NCs (e.g. gold and silver NCs) possess superior ROS generation efficiency, good aqueous solubility, and excellent photostability. Kawasaki et al. [26] proved the feasibility of $Au_{25}(SR)_{18}$ cluster for NIR-induced PDT. Under visible/near-infrared (532, 650, and 808 nm) irradiation, a characteristic peak of $^1O_2$ was observed at near 1276 nm for the $Au_{25}(SR)_{18}$ cluster, which indicated $^1O_2$ was successfully detected. In addition, Tan et al. found that $BSA-Ag_{13}$ NC also possessed excellent $^1O_2$ generation capability [27]. Metal nanostructures could also enhance the generation of $^1O_2$.

12.2.2 Transition Metal Carbide Nanomaterials

As interstitial alloys, transition metal carbides (TMCs) are synthesized by incorporating carbon atoms into the lattices of transition metals. TMC surface properties and catalytic activity are similar to those of platinum due to the enhanced D-band electron density of metals at Fermi level via electron transfer between metal atom and carbon atom [28]. Electrons and holes created by electron interband transitions in TMCs can reduce or oxidize substrates to generate ROS [29]. For example, Li et al. reported both ·OH and $^1O_2$ were formed by tungsten carbide NPs (W_2C NPs) under 1064 nm laser-activated [30]. Combined PDT and PTT of W_2C NPs will be described in the following section.

12.2.3 Carbon-based Nanomaterials

Carbon dots (CDs) are ultra-small carbon NPs, whose sizes are below 10 nm. With the advantages of excellent water dispersibility, good biocompatibility, wide absorption spectrum, and photostability, CDs have been extensively used in biomedicine including bioimaging, drug delivery, and cancer therapy. CDs are similar to organic molecules with large π-conjugated systems, i.e. the core of CDs is mainly graphitized sp^2 carbon and the shell contains abundant organic functional groups, such as –COOH, –NH$_2$, and –OH. CDs can convert light energy into heat as well as ROS, which means they can be both applied in PDT and PTT. The methods for preparing CDs are generally divided into two categories, one is top-down and the other is bottom-up. For top-down method, strong acid is used to oxidize the bulk carbon materials, which will lead to destroy the π-conjugated structure in CDs and pollute the environment while the bottom-up method is to carbonate and polymerize organic molecules into CDs, which have higher reaction yield than the top-down methods [31].

Markovic et al. obtained graphene quantum dots (GQDs, a type of CDs) by electrochemical oxidation of graphite. Under the irradiation of blue light (470 nm, 1 W), GQDs generate ROS, including $^1O_2$, and kill U251 human glioma cells, which indicates potential usefulness of GQD in PDT. Bi et al. prepared Cu-doped CDs (Cu-CDs) using acrylic acid and $Cu(NO_3)_2$ as raw materials through in-situ polymerization and subsequently pyrolysis. They showed that Cu-CDs were able to produce $^1O_2$ when exposed to an LED lamp (400–700 nm) with a quantum yield of 36%.

12.2.4 Photothermal Agents for PTT

Raising temperature in cancer treatment has been used for centuries, even dating back to India in 3000 BCE [32]. Owing to the characteristics of hypoxia and low pH in the tumor microenvironment, tumor cells are more sensitive to temperature than normal cells. In other words, tumor cells have a lower heat tolerance than normal cells [33]. High temperatures will induce thermal damage in cancer tissue. In the range of clinical hyperthermia for about 42 °C, protein denaturation may induce cell death. When temperature increases above 55 °C, the proteins in the tissue may coagulate rapidly, resulting in the cell lethality. In addition, high-temperature thermal therapy also leads to increased infiltration of host immune cells [34], inhibition of radiation repair machinery [35], and improved tumor oxygenation [36].

An ideal photothermal agent should meet the following three requirements: (i) strong near-infrared absorption. The strong light absorption and light scattering in biological tissues lead to the weak absorption wavelength of photothermal agents in the NIR region. (ii) Photothermal agents could efficiently convert the absorbed light energy into heat. (iii) It should be low-toxic or nontoxic and exhibit excellent biocompatibility.

In general, photothermal agents are composed of organic and inorganic nanomaterials. Some metal-based, carbon-based, and semiconductor-based nanomaterials can be used as inorganic photothermal agents.

12.2.5 Noble Metal-based Nanomaterials

The unique optical property exists in the interaction of metal and light. Owing to a large number of free electrons existent on the metal surface, in the field of incident light, the free electrons will oscillate collectively. Under certain conditions, the incident light resonates with the collectively oscillating electrons of the metal, and the absorption of incident light is significantly enhanced. At the same time, the amplitude of the oscillation that reaches a maximum, that termed localized surface plasmon resonance (LSPR). After absorbing the NIR light, electrons in photothermal agent will exhibit obvious thermal effects due to LSPR, causing the surrounding temperature to rise rapidly.

In particular, some noble metals, such as gold and silver, have stronger surface plasmon resonance and stronger absorption of light. Moreover, surface plasmon resonance peak can be tuned from the visible to the NIR region by adjusting size, shape, and structure of metal NPs and dielectric environment [37]. The shape and structure of particles play a major role in LSPR, while dielectric environment and particle size play a minor role.

12.2.5.1 Au NPs

For Au NPs, the shape includes gold nanospheres, nanorods, nanoshells, and nanostars dependent upon the synthetic methods adopted for their preparation.

The preparation of monodisperse gold nanospheres (particle size: 2–100 nm) is generally to reduce $HAuCl_4$ by some reducing agents. For example, trisodium citrate is a commonly used reducing agent as well as stabilizing agent. Gold NPs with different particle sizes can be produced by adjusting their concentration. The seeding growth strategy is another way to prepare gold nanospheres. $HAuCl_4$ is reduced with a strong reducing agent such as $NaBH_4$ to form seed particles, which were then added to the $HAuCl_4$ in the presence of a weak reducing agent (e.g. ascorbic acid) to produce gold nanospheres. El-Sayed et al. prepared gold nanospheres functionalized with antiepidermal growth factor receptor (EGFR) monoclonal antibodies. They immersed cells in NP solutions for 40 minutes and were exposed to laser irradiation (514 nm, 13–76 W cm^{-2}, 4 minutes). The result indicated that malignant human oral squamous cell carcinoma could be selectively killed.

Gold nanorods (GNRs) were first synthesized using the template method by Foss et al., Martin, and Perez-Juste et al., but due to their low yield, they are now widely produced via seed growth method. The preparation process is as follows: gold seeds were prepared by reducing gold salt with strong reducing agent, then added to the aqueous surfactant media containing the gold salt and the reducing agents (i.e. ascorbic acid and cationic surfactant cetyltrimethylammonium bromide [CTAB]) for the growth steps. For example, Zhang et al. used seed growth method to prepare GNRs in aspect ratio R ranging from 2.5 to 4, with CTAB as template and sodium oleate (NaOL) as stabilizer. Under irradiation of 2 W cm^{-2} NIR laser (808 nm) for 10 minutes, the temperature of GNR solution with aspect ratio of three was rinsed 41.3 °C. They proved that photothermal conversion performance of GNRs is higher than gold nanospheres (only rises 10.2 °C) under the same photothermal condition and solution concentration.

Gold nanoshells are typically made of dielectric core materials (such as silica and polystyrene) that are covered with a thin gold coating. The core is about 100 nm in diameter and the thin shell of gold is about several nanometers (~1–20 nm). Gold nanoshell precursor was prepared by seed-mediated growth method, and then gold nanoshell was generated with hydrogen peroxide as reducing agent under the catalytic action of gold NPs conforming to the particle surface. In 2003, West et al. prepared 110 nm silica cores surrounded by 10 nm gold shells and functionalized them with biocompatible polymer polyethylene glycol (PEG), which is often used to minimize NP aggregation and to increase blood circulation and half-life. The plasmon resonance wavelength is 820 nm. Gold nanoshells were incubated with SK-BR-3 human breast carcinoma cells for one hour and subsequently irradiated with laser light (820 nm, 35 W cm^{-2}, 7 minutes). The loss of cell membrane integrity and cell death were observed in irradiated region under fluorescence imaging.

12.2.5.2 Ag NPs

As one of noble metal materials, Ag NPs exhibit fascinating physical, electrical, and antimicrobial properties. Ag NPs have great advantages over other metal-based nanomaterials (e.g. Au NPs, Pd NPs) because of their simple synthesis and low cost. However, the poor biocompatibility of Ag may cause cytotoxicity, which limits its application as nanomedicine. So far, there are few reports on the separate application of Ag NPs as photothermal materials [38].

Cui and colleagues prepared silver nanodots (AgND) using the bovine serum albumin biomineralization process, which is based on the coordination reaction among metal ions and the dangling groups or ligands of amino acid residues in the protein. The BSA solution was first vigorously mixed with $AgNO_3$. Then, the pH of the mixed solution was raised above 12 by adding NaOH solution. And later the reaction of some amino acid residues of BSA and Ag+ would occur in 80 °C water bath. It was shown that when the ratio of Ag^+ ions to BSA changed from 1.32 to 52.8, the temperature increase in AgND solution changed to less than 2 and 14 °C under the irradiation of 808 nm NIR laser at $1.0\,W\,cm^{-2}$ [39].

12.2.6 Carbon-based Nanomaterials

Carbon-based nanomaterials are regarded as good PS materials for the reason of their NIR absorbance (carbon-based nanomaterials can absorb near-infrared wavelengths ranging from 700 to 1000 nm), abundant functional groups, and large specific surface area. Carbon-based nanomaterials include CDs, carbon nanotubes (CNTs), and reduced graphene oxides (rGO).

12.2.6.1 Carbon Dots

In addition to the application of CDs in PDT mentioned earlier, CDs can also be used in PTT due to their photothermal conversion ability. The room-temperature and one-pot synthesis of copper/carbon quantum dot (or CD)-crosslinked nanosheets (CuCD NSs) with high photothermal conversion efficiency of 41.3% was proposed by Wu group. After coating with thiol-PEG and fluorescent molecules, CuCD NSs could selectively target tumor tissues and contribute to triple-modal imaging-guided phototherapy.

12.2.6.2 Graphene Oxide and Reduced Graphene Oxide

Graphene is widely regarded as the most attractive carbon allotrope for its unique intrinsic properties. Graphene-based nanomaterials are generally divided into graphene with varied layers, graphene oxide (GO) and reduced GO (rGO) [40]. Due to possessing much more oxygen-containing groups, GO has a higher wettability and water solubility than rGO. By contrast, rGO maintains the sp^2 bonding networks, which makes the photothermal efficiency of rGO higher than that of GO. Both GO and rGO have great potential for use in cancer therapy as drug carriers or photothermal materials because of their large surface area, high photothermal conversion efficiency, and modifiable functional groups [41].

Sun et al. mixed polystyrene sulfonate (PSS) functionalized GO, secondary growth solution, and gold seeds to prepare GO/GNR nanohybrids at room temperature. Compared with only laser irradiation, the combination of GO/GNRs and laser leads to killing 60% of the SW1990 cancer cells [42].

In the work of Otari et al., thermostable antimicrobial nisin peptides were used to synthesize gold-NPs-reduced GO (NAu-rGO) nanocomposite by a synthesis method for one-step reduction and decoration of GO with gold NPs (NAuNPs). Under irradiation of NIR light with a power intensity of 0.5 W cm^{-2}, more than 80% of cells were inhibited after 24 hours incubation in NAu-rGO nanocomposite-treated MCF7 breast cancer cells [43].

12.2.6.3 Carbon Nanotubes

CNTs are hollow cylinders consisting of graphitic sheets and are classified into single or multiwalled CNT (SWCNT/MWCNT) [fphar]. The diameter of SWNT is around 0.4 nm, while MWNTs are about 100 nm. Their length varies from hundreds of nanometers to tens of micrometers [44]. CNTs are widely applied in biomedical fields, including biosensors, bioelectrochemistry, bioimaging, gene or drug delivery systems, etc. Specially, in antitumor therapy, as an effective NIR absorber, CNT can rapidly convert electronic excitations into molecular vibration energies and generate heat, having been attractive candidates for photothermal materials [45, 46].

Functionalized SWCNTs were mixed with curcumin (natural anti-tumor compound) and PVP K30 in methanol using an ultrasonic bath for 10 minutes, then rotary evaporated to dryness to prepare SWCNT-curcumin (SWCNT-Cur) by Li et al. SWCNT-Cur increased the inhibition effect instead of curcumin alone on PC-3 cells. In particular, the photothermal cytotoxicity of SWCNT-Cur under 808 nm laser irradiation resulted in killing more than 70.1% of PC-3 cells compared to 55.8% without laser irradiation.

A hybrid nanomaterial of gold nanostars/MWCNTs was prepared by Zhu et al. [47]. They dissolved MWCNTs into H_2SO_4/HNO_3 solution with a volumetric ratio of 1 : 3 under ultrasonication at room temperature and then diluted the solution to neutrality by using deionized water and NaOH aqueous solution. Subsequently, the solution was filtered and vacuum-dried. The photothermal conversion efficiency of MWCNTs/gold nanostars hybrid material prepared by traditional seed-mediated growth method and a two-step reduction method without CTAB were compared. They showed that the new preparation method was more convenient with low toxicity, the gold nanostars have an improved morphology and there is a small red shift absorption. The MWCNT/gold nanostars presented a 24% photothermal efficiency higher than gold nanostars alone and induced 80% B16F10 melanoma cell death, which showed a better effect on PTT.

12.2.7 Semiconductor-based Nanomaterials

Semiconductor-based nanomaterials also are ideal photothermal agents with strong absorption in NIR region, including Cu-based (CuS, Cu_9S_5, CuSe), W-based ($W_{18}O_{49}$, WS_2), and Mo-based (MoS_2) nanomaterials. Compared to noble metal nanomaterials, semiconductor nanomaterials hold extremely low costs and cytotoxicity to the organisms.

12.2.7.1 Cu-based Semiconductors

Copper sulfide (CuS) was firstly used as a photothermal agent when it was shown to possess strong NIR absorption properties among all semiconductor nanomaterials. CuS is a p-type semiconductor, which strong NIR absorption resulted from the oscillation of the valence band free carriers of this material [38].

Tian et al. obtained hydrophilic plate-like Cu_9S_5 nanocrystals (NCs) by combining a thermal decomposition and ligand exchange route. Cu_9S_5 NCs displayed a large molar extinction coefficient of 1.2×10^9 M^{-1} cm^{-1} at 980 nm and a high PCE of 25.7% which exceeded Au NRs (23.7% from 980 nm laser). They showed that exposed to NIR irradiation (980 nm, 0.51 W cm^{-2}) for 10 minutes, cancer cells in vivo can be efficiently killed by the photothermal effects of the Cu_9S_5 NCs [48].

Hessel et al. prepared copper selenide ($Cu_{2-x}Se$) nanocrystals by colloidal hot injection method and coated with amphiphilic polymer. $Cu_{2-x}Se$ exhibited a molar extinction coefficient of 7.7×10^7 M^{-1} cm^{-1} (at 980 nm) and a 22% PCE (at 808 nm), which makes it a potential PTT materials [49].

12.2.7.2 W-based Semiconductors

WS_2 is a layered material of graphene-like material with a large specific surface area. Moreover, there are many active sites on the WS_2 surface, which is helpful for the modification of the surface [50]. The lipid-modified WS_2 nanomaterials (WS_2-lipid) were synthesized by Xie and co-workers to achieve both chemo and photothermal therapeutic effect. Firstly, liposomes were produced using membrane hydration methods. Then the preparation of a lipid-modified WS_2 composite was performed by mixing. In vivo photothermal ablation result showed that 200 µg ml^{-1} WS_2-lipid can induce cancer cells death of ~80% at 100 µg ml^{-1}.

Xu and co-workers synthesized hydrophilic $W_{18}O_{49}$ nanowires (NWs) via the solvothermal synthesis-simultaneous PEGylation/exfoliation/breaking two-step route. Under 980 nm laser irradiation, the temperature of $W_{18}O_{49}$ NWs underwent a 35.2 °C increase within 5 minutes. When laser irradiated for 10 minutes, in vivo cancer cells occurred shrinking of malignant cells, eosinophilic cytoplasm, and nuclear damage by the photothermal effect of $W_{18}O_{49}$ NWs [51].

12.2.7.3 Mo-based Semiconductors

With the advantages of thermal stability, high electrocatalytic activity and adjustable bandgap, molybdenum disulfide (MoS_2) has been widely applied in PTT, drug delivery and bioimaging [38]. The bandgap of MoS_2 is adjustable: from an indirect bandgap of 1.2 eV in the bulk to a direct bandgap of 1.8 eV in single layer form. Besides, it has outstanding thermal stability and high electrocatalytic activity. It has been widely utilized in the fields of PTT, drug delivery, bioimaging, and so on [38]. For instance, Liu and co-workers fabricated hyperbranched polyglycidyl modified MoS_2 (MoS_2–HPG) nanocomposite [52]. This work utilized HPG to modify MoS_2 nanosheets by absorbing HPG on the MoS_2 surface via the multifunctional groups and the branching structure of HPG. The as-prepared MoS_2–HPG nanocomposite displayed high photothermal conversion efficiency of 29.4% at the light wavelength of 808 nm and the power of 2.0 W cm^{-2}.

12.3 Photodynamic and Photothermal Synergistic Therapy

Generally, a variety of complex pathological processes are involved in malignant tumors, leading to single treatment method can hardly effectively and thoroughly combat the tumor. Combined therapy, therefore, holds promise as a more viable strategy for tumor treatment. Nanomedicine has been shown to achieve a synergistic therapeutic effect by acting simultaneously in photodynamic and photothermal processes. Through mild hyperthermia treatments, nanomedicine can enhance blood flow in tumor tissues and improve partial oxygen pressure. At the same time, some results evidenced that heat can also contribute to the reactivity of ROS for tumor destruction. PDT mediated little damage of proteins in cells may have a significant impact on the hyperthermia sensitivity of cells.

The optical absorption peaks of photothermal materials and photodynamic materials are different, so it may need two wavelengths of laser irradiation to achieve the combined photodynamic and photothermal effect. However, sequential irradiation by two various laser sources may result in prolonged treatment time, complicated treatment process, and potential skin burns. In contrast, simultaneous PDT and PTT using a single light source is more beneficial for clinical translation. One common approach is to load photothermal materials into photodynamic materials, while another involves the use of nanomaterials that possess both PDT and PTT effects.

Reported by Li et al. [30], single 1064 nm laser-activated tungsten carbide NPs (W_2C NPs) have been shown excellent PDT/PTT synergistic effect in vitro and in vivo. Jujube core-like W_2C NPs were synthesized by the ionic self-assembly driven formation of organic–inorganic nanostructures (OI-NSs) and then were calcined and functionalized with hyaluronic acid (HA) which could improve W_2C NPs colloidal stability, biocompatibility and cancer target capability (shown in Figure 12.2). They demonstrated that HA-W_2C could produce both ·OH (type I ROS) and $^1O_2$ (type II ROS) under the irradiation of 1064 nm laser (0.8 W cm^{-2}). Even in hypoxic tumor cells (~1% O_2), HA-W_2C also enabled to generate ·OH through type I process. In vitro PTT, HA-W_2C exhibited excellent photostability and a photothermal transduction efficiency of 46.8%. Furthermore, instead of adopting PDT or PTT alone, combined PDT/PTT strategy can achieve a better treatment effect. The relevant results indicated that HA-W_2C could be served as a potential PDT/PTT synergistic nanomedicine, including a good PA/CT bioimaging contrast agent.

Furthermore, due to tunable LSPR, transition metal oxide semiconductor nanomaterials, such as $W_{18}O_{49}$ and MoO_x, are promising candidates for both PDT and PTT. Oxygen-deficient molybdenum oxide NPs (MoO_x NPs) can be employed in NIR-II based synergetic PTT and PDT by changing chemical composition to tuned its LSPR. For instance, PEG-MoO_x NPs prepared by Yin et al. [53] was used facile, green, one-pot hydrothermal method presented a remarkable NIR absorption in 800–1200 nm and showed a tumor growth inhibition ratio of 82.44% under 1064 nm laser irradiation.

Activated by NIR, copper chalcogenide ($Cu_{2-x}E$, with E = S, Se, Te) nanocrystals (NCs) were proved that not only have photothermal therapeutic effect, but also could also exploitable for PDT [54]. Wang et al. used a noninjection approach to synthesize $Cu_{2-x}S$ NCs and induced oxidation by exposing the NCs to air during the ligand exchange process. LSPR of NCs can be tuned by their chemical composition, to be specific, Cu/S ratio was changing from 1.8 (most of the reported values for copper sulfide NCs) to 1.25. When exposed to 808 nm NIR light (0.6 W cm^{-2}), aqueous $Cu_{2-x}S$ NCs solutions (2 mg ml^{-1} in Cu) exhibited up to 83.5% ·OH levels higher than untreated solutions.

Besides, ultrathin Ti_3C_2 MXene nanosheets (~100 nm) were synthesized by supplying additive Al^{3+} to avoid Al loss and employed as a photothermal/photodynamic agent for cancer therapy. In Dong et al. work, the nanosheets obtained demonstrate remarkable mass extinction coefficient (28.6 lg^{-1} cm^{-1} at 808 nm), excellent photothermal conversion efficiency (~58.3%), and effective generation of singlet oxygen ($^1O_2$) when exposed to 808 nm laser irradiation. Then, based on these Ti_3C_2 nanosheets, a multifunctional nanoplatform (Ti_3C_2-DOX) is established via layer-by-layer surface modification with DOX and HA. Both in vitro and in vivo experiments demonstrate that Ti_3C_2-DOX exhibits improved biocompatibility, targeted accumulation in tumor cells, and responsive release of drugs and achieves effective cancer cell killing and tumor tissue destruction through photothermal/photodynamic/chemosynergistic therapy [55].

12.4 Opportunities and Challenges

Although PDT and PTT have been developed wildly and exhibited great potential for their noninvasive nature and high efficiency, there are still numerous challenges waiting to be solved.

12.4.1 Light Source

Under visible and the first near-infrared region (NIR-I: 380–900 nm) laser irradiation, the penetration depth is limited by light scattering and absorption by biological tissues. For instance, hemoglobin is able to absorb light less than 650 nm, limiting the penetrability of NIR-I. Despite the effectiveness of phototherapy, it remains challenging for PTT and PDT to overcome the limitations of NIR light penetration. PTT typically employs near-infrared light at 808 nm, which penetrates only 1–2 mm and is suitable mainly for treating superficial tumor tissues such as skin cancer. It is crucial to enable the excitation power to reach the tumor's location for deep tumors. Therefore, the PSs and photothermal agents need to exhibit strong near-infrared absorption, with an absorption wavelength in the second near-infrared region (NIR-II, 1000–1700 nm).

Moreover, low-power single-laser activatable PDT and PTT are still a tough challenge. The penetration depth is deemed to be maximal for 1000–1100 nm light, and the maximum permissible exposure for skin is 1.0 W cm^{-2} in this spectral region and for 800 nm laser irradiation, the maximum permissible exposure for skin is 0.3 W cm^{-2}. In some of the above studies, the laser power exceeded the threshold. Thus, more research will be needed for further clinical transformation.

12.4.2 Hypoxia of Tumor Microenvironment

A common feature of solid tumors is hypoxic microenvironment. Hypoxic tumor microenvironment is caused by increased oxygen consumption and decreased oxygen transport. Hypoxia-inducible factor (HIF) primarily drives a range of physiological responses triggered by hypoxia. During hypoxia, the level of HIF1 protein increases, and the HIF-mediated pathway affects metabolic adaptation through mammalian target of rapamycin (mTOR) signaling, erythropoiesis, angiogenesis, cell growth, vascular tone, and differentiation. Oxygen is thus an essential factor in PDT, along with PSs and NIR light. However, hypoxia is a prominent feature of tumor microenvironment. Current research is focused on two aspects to address this issue: catalyzing excess hydrogen peroxide within the tumor to generate oxygen, and transporting oxygen from other regions to the tumor site [56].

12.4.3 Combination with Other Therapy

The adaptive immune system, activated by innate immune cells, can specifically target tumor cells and is considered the most effective approach for eradicating tumors. However, using single cancer immunotherapy to completely eradicate primary and distant tumors remains challenging due to difficulties in identifying high-efficiency biomarkers and significant patient-specific differences. To improve therapeutic efficacy, combining multiple treatments to achieve synergistic outcomes is a promising strategy, with phototherapy-synergized cancer immunotherapy occupying an essential role. It has been reported that phototherapy could promote the antitumor immune response and activate the memory of immune cells during the process of treatment. In addition to the combination with immunotherapy, the combined use of PDT and PTT with other treatment approaches has also been developed to show tremendous antitumor effects (Figure 12.2).

12.5 Summary

In this chapter, we have summarized the advantages and challenges of nanomedicines for phototherapy from the exploration of the first-generation PSs to the recent-generation nanomaterials. Then we focus on the mechanism of PDT and PTT, and various PSs and photothermal agents, particularly on the mechanism of ROS generation for various nanomaterials. Furthermore, we discussed photodynamic and photothermal synergistic therapy, including some applications. The challenges and future perspectives of phototheranostic nanomedicine for advanced cancer treatment are presented and more possible solutions should be discovered to solve these issues.

References

1 Sung, H., Ferlay, J., Siegel, R.L. et al. (2021). Global cancer statistics 2020: GLOBOCAN estimates of incidence and mortality worldwide for 36 cancers in 185 countries. *CA: A Cancer Journal for Clinicians* 71 (3): 209–249.
2 Diederich, C.J. (2005). Thermal ablation and high-temperature thermal therapy: overview of technology and clinical implementation. *International Journal of Hyperthermia* 21: 745–753.
3 Agostinis, P., Berg, K., Cengel, K.A. et al. (2011). Photodynamic therapy of cancer: an update. *CA: A Cancer Journal for Clinicians* 61 (4): 250–281.
4 Tavakkoli Yaraki, M., Liu, B., and Tan, Y. (2022). Emerging strategies in enhancing singlet oxygen generation of nano-photosensitizers toward advanced phototherapy. *Nano-Micro Letters* 14: 123.
5 Birks, J.B. Photophysics of aromatic molecules (London, 1970). *Bunson Society Reports on Physical Chemistry* 74 (12): 1294–1295.
6 Wang, A.Z., Langer, R., and Farokhzad, O.C. (2012). Nanoparticle delivery of cancer drugs. *Annual Review of Medicine* 63 (1): 185–198.
7 Spikes, J.D. (1985). The historical development of ideas on applications of photo-sensitised reactions in health sciences. In: *Primary Photoprocesses in Biology and Medicine* (ed. R.V. Bergasson, G. Jori, E.J. Land, and T.G. Truscott), 209–227. New York: Plenum Press.
8 Epstein, J.M. (1990). Phototherapy and photochemotherapy. *The New England Journal of Medicine* 32: 1149–1151.
9 Dolmans, D., Fukumura, D., and Jain, R. (2003). Photodynamic therapy for cancer. *Nature Reviews. Cancer* 3: 380–387.
10 Von Tappeiner, H. and Jesionek, A. (1903). Therapeutic tests with fluorescent substances. *German Medical Weekly.* 47: 2042–2044.
11 Von Tappeiner, H. and Jodlbauer, A. (1904). Effects of photodynamic (fluorescent) substances on protozoa and enzymes. *German Archives of Clinical Medicine* 80: 427–487.
12 Gao, D., Guo, X., Zhang, X. et al. (2019). Multifunctional phototheranostic nanomedicine for cancer imaging and treatment. *Materials Today Bio* 5: 100035.
13 Fabris, C., Valduga, G., Miotto, G. et al. (2001). Photosensitization with zinc (II) phthalocyanine as a switch in the decision between apoptosis and necrosis. *Cancer Research* 61: 7495–7500.
14 Donohoe, C., Senge, M.O., Arnaut, L.G. et al. (2019). Cell death in photodynamic therapy: from oxidative stress to anti-tumor immunity. *Biochimica et Biophysica Acta (BBA) – Reviews on Cancer* 1872 (2): 188308.
15 Rubartelli, A. and Lotze, M.T. (2007). Inside, outside, upside down: damage-associated molecular-pattern molecules (DAMPs) and redox. *Trends in Immunology* 28: 429–436.
16 Preise, D., Oren, R., Glinert, I. et al. (2009). Systemic antitumor protection by vascular-targeted photodynamic therapy involves cellular and humoral immunity. *Cancer Immunology, Immunotherapy* 58: 71–84.

17 Matzinger, P. (2002). The danger model: a renewed sense of self. *Science* 296: 301–305.
18 Castano, A.P., Mroz, P., Wu, M.X., and Hamblin, M.R. (2008). Photodynamic therapy plus low dose cyclophosphamide generates antitumor immunity in a mouse model. *Proceedings of the National Academy of Sciences of the United States of America* 105: 5495–5500.
19 Schweitzer, V.G. and Somers, M.L. (2010). PHOTOFRIN-mediated photodynamic therapy for treatment of early stage (Tis-T2N0M0) SqCCa of oral cavity and oropharynx. *Lasers in Surgery and Medicine* 42 (1): 1–8.
20 Vankayala, R., Sagadevan, A., Vijayaraghavan, P. et al. (2011). Metal nanoparticles sensitize the formation of singlet oxygen. *Angewandte Chemie, International Ed. in English* 50 (45): 10640–10644.
21 Vankayala, R., Kuo, C.L., Sagadevan, A. et al. (2013). Morphology dependent photosensitization and formation of singlet oxygen ($1\Delta g$) by gold and silver nanoparticles and its application in cancer treatment. *Journal of Materials Chemistry B* 1.
22 Vankayala, R., Huang, Y.K., Kalluru, P. et al. (2014). First demonstration of gold nanorods-mediated photodynamic therapeutic destruction of tumors via near infra-red light activation. *Small*.
23 Pasparakis, G. (2013). Light-induced generation of singlet oxygen by naked gold nanoparticles and its implications to cancer cell phototherapy. *Small* 9 (24): 4130–4134.
24 Wardman, P. and Candeias, L.P. (1996). Fenton chemistry: an introduction. *Radiation Research* 145: 523e53.
25 He, W., Zhou, Y.T., Wamer, W.G. et al. (2012). Mechanisms of the pH dependent generation of hydroxyl radicals and oxygen induced by Ag nanoparticles. *Biomaterials* 33 (30): 7547–7555.
26 Kawasaki, H., Kumar, S., Li, G. et al. (2014). Generation of singlet oxygen by photoexcited $Au_{25}(Sr)_{18}$ clusters. *Chemistry of Materials* 26 (9): 2777–2788.
27 Yu, Y., Geng, J., Ong, E.Y.X. et al. (2016). Bovine serum albumin protein-templated silver nanocluster (BSA-Ag_{13}): an effective singlet oxygen generator for photodynamic cancer therapy. *Advanced Healthcare Materials* 5 (19): 2528–2535.
28 Pang, J.S., JunmingZheng, M.L., HouqianWang, Y.Z., and Tao. (2019). Transition metal carbide catalysts for biomass conversion: a review. *Applied Catalysis, B. Environmental: An International Journal Devoted to Catalytic Science and Its Applications* 254: 510–522.
29 Gilson, R.C., Black, K.C.L., Lane, D.D., and Achilefu, S. (2017). Hybrid TiO_2-ruthenium nano-photosensitizer synergistically produces reactive oxygen species in both hypoxic and normoxic conditions. *Angewandte Chemie, International Edition* 129: 10857–10860.
30 Li, S.H., Yang, W., Liu, Y. et al. (2018). Engineering of tungsten carbide nanoparticles for imaging-guided single 1,064 nm laser-activated dual-type photodynamic and photothermal therapy of cancer. *Nano Research* 11 (9): 15.

31 Li, B., Zhao, S., Huang, L. et al. (2021). Recent advances and prospects of carbon dots in phototherapy. *Chemical Engineering Journal* 408 (1): 127245.

32 Seegenschmiedt, M.H. and Vernon, C.C. (1995). A historical perspective on hyperthermia in oncology. In: *Thermoradiotherapy and Thermochemotherapy*, vol. 1 (ed. M.H. Seegenschmiedt, P. Fessenden, and C.C. Vernon), 3–44. Berlin: Springer Verlag.

33 Zee, V.D.J. (2002). Heating the patient: a promising approach? *Annals of Oncology* 13 (8): 1173–1184. https://doi.org/10.1093/annonc/mdf280.

34 Burd, R., Dziedzic, T.S., Xu, Y. et al. (1998). Tumor cell apoptosis, lymphocyte recruitment and tumor vascular changes are induced by low temperature, long duration (fever-like) whole body hyperthermia. *Journal of Cellular Physiology* 177: 137–147.

35 Raaphorst, G.P., Azzam, E.I., and Feeley, M. (1988). Potentially lethal radiation damage repair and its inhibition by hyperthermia in normal hamster cells, mouse cells, and transformed mouse cells. *Radiation Research* 113: 171–182.

36 Jones, E.L., Prosnitz, L.R., Dewhirst, M.W. et al. (2004). Thermochemoradiotherapy improves oxygenation in locally advanced breast cancer. *Clinical Cancer Research* 10: 4287–4293.

37 Cobley, C.M., Chen, J.Y., Cho, E.C. et al. (2011). Gold nanostructures: a class of multifunctional materials for biomedical applications. *Chemical Society Reviews* 40 (1): 44–56.

38 Wei, W., Zhang, X., Zhang, S. et al. Biomedical and bioactive engineered nanomaterials for targeted tumor photothermal therapy: a review. *Materials Science and Engineering: C* 104: 109891.

39 Cui, Y., Yang, J., Zhou, Q. et al. (2017). Renal clearable Ag nanodots for in vivo computer tomography imaging and photothermal therapy. *ACS Applied Materials and Interfaces* 9: 5900–5906.

40 Yao, J., Sun, Y., Yang, M., and Duan, Y. (2012). Chemistry, physics and biology of graphene based nanomaterials: new horizons for sensing, imaging and medicine. *Journal of Materials Chemistry* 22: 14313.

41 Jinzhao, L., Jia, D., Ting, Z. et al. (2018). Graphene-based nanomaterials and their potentials in advanced drug delivery and cancer therapy. *Journal of Controlled Release* 286: 64–73.

42 Sun, B., Wu, J., Cui, S. et al. (2017). In situ synthesis of graphene oxide/gold nanorods theranostic hybrids for efficient tumor computed tomography imaging and photothermal therapy. *Nano Research* 10 (1): 37–48.

43 Otari, S.V., Kumar, M., Anwar, M.Z. et al. (2017). Rapid synthesis and decoration of reduced graphene oxide with gold nanoparticles by thermostable peptides for memory device and photothermal applications. *Science Report UK* 7: 10980.

44 Junloh, X., Lee, T.-C., Dou, Q. et al. Utilising inorganic nanocarriers for gene delivery. *Biomaterials Science* 4: 70–86.

45 Shen, Y., Skirtach, A.G., Tomohiro, S. et al. (2010). Assembly of fullerene–carbon nanotubes: temperature indicator for photothermal conversion. *Journal of the American Chemical Society* 132: 8566–8568.

46 Wang, H., Li, J., Zhang, X. et al. (2013). Synthesis, characterization and drug release application of carbon nanotube–polymer nanosphere composites. *RSC Advances* 3: 9304–9310.

47 Zhu, Y., Sun, Q., Liu, Y. et al. (2018). Decorating gold nanostars with multi-walled carbon nanotubes for photothermal therapy. *Royal Society Open Science* 5 (8): 180159.

48 Wang, Yang, Jiang et al. (2011). Hydrophilic Cu_9S_5 nanocrystals: a photothermal agent with a 25.7% heat conversion efficiency for photothermal ablation of cancer cells in vivo. *ACS Nano* 5 (12): 9761–9771.

49 Hessel, C.M., Pattani, V.P., Rasch, M. et al. (2011). Copper selenide nanocrystals for photothermal therapy. *Nano Letters* 11 (6): 2560–2566.

50 Xie, M., Yang, M., Sun, X. et al. (2020). WS_2 nanosheets functionalized by biomimetic lipids with enhanced dispersibility for photothermal and chemo combination therapy. *Journal of Materials Chemistry B* 8: 2331.

51 Xu, W., Tian, Q., and Chen, Z. (2014). Optimization of photothermal performance of hydrophilic $W_{18}O_{49}$ nanowires for the ablation of cancer cells in vivo. *Journal of Materials Chemistry B: Materials for Biology and Medicine* 2 (34): 5594–5601.

52 Wang, K., Chen, Q., Xue, W. et al. (2017). Combined chemo-photothermal antitumor therapy using molybdenum disulfide modified with hyperbranched polyglycidyl. *ACS Biomaterials Science & Engineering* 3: 2325–2335.

53 Yin, W., Bao, T., Zhang, X. et al. (2017). Biodegradable MoO_x nanoparticles with efficient near-infrared photothermal and photodynamic synergetic cancer therapy at the second biological window. *Nanoscale* https://doi.org/10.1039/C7NR07927C.

54 Wang, S.H., Riedinger, A., Li, H.B. et al. (2015). Plasmonic copper sulfide nanocrystals exhibiting near-infrared photothermal and photodynamic therapeutic effects. *ACS Nano* 9 (2): 1788–1800.

55 Liu, G., Zou, J., Tang, Q. et al. (2017). Surface modified Ti_3C_2 MXene nanosheets for tumor targeting photothermal/photodynamic/chemo synergistic therapy. *ACS Applied Materials & Interfaces* 9 (46): 40077–40086. https://doi.org/10.1021/acsami.7b13421.

56 Deng, X., Shao, Z., and Zhao, Y. (2021). Solutions to the drawbacks of photothermal and photodynamic cancer therapy. *Advanced Science* 8 (3): 2002504.

13

Nanomedicine-mediated Pulsed Electric Field Ablation Therapy

Xinzhu Yang[1], Qiong Wu[1], Ruixue Zhu[1], and Yujun Song[1,2,3]

[1] University of Science and Technology Beijing, Center for Modern Physics Technology, School of Mathematics and Physics, 30 Xueyuan Road, Haidian District, Beijing 100083, China
[2] Zhengzhou Tianzhao Biomedical Technology Company Ltd., Zhengzhou New Technology Industrial Development Zone, 7B-1209 Dongqing Street, Zhengzhou 451450, China
[3] Key Laboratory of Pulsed Power Translational Medicine of Zhejiang Province, Hangzhou Ruidi Biotechnology Company Ltd., Room 803, Bldg. 4, 4959 Yuhangtang Road, Cangqian Street, Hangzhou 310023, China

13.1 Introduction

Cardiovascular diseases, neurological disorders, and cancer pose significant threats to human health, profoundly impacting patients' quality of life and lifespan [1–3]. Among them, cardiovascular diseases, such as heart disease and stroke, rank among the leading causes of death worldwide. These diseases can result in impaired cardiac function, vascular blockages, inadequate blood supply, and even severe consequences like myocardial infarction or stroke. Symptoms may include chest pain, shortness of breath, irregular heart rhythms, paralysis, and in severe cases, disability or death.

Neurological disorders, such as Alzheimer's disease (AD), Parkinson's disease, and epilepsy, cause impairments in patients' neurological function. These conditions can lead to a decline in memory, impaired cognitive function, movement disorders, and seizures. Patients' ability to perform daily activities and self-care may be severely affected, placing a significant burden on both them and their families.

Cancer, a malignant tumor, can develop in various organs and tissues of the body. The harm of cancer primarily arises from the uncontrolled growth and spread of tumors, resulting in erosion and dysfunction of normal tissues. Cancer patients may experience symptoms such as pain, fatigue, nausea, and weight loss, with severe cases potentially leading to organ failure and life-threatening situations.

Traditional treatment methods have achieved certain successes in addressing these diseases, including drug therapy, surgical interventions, and radiation therapy. Drug therapy involves the use of chemical agents to intervene in the disease process, surgical interventions aim to remove abnormal tissues or implant devices

Xinzhu Yang and Qiong Wu are contributed equally to this article.

Nanomedicine: Fundamentals, Synthesis, and Applications, First Edition. Edited by Yujun Song.
© 2025 WILEY-VCH GmbH. Published 2025 by WILEY-VCH GmbH.

to correct anomalies, while radiation therapy employs high-energy radiation to kill cancer cells.

However, traditional treatment methods have certain limitations. Drug therapy may have side effects since drugs are distributed widely in the body, impacting not only the affected tissues but also causing damage to normal tissues [4, 5]. Surgical interventions and radiation therapy can induce trauma and complications, and for some advanced diseases, complete removal or cure may be challenging [6–9].

Therefore, there is an urgent need to explore new treatment methods to overcome the limitations of traditional approaches. New treatment modalities should exhibit higher efficacy and lower side effects, aiming to improve patients' quality of life and treatment outcomes. Researchers are actively exploring novel therapeutic strategies, including gene therapy, immunotherapy, nanomedicine, and pulsed electric field (PEF) technology [10–12], to provide more effective and personalized treatment options for patients.

The combined application of nanosecond pulsed electric fields (nsPEFs) and nanomedicine has emerged as a synergistic therapeutic approach for treating various diseases, including tumors, demonstrating tremendous potential. Nanomedicine offers targeted drug delivery and enhanced therapeutic effects, while nsPEFs provide a nonthermal means of modulating cell membrane permeability and intracellular regulation. This synergistic combination holds promise in overcoming the limitations of traditional therapies and improving treatment outcomes.

In recent years, nanomedicine has revolutionized drug delivery. By achieving controlled release and targeted delivery of drugs, nanoparticles (NPs) such as liposomes, polymer micelles, and inorganic nanomaterials (Figure 13.1) have been extensively studied for their ability to encapsulate and deliver drugs to specific sites within the body [13, 29]. This targeted delivery approach reduces nonspecific effects, lowers systemic toxicity, and improves drug bioavailability. Moreover, nanomedicine platforms can be tailored through surface modifications to achieve active targeting and stimuli-responsive drug release, enabling personalized and precise treatment strategies.

On the other hand, nsPEFs have garnered significant attention as a nonthermal biomedical application. Characterized by ultra-short pulses (nanosecond scale) and high electric field intensities, nsPEFs induce reversible and irreversible changes in cell membrane permeability when applied to cells or tissues, thereby modulating intracellular processes [30–35]. The unique ability of nsPEFs to selectively induce cell membrane permeabilization without causing significant thermal damage provides a novel approach to therapeutic interventions. In the context of tumor treatment, nsPEFs have shown promising results in inducing tumor cell death, inhibiting tumor growth, and triggering immune responses [36, 37].

The potential of combining nanosecond PEFs with nanomedicine is enormous. By leveraging the precisely targeted drug delivery capabilities of nanomedicine and the cell membrane permeabilization induced by nsPEFs, enhanced uptake of drugs by tumor cells and improved therapeutic effects can be achieved. Furthermore, the localized controllability of nsPEFs allows for selective treatment of tumor tissue while minimizing damage to surrounding healthy tissues.

13.1 Introduction | 301

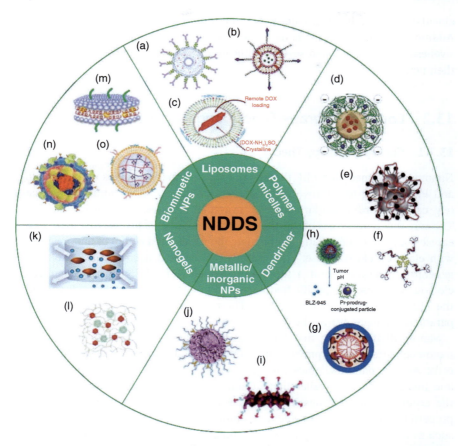

Figure 13.1 Summary of NDDS for "co-encapsulation" approach in chemoimmunotherapy. Source: Reproduced from [13] with permission under CC BY 4.0 license, Copyright the Nano-Micro Letters 2020. Liposomes (Source: (a) Reproduced with permission from Ref. [14], © 2018, Elsevier; (b) with permission from Ref. [15], © 2019, American Chemical Society; (c) with permission from Ref. [16], © 2018, American Chemical Society); Polymer micelles (Source: (d) Reproduced with permission from Ref. [17], © 2020, John Wiley and Sons; (e) with permission from Ref. [18], © 2015, John Wiley and Sons); Dendrimer (Source: (f) With permission from Ref. [19], © 2011, Elsevier; (g) with permission from Ref. [20], © 2019, Ivyspring International Publisher; (h) with permission from Ref. [21], © 2017, American Chemical Society); Inorganic NPs and metallic NPS (Source: (i) Reproduced with permission from Ref. [22], © 2017, American Chemical Society; (j) with permission from Ref. [23], © 2019, John Wiley and Sons); Nanogels (Source: (k) Reproduced with permission from Ref. [24], © 2017, Elsevier; (l) with permission from Ref. [25], © 2020, Elsevier); Biomimetic NPs (Source: (m) Reproduced with permission from Ref. [26], © 2019, American Chemical Society; (n) with permission from Ref. [27], © 2019, American Chemical Society; (o) with permission from Ref. [28], © 2017, American Chemical Society).

This chapter is structured into three main sections, providing a comprehensive introduction to the concept, treatment theory, and equipment associated with this innovative therapeutic approach. The article thoroughly explores the numerous merits of nanomedicine and PEF technology in the treatment of cardiovascular and cerebrovascular diseases, neurological disorders, and cancer, while also elucidating the advantages derived from their combined therapeutic strategy. Additionally, the synthesis methods of nanomedicines with electric field responsiveness are discussed, and several application examples are provided to illustrate their potential.

13.2 Concept, Therapy Theory, and Devices

13.2.1 Concept, Therapy Theory

Nanomedicine-mediated PEF therapy is a therapy that combines nanotechnology and PEF, aiming to improve the treatment efficacy of some intractable diseases (e.g, cancers, cardiovascular and cerebrovascular diseases, neurological disorders). The core theoretical basis of this therapeutics is to achieve precise treatment of diseasesby applying PEF ablation to regulate the release and action of nanomedicines with electric field response. The treatment equipment usually consists of a PEF generator, electrodes, and a monitoring system. The PEF generator generates electric field pulses of specific frequency and intensity, and the electrode transmits the PEF to a specific treatment area. The monitoring system is used to monitor parameter changes during the treatment process.

Nanomedicines have the following advantages in treating tumors: first, nanomedicines can improve drug targeting and reduce damage to normal cells; second, nanomedicines can improve the solubility and stability of drugs, and increase their bioavailability; In addition, nanomedicines can also improve the accumulation of drugs in tumor tissues by altering their pharmacokinetic properties. However, nanomedicine therapy for tumors also has some drawbacks, such as the complexity of synthesis and preparation processes, potential toxic side effects, and drug resistance issues [29].

The advantages of PEF therapy for tumors include: first, this treatment method can achieve local and noninvasive treatment, reducing damage to surrounding normal tissues; second, PEFs can increase the permeability of cell membranes and improve the internalization efficiency of drugs. In addition, PEFs can also activate cell apoptosis pathways and promote tumor cell death. However, PEF therapy for tumors also has some drawbacks, such as pain during the treatment process, instability of treatment effects, and difficulty in selecting treatment parameters.

Combined therapy, as a strategy that comprehensively utilizes multiple treatment methods, has shown significant advantages in cancer treatment. Nanodrug-mediated PEF therapy, as an emerging treatment method, combines the advantages of nanotechnology and PEF, and its combined application with other treatment methods has the potential to further improve the therapeutic effect [38, 39]. This article discusses in detail the advantages of nanodrug-mediated PEF therapy combination therapy, including enhancing efficacy, overcoming drug resistance, reducing side effects, and promoting personalized treatment.

13.2.1.1 Enhanced Therapeutic Effects

The combination of nanomedicine-mediated PEF therapy can enhance efficacy through multiple mechanisms. The targeting and drug delivery ability of nanomedicines can increase the accumulation of therapeutic substances in tumor tissue and improve treatment effectiveness. PEFs can increase cell membrane permeability, promote the internalization of nanomedicines, and enhance the killing effect of drugs on tumor cells. The combined application can mutually enhance the effectiveness of the two treatment methods, achieve synergistic effects, and further improve the therapeutic effect.

13.2.1.2 Overcoming Drug Resistance

Drug resistance is a common problem in tumor treatment, and the combination of nanodrug-mediated PEF therapy can effectively overcome this problem. The targeting and drug delivery ability of nanomedicines can increase the concentration of drugs in tumor cells and reduce the occurrence of cell resistance. PEFs can alter the structure and function of cell membranes, reducing cell tolerance to drugs. Combined therapy can comprehensively utilize the advantages of both treatment methods, effectively overcome tumor resistance, and improve treatment effectiveness.

13.2.1.3 Reduce Side Effects

The combination therapy of nanomedicine-mediated PEF therapy can reduce the side effects during the treatment process. The targeting of nanomedicines can reduce damage to normal cells and improve the safety of treatment. PEF therapy is a local and noninvasive treatment method that can reduce damage to surrounding normal tissues. Combined therapy can improve the quality of life of patients while reducing side effects.

13.2.1.4 Promote Personalized Treatment

The combination of nanomedicine-mediated PEF therapy can achieve the goal of personalized treatment. Nanomedicines can meet the needs of personalized treatment by adjusting the composition, dosage, and release mode of drugs. The parameters of the PEF can be adjusted according to the specific situation of the patient to achieve the best treatment effect. Combined therapy can be customized based on the patient's tumor type, condition, and individualized characteristics, improving the targeted and effective treatment.

In summary, the combination therapy of nanomedicine-mediated PEF therapy has significant advantages in tumor treatment. By enhancing efficacy, overcoming drug resistance, reducing side effects, and promoting personalized treatment, combination therapy can further improve treatment effectiveness and provide patients with more effective and safe treatment options. However, despite the enormous potential of this combined therapy strategy, further research and clinical practice are needed to validate its application prospects in different tumor types and clinical contexts.

Nanomedicine-mediated PEF therapy combined with therapy is an innovative treatment strategy that achieves precise control of drug release and enhanced therapeutic effect by combining nanomedicine carriers and PEF stimulation. This article will provide a detailed introduction to the specific instruments and related parameters used in the combination therapy of nanomedicine-mediated PEF therapy, including pulse generator, electrode system, and pulse parameter adjustment.

13.2.1.5 Pulse Generator

Pulse generator is a key device in the combination therapy of nanomedicine-mediated PEF therapy, used to generate PEF stimulation. The pulse generator derived from this technology is used to generate the signal system and generate the electrical test signal parameters required by the instrument. According to its signal waveform, it can be divided into four categories: sinusoidal signal generator, function (waveform) signal generator, pulse signal generator, and random signal generator. The typical Blumlein pulse generator is the hardware basis of biological effects. Martin's team at the British Atomic Weapons Research Center used Marx generator and transmission line technology to generate high-power nanosecond pulses in 1962. The main types of pulse generators are capacitor discharge circuits, square wave pulse generators, analog circuit generators, brumlin generators, and diode short-circuit switch generators. Typical brumlin generators are mainly divided into single transmission line type and brumlin type [40]. The specific structure is shown in Figure 13.2.

A single transmission line pulse generator, as shown in Figure 13.2, consists of a main discharge switch S, a charging resistor R_1, a load resistor R_2, and a transmission line wave impedance L. Once the switch is closed, a square wave pulse with an amplitude equal to the charging voltage and a width twice the propagation time of the electromagnetic wave in the transmission line is obtained on the load.

To overcome the drawback of a single transmission line pulse generator, which generates a pulse amplitude of only half of the charging voltage, Blumlein proposed a dual transmission line pulse generator known as the Blumlein transmission line pulse generator (referred to as the Blumlein pulse generator). The structure of this generator, as illustrated in Figure 13.3, consists of two transmission lines connected

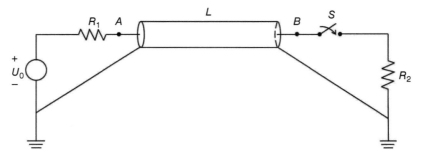

Figure 13.2 Schematic of pulse generator based on single transmission line.

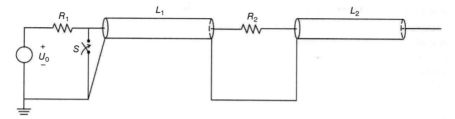

Figure 13.3 Schematic diagram of Blumlein pulse generator.

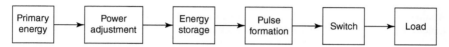

Figure 13.4 Schematic of pulse power system.

in series, with the load R_L connected between the two transmission lines. The length of the two transmission lines is denoted as l, and their wave impedance is represented by L_1 and L_2, respectively. The electromagnetic wave propagates along the transmission lines at a velocity of v [41].

Capacitor discharge device: A capacitor discharge device is a common pulse generator that generates high-voltage pulses through the charging and discharging processes. It has the characteristics of simplicity and easy control, and is commonly used in small-scale research and clinical applications.

Pulse power amplifier: Pulse power amplifier can provide higher power and more complex pulse waveforms. It is usually composed of amplifiers, pulse formers, and control systems (Figure 13.4), suitable for large-scale research and clinical treatment.

Different pulse generators have different parts assembly and have their own advantages and disadvantages. Next, we will briefly introduce the advantages and disadvantages of each representative pulse generator, as shown in Table 13.1.

13.2.1.6 Electrode System

The electrode system is used to transmit PEFs to the treatment area for precise stimulation and treatment. Common electrode systems include:

Parallel electrode: The parallel electrode is composed: of two parallel arranged conductive plates, which can be used to generate a uniform electric field distribution. It is suitable for larger treatment areas, such as tumor areas.

Needle-shaped electrode: Needle-shaped electrode is composed of multiple needle-shaped conductive materials, which can be used to generate local electric field stimulation. It is suitable for smaller treatment areas, such as lesions on the skin surface.

13.2.1.7 Pulse Parameter Adjustment

Pulse parameters are key factors that determine the therapeutic effect, including pulse voltage, pulse width, pulse frequency, and pulse number. These parameters

Table 13.1 Advantages and disadvantages of various pulse waveform generators.

Type of pulse generator	Advantages	Disadvantages
Capacitor discharge circuit	Simple structure, low price, simple control system, and high voltage	Flexibility parameter control is not high, cell survival rate is low
Square wave pulse generator	High current, simple control system, accurate, and flexible time parameter control	During the pulse period, the voltage decreases rapidly and the voltage amplitude control is not flexible
Analog circuit generator	The pulse parameter control is flexible and accurate, and the electric hole waveform is adjustable	The control system is complex and power consumption is limited
Blumlein type generator	The design is simple, high voltage and high current are obtained at the same time, and the duration and polarizability are adjustable	Switch elements are complex and need impedance matching design
Diode short circuit switch generator	The electrical components are easy to obtain and the load impedance is adjustable	Complex design, low output power

need to be adjusted and optimized according to specific treatment needs. Common pulse parameters include:

Pulse voltage: Pulse voltage refers to the peak voltage of the pulse waveform, usually measured in kilovolts (kV). Its selection should take into account factors such as the size of the treatment area, the demand for nanomedicine release, and safety.

Pulse width: Pulse width refers to the duration of the pulse waveform, usually in microseconds (μs) in units. Its choice affects the rate of nanomedicine release and the effectiveness of drug delivery.

Pulse frequency: Pulse frequency refers to the repetition frequency of a pulse waveform, usually measured in Hertz (Hz). Its selection should take into account the response time of biological tissues in the treatment area to stimuli and the need for nanomedicine release.

Pulse count: Pulse count refers to the total number of pulses in a sequence. Its selection can be adjusted based on the duration of treatment and the requirements for treatment effectiveness.

13.2.2 Devices

Currently, PEF technology has advanced to three generations. Based on the pulse width of the electric field, its biomedical applications are divided into the first generation of millisecond PEF technology, the second generation of microsecond PEF technology, and the third generation of nanosecond PEF technology. Each generation has unique features and is extensively utilized in the medical field. Its representative instruments mainly include:

Figure 13.5 MedPulser® system. Source: AngioDynamics, Inc.

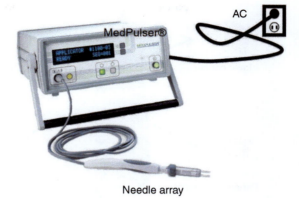

The MedPulser®, produced by Genetronics in the United States and shown in Figure 13.5 [42], is currently a representative electroporation therapeutic instrument. Its system comprises a remote (pedal) activated pulse generator MedPulser® and a sterile disposable needle array electrode, which can provide a fixed voltage and duration of µs-level square wave electric pulses. The working principle involves administering medication to the patient, inserting the needle array into the tumor, and activating MedPulser®. The device automatically activates the corresponding needle electrode pairs and releases PEFs between them. The polarity of the pulse between the electrodes changes once every two cycles. Clinicians control the treatment process through foot pedals, eliminating the need for manual operation. The voltage and pulse duration for each needle array applicator in MedPulser® are determined based on theoretical calculations, cell experiments, preclinical animal experiments, and clinical experiments. The Cliniporator, produced by Carpy Igea in Italy, can deliver electric pulses with a pulse width of 30–5000 µs, a peak voltage of 100–1000 V, and a frequency of up to 5000 Hz, demonstrating favorable therapeutic effects in clinical applications [33].

Due to the significant advantages of using applied electric fields to destroy cancer cells and other harmful tissues or cells, IRE has become an important medical tool. With the continuous updates and maturation of the theory and technology of PEFs on biological effects, many therapeutic instruments for irreversible electroporation have been designed, manufactured, and applied to the study of diseases of various types and stages. The most typical IRE treatment instrument is the nanoknife system (Nanoknife [43]), as shown in Figure 13.6. The system consists of an IRE generator and an electrode probe. The IRE generator can output high-voltage electrical pulses ranging from 100 to 3000 V, with a maximum duration of 100 ms. The electrode probe has a diameter of 16–19G and a length of 15 cm. To use the system, two electrode probes should be placed parallel and vertically inserted into the target area or surrounding tissue to be ablated. To induce cell death, two or more unipolar probes or one bipolar probe must be used simultaneously, releasing a series of microsecond PEFs between the electrodes. According to reports, the bipolar probe of the nanoknife system can create an elliptical ablation zone with a long axis of approximately 30 mm and a short axis of about 15 mm in isolated pig lesion tissue [35].

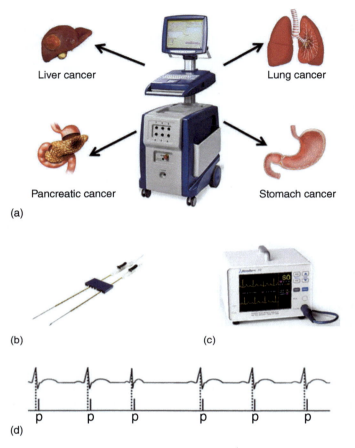

Figure 13.6 Nanoknife system: (a) IRE generator; (b) bipolar IRE prob. Source: AngioDynamics, Inc. (c) Oscilloscope. Source: Medic Exchange. (d) Pulse waveform.

This device received approval from the US FDA in April 2012 to be used in clinical applications for tumor treatment. In June 2015, China approved the use of this technology for clinical applications with CFDA approval, and it has successfully treated many patients with liver and pancreatic cancer.

Unlike millisecond and microsecond PEFs, the emerging nanosecond pulse electric field technology has only been mastered by a few companies worldwide. Representative examples include the nanosecond pulse tumor ablation system of Hangzhou Ruidi Biotechnology Co. Ltd. in China (Figure 13.7a), Pulse electric field ablation equipment (REMD-G5) independently innovatively developed by Shanghai Ruidao Medical Technology Co. Ltd. (Ruidao Medical) (Figure 13.7b) and the world's first third-generation pulse electric field cardiac ablation system, nsPFA, of Shenzhen Maiwei Medical Technology Co. Ltd. The high-pressure steep pulse third-generation machine developed by Ruidi Biotechnology Co. Ltd. has been launched in a multicenter prospective clinical trial in China and has safely completed over 100 tumor ablation surgeries at dangerous anatomical sites

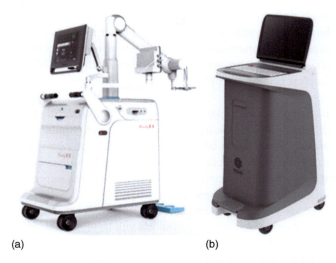

Figure 13.7 (a) RD® steep nanosecond pulse generator. Source: Hangzhou Ruidi Biotechnology Co. Ltd. (b) Ruidao Medical's "Composite Steep Pulse Treatment Equipment." Source: Tech in Asia.

near large blood vessels. This device is based on the principle of next-generation nanosecond PEF ablation, which can selectively ablate tumor tissue and treat tumors that cannot be ablated by other devices, such as tumors near important organs like large blood vessels and nerves. Ablation therapy is safe and fast with 3 cm solid soft tissue tumors being ablated within 10 minutes. Ablation plans can be customized based on the tumor's location, size, and shape, and precise pulse therapy can be achieved. It does not depend on temperature effects. Most importantly, it addresses the shortcomings of other PEF devices by dynamically adjusting pulse energy release during the treatment process, minimizing adverse effects on the patient's heart rate. Moreover, Ruidi Biological has taken the lead in original innovation with nanosecond 10 000 V high-voltage steep pulse equipment, breaking through the parameter dilemma of existing pulse electric field technology. The self-developed nanosecond pulse tumor ablation system has realized super electroporation and transmembrane nuclear penetration, which is more precise, high-energy, and sharp. It is the next generation of innovative electric field ablation therapy in China compared with foreign irreversible electroporation tumor ablation equipment with microsecond PEF.

Atrial fibrillation (hereinafter referred to as atrial fibrillation) is the most common arrhythmia. In 2022, there will be about 20 million [44] patients with atrial fibrillation in China, with a total prevalence rate of about 1.6%, and its incidence rate will gradually increase with age [45]. At present, catheter ablation of atrial fibrillation usually uses radio frequency or freezing as the energy source, and realizes tissue destruction through heat or low-temperature conduction, thus achieving pulmonary vein isolation. On the one hand, cold/hot ablation lacks selectivity in the destruction of tissue in the ablation area, causing damage to adjacent esophagus, coronary artery, and phrenic nerve, and because of local temperature, there is also the risk of pulmonary vein stenosis and thromboembolism. In addition, traditional ablation requires full attachment of the catheter to the tissue in order to play a role. In some special anatomical parts, accompanied by cardiac pulsation, respiratory movement and other interference, poor attachment of the catheter or insufficient ablation energy will lead to incomplete isolation of pulmonary vein, resulting in recurrence of atrial fibrillation.

In 2022, the clinical trial of pulse field ablation (REMD-G5) independently innovated and developed by Shanghai Ruidao Medical Technology Co. Ltd. (Ruidao Medical) in Shanxi Cardiovascular Hospital officially started and successfully completed the pulsed field ablation (PFA) operation for two patients with atrial fibrillation. The electrophysiological three-dimensional mapping showed that the pulmonary veins of both patients were successfully isolated in real time.

PFA is the use of high-voltage PEF to act on myocardial tissue. Under the action of PEF, the double molecules of cell membrane phospholipid will be abnormal, forming irreversible electroporation, which will cause necrosis and apoptosis of myocardial cells, so as to achieve the purpose of abnormal electrical signal transmission in ablated tissue. PFA has become the third new method for atrial fibrillation ablation treatment after radiofrequency ablation and cryoballoon ablation. Pulse field ablation has two main advantages: first, it has tissue selectivity, does not damage the surrounding blood vessels, nerves and esophagus, and uses nonthermal ablation to reduce other tissue damage caused by high temperature or freezing, does not cause pulmonary vein stenosis, and has low requirements for catheter attachment; the second is that the discharge time is short, which can significantly improve the safety of surgery and shorten the surgical time. Therefore, PFA for atrial fibrillation is the future direction of catheter ablation.

Deep brain stimulation (DBS) technology is a reversible method of regulating the brain. It involves the surgical implantation of electrodes into specific brain regions, which then deliver electrical pulses at a certain frequency to modulate abnormal brain discharges. This technique aims to improve or treat various diseases. Due to the similarity of the entire DBS system to pacemakers, which includes the electrodes, pulse generator, and wires, DBS is also referred to as a "brain pacemaker."

The primary approach of DBS therapy is to implant electrodes into the patient's brain and utilize a pulse generator to stimulate specific deep brain nuclei, thereby correcting abnormal brain electrical circuits and alleviating associated neurological symptoms. Unlike certain treatment methods such as lesioning or radiation therapy, which cause permanent, unadjustable, and irreversible brain damage, DBS does not disrupt brain structures and allows for potential future treatments. Currently, the most widely used clinical application of DBS is in the treatment of advanced-stage Parkinson's disease patients.

Parkinson's disease is a common neurodegenerative disorder characterized by motor dysfunction in patients. Symptoms include tremors, muscle rigidity, mask-like facial expression, and short shuffling steps, with an average onset age of around 60 years. Fifteen years ago, there were already 1.7 million Parkinson's disease patients in China, with an estimated prevalence of 1% among individuals over 55 years old. It is estimated that the number of new patients in China reaches hundreds of thousands annually. Although Parkinson's disease cannot be cured, proactive treatment can effectively control the condition and improve patients' quality of life.

In 2008, during a DBS weight loss surgery on a 50-year-old, 190 kg subject, conducted by Dr. Andres Lozano, a neurosurgeon at the University of Toronto, a remarkable discovery was made. When the electrode was implanted in the ventral part of the hypothalamus and stimulated, the patient reported a strong sense of familiarity, recalling a pleasant experience with friends in a park from 30 years prior. As the electrical current was intensified, the subject was able to recall even more details. Dr. Lozano speculated that this outcome was a result of stimulating the fornix, which activated the hippocampal memory circuitry [46]. Research investigating memory deficits resulting from lesions in the fornix in animal experiments and humans has revealed the significant role of the fornix in memory function [47]. Therefore, this unexpected finding has brought hope for improving memory function in patients with AD [46].

In 2010, the results of a phase I clinical trial (NCT00658125) investigating DBS-f treatment for AD demonstrated the activity of DBS-driven neurons in the memory circuitry, including the entorhinal cortex and hippocampal regions, as well as activation of the brain's default mode network. PET scans revealed that DBS stimulation could reverse glucose utilization impairments in the temporal and frontal lobes at an early stage, and this reversal effect could be sustained after 12 months of continuous stimulation. Assessment using the AD Assessment Scale (DAS) and Mini-Mental State Examination (MMSE) showed symptom improvement in some patients at 6 and 12 months, with a slower decline in cognitive function. No severe adverse events were observed during the study. Therefore, further research on the application of DBS for modulating pathological brain activity in AD patients is warranted [47].

In 2015, the ADvance Study research team confirmed the safety of bilateral DBS-f treatment for AD [48].

In 2016, the results of a phase II clinical trial (NCT01608061) investigating DBS-f treatment for mild AD demonstrated good safety and tolerability of the surgical procedure and electrical stimulation [49].

In addition to treating movement disorders such as dystonia and Parkinson's disease, there have been clinical attempts to use DBS to aid in the recovery of consciousness in patients with brain injuries. Furthermore, there is an increasing amount of research on using DBS to treat psychiatric disorders, including depression, obsessive-compulsive disorder, and anorexia nervosa. However, due to significant interindividual variability and potential complications, the effectiveness of surgical treatment for these conditions is still being evaluated.

It should be noted that specific instruments and parameters may vary depending on research or clinical applications. In the combination therapy of nanomedicine-mediated PEF therapy, the selection of instruments and parameters should be based on the treatment objectives, disease characteristics, and the needs of the treatment area, and undergo strict experimental verification and safety evaluation.

In summary, the specific instruments for the combination therapy of nanomedicine-mediated PEF therapy usually include pulse generators and electrode systems, and the pulse parameters need to be adjusted and optimized according to the treatment needs. The development of this treatment strategy provides new avenues for precise control of drug release and enhancement of treatment efficacy, but further research and validation are still needed in practical applications.

13.3 Synthesis of Nanomedicine with Electric Field Response

Electric field-responsive nanomedicine refers to a drug release system that can be designed as a drug carrier of NPs that is sensitive to external electric field stimuli. One common electric field-responsive nanodrug is a polymer-modified nanoparticle. Among them, drugs can include: using polymer-modified NPs as an example, chemotherapy drugs such as doxorubicin or paclitaxel can be loaded into the NPs.

(1) *Synthesis process*: The synthesis of electric field-responsive nanomedicine generally includes the following steps:
(2) *Synthesis of nanoparticle carriers*: Common methods include solution method, precipitation method, emulsification method, etc. By controlling the reaction conditions and formula, NPs with the required size and surface properties can be obtained.
(3) *Surface modification*: Introduce a polymer that can respond to electric fields on the surface of NPs. For example, electroactive polymers can be modified onto the surface of NPs through covalent bonding or physical adsorption.
(4) *Drug loading*: Loading the target drug into the interior or surface of NPs through physical adsorption, covalent bonding, or other methods.
(5) *Electric field responsiveness*: Design appropriate polymer structures to undergo morphological or structural changes under external electric field stimulation, thereby achieving drug release or regulation.
(6) *Efficacy*: Electric field-responsive nanomedicine has adjustable drug release performance in treatment. When external electric field stimulation is applied, the polymer structure of drug carriers will change, leading to drug release or controlled release, thereby achieving precise drug delivery and regulation.

13.3.1 Synergistic Anticancer Research of Multimodal Iron-based Nanodrugs and Nanosecond Pulse Technology

Nanomedicines such as Fe@Fe_3O_4 NPs [50] and Au@CoFeB–Rg3 [51], developed by the research group led by Song Yujun, have played a significant role in treating cancer, particularly liver cancer, when combined with nsPEF technology. In addition to their effective therapeutic properties, these nanomaterials possess magnetic characteristics that allow for visualization of the treatment process. It is possible to observe the accumulation of nanomedicines in the liver region with clarity [52].

13.3.1.1 Design and Synthesis of Multimodal Iron-based Nanodrugs

To demonstrate the multimodal imaging capabilities for medical diagnostics and exhibit significant therapeutic effects on cancer cells, a multimodal iron-based nanodrug, Au@CoFe–Rg3, was designed. When used for labeling biomolecules to precisely study cell–drug interactions, these multimodal nanodrugs can be optically observed under dark-field microscopy. Simultaneously, they can also be utilized for MRI and CT imaging, facilitating future clinical applications and enhancing the accuracy of clinical diagnosis/imaging. Additionally, the multimodal iron-based nanomaterials exhibit excellent antitumor properties, exerting significant toxicity on cancer cells and significantly inhibiting their proliferation.

In this chapter's study, as shown in Figure 13.8 [51], Au@CoFe NPs were first synthesized and then conjugated with ginsenoside Rg3 to form the multimodal nanodrug Au@CoFe–Rg3, intended for future multimodal diagnostics and cancer therapy.

The multimodal nanodrug Au@CoFe–Rg3 was developed by conjugating Au@CoFe NPs with ginsenoside Rg3. The Au@CoFe NPs were synthesized in a two-step programmable microfluidic process, where CoFe NPs were first synthesized using a simple programmable microfluidic process, followed by the synthesis of Au@CoFe NPs by coating the presynthesized CoFe NPs onto the surface of Au NPs.

The first step involved the synthesis of CoFe NPs using a simple programmable microfluidic process. Figure 13.9 illustrates the programmable microfluidic process used for the synthesis of CoFe NPs and Au@CoFe NPs. This straightforward process allows for convenient control of the size, crystal structure, and shape of the products. Polyvinylpyrrolidone (PVP), $FeCl_2 \cdot 4H_2O$, and $CoCl_2 \cdot 6H_2O$ were dissolved in N-methyl-2-pyrrolidone (NMP) to form a salt solution. Subsequently, sodium borohydride ($NaBH_4$) was dissolved in NMP to create a reducing solution, under nitrogen protection. The reaction was carried out at a temperature of 120 °C and a flow rate of 1 ml/min under nitrogen protection. After the completion of the reaction, the resulting dispersion solution of fresh CoFe NPs was collected in a product receiver.

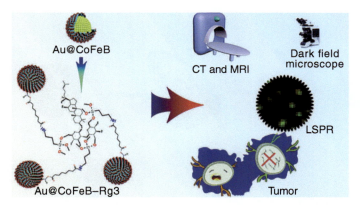

Figure 13.8 Schematic diagram of e synthesis of multimode nanomedicine and their application. Source: Reproduced with permission from Ref. [51], © 2020, American Chemical Society.

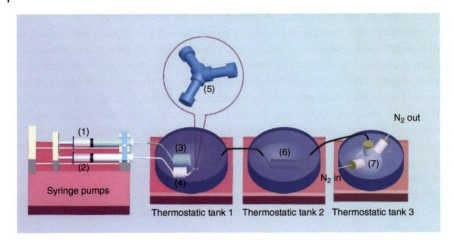

Figure 13.9 Temperature-programmed microfluidic process for the synthesis of NPs. Source: Reproduced with permission from Ref. [51], © 2020, American Chemical Society.

Next, the second step of the experiment was carried out by dissolving $HAuCl_4$ in NMP. The $HAuCl_4$ solution and the dispersion solution of presynthesized CoFe NPs were separately delivered into the microfluidic syringe pump at a flow rate of 1 ml/min to complete the reduction reaction and rapid nucleation. In this step, a small amount of $NaBH_4$ was dissolved in NMP and placed in the receiver. After the completion of the reaction, the collector was shaken and left undisturbed for 30 minutes to accomplish the displacement, re-reduction, and surface rearrangement of CoFe NPs.

The schematic diagram of the main reaction process is shown in Figure 13.10, where Au atoms are individually reduced from $HAuCl_4$ by $NaBH_4$ and CoFe, resulting in the rapid formation of Au clusters after stabilization by PVP. Subsequently, the generated iron and cobalt are further reduced by $NaBH_4$. The re-reduced Co

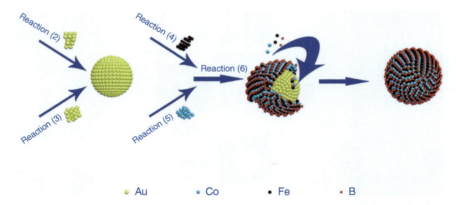

Figure 13.10 Chemical reaction process of the synthesis of the Au@CoFeB NPs. Source: Reproduced with permission from Ref. [51], © 2020, American Chemical Society.

and Fe atoms then grow around the surface of Au clusters, forming a heterogeneous structure of Au@CoFe. During this process, the atoms on the surface of CoFe particles dissolve into individual atoms in the reducing environment and deposit on the surface of gold particles to form a CoFe shell. After a rapid cooling process, both Au, Co, and Fe can effectively deposit into our particles, leading to a significant improvement in the overall stability of the particles.

13.3.1.2 Characterization of Physical Properties and Structural Analysis

Firstly, Figure 13.11 displays the morphology, structure, and composition of the CoFe NPs synthesized in the first step. Wide-angle TEM images reveal well-dispersed spherical CoFe NPs. Statistical analysis of multiple randomly selected NPs shows an average size of 2.4 ± 0.6 nm. Highly crystalline NPs are observed in the high-resolution TEM (HRTEM) images, with a measured interplanar spacing of 2.02 Å. XRD analysis was conducted to further confirm the composition of the CoFe NPs. The results indicate that the two main peaks in the XRD pattern can be attributed to the (100) and (110) planes of a body-centered cubic (bcc) CoFe alloy (PDF No. 441433), with the lattice spacing of the (110) plane determined to be in excellent agreement with the measurement from the HRTEM image mentioned earlier. Additionally, the peaks at 58.06°, 66.33°, 75.25°, and 84.17° can be assigned to the (111), (200), (210), and (211) planes, respectively, of the bcc CoFe alloy (PDF No. 44-1433).

Subsequently, Figure 13.12 illustrates the final synthesized Au@CoFe NPs, with electron microscopy images clearly depicting mostly spherical NPs with good

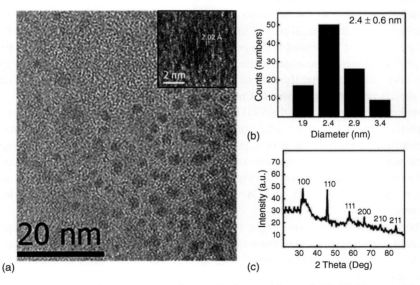

Figure 13.11 (a) Wide-viewed TEM image (the inserted image is HR-TEM image with a lattice fringe of 2.02 Å). (b) Histogram of the size distribution and (c) XRD pattern of the precursor CoFeB NPs. Source: Reproduced with permission from Ref. [51], © 2020, American Chemical Society.

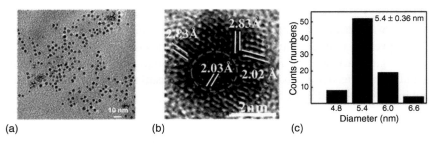

Figure 13.12 (a) Wide-viewed TEM image of the Au@CoFeB NPs. (b) HR-TEM image of one single NP showing the lattice fringe of 2.03 Å in the core and the lattice fringe of 2.02 and 2.83 Å in the shell. (c) The histogram of the size distribution of the Au@CoFeB NPs. Source: Reproduced with permission from Ref. [51], © 2020, American Chemical Society.

uniform dispersion (Figure 13.12a). The HRTEM image (Figure 13.12b) reveals a distinct core–shell structure and excellent crystallinity of the NPs, with lattice spacings of 2.03 Å in the core (corresponding to the (200) plane of face-centered cubic (fcc) Au) and lattice spacings of 2.83 and 2.02 Å (corresponding to the (100) and (110) planes of bcc CoFe, respectively). Subsequently, 100 randomly selected NPs were analyzed for size statistics, and the results indicated an average diameter of 5.4 ± 0.4 nm for the Au@CoFe NPs (Figure 13.12c).

To further confirm the core–shell structure, the compositional distribution of the NPs was characterized using energy-dispersive X-ray spectroscopy (EDS). Figure 13.13 presents representative point scan results, where the EDS spectra show strong peaks for Au and weaker peaks for Co and Fe in the central region, confirming that the central part of the NPs primarily consists of Au with some Co and Fe. However, compared to the peaks in the central region, the intensity of the Au peak decreases at the edges, while the Co and Fe peaks become more prominent. This result indicates that the core of the sample is mainly composed of Au, while the shell is primarily composed of CoFe.

As shown in Figure 13.14, the XRD spectrum of Au@CoFe NPs exhibits peaks corresponding to CoFe and Au, confirming the formation of the Au@CoFe NPs structure during microfluidic reduction. Subsequently, X-ray photoelectron spectroscopy (XPS) was conducted to determine the elemental composition as well as chemical and electronic states, as shown in Figure 13.14b. The Au@CoFe NPs are primarily composed of Au, B, C, O, Fe, and Co, which is consistent with the proposed reaction process.

Figure 13.15 displays wide-angle TEM images of the nanodrug Au@CoFe-Rg3, indicating good dispersion of these nanodrugs as well. However, the lattice fringes are less distinct compared to the original state due to the coupling of ginsenoside Rg3. Nevertheless, the core–shell nanostructure remains clear. After conducting size statistics analysis, it was found that these nanodrugs exhibit uniform size distribution with an average diameter of 6.6 ± 0.7 nm.

Subsequently, a comparison was made between the pure Au@CoFe NPs and the nanodrugs after ginsenoside coupling, as shown in Figure 13.15b. Compared to the

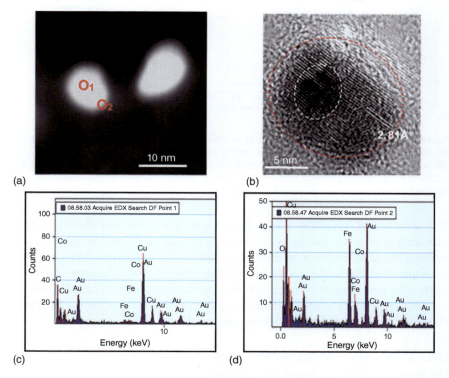

Figure 13.13 (a) The high-angle annular dark field (HAADF) STEM image and (b) the corresponding HR-TEM image of one typical NP. (c) The EDS spectra of point 1 in the center of the single NP and (d) the EDS spectra of point 2 in the outer shell of the single NP to show distinctly increased Au content and decreased Co and Fe contents in the center of the single NP. Source: Reproduced with permission from Ref. [51], © 2020, American Chemical Society.

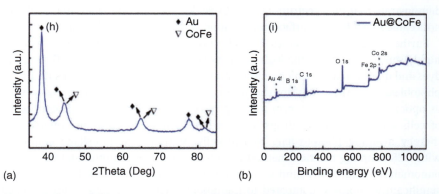

Figure 13.14 (a) XRD pattern of the Au@CoFeB NPs suggesting the fcc Au and bcc CoFe alloy and (b) full X-ray photoelectron spectroscopy (XPS) spectra. Source: Reproduced with permission from Ref. [51], © 2020, American Chemical Society.

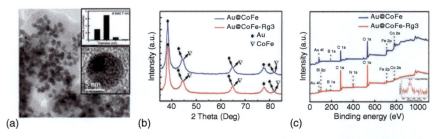

Figure 13.15 (a) Wide view TEM image of Au@CoFe(B)–Rg3 NMs (the inserted are the HR-TEM image of one single NM and the histogram of the size distribution). (b) X-ray diffraction and (c) full X-ray photoelectron spectroscopy (XPS) spectra (the inserted image is XPS of boron binding energy of the NMs). Source: Reproduced with permission from Ref. [51], © 2020, American Chemical Society.

XRD spectrum of Au@CoFe, the peaks in Au@CoFe–Rg3 appear slightly sharper, indicating improved crystallinity due to surface modification and conjugation processes. XPS characterization, as shown in Figure 13.15c, reveals that Au@CoFe–Rg3 is composed of Au, Si, B, C, N, O, Fe, and Co, consistent with the proposed mechanism. The new element N originates from the surface modifier (APTMS) and the coupling agent (DSS), while Si evidently comes from the surface modifier (APTMS). These results preliminarily confirm the success of the coupling process.

13.3.1.3 Research on Multimodal Physical Imaging Applications

The plasmonic colors of Au@CoFe NPs and Au@CoFe–Rg3 were observed using dark-field microscopy. As shown in Figure 13.16, the blue-green color of Au@CoFe NPs changed to yellow-green in Au@CoFe–Rg3. Based on the peak position statistics of the LSPR spectra of 50 randomly selected particles, Au@CoFe–Rg3 exhibited a redshift of approximately 36 nm after surface modification and conjugation processes. The typical LSPR spectra of individual Au@CoFe NPs and Au@CoFe–Rg3 (i.e. bright spots in Figure 13.16) are shown in Figure 13.16. Their LSPR spectra demonstrate that the spectrum of Au@CoFe–Rg3 becomes broader and exhibits a noticeable redshift after surface modification and conjugation with Rg3.

Furthermore, in certain cases, they also extend potential applications as various highly sensitive chemical/biological sensor components. Compared to fluorescent dyes and quantum dots, which require strong laser excitation and often suffer from photobleaching, these multimodal iron-based nanomaterials can be easily used as optical probes for long-term tracking of cell–cell interactions and observation of cells and biomolecules/drugs under white light using dark-field microscopy, thereby avoiding phototoxicity to live cells. With the amino-silane combined with DSS coupling reaction method invented by us, these multimodal iron-based nanomaterials can easily bind with biomolecules and drugs, and their stability is significantly improved compared to traditional dye-based probes or laser-excited quantum dots [53–55].

Previous studies have demonstrated that magnetic iron-based nanomaterials can be used as MRI contrast agents by significantly improving proton relaxation in the

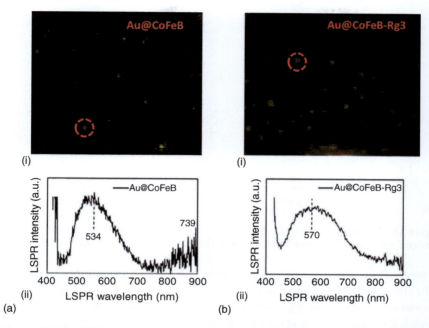

Figure 13.16 LSPR real color images of (a, i) Au@CoFeB NPs and (b, i) Au@CoFeB–Rg3 NMs and their LSPR spectra (a, ii, Au@CoFeB; b, ii, Au@ CoFeB–Rg3) measured by dark-field microscopy and spectroscopy. MRI effects of Au@CoFeB and Au@CoFeB–Rg3 solutions. Source: Reproduced with permission from Ref. [51], © 2020, American Chemical Society.

tissue environment [56]. Therefore, considering the unique magnetic properties of Au@CoFe NPs and Au@CoFe-Rg3, in addition to the aforementioned optical imaging applications, we further extended their applications to clinical settings as contrast agents for MRI and CT imaging. Molecular imaging, due to its noninvasive nature, high spatial resolution, and tissue sensitivity, is applicable for the diagnosis of various diseases and enables visualization of cellular functions within biological organisms [57]. In molecular imaging, MRI is used for morphological and functional imaging of anatomical structures and physiological processes in the body. To achieve high specificity, appropriate MRI contrast agents are required based on relaxation and biocompatibility during the imaging process. Therefore, the MRI imaging effects of Au@CoFe and Au@CoFe–Rg3 were evaluated. The T2-weighted spin-echo imaging effects and corresponding T2 relaxation rates, dependent on the effective metal concentrations (Co and Fe), in Au@CoFe NPs and Au@CoFe–Rg3 are shown in Figure 13.17.

Furthermore, the combination of MRI and CT enables highly accurate diagnosis of diseases. Therefore, the potential of Au@CoFe–Rg3 as a dual contrast agent in CT imaging was further investigated. These findings indicate that Au@CoFe–Rg3 is a multimodal clinical nanodrug, allowing for MRI and CT imaging and enabling three-dimensional deep tracking of nanodrugs in tissues or opaque biological organisms with enhanced resolution, potentially reaching the nanoscale.

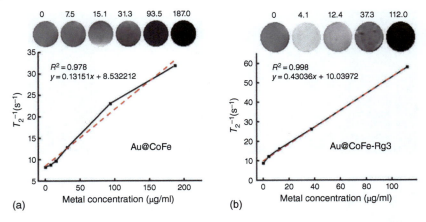

Figure 13.17 Au@CoFe NPs (a) and Au@CoFe–Rg3 (b) in vitro MRI imaging effect. Source: Reproduced with permission from Ref. [51], © 2020, American Chemical Society.

First, cell viability analysis was conducted to assess the cytotoxicity of Au@CoFe NPs and Au@CoFe–Rg3 on different types of cells for future in vivo animal studies and eventual clinical applications. 3T3 cells and Jurkat-CT cells were chosen as models for normal healthy cells, while K562-CT cells and HepG2 cells were selected as tumor cell models. 3T3 cells belong to vascular endothelial cells and are commonly used for the culture of keratinocytes, which secrete growth factors favorable for the growth of these cells in tumor tissues where vascular permeability factors are more abundant than in healthy tissues. Traditionally, these cells are targeted for treatment through injection methods, and the toxicity toward these healthy cells is worth investigating. Additionally, the growth of solid tumors and angiogenesis also require these cells, and their interaction directly affects the enhancement of penetration and retention in tumor sites, delivering more drugs to the tumor and even deep tissues. Jurkat-CT cells are immune T cells that typically exist in the form of suspended cells in the blood. K562-CT cells are chronic myeloid leukemia cancer cells that exist in the blood in the form of suspended cells. HepG2 cells are adherent liver cancer cells capable of forming solid tumors.

13.3.1.4 Study on In Situ Anticancer Effects in Liver Cancer

The cytotoxicity data indicated that the NPs at lower concentrations (<95 μg/ml) exhibited low toxicity toward all types of cells (Figure 13.18), suggesting their potential as nanoprobes. It was observed that as the concentration increased, the NPs showed significant toxicity toward 3T3 cells and HepG2 cells, indicating that the NPs had greater toxicity toward adherent cells that can grow in tissues compared to suspended cells. The cytotoxicity of Au@CoFe–Rg3 also showed that at lower concentrations (<95 μg/ml), Au@CoFe–Rg3 could be used as a nanoprobe for cells. By comparing the cytotoxicity of Au@CoFe–Rg3 with that of NPs, it was evident that the coupling of Rg3 with NPs in the form of Au@CoFe–Rg3 significantly increased the toxicity toward cells. For cancer cells K562 and HepG2, Au@CoFe–Rg3 at concentrations >195 μg/ml exhibited good therapeutic effects.

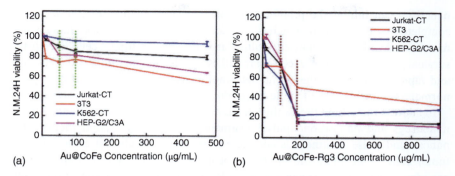

Figure 13.18 Au@CoFe NPs (a) and Au@CoFe–Rg3 (b) study on cytotoxicity of different cells. Source: Reproduced with permission from Ref. [51], © 2020, American Chemical Society.

13.3.2 Progresses of Innovative Combination Therapy

Furthermore, the in vivo antitumor effect of the multimodal iron-based nanodrug was assessed in situ in liver cancer mice. The results demonstrated that Au@CoFe–Rg3 combines the advantages of Au@CoFe NPs and ginsenoside Rg3, synergistically exerting excellent antitumor effects. Long-term toxicity evaluation of the multimodal iron-based nanodrug confirmed that Au@CoFe–Rg3 avoids the potential toxicity associated with traditional NPs, thereby enabling safe and efficient antitumor therapy.

Combination therapy represents a further advancement in nanomedicine. Over the years, significant progress has been made in cancer treatment with various modalities such as chemotherapy, radiation therapy, targeted therapy, and immunotherapy. However, tumors are complex and heterogeneous, and single-therapy approaches still have significant limitations. Certain therapies may be effective for a specific part of the body, while others may be more effective for different areas. The combination of traditional and modern therapies can help prolong a patient's life, overcome drug resistance, and alleviate symptoms. Combination therapy for cancer treatment has shown to be more effective and can achieve higher overall survival rates compared to single treatments in different types of cancer, including liver cancer. Therefore, in this chapter, we propose a combination therapy using nanosecond pulse treatment in conjunction with nanodrugs. Nanosecond pulse combined with nanodrugs is a new approach in the field of tumor treatment that is still in the research stage. Several chemotherapeutic drugs and traditional drugs can be combined with nanosecond pulse and exhibit different synergistic effects, thereby enhancing the therapeutic effect on tumors. As mentioned in previous chapters, the superiority of the multimodal iron-based nanodrug over traditional chemotherapy drugs has been confirmed.

In this chapter, we combine the advantages and limitations of the multimodal iron-based nanodrug and nanosecond pulse ablation, targeting the characteristics of tumor cells, to design a combination therapy using nanodrugs and nanosecond pulse treatment. First, as mentioned earlier, although the multimodal iron-based nanodrug can inhibit the proliferation and metastasis of tumor cells, it cannot completely eradicate them, which is a clinical challenge. On the other hand, while nanosecond pulse treatment can kill a large number of tumor cells during the surgical process, some tumor cells may still escape, leading to tumor cell metastasis and rapid regrowth in other locations. Another issue is the potential damage to normal tissue cells adjacent to the tumor during the nanosecond pulse process.

To address these limitations, this chapter presents a combination therapy using nanodrugs and nanosecond pulse treatment. In this combination therapy, the nanosecond pulse treatment is conducted first to effectively kill a high dose of tumor cells, overcoming the limitation of incomplete tumor cell eradication by nanodrugs alone. After the nanosecond pulse treatment, although some tumor cells may escape, we administer the nanodrug, which can rapidly target the liver. Due to the high concentration of H_2O_2 in tumor cells, the nanodrug can specifically recognize and kill the escaped tumor cells. Moreover, Rg3 in the nanodrug can effectively enhance and protect liver and kidney function, preserving and repairing the normal cells damaged by nanosecond pulse treatment. In addition to these obvious synergistic effects, nanosecond pulse treatment can generate reversible electroporation on the cell surface, helping more nanodrugs enter tumor cells and further enhancing the synergistic effect. The study of the interaction between nanodrugs and nanosecond pulse therapy and the apoptotic mechanism of tumor cells can provide a theoretical basis and clinical guidance for the discovery of new methods and approaches for tumor treatment.

13.4 Application Examples

Nanodrug-mediated PEF combined therapy, as an emerging treatment strategy, has made significant progress in the treatment of various diseases. This article presents several application examples in detail to demonstrate the potential application of nanodrug-mediated PEF therapy combined with therapy in the fields of cancer treatment, neurological disease treatment, and bacterial infection treatment.

13.4.1 Cancer Treatment

Example 1: Nanodrug-mediated PEF therapy combined with chemotherapy in the treatment of breast cancer.

The research team used targeted NPs as carriers to load docetaxel drugs into the NPs. By adjusting the parameters of the external PEF, drug release within NPs can be achieved. The results showed that the combined treatment group was significantly better than the chemotherapy alone in inhibiting the growth of breast cancer cells and inducing apoptosis, while reducing the damage of chemotherapy drugs to normal tissues.

Example 2: Application of nanomedicine-mediated PEF therapy combined with photodynamic therapy in the treatment of malignant melanoma of the skin.

Researchers have designed a nanoparticle with electric field responsiveness to load photosensitizers such as porphyrin compounds into the nanoparticle. The combination therapy of photodynamic therapy is achieved by applying PEFs while illuminating. The results showed that the combination therapy group showed significant efficacy in inducing tumor cell apoptosis, inhibiting metastasis, and promoting local tumor control.

13.4.2 Treatment of Neurological Diseases

Application of nanodrug-mediated PEF therapy combined with gene therapy in the treatment of Parkinson's disease.

The research team has developed electric field-responsive NPs that load gene therapy drugs (such as adeno-associated virus vectors) into the NPs. Nanodrugs can be designed as NPs that can cross the blood–brain barrier and be delivered to specific nerve cells through the action of nanosecond PEFs, thereby achieving therapeutic effects. By applying a PEF, gene therapy drugs are released and transfected. The experimental results showed that the combined treatment group showed significant therapeutic effects in alleviating motor disorders, enhancing the survival of dopamine neurons, and improving behavioral function in Parkinson's disease animal models.

13.4.3 Bacterial Infection Treatment

Application of nanodrug-mediated PEF therapy combined with antibiotic therapy in the treatment of drug-resistant bacterial infections.

Researchers have developed electric field-responsive NPs that load antibiotics such as vancomycin into the NPs. By applying a PEF, the release of antibiotics within NPs and the killing of bacteria are achieved. The research results showed that the combination therapy group showed superior efficacy in treating drug-resistant bacterial infections compared to individual antibiotic therapy while reducing the dosage of antibiotics and their impact on normal microorganisms.

13.5 Conclusion

The combination therapy via nanomedicine-mediated PEF ablation has shown broad potential as an innovative treatment strategy. In the fields of cancer treatment, nervous system disease treatment, and bacterial infection treatment. This combined therapy strategy achieves precise drug release and enhanced therapeutic effect through the design of nanomedicine carriers and the regulation of PEFs. However, further research still requires an indepth exploration of the coherent biological mechanism in the treatment of different intractable diseases, operation and biological safety, and clinical application, for the promotion of its clinical transformation and advanced application.

References

1 de Boer, R.A., Meijers, W.C., van der Meer, P., and van Veldhuisen, D.J. (2019). Cancer and heart disease: associations and relations. *European Journal of Heart Failure* 21: 1515–1525, https://doi.org/10.1002/ejhf.1539.

2 Mancusi, R. and Monje, M. (2023). The neuroscience of cancer. *Nature* 618: 467–479.

3 IARC Biennial Report 2020–2021. (2022). Available from: https://www.iarc.who.int/news-events/iarc-biennial-report-2020-2021/

4 Stewart, M.P., Sharei, A., Ding, X. et al. (2016). In vitro and ex vivo strategies for intracellular delivery. *Nature* 538 (7624): 183–192.

5 Ding, X., Stewart, M., Sharei, A. et al. (2017). High-throughput nuclear delivery and rapid expression of DNA via mechanical and electrical cell-membrane disruption. *Nature Biomedical Engineering* 1: 0039.

6 Badgley, M.A., Kremer, D.M., Maurer, H.C. et al. (2020). Cysteine depletion induces pancreatic tumor ferroptosis in mice. *Science* 368 (6486): 85–89.

7 Hao, W., Wang, H., Mu, X. et al. (2018). Cancer radiosensitizers. *Trends in Pharmacological Sciences* 39: 1, 24–48.

8 Gill, M.R. and Vallis, K.A. (2019). Transition metal compounds as cancer radiosensitizers. *Chemical Society Reviews* 48 (2): 540–557.

9 Gabizon, A.A., Patil, Y., and La-Beck, N.M. (2016). New insights and evolving role of pegylated liposomal doxorubicin in cancer therapy. *Drug Resistance Updates* 29: 90–106.

10 Steuer, C.E., Behera, M., Ernani, V. et al. (2017). Comparison of concurrent use of thoracic radiation with either carboplatin-paclitaxel or cisplatin-etoposide for patients with stage III non-small-cell. *Lung Cancer* 3 (8): 1120–1129.

11 Chovanec, M., Zaid, M.A., Hanna, N. et al. (2017). Long-term toxicity of cisplatin in germ-cell tumor survivors. *Annals of Oncology* 28 (11): 2670–2679.

12 Gottschalk, A., Sharma, S., Ford, J. et al. (2010). Review article: the role of the perioperative period in recurrence after cancer surgery. *Anesthesia & Analgesia* 110 (6): 1636–1643.

13 Weiwei, M., Chu, Q., Liu, Y., and Zhang, N. (2020). A review on nano-based drug delivery system for cancer chemoimmunotherapy. *Nano-Micro Letters* 12: 1–24.

14 Gu, Z., Wang, Q., Shi, Y. et al. (2018). Nanotechnology-mediated immunochemotherapy combined with docetaxel and PD-L1 antibody increase therapeutic effects and decrease systemic toxicity. *Journal of Controlled Release* 286: 369–380.

15 Liu, Y., Chen, X.G., Yang, P.P. et al. (2019). Tumor microenvironmental pH and enzyme dual responsive polymer-liposomes for synergistic treatment of cancer immuno-chemotherapy. *Biomacromolecules* 20: 882–892.

16 Lu, J.Q., Liu, X.S., Liao, Y.P. et al. (2018). Breast cancer chemo-immunotherapy through liposomal delivery of an immunogenic cell death stimulus plus interference in the IDO-1 pathway. *ACS Nano* 12: 11041–11061.

17 Su, Z.W., Xiao, Z.C., Wang, Y. et al. (2020). Codelivery of anti-PD-1 antibody and paclitaxel with matrix metalloproteinase and pH dual-sensitive micelles for enhanced tumor chemoimmunotherapy. *Small* 16: 1906832.

18 Hernandez-Gil, J., Cobaleda-Siles, M., Zabaleta, A. et al. (2015). An iron oxide nanocarrier loaded with a Pt(IV) prodrug and immunostimulatory dsRNA for combining complementary cancer killing effects. *Advanced Healthcare Materials* 4: 1034–1042.

19 Lee, I.H., An, S., Yu, M.K. et al. (2011). Targeted chemoimmunotherapy using drug-loaded aptamer-dendrimer bioconjugates. *Journal of Controlled Release* 155: 435–441.

20 Xia, C., Yin, S., Xu, S. et al. (2019). Low molecular weight heparin-coated and dendrimer-based core-shell nanoplatform with enhanced immune activation and multiple anti-metastatic effects for melanoma treatment. *Theranostics* 9: 337–354.

21 Shen, S., Li, H.J., Chen, K.G. et al. (2017). Spatial targeting of tumor-associated macrophages and tumor cells with a pH-sensitive cluster nanocarrier for cancer chemoimmunotherapy. *Nano Letters* 17: 3822–3829.

22 Ou, W., Byeon, J.H., Thapa, R.K. et al. (2018). Plug-and-play nanorization of coarse black phosphorus for targeted chemo-photoimmunotherapy of colorectal cancer. *ACS Nano* 12: 10061–10074.

23 Chen, L., Zhou, L.L., Wang, C.H. et al. (2019). Tumor-targeted drug and CpG delivery system for phototherapy and docetaxel-enhanced immunotherapy with polarization toward M1-type macrophages on triple negative breast cancers. *Advanced Materials* 31: 1904997.

24 Wu, X., Wu, Y., Ye, H. et al. (2017). Interleukin-15 and cisplatin co-encapsulated thermosensitive polypeptide hydrogels for combined immuno-chemotherapy. *Journal of Controlled Release* 255: 81–93.

25 Dong, X., Yang, A., Bai, Y. et al. (2020). Dual fluorescence imaging-guided programmed delivery of doxorubicin and CpG nanoparticles to modulate tumor microenvironment for effective chemo-immunotherapy. *Biomaterials* 230: 119659.

26 Kadiyala, P., Li, D., Nunez, F.M. et al. (2019). High-density lipoprotein-mimicking nanodiscs for chemo-immunotherapy against glioblastoma multiforme. *ACS Nano* 13: 1365–1384.

27 Wu, M., Liu, X., Bai, H. et al. (2019). Surface-layer protein-enhanced immunotherapy based on cell membrane-coated nanoparticles for the effective inhibition of tumor growth and metastasis. *ACS Applied Materials & Interfaces* 11: 9850–9859.

28 Song, Q., Yin, Y., Shang, L. et al. (2017). Tumor microenvironment responsive nanogel for the combinatorial antitumor effect of chemotherapy and immunotherapy. *Nano Letters* 17: 6366–6375.

29 He, H., Liu, L., Morin, E.E. et al. (2019). Survey of clinical translation of cancer nanomedicines – lessons learned from successes and failures. *Accounts of Chemical Research* 52 (9): 2445–2461.

30 Beebe, S.J., Blackmore, P.F., White, J. et al. (2004). Nanosecond pulsed electric fields modulate cell function through intracellular signal transduction mechanisms. *Physiological Measurement* 25 (4): 1077.

31 Yin, S., Chen, X., Hu, C. et al. (2014). Nanosecond pulsed electric field (nsPEF) treatment for hepatocellular carcinoma: a novel locoregional ablation decreasing lung metastasis. *Cancer Letters* 346 (2): 285–291.

32 He Tianshuai, T.K., Sun, Q. et al. (2021). Development status of irreversible electric perforated tumor ablation device. *Chinese Journal of Medical Instrumentation* 656–658.

33 Cemazar, M., Miklavcic, D., Mir, L.M. et al. (2001). Electrochemotherapy of tumours resistant to cisplatin: a study in a murine tumour model. *European Journal of Cancer* 37 (9): 1166–1172.

34 Yasir, H.E., Tahir, O.A., Leena, B.M. et al. (2016). Metabolic syndrome and its association with obesity and lifestyle factors in Sudanese population. *Diabetes & Metabolic Syndrome* 10 (3): 128–131.

35 Pech, M., Janitzky, A., Wendler, J.J. et al. (2011). Irreversible electroporation of renal cell carcinoma: a first-in-man phase I clinical study. *Cardiovascular and Interventional Radiology* 34 (1): 132–138.

36 Wang, Y., Yin, S., Zhou, Y. et al. (2019). Dual-function of Baicalin in nsPEFs-treated hepatocytes and hepatocellular carcinoma cells for different death pathway and mitochondrial response. *International Journal of Medical Sciences* 16 (9): 1271–1282.

37 Qian, J., Chen, T., Wu, Q. et al. (2020). Blocking exposed PD-L1 elicited by nanosecond pulsed electric field reverses dysfunction of CD8+ T cells in liver cancer. *Cancer Letters* 495: 1.

38 Shi, J., Kantoff, P.W., Wooster, R., and Farokhzad, O.C. (2017). Cancer nanomedicine: progress, challenges and opportunities. *Nature Reviews Cancer* 17 (1): 20–37.

39 Alric, C., Taleb, J., Le Duc, G. et al. (2008). Gadolinium chelate coated gold nanoparticles as contrast agents for both X-ray computed tomography and magnetic resonance imaging. *Journal of the American Chemical Society* 136 (18): 5908–5915.

40 Robledo-Martinez, A., Vega, R., Cuellar, L.E. et al. (2007). Reversible, high-voltage square-wave pulse generator for triggering spark gaps. *Review of Scientific Instruments* 78 (5): 056104-2.

41 Dengbin, M. (2007). *Development of Nanosecond Pulse Generator Based on Blumlein Transmission Line*. Chongqing: Chongqing University College of Electrical Engineering.

42 Hofmann, G.A., Dev, S.B., Dimmer, S., and Nanda, G.S. (1999). Electroporation therapy: a new approach for the treatment of head and neck cancer. *IEEE Transactions on Biomedical Engineering* 46 (6): 752–759.

43 Yasir, H.E., Tahir, O.A., Leena, B.M., and Imam, S.N. (2016). Metabolic syndrome and its association with obesity and lifestyle factors in Sudanese population. *Diabetes and Metabolic Syndrome: Clinical Research and Reviews* 10 (3): 128–131.

44 Shi, S., Tang, Y., Zhao, Q. et al. (2022). Prevalence and risk of atrial fibrillation in China: a national cross-sectional epidemiological study. *The Lancet Regional Health–Western Pacific* 23: 100439.

45 Liqun, W., Congxin, H., and Dejia, J.H. (2020). Cryoballoon-based catheter ablation for atrial fibrillation: expert consensus in China. *Chinese Journal of Cardiac Pacing and Electrophysiology* 34: 95–97.

46 Hamani, C., McAndrews, M.P., Cohn, M. et al. (2008). Memory enhancement induced by hypothalamic/fornix deep brain stimulation. *Annals of Neurology* 63 (1): 119–123.

47 Laxton, A.W., Tang-Wai, D.F., McAndrews, M.P. et al. (2010). A phase I trial of deep brain stimulation of memory circuits in Alzheimer's disease. *Annals of Neurology* 68 (4): 521–534.

48 Ponce, F.A., Asaad, W.F., Foote, K.D. et al. (2016). Bilateral deep brain stimulation of the fornix for Alzheimer's disease: surgical safety in the ADvance trial. *Journal of Neurosurgery* 125 (1): 75–84.

49 Lozano, A.M., Fosdick, L., Chakravarty, M.M. et al. (2016). A phase II study of fornix deep brain stimulation in mild Alzheimer's disease. *Journal of Alzheimer's Disease* 54 (2): 777–787.

50 Zhao, X., Wang, J., Song, Y. et al. (2018). Synthesis of nanomedicines by nanohybrids conjugating ginsenosides with auto-targeting and enhanced MRI contrast for liver cancer therapy. *Drug Development and Industrial Pharmacy* 44 (8): 1307–1316.

51 Zhang, W., Zhao, X., Yuan, Y. et al. (2020). Microfluidic synthesis of multimode Au@CoFeB–Rg3 nanomedicines and their cytotoxicity and anti-tumor effects. *Chemistry of Materials* 32 (12): 5044–5056.

52 Xiaoxiong, Z. (2022). *Study on the Antitumor Efficacy and Mechanism of Multimode Iron-based Nanodrugs Combined with Nanosecond Pulse*. Beijing: School of Mathematics and Physics University of Science and Technology Beijing.

53 Xu, X., Song, Y., and Nallathamby, P. (2006). Probing membrane transport of single live cells using single-molecule detection and single nanoparticle assay. *Chemical Analysis* 172: 41.

54 Yu, J.S., Kim, S.H., Man, M.T. et al. (2019). Synthesis and dual-channel optical properties of Mn-doped ZnSe quantum dots. *Materials Letters* 253: 367–371.

55 Zhao, X., Liang, H., Chen, Y. et al. (2021). Magnetic field coupling microfluidic synthesis of diluted magnetic semiconductor quantum dots: the case of Co doping ZnSe quantum dots. *Journal of Materials Chemistry C* 9 (13): 4619–4627.

56 Yujun, S., Ruixue et al. (2011). Synthesis of well-dispersed aqueous-phase magnetite nanoparticles and their metabolism as an MRI contrast agent for the reticuloendothelial system. *European Journal of Inorganic Chemistry* (22): 3303–3313.

57 Tang, W., Zhen, Z., Yang, C. et al. (2014). Fe_5C_2 nanoparticles with high MRI contrast enhancement for tumor imaging. *Small* 10 (7): 1245–1249.

14

Nanomedicine-mediated Magneto-dynamic and/or Magneto-thermal Therapy

Yangfei Wang[1] and Yujun Song[1,2,3]

[1] University of Science and Technology Beijing, Center for Modern Physics Technology, School of Mathematics and Physics, 30 Xueyuan Road, Haidian District, Beijing 100083, China
[2] Zhengzhou Tianzhao Biomedical Technology Company Ltd., Zhengzhou New Technology Industrial Development Zone, 7B-1209 Dongqing Street, Zhengzhou 451450, China
[3] Key Laboratory of Pulsed Power Translational Medicine of Zhejiang Province, Hangzhou Ruidi Biotechnology Company Ltd., Room 803, Bldg. 4, 4959 Yuhangtang Road, Cangqian Street, Hangzhou 310023, China

14.1 Concepts, Treatment Theories, and Methods

14.1.1 Overview of Magnetic Hyperthermia Therapy

Magnetic-mediated hyperthermia (MMH), as a new type of tumor hyperthermia method, has become a hot topic in hyperthermia research in recent years. Hyperthermia has always been widely used in clinical practice. It uses physical "thermal effects" to improve blood supply and local pathological conditions in diseased areas. However, traditional thermal therapy methods often have problems such as complicated operations, inability to effectively control heat, and low heat conduction efficiency. The core concept of MMH is to position magnetic particles, especially nanoscale magnetic particles, in tumor tissue, and then apply an external alternating magnetic field to cause the material to be heated due to hysteresis, relaxation, or induced eddy currents. This heat is then transferred to the tumor tissue surrounding the material, causing the temperature of the tumor tissue to exceed 41 °C and also causing cell apoptosis and necrosis, thereby achieving a therapy for tumor treatment. Compared with traditional hyperthermia, magnetic hyperthermia has high safety, strong tissue penetration, controllable temperature and easy maintenance, and can carry out targeted and precise treatment of local lesions, which are not available in other traditional hyperthermia methods. With special advantages, tumor-related clinical research was carried out in the United States and Europe a few years ago, and it has been widely used in many biomedical fields such as tissue engineering, brain nerve regulation, and immunotherapy. At present, two main methods, arterial embolization magnetic hyperthermia and direct injection magnetic hyperthermia, have been developed. In addition, intracellular magnetic hyperthermia technology has also begun to develop rapidly. Because magnetic liquid has the advantages of targeting, minimally invasive, and

Nanomedicine: Fundamentals, Synthesis, and Applications, First Edition. Edited by Yujun Song.
© 2025 WILEY-VCH GmbH. Published 2025 by WILEY-VCH GmbH.

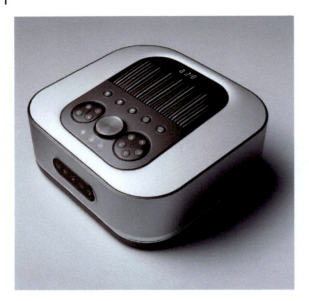

Figure 14.1 Magnetic heat therapy apparatus.

low toxic and side effects, MMH is gradually attracting people's attention as a new method of treating tumors [1].

A desktop magnetic vibration heat therapy instrument (dual-channel) is shown in Figure 14.1, which adopts a combination of three physical factors, namely, magnetic field, vibration, and warming. There are s two treatment modes, which can be selected by oneself according to the disease condition. The instrument has a certain effect on sensing nerve signals, promoting local blood circulation and revitalizing human cells. It is small in size and easy toperation, suitable for medical institutions and family use.

14.2 Treatment Theories and Methods

14.2.1 Magneto-thermal Conversion Mechanism

Under the action of an alternating magnetic field, magnetic nanoparticles absorb magnetic field energy to generate heat due to the influence of eddy current effects, hysteresis effects, magnetic aftereffects, and other factors.

The mechanism of the magnetothermal effect is schematically illustrated in Figure 14.2. When the frequency of the external magnetic field cannot cause significant eddy current heat generation, that is, it cannot reach the frequency at which the particles naturally resonate and generate significant heat, the magnetic nanoparticles in the alternating magnetic field absorb a large amount of the magnetic field mainly through hysteresis loss. Energy generates heat because magnetic nanoparticles have superparamagnetic properties. For hysteresis loss

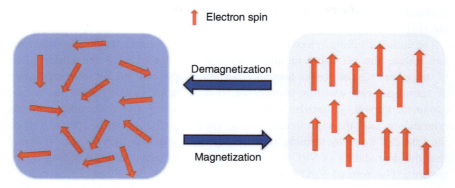

Figure 14.2 Magnetothermal effect: spin entropy changes in excitation and demagnetization states cause changes in system temperature.

heat generation, the smaller the size of the nanoparticles, the greater their coercive force and remanence, and thus the more heat they generate. Through the alternating magnetic field, when the magnetic energy of the magnetic liquid in the cells is converted into heat energy, the effect of hyperthermia can be achieved. The heating process of magnetic materials under an alternating magnetic field is actually a process of energy conversion and loss, that is, the material absorbs the energy of the alternating magnetic field and converts it into thermal energy. Magnetic materials are magnetized in an alternating magnetic field on the one hand, and on the other hand, they also consume energy in various ways.

14.2.2 The Killing Effect of Magnetic Hyperthermia on Tumor Cells

Research shows that the cell-killing effect directly caused by heat energy is essentially different from the cell death caused by radiotherapy. Its mechanism of action is reflected in the direct killing effect and indirect killing effect of heat therapy on tumor cells.

The direct killing effect is caused by hyperthermia treatment of the following target areas of tumor cells, mainly cell membranes, cytoskeleton, cellular proteins, and nucleic acids. Hyperthermia can change the fluidity and permeability of cell membranes, affect the intracellular environment, and hinder the functions of transmembrane transport proteins and cell surface receptors. Furthermore, the cholesterol content of tumor cell membranes is lower than that of normal cells, and the membrane fluidity is stronger, so more susceptible to the influence of heat; damage to the cytoskeleton mainly destroys cell morphology, mitotic apparatus, intracellular plasma membrane, etc. Hyperthermia inhibits the synthesis and polymerization of DNA and RNA, but RNA and protein synthesis recover quickly after hyperthermia stops, while DNA synthesis is inhibited [2]. It can also damage chromosomal proteins that bind to DNA, causing degeneration within the nuclear matrix. The aggregation of proteins further affects the functions of several molecules (including DNA replication, transcription, repair, hnRNA processing, etc.) [3].

The indirect killing effect is the first mean to kill tumor cells. Hyperthermia can reduce the pH value in tumor cells, which can increase the killing effect of hyperthermia on cells. Second, thermotherapy can inhibit the expression of tumor-derived vascular endothelial growth factor (VEGF) and the related products, impeding vascular endothelial proliferation and extracellular matrix remodeling and then inhibiting tumor growth and metastasis; Third, Hyperthermia can favor to forming more superoxide and free radicals. Li [4] found that the degradation of VitC and VitC sodium salts was enhanced during hyperthermia. The formed VitC free radicals increase, and the elevated free radicals also strengthen the anti-tumor effect of hyperthermia; then, high temperature promotes the activity of immune cells and the synthesis of cytokines, improving the body's immune function; finally, hyperthermia enhances apoptosis. Regulate the expression of genes, such as p53, FAS, Bax, Bcl-x, Bcl-2, etc., and induce cell apoptosis.

Figure 14.3 shows the application of magnetothermal effect in the medical field. Figure 14.3a shows the use of magnetothermal to destroy cancerous cells, while normal cells are not affected. Figure 14.3b is the use of magnetoheat to trigger drug delivery; Figure 14.3c is the control of material entry and exit through magnetothermal, thus controlling cell activity.

14.2.3 Cytotoxicity

Intracellular magnetic hyperthermia therapy must be carried out on the premise that the nanomagnetic particles are nontoxic. To this end, it is necessary to determine the applicable concentration range, action time, and coating modification method of nanomagnetic particles. In 2003, Hilger et al. pointed out that unmodified 8 nm iron oxide particles have no cytotoxic effect at low concentrations. When the particle concentration is increased and the culture time is prolonged, the cell survival rate decreases. After cationic modification of the surface of the ferromagnetic fluid, the chemical composition of the cell membrane was affected or its physiological pH

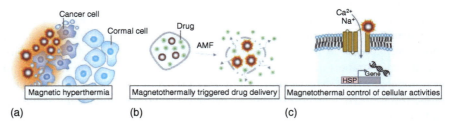

Figure 14.3 Application of magnetic heating in biomedicine. (a) Magnetothermal control of cancer cell apoptosis; (b) magnetothermal-triggered drug delivery; and (c) magnetothermal control of specific cell activity.

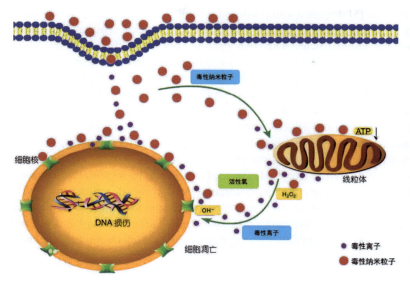

Figure 14.4 Toxic nanoparticles cause cell damage.

was changed due to its strong binding force with N-acetylceramide on the human adenocarcinoma cell membrane BT-20. This results in cytotoxicity, which is not limited to tumor cells but also causes toxicity to normal cells. Therefore, nanomagnetic particles are nontoxic at a certain concentration and for a certain action time, and a simple magnetic field will not affect the survival of cells. When coating and modifying cells, materials with good biocompatibility should be selected to avoid toxicity to cells. The mechanism of action of toxic ions or toxic nanoparticles causing apoptosis is shown in Figure 14.4.

14.3 Magnetic Field-responsive Nanomedicine Synthesis

14.3.1 Fe_3O_4 Materials

As the most widely used magnetic nanomaterial in the biomedical community, Fe_3O_4 nanoparticles have many attractive properties, such as low production cost, high biocompatibility, and strong magnetic response, and can be modified to increase its selectivity to the target tissue. Fe_3O_4 nanoparticles can produce different contrasts and can also be used in magnetic resonance imaging (MRI) detection. Its magnetism can also be used for stem cell tracking by labeling stem cells. The magnetic nanoparticles can be guided to the desired location by applying a magnetic field, and heating can be achieved through the magnetic field to accelerate the drug release. Therefore, Fe_3O_4 nanoparticles can be used as integrated nanoprobes for cancer diagnosis and treatment. While providing simple diagnosis and treatment, they can also customize more functions through surface functionalization.

Guldris et al. [5] designed and produced a multifunctional Fe_3O_4 nanoparticle platform that can perform clickable orthogonal reactions. They simply functionalized the nanoparticles with two ligands containing isocyanate moieties, providing high surface coverage with maleimide and alkynyl groups. Due to limitations in the characterization and purification of Fe_3O_4 nanoparticles, click reactions are an ideal solution for functionalization for biomedical applications. They loaded fluorophores as drug models via the Diels–Alder reaction and reacted with polyethylene glycol (PEG)-containing biotin ligands via an azide–alkyne cycloaddition reaction to demonstrate the applicability of the platform. Through the thermally reversible nature of the Diels–Alder reaction, application of external AMF enables temperature-triggered release of fluorescent molecules. Furthermore, the ability of Fe_3O_4 nanoparticles to enhance MRI contrast, which was key to verifying their tumor cell-targeting capabilities, was found to be enhanced by biotin. Validated by MRI, this hybrid platform shows potential to further exploit combined imaging and magnetothermal therapy applications to develop theranostic agents.

14.3.2 Ferrite Materials

Ferrite is an oxo acid salt, which generally includes oxo acid salts in four valence states: iron (III), iron (IV), iron (V), and iron (VI). Unless otherwise specified, it generally refers to iron (the oxygen-containing acid salt of III)is called ferrate while the salt of iron (VI) is called ferrate. The other two valence states are relatively rare. Ferrite is widely used in magnetic thermal therapy due to its high heat production rate and high saturation magnetic susceptibility. Because they are iron-based nanoparticles, they also have good MRI imaging effects. There have been many experiments to synthesize a variety of different ferrites for use in magnetothermal therapy and imaging, or for synergistic treatment with other therapeutic methods. Fe_3O_4 nanoparticles, especially magnetite, are the most commonly used magnetic nanoparticles for hyperthermia due to their lower toxicity than other magnetic materials, adjustable magnetism, relatively simple preparation, and good stability. Magnetite nanoparticles with an inverse spinel structure are the most studied ferrites in biomedicine. Currently, research is focused on developing hybrid ferrites by partially replacing Fe ions with other cations such as Zn, Ca, Mg, Mn, Co, Cu, Ca, and Ni. For example, $CoFe_2O_4$ has higher magnetocrystalline anisotropy than Fe_3O_4, which will lead to greater thermal energy dissipation during hyperthermia [6].

Demirci et al. [7] studied how to perform effective magnetic hyperthermia by adjusting the K value of Co-ferrite nanoparticles by replacing Fe^{3+} particles with La^{3+} particles. They studied the structure, composition, magnetism, and magnetocaloric reaction of $CoFe_{2-x}La_xO_4$ (x = 0.0, 0.2, 0.5) nanoparticles. Compared to unsubstituted Co-ferrite, the saturation magnetization (Ms) and K increase with increasing substitution concentration of La^{3+}. No change in SLP was observed under a magnetic field of 50 kA m^{-1} at 195 kHz when the particles were dispersed in deionized H_2O. But contrary to expectations, SLP enhanced with increasing viscosity. Explore the optimal alternative concentration of La^{3+} particles to obtain the best heating efficiency based on the K and Ms values obtained by tracking. Compared with the unsubstituted and 25% La^{3+} substituted Co-ferrate nanoparticles, the 10% La^{3+} ion-substituted Co-ferrite nanoparticles have higher SLP values, which may be due to the Néel relaxation account dominant position. By replacing different types of rare earth ions with Co-ferrite nanoparticles, it is very important to explore the optimal La ion concentration and the effect of adjusting K.

14.3.3 Alloy Materials

As one of the best magnetocaloric materials, Fe-containing nanoalloys are also an important type of magnetocaloric material. As a representative among them, FeCo has high saturation Ms, high Curie temperature, high permeability, and good physical properties. So, it has attracted widespread attention and applications. Çelik and Fırat [8] synthesized FeCo colloidal magnetic nanoalloys of different sizes through surfactant-assisted ball milling, with controllable sizes ranging from 11.5 to 37.2 nm. The particle distribution was measured by TEM, and the FeCo alloy was verified to have a BCC (body-centered cubic) structure by XRD measurement. The magnetic properties of the nanoalloys were then studied through the vibrating sample magnetometer (VSM) method, which confirmed that the magnetic properties of the nanoparticles are related to size. The saturation Ms intensities of FeCo nanoparticles with average sizes of (11.2 ± 2.1), (13.7 ± 3.3), (17.2 ± 3.6), and (37.4 ± 2.4) nm are 145, 156, 165, and 172 $Am^2\,kg^{-1}$, respectively. As the size decreases, the saturation Ms of the ball-milled nanoparticles decreases relative to the overall structure of the nanoparticles, and their Ms curve proves the typical single-domain ferromagnetic behavior of FeCo nanoparticles. As the size of the magnetic material becomes smaller compared to the critical size, the magnetic particles become single domains reducing their magnetic properties. The coercive force of single-domain nanoparticles strongly depends on the size. The coercive forces of the above-mentioned nanoparticles of different sizes are 17.5, 19.4, 17.8, and 16.0 mT, respectively. The coercive force is related to the magnetic anisotropy. Finally, they used LRT and SW models to simulate the heating mechanism of nanoparticles under AMF. For FeCo nanoparticles with an average size of (13.3 ± 1.6) nm, under a magnetic field frequency of 171 kHz and a magnetic field strength of 14 mT, the SAR value was (15.0 ± 0.3) $W\,g^{-1}$, which is smaller than the theoretically calculated value. From the results of magnetic hyperthermia therapy, it was found that the heat generation mechanism is similar to the Stoner–Wohlfarth method, and the theoretical prediction of the step-like form is consistent with the magnetic field dependence of SAR loss.

14.3.4 Composite Materials

Magnetic nanoparticles exhibit unique physical and chemical properties, making them widely studied in the medical field. However, their tendency to aggregate easily also results in the loss of some of their unique properties and application potential. Organic/inorganic nanocomposites combine the unique properties of nanoparticles with the advantages of polymer matrices (including lightweight and easy molding). The usual method is to embed magnetic nanoparticles in polymers to create injectable preparations and magnetic hyperthermia for local delivery. Colloidal suspensions of nanoparticles dispersed in polymer solutions can prevent flocculation in polymer solutions or polymer melts. Nanocomposite films can also be prepared by spin coating and precipitation of nanocomposites. The materials themselves are easily ground into the powder form.

Gyergyek et al. [9] proposed a simple method to prepare highly uniformly dispersed magnetic Fe_3O_4/polymethacrylic acid (PMMA) nanocomposites. Methacrylate-monomer-functionalized magnetic Fe_3O_4 nanoparticles were suspended in a colloidal suspension. Copolymerized with methyl methacrylate (MMA) monomer. The developed copolymerization method has two advantages: first, the magnetic nanoparticles are decorated with PMMA chains, which provide colloidal stability in the polymer solution and prevent aggregation of the nanoparticles during preparation; second, the PMMA matrix is synthesized in one pot. At the same time, magnetic nanoparticles were formed and evenly dispersed in the PMMA matrix, and the mass fraction reached 53%. Composite material, is simple and convenient. Nanocomposites precipitated from suspension with PMMA chains are firmly attached to the surface of the nanoparticles. Hysterisis analysis shows that it has a superparamagnetic property and exhibits a large value when the saturation Ms is as high as 36 emu g^{-1}. Nanocomposites incorporating a large number of nanoparticles generate large amounts of heat at relatively low amplitude AMF. So, it also has potential for cancer treatment applications based on magnetic induction hyperthermia.

14.3.5 Other Materials

In addition to commonly used Fe-containing materials, people are also looking for other substances with better magnetocaloric effects than Fe to prepare nanomaterials. For example, a type of perovskite structural material has recently received great attention from the hyperthermia field, namely $La_{1-x}Sr_xMnO_3$(LSMO) nanoparticles. $La_{1-x}Sr_xMnO_3$ is a ferromagnetic material that generates heat during magnetic hyperthermia therapy. Compared with iron compounds, $La_{1-x}Sr_xMnO_3$ has a lower Curie temperature, better biocompatibility, high Ms, high stability, and low coercive force. These characteristics make this material attract widespread attention. Researchers have explored many synthesis methods of $La_{1-x}Sr_xMnO_3$ nanoparticles, including sol–gel, hydrothermal synthesis, and molten salt methods. Lotfi et al. [10] proposed a simple synthesis method of $La_{1-x}Sr_xMnO_3$ nanoparticles through sol–gel technology, and successfully prepared four kinds of $La_{1-x}Sr_xMnO_3$ nanoparticles with $x = 0.2, 0.3, 0.4$, and 0.5, and A series of investigations were conducted into their magnetic properties. The structure, morphology and magnetism of the nanoparticles were studied using FTIR, XRD, SEM, TEM, and VSM characterization methods, which proved that the $La_{1-x}Sr_xMnO_3$ crystal structure is in line with expectations, most of the nanoparticles are in an aggregated state, and when $x = 0.2$, The Ms value of $La_{1-x}Sr_xMnO_3$ nanoparticles is higher than that of other samples, which proves that the concentration of Sr affects the Ms of nanoparticles. These experimental results indicate that superparamagnetic perovskite $La_{1-x}Sr_xMnO_3$ nanoparticles with low toxicity have high potential in biomedical applications, especially magnetic hyperthermia applications, and are strong competitors as magnetocaloric materials.

14.4 Applications

14.4.1 Instance 1

In 2021, the Maastricht University Medical Center in the Netherlands published an article using noninvasive magnetic thermal deep brain stimulation technology to improve the motor function symptoms of Parkinson's model mice. The researchers optimized the alternating magnetic field coil through experiments, so that these magnetic nanoparticles can exhibit high SLP in a short time and improve the magnetocaloric effect.

First, the researchers used genetic engineering to package TRPV1 in a viral vector with a neuronal promoter, so that neurons can express heat-activated ion channel proteins. Specially prepared magnetic nanoparticles are then injected into the same area of the brain. These nanoparticles adhere to the surface of the target neurons. When the mouse is exposed to an alternating magnetic field, the magnetic field will cause the magnetized nanoparticles to rapidly flip, generating heat energy. The heat forces temperature-sensitive ion channels to open, stimulating neurons.

Through the above experimental operations (Figure 14.5), the researchers demonstrated that magnetic thermal stimulation of the subthalamic nucleus could promote the rotational movement function of mice. Subsequent immunofluorescence experiments found that primary motor cortex neurons were activated, but nonmotor-related amygdala brain areas did not appear Subsequently, in two Parkinson's mouse models, magnetic stimulation of the subthalamic nucleus could rapidly and persistently improve the motor dysfunction of the mice. Previous optogenetic experiments (activating or inhibiting) subthalamic nucleus neurons did not affect Parkinson's dyskinesia. This article reveals that magnetic activation of the neural circuits of the subthalamic nucleus significantly improves movement disorders. At the same time, neurons in the motor cortex are also activated. This shows that the effect of magnetic stimulation of the subthalamic nucleus is not limited to this brain area, but will spread to the brain area associated with the subthalamic nucleus to exert the above therapeutic effects [11].

14.4.2 Instance 2

In recent years, a new iron oxide nanoring with a vortex magnetic domain structure has been developed. Due to its small size boundary effect and special structure, its magnetic moment is distributed in a vortex in the clockwise or counterclockwise

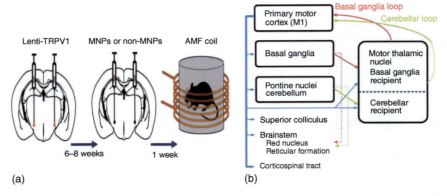

Figure 14.5 (a) In vivo experimental scheme. (b) Major central nervous system regions involved in control of the motor system. The primary motor cortex (M1) has subcortical descending pathways to the corticospinal tract. Axons of these pathways also target subcortical motor nuclei, such as the red nucleus and brainstem reticular formation, which serve as the origin of the descending rubrospinal and reticulospinal tracts. Axonal collaterals target structures including the pontine nuclei, the superior colliculus, the basal ganglia, and the thalamus. Basal ganglia (red) and cerebellum (gold) form two major loops which can influence M1 and thus motor function.

direction in the plane, forming unique multi-domain structure of closed distribution. It has the sole dispersion of superparamagnetic iron oxide nanoparticles and the external field-induced vortex-onion state. The magnetization reversal will result in a large hysteresis loss and the excellent magnetic properties are close to bulk materials. Under the action of an alternating magnetic field, its magnetothermal conversion efficiency is an order of magnitude higher than that of superparamagnetic nanoparticles. Using tumor-bearing mice as a model, the magnetothermal treatment time is shortened from 60 minutes in clinical practice to 10 minutes, and the dose is reduced from 5 to 10 mg. Fe/kg tumor tissue was reduced to 0.5 mg Fe/kg tumor tissue, proving that materials with high magnetothermal conversion efficiency can significantly improve tumor magnetic hyperthermia therapy technology, and reduce dosage and toxic side effects.

14.4.3 Instance 3

Gao Fei and others from the Institute of Integrative Medicine of Shaanxi University of Traditional Chinese Medicine combined traditional hyperthermia concepts with magnetic induction technology to originalize the use of magnetic thermal therapy to treat knee osteoarthritis (KOA) and explore its mechanism to establish a system with modern medical characteristics A primary KOA model in rats was established and a high-efficiency magnetic hyperthermia therapy medium was then constructed based on vortex magnetic iron oxide nanorings. Under low-intensity magnetic field conditions, precise temperature-controlled heating of the KOA lesion site can be achieved through pain and knee joint tissue. Pathology, knee joint morphology, MicroCT microscopic bone structure changes, and other clinical indicators were detected, as well as serum inflammatory factor levels were measured to observe the efficacy and mechanism of magnetic heat treatment of KOA. After magnetic heat treatment, the mechanical pain threshold of rats increased by approximately 48.9% compared with the model group; the degree of congestion, edema, and cartilage surface wear of the joint capsule and synovial tissue were significantly reduced compared with the model group; the Mankin and OARSI scores were reduced by approximately 33% and 20%; MicroCT results showed that the degree of sclerosis of subchondral bone was improved; the levels of inflammatory factors in serum were reduced. Vortex magnetic iron oxide nanorings are used to design and construct high-efficiency magnetic hyperthermia nanopreparations, which have definite efficacy in the treatment of KOA, and the mechanism is related to the inhibition of inflammatory factors [12].

References

1 Laha, S.S., Thorat, N.D., Singh, G. et al. (2022). Rare-earth doped iron oxide nanostructures for cancer theranostics: magnetic hyperthermia and magnetic resonance imaging. *Small* 18 (11): 2104855.
2 Hildebrandt, B., Wust, P., Ahlers, O. et al. (2002). The cellular and molecular basis of hyperthermia. *Critical Reviews in Oncology/Hematology* 43 (1): 33–56.
3 Cummings, M. (1995). Increased c-Fos expression associated with hyperthermia-induced apoptosis of a Burkitt lymphoma cell line. *International Journal of Radiation Biology* 68 (6): 687–692.
4 Li, J. and Oberley, L.W. (1997). Overexpression of manganese-containing superoxide dismutase confers resistance to the cytotoxicity of tumor necrosis factor α and/or hyperthermia. *Cancer Research* 57 (10): 1991–1998.
5 Guldris, N., Gallo, J., García-Hevia, L. et al. (2018). Orthogonal clickable iron oxide nanoparticle platform for targeting, imaging, and on-demand release. *Chemistry: A European Journal* 24 (34): 8624–8631.
6 Rosensweig, R.E. (2002). Heating magnetic fluid with alternating magnetic field. *Journal of Magnetism and Magnetic Materials* 252: 370–374.
7 Demirci, Ç.E., Manna, P.K., Wroczynskyj, Y. et al. (2018). Lanthanum ion substituted cobalt ferrite nanoparticles and their hyperthermia efficiency. *Journal of Magnetism and Magnetic Materials* 458: 253–260.
8 Çelik, Ö. and Fırat, T. (2018). Synthesis of FeCo magnetic nanoalloys and investigation of heating properties for magnetic fluid hyperthermia. *Journal of Magnetism and Magnetic Materials* 456: 11–16.
9 Gyergyek, S., Pahovnik, D., Žagar, E. et al. (2018). Nanocomposites comprised of homogeneously dispersed magnetic iron-oxide nanoparticles and poly (methyl methacrylate). *Beilstein Journal of Nanotechnology* 9 (1): 1613–1622.
10 Lotfi S, Bahari S, Bahari A, et al., Magnetic performance and evaluation of radiofrequency hyperthermia of perovskite $La_{1-x}Sr_xMnO_3$ *Journal of Superconductivity and Novel Magnetism*, 2018, 31(1): p. 2187–2193.

11 Hescham, S., Chiang, P.H., Gregurec, D. et al. (2021). Magnetothermal nanoparticle technology alleviates parkinsonian-like symptoms in mice. *Nature Communications* 12 (1): 5569.

12 Gao, F., Longlong, D., Wang, T. et al. (2023). Clinical observation and mechanism of magnetothermal therapy in the treatment of knee osteoarthritis. *Journal of Xi'an Jiaotong University (Medical Edition)* 44 (5): 784–793.

15

Nanomedicine-mediated Radiotherapy for Cancer Treatment

Ruixue Zhu[1] and Yujun Song[1,2,3]

[1] University of Science and Technology Beijing, Center for Modern Physics Technology, School of Mathematics and Physics, 30 Xueyuan Road, Haidian District, Beijing 100083, China
[2] Zhengzhou Tianzhao Biomedical Technology Company Ltd., Zhengzhou New Technology Industrial Development Zone, 7B-1209 Dongqing Street, Zhengzhou 451450, China
[3] Key Laboratory of Pulsed Power Translational Medicine of Zhejiang Province, Hangzhou Ruidi Biotechnology Company Ltd., Room 803, Bldg. 4, 4959 Yuhangtang Road, Cangqian Street, Hangzhou 310023, China

15.1 Concept, Therapy Theory, and Devices

Cancer, or termed malignant tumors and neoplasms, is a genetic disorder manifested as uncontrollable growth of abnormal cells beyond the usual boundaries, which can invade and spread to adjoining parts and other organs of the body, referred to as metastasis. Cancer is one of the major causes of death globally recognized by the World Health Organization (WHO), accounting for nearly 10 million deaths in 2020. If detected early and effectively treated, many cancers can be cured. Despite the great progress in cancer clinical treatments including operation, chemical medication, RT, and immunotherapy, the overall five-year survival rate of cancer has only risen from 49% in the mid-1970s to 68% between 2012 and 2018 [1]. Radiation therapy or radiotherapy, abbreviated RT, is one of the major cancer treatment entities using ionizing radiation to destroy, inactivate, or control the growth of malignant cells by damaging the DNA of cancerous tissues while controlling the dosage absorbed by healthy tissues to maintain their integrity [2]. It is estimated that 31% of cancer patients utilize RT as the primary treatment option for the first course of treatment, and over 50% of patients receive RT for disease management during the treatment period worldwide [3]. In China, approximately 50% of cancer incidence require RT as part of their definitive treatment [4]. Considering the statistics of about 4.29–4.82 million Chinese newly diagnosed with cancer each year between 2015 and 2022, approximately 2.15–2.41 million cancer patients receive RT every year [4, 5]. This number is likely to be higher as it does not include data on cancer recurrence and palliative treatment. As of 2019, there were a total of 1463 hospitals or research institutes registered in China engaged in the RT, with about 1.5 accelerators per million people [4].

The principle of radiation therapy can be understood from the nature of ionizing radiation and its interaction with living matter. Ionizing radiation is a form of high-energy radiation that is sufficiently strong to break down the chemical bonds in atoms and molecules, and displace or detach electrons from them [6]. This process creates ions (atoms with a positive charge) and free radicals, which can damage cells and tissues through their interaction with biological objects. Ionizing radiation can be in electromagnetic (EM) form, such as high-energy photons with no mass, or in particulate form, such as charged particles including electrons, protons, alpha (positively charged), and beta (negatively charged) particles as well as uncharged particles like neutrons [7]. The high-energy photons including X-rays and gamma-rays in high-frequency range of EM radiation are the most common types of ionizing radiation used in practice, and can be generated electrically by clinical linear accelerators or through spontaneous fission or fusion reactions of certain radioactive nuclei, respectively [7]. Particulate ionizing radiation consisting of atomic or subatomic particles often results from the radioactive decay of the nuclei of radioactive atoms [8]. When these high-energy photons or particles collide with biological objects such as cells, tissues, and organs, they can cause irreparable or mis-repaired damage to the DNA and other molecules in the object. Ionization of molecules can lead to radioactive decomposition (breakdown of chemical bonds) and the formation of highly reactive free radicals, which may react chemically with neighboring materials even after the original radiation has ceased [9]. The damage to the DNA can lead to mutations in the DNA, which can result in changes in the genetic code of the cell and potentially lead to a series of severe biological consequences including cell death, mutagenesis, chromosomal aberrations, carcinogenesis, or other diseases [10]. Radiation effects on tissue are divided into early and late effects, a brief summary of normal cellular responses to ionizing radiation is shown in Table 15.1, and more specific information can be found in Refs. [11–16].

The interaction between ionizing radiation and biological objects depends on several factors, including the energy of the radiation, the types, density, and composition of the target organs and cells, and the distance between the object and the source of the radiation. The linear energy transfer (LET) is the most commonly accepted characteristic quality parameter for the RT, and ranges from low-LET (about $1\,keV\,\mu m^{-1}$, such as beta particles and X- and gamma-rays) to slightly high (about $10\,keV\,\mu m^{-1}$ for RT protons, or around 10–$100\,keV\,\mu m^{-1}$ for carbon ions) and extremely high values ($>100\,keV\,\mu m^{-1}$ for alpha-particles or heavy charged particles as space radiation) [17]. At lower energies (below $10\,keV\,\mu m^{-1}$), ionizing radiation is typically absorbed by atoms in biological objects, causing them to become excited and ultimately decay by emitting gamma-rays. At higher energies ($>10\,keV\,\mu m^{-1}$), ionizing radiation can penetrate deeper into objects and interact with electrons in molecules, causing chemical reactions that can damage DNA [18].

In general, RT can be performed inside or outside the human body and thus be categorized as external beam RT (EBRT) and internal RIT or brachytherapy. EBRT utilizes a radiation machine to carefully deliver tightly targeted radiation beams from outside the body, and usually involves several treatment fractions over a period (from a few days to a few weeks). Photon radiation (such as X-rays and gamma-rays) and electron beams that have low penetration power are used for the treatment of superficial cancer cells, and particulate radiation (such as proton and neutron beams) is

Table 15.1 A brief summary of the response of normal cells to ionizing radiation.

Objects	Responses to ionizing radiation (IR)
Bone marrow	High doses exposure will lead to bone marrow failure and death. Sub-lethal doses will cause bone marrow suppression and will cause immune suppression due to abnormal numbers of functional blood cells
Skin	The most radiosensitive cells of the integumentary system are the basal keratinocytes, stem cells of the hair follicle, and melanocytes. Upon exposure, destroyed basal keratinocytes will disrupt the skin's ability to self-renew; the stem cells of the hair follicle are injured due to damage to nuclear and mitochondrial DNA induced by free radicals
Brain	The brain is sensitive to IR due to high density of DNA-containing cells and tissues. IR can modulate responses of microglial cells in the central nervous system (CNS) and high doses can induce reactive oxygen species (ROS) formation, oxidative stress, and neuroinflammation
Intestine	IR can promote bacterial translocation across the intestinal barrier. The small intestines exposed to high levels of IR exhibit characteristic morphological changes including decreased villi height and irregular, shortened microvilli, cytoplasmic vacuolization, and detachment of epithelial and endothelial cells from their basement membrane
Muscles	Adult and developing skeletal muscles are resistant and highly sensitive to IR, respectively; thus, RT in childhood may induce muscular atrophy. Radiation affects the activation, proliferation, and differentiation of muscle satellite cells, and damaging neuromuscular junctions by influencing the membrane permeability of Na^+/K^+ pump
Capillaries	Capillary endothelium responds to radiation by leukocyte attachment, endothelial cell swelling, and increased capillary permeability. Characteristic changes in capillary histology after irradiation include detachment of endothelial cells from the basal lamina, cell pyknosis, thrombosis, and loss of entire capillary segments, resulting in tissue ischemia or regrowth of lost vessels in some organs

very effective in treating deep tumors [19]. RIT usually involves introducing small radioactive seeds inside the body, into or adjacent to the tumor or affected area, allowing the radiation to reach the treatment area directly. The isotopes are usually administered intravenously and travel throughout the body with the bloodstream, where they accumulate in the cells targeted for treatment. Currently, the radionuclides most used in the RIT emit beta particles with a low tissue penetration range, thereby reducing the risk of damaging nearby normal tissues [20]. It is noteworthy that most modern RIT enhance the targeted binding of isotopes to tissues or specific lesions through molecular targets, such as targeting metabolic pathways, ligand receptors, or specific antibody for its corresponding antigen [20].

RT has been used for over a century to treat various types of cancer, including lung, breast, prostate, and brain tumors. The development of RT has been a major milestone in cancer treatment, as it has greatly improved the survival rates of patients with advanced cancer. The application of radiation in medical treatment can be dated back to 1896, the second year after Röntgen discovered X-rays (1895) [21].

Soon after the discovery of polonium and radium by Maria Sklodowska-Curie (1867–1934) and Pierre Curie (1859–1906) in 1898, radium was successfully used in the treatment of cutaneous lupus [22]. The invention of Snook's "interrupterless" transformer in 1907 allowed for higher energy output under better control, further promoted the practical application of X-ray [21]. The first decade of the twentieth century witnessed the invention of a series of equipment, including X-ray tubes (in 1913 by William David Coolidge), and it was until the 1920s and 1930s that RT became widely accepted as a treatment option for cancer [23]. During this time, researchers developed new techniques for delivering radiation to the body, such as the use of machines to rotate X-ray beams around the patient and the use of isotopes to enhance the effectiveness of radiation therapy. From the 1950s to the early 1980s, cobalt therapy machines (using cobalt-60 sources for emitting gamma-rays) were used in the field of radiation therapy, allowing for the treatment of deep and refractory cancers, while X-rays were only suitable for the treatment of superficial tumors [24]. On the other hand, the medical linear accelerators have gradually entered the historical stage since 1953 and replaced the cobalt units with the possibility to generate higher energies [25]. In the following decades, RT continued to evolve and improve. Researchers developed new techniques for targeting specific types of cancer cells, such as the use of proton therapy to deliver high-energy radiation directly to cancer cells without affecting healthy tissue. Technological advances in new imaging modalities (such as magnetic resonance imaging (MRI), spiral computed tomography (CT), positron emission tomography (PET), etc.), powerful computers and software, and novel delivery systems have made it possible to more accurately locate and target cancerous areas within the body with better dose distribution [26]. Nowadays, diverse RT treatment strategies have been developed and well-applied in many cancer treatment programs, including 3D conformal RT (3DCRT), intensity-modulated radiation therapy (IMRT), image-guided RT (IGRT), and stereotactic body radiation therapy (SBRT) [27]. In summary, a brief list of the major milestones in cancer RT is shown in Table 15.2.

In practice, RT can be used alone or in combination with other treatments, such as chemotherapy and surgery, and can also be used to control the spread of cancer in populations who have already received other treatments but have a high risk of recurrence. Overall, the constant effort in the development of RT has been increasingly successful in cancer treatment and has saved countless lives around the world. As research continues, it is likely that RT will continue to evolve and become even more effective in treating a wide range of cancers.

Despite the successful development and wide use of RT for cancer treatment, it still has certain shortcomings that are widely acknowledged by medical professionals and researchers, those can be summarized into five main aspects: accessibility, side effects, limited effectiveness, risk of secondary malignancies, and cost. First of all, radiation therapy machines require large equipment and professional personnel to operate, making it difficult for the treatment of patients in remote areas with insufficient medical facilities or medical services [28]. Secondly, the side effects of RT arising from the unavoidable damage delivered to the normal, noncancerous cells pose a major limitation of RT. The side effects of RT may vary depending on the type and duration of treatment and mainly include fatigue, skin changes

Table 15.2 A brief summary of the major milestones in cancer radiotherapy.

Year	Inventor	Milestones
1895	W. Röntgen	Discovery of X-rays
1896	E. Grubbe, V. Despeignes	First attempt to use X-rays to cure cancer
1898	M. Sklodowska-Curie, P. Curie	Discovery of polonium and radium
1899	Tage Sjögren	First successful treatment of cancer patient with X-rays
1900	Thor Stenbeck	Healing of skin cancer with small doses of radiation
1903	S.W. Goldberg, E.S. London	Histologically proven cure of skin cancer with radiation
1905	R. Abbe	Cure of cervical cancer with radium sources
1906	J. Bergonie, L. Tribondeau	Cell radiosensitivity law
1913	W.D. Coolidge	First X-ray tube
1922	H. Coutard	Verified the minimization of side effects of radiation
1928	R. Wideröe	Radiofrequency linear accelerator for ions
1930	R.F. Mottram	Oxygen effect on radiosensitivity
1933	H. Crabtree, W. Cramer	Oxygen effect in RT
1934	I. Joliot-Curie, F. Joliot-Curie	Discovery of artificial radioactivity
1938	R. Stone	First treatment with neutrons
1951	I. Smith, H.E. Johns	Cobalt-60 teletherapy
1951	L. Leksell	"Gamma knife" radiosurgery
1953	Metropolitan-Vickers	First medical linear accelerator (Hammersmith)
1956	H.S. Kaplan, E. Ginzton	Medical linear accelerator (Stanford)
1965	D. O'Connell, M. Wakabayashi	Cobalt-60 sources for high dose rate brachytherapy
1973	G. Hounsfield	X-ray computed tomography (CT) and first CT scanner
1980	P. Bottomley	First commercial magnetic resonance imaging (MRI) device
1994	N. Peacock	Intensity modulated radiation therapy (IMRT)
2001	A.L. Boyer	Volumetric modulated arc therapy (V-MAT)
2006	J.A. Bonner et al.	Targeted therapy and modern RT

(i.e. redness, swelling, and blistering), temporary or permanent hair loss, digestive problems (i.e. diarrhea, constipation, and abdominal pain), mouth sores and ulcers, cognitive changes (i.e. memory loss, difficulty concentrating, and mood changes), and infertility [29]. Long-term exposure to radiation may even increase the risk of developing secondary cancers such as lung cancer, head and neck cancer, and leukemia. In addition, limited effectiveness of RT as it may not be effective in all types of cancer (especially in metastatic cancer) and relatively high cost are also challenges needed to be considered in choosing radiation therapy [30]. Particularly, since only a small portion of the radiation energy is absorbed by tumors as they are soft tissues, high intensity radiation beams are usually required for EBRT to effectively kill the tumor, which poses a high risk of damaging adjacent normal tissues in the beam path of radiation (such as X-rays). On the other hand, achieving targeted delivery of radioactive isotopes to malignant tumors in the body and avoiding systemic radiotoxicity to normal organs is an important challenge for RIT.

Nanotechnology is a multidisciplinary field. Nanomaterials are two- or three-dimensional microstructures with sizes ranging from a few nanometers to 100 nm [31]. To date, depending on different dimension and geometry, diverse types of nanomaterials have been studied, including NPs, nanofibers, nanotubes, nanorods (NRs), nanofilms, nanoribbons, and bulk nanomaterials. Those nanostructures can serve as imaging reporter for tumor display [32], delivery capsules of photosensitizers in photodynamic therapy [33], or as artificial enzymes or biomarkers in biosensing, bioimaging, and tumor diagnosis [34].

The revolutionary development in nanotechnology and nanomedicine has encouraged tremendous progress in the design of in vivo/in vitro diagnostic/therapeutic tools utilizing nanostructures with various types and compositions. Nanotechnology-mediated radiation therapy has emerged as one of the most advantageous fields in the treatment of malignant tumors in recent decades as there is a great potential in utilizing nanomaterials to enhance radiation responses and overcome severe systemic toxicity, thereby achieving enhanced RT for cancer treatment [32]. Nanomaterials can not only be used as therapeutic units themselves based on their special characteristics in RT, but can also serve as in vivo delivery carriers for other therapeutic agents to enhance the targeting of treatment, thus playing a crucial role in tumor treatment [35]. For example, metallic NPs including gold NRs (GNRs) or silver nanoclusters (SNCs) that contain high atomic number (high-Z) elements (i.e. Cu, Ag, Pt, and Au) can act as radiosensitizers as they have a higher stopping power for ionizing radiation (or strong X-ray attenuation ability) than soft tissue [36]. The radiation energy during external RT can be better deposited within tumors under the action of such nanoradiosensitizers. Additionally, nanomaterials can be chemically modified as multifunctional nanocarriers for active tumor targeting, providing in vivo delivery of therapeutic radioisotopes for RIT [37]. Moreover, the special tumor microenvironment (TME) characterized by insufficient oxygenation (hypoxia), acidic pH (acidosis), and elevated levels of glutathione (GSH) and hydrogen peroxide (H_2O_2) can be modulated by various nanomedicines, thus overcoming hypoxia-related radioresistance and enhancing the efficacy of RT [35].

The application of nanotechnology in cancer treatment has been developed for more than a decade, and new advances are constantly being proposed. In this chapter, the concept, therapy theory, devices, and recent developments in the context of nanomedicine-mediated radiation therapy will be summarized and presented from the aspects of nanomaterials as radiosensitizers for RT, nanomaterials delivering radioisotopes for RIT, nanomaterials for chemo-RT, nanomaterials for thermo-RT, and nanomaterials for cancer radioimmunotherapy. Other emerging strategies for improved RT with nanomedicines will also be introduced.

15.2 Nanomaterials as Radiosensitizers for Radiation Therapy

Radiosensitizers, also known as radiation sensitizers or radioenhancers, are chemical compounds or pharmaceutical agents that can improve the lethal effects of RT on tumor cells by accelerating DNA damage or generating free radicals indirectly with minimal effects on normal tissues [38]. Traditional chemical sensitizers are usually classified into five categories by an early pioneer in this field, G.E. Adams, including: (i) inhibitors of intracellular thiol (SH) or other endogenous radioprotective substances, (ii) compounds that produce radiation-induced cytotoxic radiolytic products, (iii) suppressors of post-irradiation cellular repair processes, (iv) DNA-binding thymine analogues that can be structurally incorporated into intracellular DNA, and (v) oxygen-mimetic sensitizer—stable free radicals [38]. However, in clinical treatment trials for some diseases such as glioblastoma, small molecule radiosensitizers have mostly failed in improving patients' RT efficacy and overall survival, largely due to the inability of these radiosensitizers to cross the blood–brain barrier at effective therapeutic concentrations [39]. In recent years, NP-based radiosensitizers have shown great promise in enhancing the effectiveness of RT with their unique physical, chemical, and electrical properties. In addition, compared to free therapeutics alone, drugs loaded in nanocarriers with suitable sizes and proper surface modifications could obtain better stability and targeted releasing, as well as longer circulation half-life.

As mentioned before, nanomaterials based on high atomic number (high-Z) elements (i.e. Cu, Ag, Pt, and Au) have been developed as an important type of radiosensitizers. These high-Z-element-based drugs can enhance the local radiation dose when they accumulate in tumor sites by strongly absorbing, scattering, and emitting the radiation energy. In principle, four types of interaction processes may occur when the photon encounters the matter, including: photoelectric effect, Compton scattering, coherent scattering, and pair production [40].

In the photoelectric effect, an incident photon collides with an electron tightly bound in the atomic shell and transfers almost all of its energy to it, while the photon itself no longer exists. As a result, the electron carrying the photon energy is knocked out of the inner shell of the atom and begins to ionize surrounding molecules. This interaction depends on both the atomic number of the absorbing substance and the energy of the incoming radiation. Such an event is more likely to occur when the binding energy of electrons within the material is slightly less than the energy of the incident photon. It is noteworthy that this effect is typically proportional to the cubed power of the absorbing substance's atomic number [40].

The Compton effect is the most important process of light-material interaction in cancer RT. In this process, an incident photon collides with an electron that is loosely bound to the atom, causing the incident photon to be scattered in a different direction with lower energy, or causing the impacted loosely bound electron to be scattered away from the atomic shell. Then, the scattered photon with lower energy could continue to undergo additional interaction with matter, and the scattered electron starts to ionize surrounding molecules with the energy obtained from the photon. Typically, this effect is not dependent on the atomic number of the interacting substance, but is positively correlated with the physical density of the substance, and negatively related to the energy of the incident photon [41].

The coherent scattering occurs when the energy of the incoming radiation is lower than the binding energy of the electrons it hits. It usually causes the scattering of the photons in different directions without energy transfer between photons and electrons. Coherent scattering is usually proportional to the square of the atomic number of the absorbing substance [40].

In the process of pair production, a photon interacts with the nucleus of an atom instead of an orbital electron, transferring practically all of its energy to the nucleus and creating a pair of positively (positron) and negatively charged electrons. The positron combines with the free electron and annihilates, creating two energetic photons scattered in opposite directions. Notably, the probability of pair production is proportional to the square of the atomic number and the logarithm of the incident photon-energy [41].

It can be seen from the above analysis of the photon-matter interaction process that the high atomic number (high-Z) elements can enhance the photoelectric effect, coherent scattering, and the probability of pair production when ionizing irradiation encounters matters. It was found that the generation of secondary X-rays, photoelectrons, and Auger electrons can enhance the radiation dose when high-Z metal nanomaterials were irradiated [42]. A detailed illustration of the interactions of X-ray radiation with high-Z materials and the corresponding direct and indirect effects on cancer cells is shown in Figure 15.1.

When X-rays directly act on cells, the incoming beams of X-photons can cause excitations and ionizations of intracellular components, resulting in direct

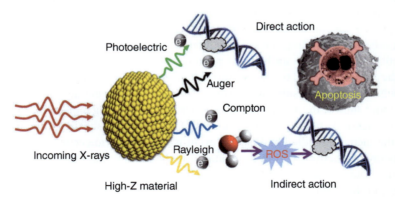

Figure 15.1 Interactions of X-rays with high-Z elements and the corresponding direct and indirect effects on cancer cells. Source: Reproduced with permission from Ref. [40], © 2017, The Royal Society of Chemistry.

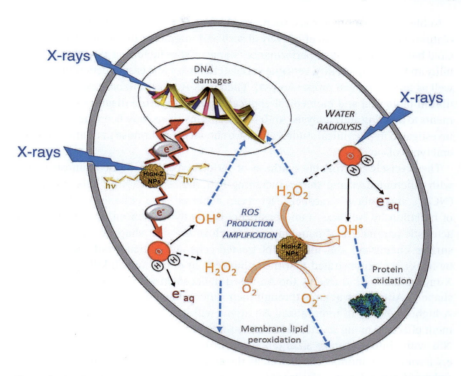

Figure 15.2 Physical and biochemical radiosensitization mechanism of high-Z-metal-based nanomaterials. Source: Pinel et al. [43]/Elsevier/Public domain/CC BY 4.0.

effects including DNA damage (e.g. breaking of the DNA single- and double-strands, cross-linking of DNA-protein) and the water radiolysis leading to production of ROS. ROS in turn alter cellular macromolecules, inducing protein oxidation, membrane lipid peroxidation, and indirect DNA damage. In the presence of high-Z-metal-based nanomaterials, the interactions between X-rays and high-Z materials leads to an increase in the production of secondary electrons such as photoelectrons, auger electrons, and Compton electrons, as well as ROS, which helps enhance the cytotoxicity on irradiated cells [43]. This process describes the physical sensitization mechanism. On the other hand, the biochemical sensitization mechanism indicates that the functionalized high-Z-metal-based nanomaterials not only enhance the ROS production but also shift the cell cycle into a radiosensitive state and inhibit the p53 signaling pathway to induce cell autophagy and lysosome dysfunction, thus improving RT sensitivity [38]. The physical and biochemical radiosensitization mechanism of high-Z-metal-based nanomaterials is shown in Figure 15.2.

Various types of high-Z-element-based nanomaterials have been investigated as potential radiosensitizers. In this section, the representative high-Z materials including some kinds of noble metal nanomaterials and heavy metal nanomaterials, as well as some of the nonhigh-Z-element-based nanomaterials including semiconductor nanomaterials, ferrite nanomaterials, and nonmetallic nanomaterials and their applications as radiosensitizers for RT will be summarized.

Noble metal nanomaterials, such as gold (Au, $Z = 79$), silver (Ag, $Z = 47$), and platinum (Pt, $Z = 78$) can effectively absorb and interact with the incoming X-rays. Gold has various excellent performance characteristics including good chemical stability and biocompatibility, versatile surface chemistry and easy functionalization, as well as unique optical properties [44]. Therefore, Au-based nanomaterials are popular in biological and biomedical applications ranging from diagnostics and treatments to imaging and catalysis. Au-based radiosensitizers have been the most widely investigated in preclinical studies and have shown great promise in treatment of several types of cancer.

The precisely controllable synthesis of various types of Au-based nanostructures with different sizes and shapes including NPs, nanospikes (NSs), NRs, nanocubes (NCbs), and hollow nanoshells, has been achieved. The enhancement efficiency of radiation by Au-based radiosensitizers is greatly influenced by both the characteristic parameters of nanomaterials (such as the size, shape, composition, and surface chemistry) and the type of treated cells and ionizing radiation. A study by Ma et al. demonstrated that the enhancement of cancer cell-killing rates upon X-ray irradiation induced by the Au-based radiosensitizers were influenced by the shape of Au-nanostructures through their different cellular uptake efficiency [45]. A higher amount of internalized Au atoms will lead to better radiation enhancement effects. Among several common Au-nanostructures including NPs, NRs, and NSs with similar average sizes (about 50 nm), the cellular uptake in human oral epidermoid carcinoma cells as well as the efficiency of cancer RT increased in the order of GNPs > GNSs > GNRs [45].

The utilization of Au NPs with low concentration levels (e.g. 7 and 18 mg cases) as radiosensitizers can achieve a significant tumor dose enhancement (over 40%) in a brachytherapy approach using low-energy (e.g. with an average energy of below 100 keV) gamma-/X-ray sources, such as ^{125}I and ^{169}Yb or 50 kVp X-rays [46]. The macroscopic dose enhancement factors (MDEF), defined as the ratio of the average dose in the tumor area in the presence and absence of Au NPs during the irradiation process of the tumor, determined by the Monte Carlo (MC) calculations increased with the concentration of Au NPs within the tumor [46]. However, the glomerular filtration threshold of metal-based NPs in the kidney is 5.5 nm [47]. Therefore, larger (>6 nm) NPs exceed the glomerular filtration threshold, making it difficult to excrete the particles through the kidneys and potentially causing liver damage.

Gold NCs (AuNCs) smaller than the kidney filtration threshold with diameters of 1–3 nm have improved renal clearance efficiency and can be designed as effective radiosensitizers with high biocompatibility. Atomically precise gold-levonorgestrel NCs consisting of $Au_8(C_{21}H_{27}O_2)_8$ (Au_8NC) were designed and investigated by Jia et al. as radiosensitizers for enhanced cancer therapy [48]. The design has three main characteristics. First, the ultrasmall Au NCs can easily accumulate in tumor areas through passive transportation and concentrate the dose of the applied radiation in the tumor area. Second, the atomically precise structure provides the possibility of optimizing functional ligands at the atomic level to regulate the Au radiosensitizer. Third, the surface modification by a typical drug molecule levonorgestrel greatly enhanced the biocompatibility of the AuNCs. A one-pot method

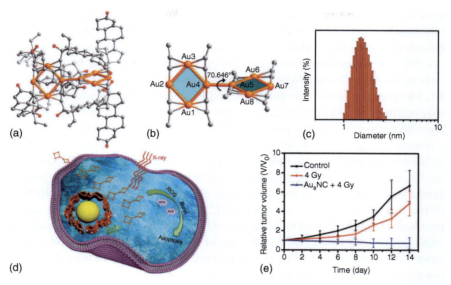

Figure 15.3 Perspective views of (a) the molecular structure of the Au_8NCs and (b) the dihedral angle formed by the planes of two tetranuclear units. (c) The sizes of the Au_8NCs measured by the dynamic light scattering method showed a relatively narrow distribution of approximately 2 nm with good dispersibility. (d) The cancer RT mechanism by the Au_8NCs via the ROS burst upon X-ray irradiation. (e) Relative tumor volume curves of the mice not treated and treated with the X-rays (4 Gy) alone or together with the Au_8NCs. Source: Reproduced with permission from Ref. [48], © 2019, American Chemical Society.

involving the reaction between $C_{21}H_{28}O_2$ (levonorgestrel) and Me_2SAuCl in CH_2Cl_2 (dichloromethane, DCM)/CH_3CN (acetonitrile) ($v/v = 1:1$) in the presence of Et_3N (triethylamine) was utilized to synthesize the $Au_8(C_{21}H_{27}O_2)_8$ NCs at a high yield. The Au_8NC demonstrated intense yellow-green luminescence at room temperature with main emission peaks at 518 and 578 nm under the excitation band with peak wavelength of around 400 nm in phosphate buffer (10 µM). The enhanced cancer cell-killing efficiency was achieved by excessive ROS production and consequent irreversible cell apoptosis. The structure and size characterization of the Au_8NCs are shown in Figure 15.3a–c, and the treatment mechanism is shown in Figure 15.3d. As shown in Figure 15.3e, the relative tumor volume was effectively controlled by the simultaneous use of Au_8NCs and X-rays compared to no treatment and X-ray treatment alone. The combined use of Au_8NCs and X-rays exhibited a tumor inhibition rate of 74.2% compared with treatment by X-rays alone.

It is worth noting that harmful side effects of RT on healthy tissues should be avoided as much as possible while enhancing the radiation dose to the tumor area through radiosensitizers. Therefore, effective monitoring of the location distribution of radiosensitizers and determination of the time delay between intravenous injection of radiosensitizers and initiation of irradiation are essential aspects of RT using radiosensitizers. The noninvasive imaging technology MRI that usually uses magnetic NPs as contrast agents can accurately monitor the distribution of NPs and has been well used to localize specially designed radiosensitizers in RT. Alric et al. designed and synthesized the ultra-small AuNPs core coated by DTDTPA (dithiolated diethylenetriamine pentaacetic acid) shell which permits the immobilization of gadolinium (Gd) ions for MRI and radiometals for nuclear imaging [49, 50]. The Gd chelate-coated gold NPs Au@DTDTPA-Gd$_{50}$ can be utilized not only as an MRI/CT dual-imaging agent but also as efficient radiosensitizers in RT. The presence of Gd ions entrapped in the organic shell provides the contrast enhancement in MRI localization of the functionalized NPs and the Au core provides a strong X-ray absorption for effective eradication of the tumor [50].

Selective targeting is an important means to enhance the tumor enrichment of nanoradiosensitizers, increase the radiation dose in tumor areas, and control healthy tissue toxicity. Luo et al. developed ultrasmall Au$_{25}$ NCs for selective prostate cancer targeting and achieved satisfactory RT enhancement and rapid clearance of the nanomaterials from the body [51]. The highly selective prostate-specific membrane antigen (PSMA) targeting ligand (CY-PSMA-1) was combined with Au$_{25}$ NCs to specifically bind to the PSMA receptor that is overexpressed in most prostate cancers. The ultrasmall Au NCs of about 1.5 nm in diameter below the threshold for renal clearance avoided the unintended and nonspecific organ uptake by the liver and spleen, and reduced the potential organ toxicity induced by the Au cluster. The CY-PSMA-1-Au$_{25}$NCs were synthesized using a one-step method through adding the HAuCl$_4$ aqueous solution (160 μL, 25 mM) dropwise to the CY-PSMA-1 peptide water solution (4 ml, 1.0 mM) under appropriate pH values as illustrated in Figure 15.4a. X-ray irradiation with a dose of 6 Gy (Joule/kilogram) was given at four hours after the intravenous injection of NCs when the peak targeted accumulation was reached in the tumor and the concentration of Au in blood was relatively low. The results showed higher targeting levels to the PSMA-receptor expressing PC3pip tumor than to the PC3flu tumor cells with no PSMA receptor, resulting in improved targeted radiation therapy as illustrated in Figure 15.4b.

Similar to noble metal nanomaterials, heavy metal nanomaterials with high atomic number including gadolinium (Gd, $Z = 64$), hafnium (Hf, $Z = 72$), tantalum (Ta, $Z = 73$), tungsten (W, $Z = 74$), and bismuth (Bi, $Z = 83$) also have large atomic coefficients and a great X-ray attenuation capability, thus being extensively explored as radiosensitizers. Verry et al. performed the first-in-man phase I trial of the theranostic Gd-based NPs through a simple intravenous injection for the treatment of multiple brain metastases [52]. This new Gd-based theranostic nanoagent was named AGuIX (activation and guidance of irradiation by X-ray), acting as a contrast agent for MRI imaging and a cytotoxic agent for whole brain RT (WBRT). These NPs with diameter of 3 ± 1.5 nm consist of a polysiloxane core surrounded by Gd ring ligands derived from DOTA (1,4,7,10-tetraazacyclododecane-1,4,7,10-tetraacetic acid) and covalently grafted onto a polysiloxane matrix. The passive accumulation of the AGuIX in the tumors was achieved through the enhanced permeability and retention (EPR) effect. The rich Gd content allows for high absorption of photons from the ionizing radiation and/or secondary ionizing species, leading to locally enhanced free radical generation. Rajaee et al. introduced a new theranostic agent combining several high-Z-elements, the multifunctional Bi Gd oxide NP surface modified with polyethylene glycol (BiGdO$_3$-PEG), for multimodal CT/MRI imaging

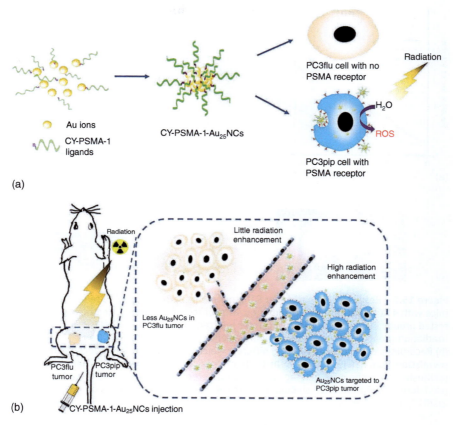

Figure 15.4 (a) Schematic illustration of synthesis of the CY-PSMA-1-Au$_{25}$NCs radiosensitizer with high affinity to the PC3pip cells overexpressing PSMA receptor. (b) Illustration of the high radiation enhancement in PC3pip tumor due to the higher targeting levels of the CY-PSMA-1-Au$_{25}$NC radiosensitizer compared to the little radiation enhancement in PC3flu tumor with low NC uptake due to the lack of PSMA receptor after the intravenous injection of NCs. Source: Reproduced with permission from Ref. [51], © 2019, John Wiley & Sons.

and enhanced RT [53]. Both Bi and Gd elements can serve as radiosensitizers to amplify radiation dose and CT-enhanced contrast agents, while Gd can also serve as MRI-enhanced contrast agent. The in vivo results of cancer treatment experiments carried out in four groups of female BALB/c mice bearing 4T1 breast tumors as shown in Figure 15.5 demonstrated that the increase of the tumor volume was significantly suppressed by RT together with the BiGdO$_3$-PEG NP injection while maintaining the health of the mice.

Some of the nonhigh-Z-element-based nanomaterials including semiconductor nanomaterials (e.g. II–VI group semiconductor quantum dots, QDs), ferrite nanomaterials, and nonmetallic nanomaterials (e.g. fullerene [C$_{60}$]) possessing unique properties have also been developed as radiosensitizers for RT. Nanoscale semiconductor materials tightly confine electrons or electron holes, and QDs are semiconductor

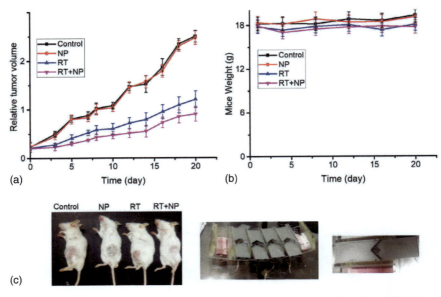

Figure 15.5 Results of in vivo cancer RT experiments carried out using female BALB/c mice with 4T1 breast tumors. (a) Tumor growth curves after different treatments of four tested groups including control group, mice injected with BiGdO$_3$-PEG NPs group, X-ray irradiation group, and X-ray irradiation in presence of BiGdO$_3$-PEG NPs group.
(b) Recordings of mice weight during different treatments. Source: Reproduced with permission from Ref. [53], © 2019, Institute of Physics and Engineering in Medicine.
(c) Illustration of the mice bearing tumor after different treatments and the lead (Pb) plate used during RT for radiation shielding. Source: Reproduced with permission from Ref. [53], © 2019, IOP Publishing.

nanocrystals in the size range of 2-100 nm. QDs have unique properties between bulk semiconductors and discrete atoms or molecules such as electronic, tunable optical, and targeting properties that vary with size and shape as a result of quantum mechanics. QDs are sometimes referred to as artificial atoms [54]. Engineered QDs exhibiting crystalline metalloid structure, quantum dimension/confinement effect, and surface effect have great promise in biomedicine applications [55]. When photons collide with QDs, several possible events can occur at different energy levels of the incident photons as shown in Figure 15.6 [55].

In situations for low-energy photons when the energy of the photon is equal to or greater than the bandgap energy of the QD (ΔE_g), such as ultraviolet (UV), visible light (VIS), and near-infrared (NIR) radiation, charge transfer and photosensitization may occur. In the process of charge transfer, the absorption of a photon generates an electron-hole pair (e$^-$ and h$^+$ pair) and transfers charge between the donor QD and neighboring acceptor molecule "A" as shown in Figure 15.6a. When the photon absorption causes the QD to be promoted from its ground state "a" to a higher energy excited state "b" and then to a "trap" state "c" as shown in Figure 15.6b, photosensitization takes place and triplet ground-state oxygen ($^3O_2$) is promoted into highly reactive singlet oxygen ($^1O_2$) by triplet energy transfer

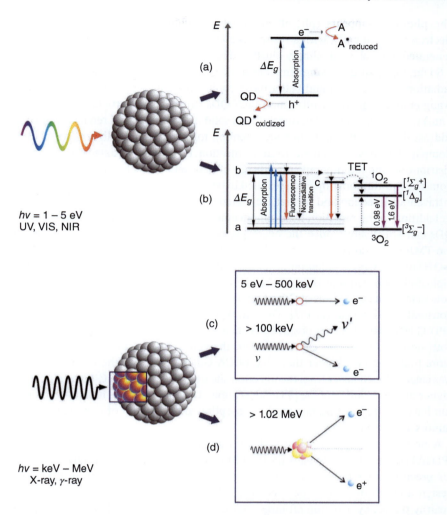

Figure 15.6 Schematic illustration of possible events that may occur when photons collide with quantum dots at different energy levels: (a and b) situations for low-energy photons (such as ultraviolet, visible light, near-infrared radiation), (c and d) situations for high-energy photons (such as X-rays and gamma rays). Source: Reproduced with permission from Ref. [55], © 2008, Elsevier.

(TET). Under this condition, the QDs can emit fluorescence like other organic fluorophores, act as photosensitizers inducing photochemical processes, and demonstrate photocatalytic properties. For high-energy incoming photons (X-rays and gamma rays), the QDs can act as radiosensitizers to enhance radiation absorption owing to their high atom and electron density. The photoelectric ionization effect (i.e. transfer of the high energy to the electron) and the Compton scattering could induce the ejection of a high-speed electron from an atom constituting the QD as shown in Figure 15.6c. When the energy of the incident photon is above the threshold energy of about 1.022 MeV, the process of pair production occurs.

The photon disappears (photon annihilation) while two charged particles (an electron-positron pair), with a rest-mass energy equivalent to 0.511 MeV each, are generated in its place, as simply illustrated in Figure 15.6d.

So far, many studies using semiconductor QDs as photo- and radiosensitizers for radiation enhancement and better biosafety in tumor therapy have been reported. Wang et al. designed the ultrasmall BiOI QDs (the average particle size is around 3 nm) with strong internal permeability of solid tumors and rapid renal clearance to address the issue of long-term body retention toxicity of nanoradiosensitizers [56]. Compared to conventional intravenous injection, this study evaluated the biological distribution and radiotherapeutic efficacy of radiosensitizers by intratumoral injection. The intratumoral injection resulted in better tumor accumulation compared to intravenous injection and further facilitated the high-efficiency radiosensitization. In addition to the radiation dose enhancement by the high-Z element Bi, the BiOI QDs can effectively reduce the overexpressed H_2O_2 (as the electron acceptor) in the TME and oxidize the surrounding H_2O (as hole acceptor) to hydroxyl radical (•OH) under X-ray irradiation for enhanced RT. Recently, Ma et al. combined the high-Z-element Gd that is also a frequently-used MRI contrast agent with carbon dots to fabricate two kinds of Gd-intercalated carbon dots (Gd@C-dots) for IGRT of nonsmall cell lung cancer [57]. The two types of Gd@C-dots, CA-Gd@C-dots and pPD-Gd@C-dots, bearing positive and neutral surface charges, respectively, were both nontoxic to H1299 cells. However, the cell uptake of pPD-Gd@C-dots was more than 10 times higher than that of CA-Gd@C-dots, and the CA-Gd@C-dots was cleared much faster from the injection site than the pPD-Gd@C-dots, indicating higher tumor retention of the pPD-Gd@C-dots. Therefore, surface functional groups can have a significant impact on the absorption and retention of nanomaterials in tumors and normal cells.

A novel design of nanoscale systems utilizing the poly(lactic-*co*-glycolic acid) (PLGA) NPs as carriers to package the ultra-small black phosphorus QDs (BPQDs) for precise tumor radiosensitization was accomplished by Chan et al. [58]. The design aims to overcome the excessive radiation that may damage the adjacent healthy tissues by avoiding off-target release and realizing rapid clearance during blood circulation. The prepared PLGA$_{SS}$-D@BPQDs with a core-shell structure and conjugated with several layers of polymers are designed to be sequentially triggered to provide precise bio-responsiveness in the complex biological environment for enhanced RT. The exquisite design of the nanoscale system and the principles of the sequential triggering for precise bio-responsive X-ray radiation-enhanced therapy are shown in Figure 15.7.

The BPQDs are adopted as radiosensitizers. PLGA is used to improve the biocompatibility of the system, cystamine and 2,3-dimethylmaleic anhydride (DMMA) serve as bio-responsive triggers and pH-responsive shell endowing the system with the surface charge-switching ability. The active accumulation in tumor site was achieved by the conjugation of cancer-targeting molecules, the polypeptide peptide motif arginyl-glycyl-aspartic acid (RGD). The polyethylenimine (PEI) layer connected to cystamine is exposed on the surface of the nanoscale system with positive charges. The introduction of this cationic polymer with rich amino

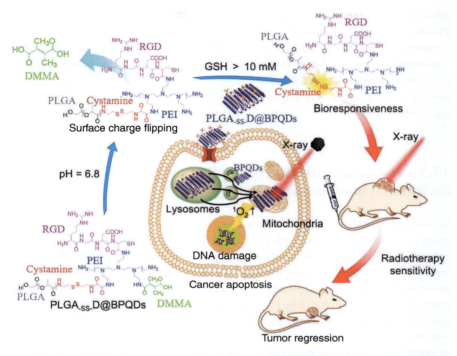

Figure 15.7 Schematic illustration of the rational design of the PLGA-ss-D@BPQDs nanoscale system with precise bio-responsiveness that can be sequentially triggered in complex biological environments, and the mechanisms of surface charge flipping, active tumor cell accumulation, and combined photodynamic effects and radiation-enhanced therapy. Source: Reproduced with permission from Ref. [58], © 2018, American Chemical Society.

groups not only improves the hydrophilicity and cellular uptake of the nanoscale system but also provides sites for further chemical modifications. Characterized by transmission electron microscopy (TEM), the synthesized BPQDs are about 3 nm in diameter and the PLGA-ss-D@BPQDs with a spherical shape were about 150 nm in diameter. The prepared smart PLGA-ss-D@BPQDs nanoscale system was proved to have surface charge-switching ability, bio-responsiveness, good stability in blood environment (pH ≈ 7.4), selective cellular uptake property, and enhanced X-ray RT effects. After being intravenously injected, materials in the nanoscale system are circulated with the blood and transported to tumor sites. Once exposed to and actively accumulated in the acidic TME (pH ≈ 6.8), the nanoscale system with the negatively charged DMMA shell expands from about 150 nm to over 2500 nm, causing the hydrolysis of the amide bonds of DMMA and dissociation of DMMA from the nanoscale system. The exposure of the positively charged PEI layer on the surface of the nanoscale system enhances the cellular uptake. At a large concentration of the reductive GSH overexpressed in the TME, the disulfide bond between PEI and PLGA will be deoxidized by GSH and the size of the nanoscale system will drop to about 139 nm, leading to efficient release of the loaded BPQDs.

After X-ray irradiation, the cytotoxicity against A375 and HeLa cells induced by the PLGA$_{SS}$-D@BPQDs nanoscale system was improved by about 14 times than that caused by the traditional radiosensitizer AuNPs and was 2 times higher than that of bare BPQDs. In addition, the enhanced generation of reactive singlet oxygen ($^1O_2$) via photodynamic effects of BPQDs further promotes DNA damage and apoptosis induction. This study has important implications for the design of intelligent nanomedicine systems.

15.3 Nanomaterials Delivering Radioisotope for Internal Radioisotope Therapy

As introduced before, internal RIT is a type of cancer treatment that uses small doses of radioactive isotopes (solid or liquid) that are placed inside the body to kill cancer cells. In clinical applications, the therapeutic radioisotopes are typically introduced into the tumor site through a minimally invasive method, such as intravenous administration or direct infusion via a catheter (also called brachytherapy) under the condition that the tumor can be reached from the main arteries [37]. Radioisotopes, also referred to as radioactive nuclides, radionuclides, or radioactive isotopes, are unstable nuclides having excess nuclear energy that can be emitted from the nucleus as gamma radiation, transferred to one of its electrons to release it as a conversion electron, or used to create and emit new particles (alpha particles or beta particles) from the nucleus.

Once inside the body, the isotopes travel to the tumor and the ionized atoms and free radicals emitted from the internal radioisotopes destroy cancer cells within a certain area while sparing healthy cells. IRT has several advantages over other types of cancer treatment, including its ability to target specific types of cancer cells, its relatively low risk of side effects, and its potential to be used in combination with other treatments. IRT is often used to treat cancers such as colon, breast, lung, and prostate cancer. However, IRT also has some disadvantages, including the potential for long-term exposure to radiation, which can increase the risk of developing secondary cancers. Additionally, not all patients may be good candidates for IRT due to factors such as their age, overall health, and the location and size of their tumor.

In order to overcome the drawbacks of IRT, improve the RT efficacy, and reduce normal tissue damage, therapeutic radioisotopes have been loaded onto biocompatible NPs characterized by a large specific surface area for high-efficiency in vivo targeted delivery. The large surface areas and various types of functional groups on the surface of NPs allow for the assembly or loading of bioactive ligands, small molecule formulations, as well as radioisotopes through various methods including encapsulating and chemical reactions such as chelating, chemical doping, and displacement [59]. Through ingenious and reasonable design, a single NP can carry multiple radionuclides to achieve a higher radioactive payload. The efficiency of active targeted binding with tumor cells can be promoted through the functional conjugation of multiple cancer-targeting molecules on the surface of the NP.

In IRT, typically used radioisotopes mainly consist of three categories including alpha (α), beta (β), and Auger-particle emitters [35], and various nanocarriers including metal NPs, inorganic and organic nanomaterials, polymer-based polymeric NPs, and liposomes have been developed as radioisotope carriers. The α-particles, also referred to as α-rays or α radiation, are formed by the combination of two protons and two neutrons identical to helium nuclei, possessing double positive charges and a large mass (compared to β-particles). The α-particles have a high LET, which denotes the average radiant energy deposited in tissue per unit track length (keV µm^{-1}), about 50-230 keV µm^{-1}, but a limited penetration depth (20-100 µm) [60]. Therefore, α-particles are suitable for the therapy of small-volume microscopic tumors or residual tumors and can lead to efficient and specific tumor cell death by irreversible DNA and chromosomal damage [59, 61]. The radiobiological properties of α-particles have been recognized since the early 1960s, but it was not until 1997 that the first clinical trial of radiolabeled antibody therapy using an α-particle emitter utilizing ^{213}Bi-conjugated anti-leukemia antibody HuM195 was reported [62]. The majority of over 100 types of radioactive isotopes emitting α-particles decay too quickly to be considered for therapeutic uses, and only a few α-emitters including astatine-211 (^{211}At), actinium-225 (^{225}Ac), terbium-149 (^{149}Tb), bismuth-213 (^{213}Bi), and lead-212 (^{212}Pb) as summarized in Table 15.3 have been developed for their therapeutic potential in RIT [70].

The β-particles, also called β-rays or β-radiation, are high-energy, high-speed charged particles emitted through the process of radioactive β decay of atomic nucleus. The β$^-$ and β$^+$ decay produce the negatively charged β-particles identical to electrons and the positively charged β-particles called positrons, respectively. The β-particles have a relatively low LET than α-particles (approximately 0-2 keV µm^{-1}) but profound tissue penetration (20–130 mm) [37]. Additionally, β-particles have lower cytotoxicity and could induce the "crossfire effect" that enables the eradication of cancer cells that are not directly bound to radioimmunoconjugates but are affected by the radiation [59]. As a result, β-particle emitters are the most widely used radioisotopes in cancer treatment as they are useful in overcoming radioresistance and are particularly suitable for the treatment of bulky or large-volume tumors [71]. The most commonly used β-emitters including iodine-131 (^{131}I), rhenium-188 (^{188}Re), copper-64 (^{64}Cu), yttrium-90 (^{90}Y), and gold-198 (^{198}Au) are summarized in Table 15.4.

Among them, the use of radionuclide ^{131}I for the treatment of thyroid diseases is one of the most successful applications of the metabolic pathway and ^{131}I has been widely applied in the treatment of well-differentiated thyroid cancer for more than 50 years [20]. ^{131}I decays with both β and γ emissions and the principal γ-ray of 364 keV can be used for gamma camera-based functional imaging of thyroid and post-treatment ^{131}I distribution monitoring. The principal β-particle with an average energy of 0.192 MeV acts as the radiation effector playing the major role in ^{131}I-based RIT. A specific target of radioactive ^{131}I in thyroid disease is an intrinsic plasma membrane protein known as the sodium iodide transporter (NIS), which mediates the active transport of iodide into the thyroid gland [20]. Therefore, ^{131}I is a highly targeted form of radiation that can be precisely delivered to the tumor site

Table 15.3 A brief summary of some therapeutic α-emitters delivered by nanomaterials for RIT.

α-Emitters	Half-life	Biomaterials	Characteristics	References
^{212}Pb	10.64 h	Monoclonal antibodies (mAbs)	^{212}Pb is the immediate parental nuclide of ^{212}Bi and can be used as an in vivo generator for ^{212}Bi to overcome its short half-life (61 min)	[63]
^{211}At	7.2 h	Gold NPs, mAbs, colloids, methylene blue (MTB), urea-based small-molecule	The α-particles of ^{211}At have high energy, short path length, and a mean LET of about 97 keV μm^{-1}. Due to its long half-life, it can be used in situations where the targeting molecules accumulate slowly in the tumor sites	[64]
^{149}Tb	4.2 h	Albumin-binding PSMA-targeting ligand, mAbs	^{149}Tb belongs to the group of rare earth elements and is a type of partial alpha-emitting nuclides. ^{149}Tb emits low-energy α-particles together with β$^+$-particles (positrons), making it suitable for on-site tracking using positron emission tomography (PET)	[65, 66]
^{225}Ac	10 d	PSMA-targeting antibody, liposomes, tumor-homing peptide	^{225}Ac can be obtained either from the decay of ^{233}U or from the neutron transmutation of ^{226}Ra. ^{225}Ac carriers can act as atomic nanogenerators delivering an α-particle cascade to a cancer cell	[67, 68]
^{213}Bi	46 min	DOTA–biotin	^{213}Bi possessing remarkable antitumor activity and tolerability is a kind of short half-life nuclide not suitable for clinical translation	[69]

without affecting healthy surrounding tissues. It does not produce α-particles or other harmful byproducts, thus causing minimal toxicity compared to other forms of radioisotopes. In addition, ^{131}I can be administered through a variety of routes, including intravenous injection, oral ingestion, or implantation into a tumor site, making it easy to deliver the radiation to the affected area without causing too much discomfort to the patients [80].

Auger-particles are low-energy (<25 keV) electrons emitted by the Auger effect (i.e. inner-shell electron transitioning) [59]. The high LET (between 1 and 26 keV μm^{-1}) Auger-particles can produce clustered damage in macromolecular targets (e.g. DNA and the cell membrane) of cancer cells and are suitable for targeting single cells [71, 81]. Auger-emitters deposit high LET at extremely short distances (≤150 nm), making them most effective when decay occurs in the nucleus and less effective when decay occurs in the cytoplasm [59]. Consequently, Auger-emitters

Table 15.4 A brief summary of some therapeutic β-emitters delivered by nanomaterials for RIT.

β-Emitters	Half-life	Biomaterials	Characteristics	References
^{131}I	193 h	Pd nanosheets, human serum albumin NPs, liposome, Au NPs, dendrimers	^{131}I is one of the frequently used radioisotopes that can be administered alone into the bloodstream or linked to antibodies targeting specific markers. It can be used for a variety of cancers, including thyroid cancer, melanoma, prostate cancer, and others	[72, 73]
^{186}Re	91 h	Liposomes, lipid NPs	The radioisotope ^{186}Re emitting β-particles of 1.07 MeV and γ-rays of 137 keV can be produced either at a nuclear reactor or at a particle accelerator (cyclotron). It has been used in treatment of metastatic cancer of the prostate, breast, colon, and lung	[74, 75]
^{64}Cu	12.7 h	Liposome, copper sulfide-loaded microspheres	The metallic radionuclide ^{64}Cu can be used for both diagnostic PET imaging and RT. It decays to ^{64}Ni by positron-emission ($β^+$-particles with average energy of 288 keV) or electron capture (EC), or to ^{64}Zn by β-decay (emitting $β^-$-particles of 190.2 keV)	[76, 77]
^{90}Y	64 h	Peptides, upconversion NPs, injectable hydrogels	^{90}Y together with ^{131}I is readily available and inexpensive radionuclides most popular in clinical applications. The pure β-emitter nature, high energy, and long residence time in tumor site of ^{90}Y make it subject to few environmental radiation restrictions	[78]
^{198}Au	2.7 d	Au NPs, Fe$_3$O$_4$@Au core-shell NPs	The superparamagnetic iron oxide (magnetite) nanoparticles coated with ^{198}Au can be used for combined targeted magnetic hyperthermia and radionuclide therapy of hepatocellular carcinoma	[79]

are most harmful to cellular structures whose nuclear component dimensions fall within the range of Auger-emitters (≤150 nm). However, the extreme toxicity of the Auger-particles greatly limits their applications. The selection of radioisotopes for cancer treatment should fully consider the specific cancer types and cell-level characteristics, the toxicity, safety, and availability of the radioisotope, as well as the special design requirements for loading and targeted release through nanocarriers.

Table 15.5 A brief summary of some therapeutic Auger-emitters delivered by nanomaterials for RIT.

Auger-emitters	Half-life	Biomaterials	Characteristics	References
^{123}I	13.2 h	Poly adenosine diphosphate-ribose polymerase (PARP) inhibitor	^{123}I releases about 14% of its decay energy in the form of Auger electrons and is efficient in inducing chromosomal damage	[82]
^{125}I	60 d	PSMA, TiO$_2$ NPs, thymidine analog	^{125}I is commonly used in studying in vivo and in vitro Auger-electron-induced DNA damage. The long half-life of ^{125}I requires the exposure of cells for weeks to accumulate detectable biological damage	[83]
^{67}Ga	78.3 h	Sphingolipid nanoemulsions, modified polymer dots, lipid NPs	^{67}Ga produces the most energetic (7–9 KeV) and longest ranging (up to 2.4 μm in water) Auger electrons amongst Auger electron emitters, and emits γ-rays for gamma camera imaging	[84, 85]
^{58m}Co	9 h	Octreotide analogs	^{58m}Co emits Auger and conversion electrons with high LET in water (2-18 keV μm^{-1}) and decays to a positron-emitting daughter ^{58g}Co	[86]

Examples of some typical therapeutic Auger-particles such as iodine-123 and -125 (^{123}I and ^{125}I), gallium-67 (^{67}Ga), and cobalt-58m (^{58m}Co), as well as delivering nanomaterials for internal RIT, are summarized in Table 15.5.

15.4 Nanomedicine Synthesis with Radio/Nuclear Radiation Response

The principles for nanomedicine synthesis with radio/nuclear radiation response involve the use of radiation-sensitive materials and techniques to generate targeted DDSs. The goal is to develop drugs that can be specifically delivered to cancer cells, thereby minimizing systemic side effects and improving therapeutic outcomes. In the design and synthesis of nanomedicines with radio/nuclear radiation response, the first consideration is the treatment method to be used, whether it is EBRT or RIT. On this basis, it is necessary to determine the type of the core therapeutic material, such as high-Z-element-based nanomaterials or radioisotopes to be delivered by nanocarriers. Other essential considerations include the collaborative design of multiple diagnostic and therapeutic modalities coupled with RT through a certain nanomaterial platform. The improvement in biocompatibility, active targeting, as well as special responsiveness of the nanoplatforms in various biological environments through functional surface modification of nanomaterials, has gradually shown great significance in design and synthesis of nanomedicines and achieved rapid development in recent years.

Typically, nanoplatforms enter the bloodstream through intravenous injection or oral administration, and their size and composition will affect their in vivo biodistribution, metabolic excretion, and nonselective aggregation in tumor sites in the body. A well-designed polymer coating on the surface of nanomaterials can improve their biocompatibility and stability, including but not limited to minimizing toxicity from metal NPs, regulating the interaction between nanomaterials and biomolecules, cells, and tissues, regulating the in vivo biodistribution and metabolic excretion, and stabilizing the aggregation and precipitation of metal or metal oxide NPs under physiological conditions [59]. Such polymer coatings functionalized by reactive functional groups such as -COOH, -NH$_2$, and -SH can provide binding sites for tumor-targeting ligands or other reagents, and allow for covalent linking or noncovalent coupling of chelates of radioactive isotopes [59]. Furthermore, integrated diagnosis and treatment can be achieved by surface modification with other functional moieties including imaging agents, photosensitizers, and chemotherapy drugs [61]. The idea and possible designs of multifunctional NP platforms that can integrate tumor targeting, diagnosis (multiple functional imaging ability), and therapy (delivery and release of drugs and/or radioisotopes) are schematically demonstrated in Figure 15.8.

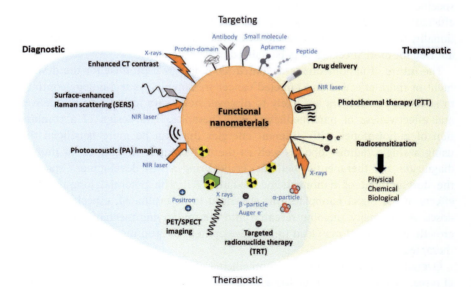

Figure 15.8 Schematic illustration of the design of multifunctional integrated theranostic nanoparticle platforms for tumor targeting, diagnostic, and delivery of therapeutic drugs and/or radioisotopes. Source: Reproduced with permission from Ref. [61], © 2021, Elsevier.

It can be seen from Figure 15.8 that the core of the functional nanomaterials could be the high-Z-element-based or other radiosensitizing nanomaterials, as well as the various nanodelivery platforms carrying radionuclides such as α-, β-, and Auger-emitters. The active targeting of nanomedicine involves the use of specific molecules or antibodies that specifically bind to cancer cells, allowing for targeted delivery of radioactive and other therapeutic agents. The increased specificity and improved therapeutic outcomes give advantages to active targeting nanoplatforms over traditional drugs. One example of active targeting is the use of antibody-drug conjugates (ADCs), which are composed of an antibody that binds to a tumor-specific antigen and a drug that is designed to destroy cancer cells [87]. The antibody acts as a "key" that unlocks the drug's activity within the cancer cell, resulting in targeted destruction of the cancerous tissue. The utilization of metal elements such as Au- and Gd-based nanomaterials could further endow nanoplatforms with surface plasmon resonance (SPR) and magnetic resonance-enhancing properties, making them excellent tools for multiple diagnostic and therapeutic applications, such as photothermal therapy (PTT), photoacoustic (PA) imaging, enhanced Raman scattering (SERS), and MRI [88, 89].

15.5 Conclusion and Prospects

In conclusion, the use of nanomedicine in RT has shown significant potential in improving the efficacy and safety of cancer treatment. By targeting cancer cells with specific nanomaterials, researchers have been able to increase the radiation dose efficiently delivered to tumor sites while minimizing damage to normal tissue. Additionally, the use of active targeting strategies, such as ADCs and proteomics, has allowed for more precise and personalized cancer therapy.

The future of nanomedicine-mediated RT holds great promise for the development of more effective and targeted cancer treatments. As technology continues to advance, researchers are likely to discover new ways to improve the delivery and effectiveness of nanomedicine-based therapies. The design of a combination of radioisotopes with different energies that can be more beneficial than using a single radioisotope, as well as the collaborative integration of multiple diagnostic and therapeutic methods, could bring a vast development space for the development of nanomedicine-mediated RT. The potential long-term side effects of the novel nanoplatforms still need to be evaluated, especially at high doses. Additionally, ongoing research into the underlying mechanisms of cancer growth and survival may lead to the development of even more effective targeted therapies.

Overall, the combination of nanomedicine and RT represents an exciting area of research with the potential to significantly improve cancer treatment outcomes. With continued investment and collaboration between researchers, clinicians, and industry partners, it is likely that these technologies will become increasingly integrated into standard cancer treatment regimens in the years to come.

References

1 Siegel, R.L., Miller, K.D., Wagle, N.S., and Jemal, A. (2023). Cancer statistics, 2023. *CA: A Cancer Journal for Clinicians* 73 (1): 17–48.

2 Dunne-Daly, C.F. (1999). Principles of radiotherapy and radiobiology. *Seminars in Oncology Nursing* 15 (4): 250–259.

3 Noy, M.A., Rich, B.J., Llorente, R. et al. (2022). Levels of evidence for radiation therapy recommendations in the national comprehensive cancer network (NCCN) clinical guidelines. *Advances in Radiation Oncology* 7 (1): 100832.

4 Yan, H., Hu, Z., Huang, P. et al. (2021). The status of medical physics in radiotherapy in China. *Physica Medica* 85: 147–157.

5 Xia, C., Dong, X., Li, H. et al. (2022). Cancer statistics in China and United States, 2022: profiles, trends, and determinants. *Chinese Medical Journal* 135 (5): 584–590.

6 Goodhead, D.T. (1994). Initial events in the cellular effects of ionizing radiations: clustered damage in DNA. *International Journal of Radiation Biology* 65 (1): 7–17.

7 Winiecki, J. (2022). Principles of radiation therapy. *Physical Sciences Reviews* 7 (12): 1501–1528.

8 Snider, J.W. and Mehta, M. (2016). Principles of radiation therapy. In: *Handbook of Clinical Neurology*, vol. 134 (ed. M.S. Berger and M. Weller), 131–147. Elsevier.

9 Desouky, O., Ding, N., and Zhou, G. (2015). Targeted and non-targeted effects of ionizing radiation. *Journal of Radiation Research and Applied Science* 8 (2): 247–254.

10 Byakov, V.M. and Stepanov, S.V. (2006). The mechanism for the primary biological effects of ionizing radiation. *Physics-Uspekhi* 49 (5): 469.

11 Rodemann, H.P. and Blaese, M.A. (2007). Responses of normal cells to ionizing radiation. *Seminars in Radiation Oncology* 17 (2): 81–88.

12 Green, D.E. and Rubin, C.T. (2014). Consequences of irradiation on bone and marrow phenotypes, and its relation to disruption of hematopoietic precursors. *Bone* 63: 87–94.

13 Jalili-Firoozinezhad, S., Prantil-Baun, R., Jiang, A. et al. (2018). Modeling radiation injury-induced cell death and countermeasure drug responses in a human Gut-on-a-Chip. *Cell Death & Disease* 9 (2): 223.

14 Bennardo, L., Passante, M., Cameli, N. et al. (2021). Skin manifestations after ionizing radiation exposure: a systematic review. *Bioengineering* 8 (11): 153.

15 Betlazar, C., Middleton, R.J., Banati, R.B., and Liu, G.-J. (2016). The impact of high and low dose ionising radiation on the central nervous system. *Redox Biology* 9: 144–156.

16 Jurdana, M. (2008). Radiation effects on skeletal muscle. *Radiology and Oncology* 42 (1): 15–22.

17 Mavragani, I.V., Nikitaki, Z., Kalospyros, S.A., and Georgakilas, A.G. (2019). Ionizing radiation and complex DNA damage: from prediction to detection challenges and biological significance. *Cancers (Basel)* 11 (11): 1789.

18 Turner, J.E. (2005). Interaction of ionizing radiation with matter. *Health Physics* 88 (6): 520–544.

19 Koka, K., Verma, A., Dwarakanath, B.S., and Papineni, R.V. (2022). Technological advancements in external beam radiation therapy (EBRT): an indispensable tool for cancer treatment. *Cancer Management and Research* 14: 1421–1429.

20 Yansong, L. (2015). Internal radiation therapy: a neglected aspect of nuclear medicine in the molecular era. *The Journal of Biomedical Research* 29 (5): 345.

21 Lederman, M. (1981). The early history of radiotherapy: 1895–1939. *International Journal of Radiation Oncology – Biology – Physics* 7 (5): 639–648.

22 Mould, R.F. (1998). The discovery of radium in 1898 by Maria Sklodowska-Curie (1867–1934) and Pierre Curie (1859–1906) with commentary on their life and times. *The British Journal of Radiology* 71 (852): 1229–1254.

23 Nascimento, M.L.F. (2014). Brief history of X-ray tube patents. *World Patent Information* 37: 48–53.

24 Martins, P.N. (2018). A brief history about radiotherapy. *International Journal of Latest Research in Engineering and Technology* www.ijlret.com 04: 8–11.

25 Thwaites, D.I. and Tuohy, J.B. (2006). Back to the future: the history and development of the clinical linear accelerator. *Physics in Medicine and Biology* 51 (13): R343–R362.

26 Svensson, H. and Möller, T.R. (2003). Developments in radiotherapy. *Acta Oncology (Madr)* 42 (5–6): 430–442.

27 Baskar, R., Lee, K.A., Yeo, R., and Yeoh, K.-W. (2012). Cancer and radiation therapy: current advances and future directions. *International Journal of Medical Sciences* 9 (3): 193–199.

28 Welz, S., Nyazi, M., Belka, C., and Ganswindt, U. (2008). Surgery vs. radiotherapy in localized prostate cancer. Which is best? *Radiation Oncology* 3 (1): 23.

29 Löbrich, M. and Kiefer, J. (2006). Assessing the likelihood of severe side effects in radiotherapy. *International Journal of Cancer* 118 (11): 2652–2656.

30 Norlund, A. (2003). Costs of radiotherapy. *Acta Oncology (Madr)* 42 (5–6): 411–415.

31 Rao, C.N.R. and Cheetham, A.K. (2001). Science and technology of nanomaterials: current status and future prospects. *Journal of Materials Chemistry* 11 (12): 2887–2894.

32 Wang, Y., Liang, R., and Fang, F. (2015). Applications of nanomaterials in radiotherapy for malignant tumors. *Journal of Nanoscience and Nanotechnology* 15 (8): 5487–5500.

33 Abrahamse, H. and Hamblin, M.R. (2016). New photosensitizers for photodynamic therapy. *Biochemical Journal* 473 (4): 347–364.

34 Zhu, R., Avsievich, T., Popov, A. et al. (2021). In vivo nano-biosensing element of red blood cell-mediated delivery. *Biosensors & Bioelectronics* 175: 112845.

35 Song, G., Cheng, L., Chao, Y. et al. (2017). Emerging nanotechnology and advanced materials for cancer radiation therapy. *Advanced Materials* 29 (32): 1700996.

36 Crapanzano, R., Secchi, V., and Villa, I. (2021). Co-adjuvant nanoparticles for radiotherapy treatments of oncological diseases. *Applied Sciences* 11 (15): 7073.

37 Pei, P., Liu, T., Shen, W. et al. (2021). Biomaterial-mediated internal radioisotope therapy. *Materials Horizons* 8 (5): 1348–1366.

38 Gong, L., Zhang, Y., Liu, C. et al. (2021). Application of radiosensitizers in cancer radiotherapy. *International Journal of Nanomedicine* 16: 1083–1102.

39 Ali, M.Y., Oliva, C.R., Noman, A.S.M. et al. (2020). Radioresistance in glioblastoma and the development of radiosensitizers. *Cancers (Basel)* 12 (9): 2511.

40 Goswami, N., Luo, Z., Yuan, X. et al. (2017). Engineering gold-based radiosensitizers for cancer radiotherapy. *Materials Horizons* 4 (5): 817–831.

41 Choi, J., Kim, G., Cho, S.B., and Im, H.-J. (2020). Radiosensitizing high-Z metal nanoparticles for enhanced radiotherapy of glioblastoma multiforme. *Journal of Nanobiotechnology* 18 (1): 122.

42 Kolyvanova, M.A., Belousov, A.V., Krusanov, G.A. et al. (2021). Impact of the spectral composition of kilovoltage X-rays on high-Z nanoparticle-assisted dose enhancement. *International Journal of Molecular Sciences* 22 (11).

43 Pinel, S., Thomas, N., Boura, C., and Barberi-Heyob, M. (2019). Approaches to physical stimulation of metallic nanoparticles for glioblastoma treatment. *Advanced Drug Delivery Reviews* 138: 344–357.

44 Hainfeld, J.F., Dilmanian, F.A., Slatkin, D.N., and Smilowitz, H.M. (2010). Radiotherapy enhancement with gold nanoparticles. *Journal of Pharmacy and Pharmacology* 60 (8): 977–985.

45 Ma, N., Wu, F.-G., Zhang, X. et al. (2017). Shape-dependent radiosensitization effect of gold nanostructures in cancer radiotherapy: comparison of gold nanoparticles, nanospikes, and nanorods. *ACS Applied Materials & Interfaces* 9 (15): 13037–13048.

46 Cho, S.H., Jones, B.L., and Krishnan, S. (2009). The dosimetric feasibility of gold nanoparticle-aided radiation therapy (GNRT) via brachytherapy using low-energy gamma-/X-ray sources. *Physics in Medicine and Biology* 54 (16): 4889–4905.

47 Du, B., Yu, M., and Zheng, J. (2018). Transport and interactions of nanoparticles in the kidneys. *Nature Reviews Materials* 3 (10): 358–374.

48 Jia, T.-T., Yang, G., Mo, S.-J. et al. (2019). Atomically precise gold–levonorgestrel nanocluster as a radiosensitizer for enhanced cancer therapy. *ACS Nano* 13 (7): 8320–8328.

49 Alric, C., Taleb, J., Le Duc, G. et al. (2008). Gadolinium chelate coated gold nanoparticles as contrast agents for both X-ray computed tomography and magnetic resonance imaging. *Journal of the American Chemical Society* 130 (18): 5908–5915.

50 Miladi, I., Alric, C., Dufort, S. et al. (2014). The in vivo radiosensitizing effect of gold nanoparticles based MRI contrast agents. *Small* 10 (6): 1116–1124.

51 Luo, D., Wang, X., Zeng, S. et al. (2019). Targeted gold nanocluster-enhanced radiotherapy of prostate cancer. *Small* 15 (34).

52 Verry, C., Sancey, L., Dufort, S. et al. (2019). Treatment of multiple brain metastases using gadolinium nanoparticles and radiotherapy: NANO-RAD, a phase I study protocol. *BMJ Open* 9 (2): e023591.

53 Rajaee, A., Wang, S., Zhao, L. et al. (2019). Multifunction bismuth gadolinium oxide nanoparticles as radiosensitizer in radiation therapy and imaging. *Physics in Medicine and Biology* 64 (19): 195007.

54 Hada, Y. and Eto, M. (2003). Electronic states in silicon quantum dots: multivalley artificial atoms. *Physical Review B* 68 (15): 155322.

55 Juzenas, P., Chen, W., Sun, Y.-P. et al. (2008). Quantum dots and nanoparticles for photodynamic and radiation therapies of cancer. *Advanced Drug Delivery Reviews* 60 (15): 1600–1614.

56 Wang, X., Guo, Z., Zhang, C. et al. (2020). Ultrasmall BiOI quantum dots with efficient renal clearance for enhanced radiotherapy of cancer. *Advanced Science* 7 (6): 1902561.

57 Ma, X., Lee, C., Zhang, T. et al. (2021). Image-guided selection of Gd@C-dots as sensitizers to improve radiotherapy of non-small cell lung cancer. *Journal of Nanobiotechnology* 19 (1): 284.

58 Chan, L., Gao, P., Zhou, W. et al. (2018). Sequentially triggered delivery system of black phosphorus quantum dots with surface charge-switching ability for precise tumor radiosensitization. *ACS Nano* 12 (12): 12401–12415.

59 Mao, H. (2010). Delivery of therapeutic radioisotopes using nanoparticle platforms: potential benefit in systemic radiation therapy. *Nanotechnology, Science and Applications* 3 (1): 159.

60 Silindir-Gunay, M., Karpuz, M., and Ozer, A.Y. (2020). Targeted alpha therapy and nanocarrier approach. *Cancer Biotherapy & Radiopharmaceuticals* 35 (6): 446–458.

61 Daems, N., Michiels, C., Lucas, S. et al. (2021). Gold nanoparticles meet medical radionuclides. *Nuclear Medicine and Biology* 100–101: 61–90.

62 Sgouros, G. (2008). Alpha-particles for targeted therapy. *Advanced Drug Delivery Reviews* 60 (12): 1402–1406.

63 Yong, K. and Brechbiel, M. (2015). Application of 212Pb for targeted α-particle therapy (TAT): pre-clinical and mechanistic understanding through to clinical translation. *AIMS Medical Science* 2 (3): 228–245.

64 Zalutsky, M. and Vaidyanathan, G. (2000). Astatine-211-labeled radiotherapeutics an emerging approach to targeted alpha-particle radiotherapy. *Current Pharmaceutical Design* 6 (14): 1433–1455.

65 Umbricht, C.A., Köster, U., Bernhardt, P. et al. (2019). Alpha-PET for prostate cancer: preclinical investigation using 149Tb-PSMA-617. *Scientific Reports* 9 (1): 17800.

66 Beyer, G.-J., Miederer, M., Vranješ-Đurić, S. et al. (2004). Targeted alpha therapy in vivo: direct evidence for single cancer cell kill using 149Tb-rituximab. *European Journal of Nuclear Medicine and Molecular Imaging* 31 (4): 547–554.

67 Kratochwil, C., Haberkorn, U., and Giesel, F.L. (2020). 225Ac-PSMA-617 for therapy of prostate cancer. *Seminars in Nuclear Medicine* 50 (2): 133–140.

68 Miederer, M., Scheinberg, D.A., and McDevitt, M.R. (2008). Realizing the potential of the Actinium-225 radionuclide generator in targeted alpha particle therapy applications. *Advanced Drug Delivery Reviews* 60 (12): 1371–1382.

69 Song, E.Y., Rizvi, S.M.A., Qu, C.F. et al. (2007). Pharmacokinetics and toxicity of 213Bi-labeled PAI2 in preclinical targeted alpha therapy for cancer. *Cancer Biology & Therapy* 6 (6): 898–904.

70 Mulford, D.A., Scheinberg, D.A., and Jurcic, J.G. (2005). The promise of targeted α-particle therapy. *Journal of Nuclear Medicine* 46 (Suppl 1): 199S–204S.

71 Gill, M.R., Falzone, N., Du, Y., and Vallis, K.A. (2017). Targeted radionuclide therapy in combined-modality regimens. *The Lancet Oncology* 18 (7): e414–e423.

72 Le Goas, M., Paquet, M., Paquirissamy, A. et al. (2019). Improving 131I radioiodine therapy by hybrid polymer-grafted gold nanoparticles. *International Journal of Nanomedicine* 14: 7933–7946.

73 Degoul, F., Borel, M., Jacquemot, N. et al. (2013). In vivo efficacy of melanoma internal radionuclide therapy with a 131I-labelled melanin-targeting heteroarylcarboxamide molecule. *International Journal of Cancer* 133 (5): 1042–1053.

74 Kinuya, S., Yokoyama, K., Izumo, M. et al. (2005). Locoreginal radioimmunotherapy with 186Re-labeled monoclonal antibody in treating small peritoneal carcinomatosis of colon cancer in mice in comparison with 131I-counterpart. *Cancer Letters* 219 (1): 41–48.

75 Argyrou, M., Valassi, A., Andreou, M., and Lyra, M. (2013). Dosimetry and therapeutic ratios for rhenium-186 HEDP. *ISRN Molecular Imaging* 2013: 1–6.

76 Obata, A., Kasamatsu, S., Lewis, J.S. et al. (2005). Basic characterization of 64Cu-ATSM as a radiotherapy agent. *Nuclear Medicine and Biology* 32 (1): 21–28.

77 Holland, J.P., Ferdani, R., Anderson, C.J., and Lewis, J.S. (2009). Copper-64 radiopharmaceuticals for oncologic imaging. *PET Clinics* 4 (1): 49–67.

78 Goffredo, V., Paradiso, A., Ranieri, G., and Gadaleta, C.D. (2011). Yttrium-90 (90Y) in the principal radionuclide therapies: an efficacy correlation between peptide receptor radionuclide therapy, radioimmunotherapy and transarterial radioembolization therapy. Ten years of experience (1999–2009). *Critical Reviews in Oncology/Hematology* 80 (3): 393–410.

79 Gharibkandi, N.A., Żuk, M., Muftuler, F.Z.B. et al. (2023). 198Au-coated superparamagnetic iron oxide nanoparticles for dual magnetic hyperthermia and radionuclide therapy of hepatocellular carcinoma. *International Journal of Molecular Sciences* 24 (6): 5282.

80 Liang, C., Chao, Y., Yi, X. et al. (2019). Nanoparticle-mediated internal radioisotope therapy to locally increase the tumor vasculature permeability for synergistically improved cancer therapies. *Biomaterials* 197: 368–379.

81 Ku, A., Facca, V.J., Cai, Z., and Reilly, R.M. (2019). Auger electrons for cancer therapy – a review. *EJNMMI Radiopharmacy and Chemistry* 4 (1): 27.

82 Fourie, H., Nair, S., Miles, X. et al. (2020). Estimating the relative biological effectiveness of auger electron emitter 123I in human lymphocytes. *Frontiers of Physics* 8: 519.

83 Grudzinski, J., Marsh, I., Titz, B. et al. (2018). CLR 125 Auger electrons for the targeted radiotherapy of triple-negative breast cancer. *Cancer Biotherapy & Radiopharmaceuticals* 33 (3): 87–95.

84 Othman, M.F.b., Mitry, N.R., Lewington, V.J. et al. (2017). Re-assessing gallium-67 as a therapeutic radionuclide. *Nuclear Medicine and Biology* 46: 12–18.

85 Díez-Villares, S., Pellico, J., Gómez-Lado, N. et al. (2021). Biodistribution of 68/67Ga-radiolabeled sphingolipid nanoemulsions by PET and SPECT imaging. *International Journal of Nanomedicine* 16: 5923–5935.

86 Thisgaard, H., Olsen, B.B., Dam, J.H. et al. (2014). Evaluation of cobalt-labeled octreotide analogs for molecular imaging and Auger electron-based radionuclide therapy. *Journal of Nuclear Medicine* 55 (8): 1311–1316.

87 Arslan, F.B., Ozturk, K., and Calis, S. (2021). Antibody-mediated drug delivery. *International Journal of Pharmaceutics* 596: 120268.

88 Kouri, M.A., Polychronidou, K., Loukas, G. et al. (2023). Consolidation of gold and gadolinium nanoparticles: an extra step towards improving cancer imaging and therapy. *Journal of Nanotheranostics* 4 (2): 127–149.

89 Perry, H.L., Botnar, R.M., and Wilton-Ely, J.D.E.T. (2020). Gold nanomaterials functionalised with gadolinium chelates and their application in multimodal imaging and therapy. *Chemical Communications* 56 (29): 4037–4046.

16

Nanomedicine Conjugating with AI Technology and Genomics for Precise and Personalized Therapy

Lin Liu, Siyu Chen, and Sen Zhang

School of Mathematics and Physics, University of Science and Technology Beijing, 30 Xueyuan Road, Beijing 100083, China

16.1 Concept

Personalized and precise nanomedicine, integrating artificial intelligence (AI) and genomics-related technologies, is an approach to nanomedicine design that utilizes individual genomic information to predict and optimize the characteristics of designed nanomedicines, thereby achieving accurate and personalized nanomedicine design. A growing number of studies pertaining to genomic detection are being driven by big data research. Examples of these studies include locating disease-related genes in noncoding regions of the genome [1], evaluating the pharmacogenomic effects of polymeric materials [2, 3], and figuring out how genetic and environmental factors interact to influence the onset and course of diseases [4].

Nanotechnology refers to the scientific and technological research and application of materials at the nanoscale (typically in the range of 1–100 nm), encompassing nanomaterials, nanodevices, and nanosystems. In recent years, nanotechnology has made a significant impact in various fields, from electronics to cosmetics and pharmaceuticals. The emerging field of nanomedicine involves applying nanotechnology to biomedical and pharmaceutical sciences for effective disease diagnosis, treatment, and management. Nanomedicine involves the processing of drug molecules or drug carriers into nanoscale drug particles that can be administered through various routes, such as oral, topical, or injectable. Notably, nanocarriers play a pivotal role in loading small-molecule drugs or biopharmaceutical agents in nanomedicine development. Due to their small size, nanocarriers exhibit unique physicochemical properties that can enhance the therapeutic effectiveness of active drug components. Nanomedicines also possess excellent dispersibility and solubility, enabling rapid distribution within the body, and increasing drug bioavailability, thus providing a larger surface area for drug-cell interactions, thereby enhancing drug bioactivity. Additionally, nanomedicines are characterized by low toxicity and precision.

AI is a technology that extends and simulates human intelligence through computer programs. The goal of AI development is to enable computers to think,

Nanomedicine: Fundamentals, Synthesis, and Applications, First Edition. Edited by Yujun Song.
© 2025 WILEY-VCH GmbH. Published 2025 by WILEY-VCH GmbH.

perceive, learn, and communicate like humans, achieving human–computer interaction and autonomous decision-making capabilities. In recent years, the integration of AI and drug design has become increasingly close. AI can help drug development researchers discover and design new drug molecules faster and more accurately, improving the efficiency and success rate of drug development while reducing costs and time. AI technologies encompass machine learning and deep learning, allowing researchers to predict the pharmacokinetics and toxicity of nanomedicines, as well as their effects and side effects in different individuals.

Machine learning, as a subset of AI, involves using computational models to extract information and learn from data, making predictions and optimizing target variables through model training. Machine learning algorithms essentially discover patterns in data for prediction, related to statistical methods but different in the approach. In brief, there are two primary types of machine learning techniques: supervised learning and unsupervised learning. In supervised learning, the output values are known for corresponding input data used to train the model. Supervised learning is employed in classification or regression tasks, mapping input to discrete output labels or continuous output values, respectively. In contrast, unsupervised learning aims to discover patterns in unlabeled data, commonly used for exploratory analysis and clustering to identify inherent structures and meaningful groups.

Deep learning, another subset of AI, is a method of predicting and optimizing target variables through multilayered neural networks. Deep learning, compared to machine learning, employs more complex model structures, capable of handling more intricate data and tasks, and can automatically learn features and representations from data without the need for manual feature design. This greatly enhances learning efficiency and effectiveness. Deep learning can be employed in the analysis of large amounts of nanomedicine data to automatically learn features and representations, thereby predicting and optimizing nanomedicine structures and properties to enhance bioavailability, selectivity, and stability. Deep learning can also be used to predict the dose response of nanomedicines in different individuals for personalized therapy. Furthermore, deep learning can be used to predict the safety and toxicity of nanomedicines, improving drug development efficiency and success rates. AI has led to many groundbreaking discoveries in the fields of biology and medicine. AI dramatically facilitates clinical decision making, big data in medicine, and medical efficiency. The field of nanomedicine is currently receiving unprecedented attention in order to treat and cure precision medicine and therapeutic diagnostic imaging, which can improve biosensing of cellular and system processes, diagnose disease states, deliver medications or target therapies, and control the chemical activity of neuronal signals [5–9]. Some AI systems of these working are shown in Figure 16.1.

Genomics is a multidisciplinary biology field that collectively characterizes, quantifies, and compares all the genes of an organism. It encompasses the analysis of genomic phenotype and gene function, as well as comparative studies of genomic structure and function. Genomics helps researchers understand genetic variations and gene expression in individuals, providing crucial information for nanomedicine design. In genomics, researchers often use sequencing techniques to obtain an individual's genetic sequence, which can be used to study genetic variations and gene expression. Genetic variation refers to changes in the genetic sequence, which

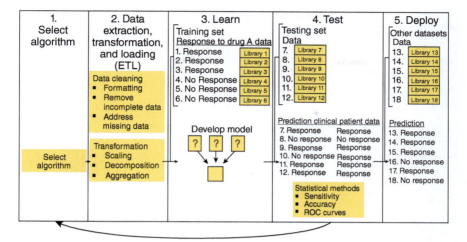

Figure 16.1 Schematic of machine learning steps for supervised learning. Machine learning is conducted in five phases.

may affect gene function. Gene expression involves the transcription of genes into RNA, which can further be translated into proteins, affecting an individual's physiology and pathology. By analyzing genetic sequences, researchers can understand genetic variations and gene expression in individuals. This information can be used to predict the effects and side effects of nanomedicines in different individuals. For instance, if an individual has a genetic variation in a specific gene, it may impact the absorption, distribution, metabolism, and excretion of nanomedicines, consequently influencing their effects and side effects. These studies are based on advancements in machine learning and deep learning technologies. AI technologies can also produce risk assessment models to forecast illness risk [10]. Figures 16.2 and 16.3 show the flow diagram of some of the technologies.

The integration of genomics and AI in precise and personalized nanomedicine design can assist scientists in designing safer and more effective nanomedicines. This approach not only enhances the therapeutic effects of nanomedicines but also helps scientists better comprehend the interactions between genes and drugs, promoting nanomedicine research and development.

One of the primary diseases targeted by nanomedicines is cancer. Scientists can use genomics and AI to design nanomedicines for cancer treatment. Cancer is a disease caused by genetic mutations and abnormal gene expression. By analyzing a patient's genomic information, scientists can understand the genetic variations and gene expression abnormalities present in the patient. They can then use AI to predict the effects and side effects of nanomedicines in the patient's body, optimizing the design of nanomedicines. This design approach can enhance the therapeutic effectiveness of nanomedicines and reduce side effects, thereby improving the survival rates of cancer patients. Additionally, the precise and personalized nanomedicine design, combining genomics and AI, can also be applied to treat other diseases, such as heart disease, diabetes, and neurological disorders. By analyzing a patient's genomic information, scientists can understand the genetic

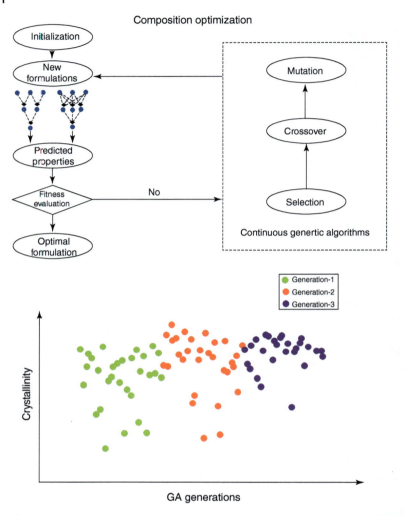

Figure 16.2 Schematic representation of the simultaneous multiobjective optimization processes including ANN for the formulation modeling and the continuous genertic algorithms.

variations and gene expression abnormalities in the patient's body. AI can be then used to predict the therapeutic effects and side effects of nanomedicines in the patient's body, optimizing the nanomedicine design. This approach can enhance the therapeutic effectiveness of nanomedicines while reducing side effects, thereby improving the patient's survival rate.

16.2 Genomics of Nanomedicine

Drug genomics is the study of the role of genomics in drug response, analyzing how an individual's genetic makeup affects their response to drugs. This involves

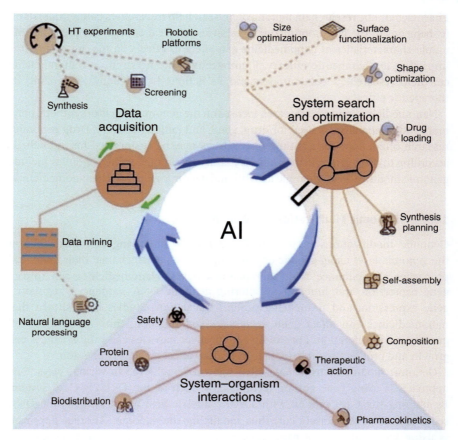

Figure 16.3 Artificial intelligence-assisted real-time dynamic computation and auxiliary data acquisition in nanomedicine.

linking gene expression or single nucleotide polymorphisms to pharmacokinetics, pharmacodynamics, and drug receptor target effects, and understanding the impact of both acquired and innate genetic variations on drug responses. Pharmacokinetics primarily studies the absorption, distribution, metabolism, and excretion of drugs in the body, while pharmacodynamics primarily examines the effects of drugs on living organisms.

Nanomedicine genomics is an important branch of nanomedicine research and a subfield of drug genomics. It aims to study the interactions between nanomedicine and genomics, as well as the role of genomics in the design and application of nanomedicine. Nanomedicine genomics can help researchers better understand the interactions between nanomedicine and genomics, as well as the role of genomics in nanomedicine design and application. Furthermore, nanomedicine genomics can provide new insights and methods for optimizing nanomedicine.

The pharmacokinetic characteristics of nanomedicine can affect its distribution and metabolism in the body, thereby influencing its pharmacological effects.

Pharmacokinetic research on nanomedicine can help researchers better understand its behavior in the body, optimizing its pharmacological effects. The pharmacodynamic characteristics of nanomedicine can affect its effects on the organism, thus impacting its therapeutic efficacy. Pharmacodynamic research on nanomedicine can help researchers better understand its actions in the body, optimizing its therapeutic efficacy.

Nanomedicine genomics research focuses on the genomic effects of synthetic polymers and nanomaterials used for drug, gene, and antigen delivery. Drug genomics research offers hope for a future where drugs may be customized for individuals according to their genomic makeup. Key issues in nanomedicine genomics include genomic modifications, genomic toxicity, and genomic responses to nanomedicines.

16.2.1 Genomic Modifications in Nanomedicine

Genomic modifications in nanomedicine involve modifying nanomedicines at the genomic level using genetic engineering techniques to change their biological properties and enhance their therapeutic efficacy. Nanomedicines can impact gene expression and function by altering genomic DNA sequences, regulating gene expression, and influencing protein translation. Studies have shown that nanomedicines can affect gene expression and function, thus influencing cell growth and proliferation. For example, nanosilver particles can inhibit cancer cell growth and proliferation due to their regulatory effect on gene expression.

Genomic modifications in nanomedicine can be achieved in various ways. One approach is using gene editing technologies, such as the CRISPR–Cas9 system. By binding the Cas9 protein with specific nucleic acid adapters, genomic DNA sequences can be precisely cut, thus modifying the genome. Another approach is using nanomaterials as gene delivery vectors to introduce target genes into host cells. This method leverages the efficient delivery capability and genomic modification precision of nanomaterials. Gene expression regulation involves changing gene expression levels to alter cellular biological properties, and it can be used in the design and optimization of nanomedicines to modify their properties and performance. For instance, gene expression regulation can alter nanomedicine bioavailability and stability and enhance their therapeutic efficacy.

Genomic modifications in nanomedicine offer new perspectives and methods for the design and optimization of nanomedicines. Genomic modifications can improve nanomedicine bioavailability and stability, enhancing their therapeutic efficacy. Additionally, genomic modifications can help scientists better understand the interactions between nanomedicines and genomics, optimizing therapeutic effects.

16.2.2 Genomic Toxicity of Nanomedicines

Genomic toxicity of nanomedicines refers to the phenomenon where nanomedicines cause genomic damage or changes in living organisms. Nanomedicines can harm the genome directly or indirectly, leading to DNA double-strand breaks, chromosomal aberrations, and cell apoptosis, affecting cell proliferation and differentiation.

Genomic toxicity is a significant concern because nanomedicines come into contact with the genome once they enter the human body, which may result in genomic damage. Genomic toxicity can influence the properties of nanomedicines, such as bioavailability, biodistribution, and biodegradation, thus affecting their therapeutic efficacy. Excessive genomic toxicity may lead to severe side effects and adverse reactions. Therefore, it is crucial to consider the impact of nanomedicines on the genome and take appropriate measures to reduce genomic toxicity during the design and development of nanomedicines.

Several strategies can be employed to reduce the genomic toxicity of nanomedicines. One strategy is to use low-toxicity nanomaterials. Another approach is to employ modification techniques, such as genetic and chemical modifications, to reduce the genomic damage caused by nanomedicines. Additionally, personalized nanomedicine design can be used to customize nanomedicines based on individual patient profiles, reducing genomic toxicity.

In summary, the genomic toxicity of nanomedicines is an important component of nanomedicine research that helps researchers better understand the biological effects and safety of nanomedicines. Genomic toxicity of nanomedicines is a crucial and complex issue that should be considered during the design and development of nanomedicines.

16.2.3 Genomic Responses to Nanomedicines

Genomic responses to nanomedicines refer to the phenomenon where nanomedicines induce genomic changes or responses within living organisms. Research indicates that nanomedicines can influence the genome by altering DNA sequences, regulating gene expression, and affecting protein translation. Additionally, nanomedicines can induce stress responses within the genome, enhancing cell adaptability and survival. For example, nanomedicines can promote antioxidant reactions and repair mechanisms, thus increasing cell survival and adaptability.

Genomic responses to nanomedicines are a significant concern because nanomedicines come into contact with the genome, which may impact genome functionality. Excessive genomic responses to nanomedicines may lead to severe side effects and adverse reactions. Therefore, it is essential to consider the influence of nanomedicines on the genome and take appropriate measures to reduce genomic responses during the design and development of nanomedicines. Figure 16.2 shows some diagrams of AI-recognizing genomes.

16.3 Artificial Intelligence Technology in Nanomedicine Development

In the rapidly advancing field of nanomedicine, the role of AI technology is indispensable, and the application of AI in nanomedicine is gaining increasing attention. The primary applications of AI technology in nanomedicine encompass the following three aspects.

16.3.1 Machine Learning Techniques

Machine learning is used to analyze extensive nanomedical data, such as cell images and tissue samples, to unveil the mechanisms of action and therapeutic effects of nanodrugs. Machine learning can be categorized into classification, clustering, prediction, and reinforcement learning algorithms.

Classification algorithms are employed for classifying and screening nanodrugs based on different characteristics or functions. This assists in analyzing the behavior of nanoparticles. For example, support vector machine (SVM) algorithms can model and classify various types of nanodrugs to determine which ones exhibit optimal therapeutic effects.

Clustering algorithms help categorize nanodrugs into different groups, leading to a better understanding of their mechanisms of action and therapeutic effects. For instance, K-means clustering is used to group different types of nanodrugs and identify which ones have the best therapeutic effects. Additionally, both SVM and K-means can be utilized for analyzing nanomedical images, such as MRI or CT scans, to identify cancer cells or other pathological abnormalities. This aids nanomedical experts in accurately predicting nanodrug efficacy and safety, thereby providing better treatment options.

Prediction algorithms provide researchers with opportunities to predict the behavior of nanoparticles, enabling a better understanding of their properties. This also assists in predicting nanoparticle behavior in different environments, which in turn enhances understanding of nanoparticle design, functionality, therapeutic effects, and potential side effects. Algorithms such as neural networks or random forests are applied to predict nanodrug efficacy and safety.

Reinforcement learning algorithms can optimize the design and preparation processes of nanodrugs. Techniques like Q-learning or deep reinforcement learning are used to improve the formulation and preparation of nanodrugs, leading to more efficient and precise nanodrug design, ultimately improving treatment success rates and efficiency.

16.3.2 Computer Vision Technology

Computer vision technology aids researchers in gaining a better understanding of the properties and behavior of nanomaterials through image analysis and processing. This is mainly used for visualizing and detecting nanodrugs. Techniques like fluorescence dyes or magnetic resonance imaging are used to observe the distribution and metabolism of nanodrugs within the body, thus enhancing treatment success rates and efficiency.

16.3.3 Natural Language Processing Technology

Natural Language Processing (NLP) is utilized for analyzing extensive literature and clinical data to identify the potential therapeutic effects and side effects of nanodrugs. Text mining and information extraction techniques are employed to analyze large volumes of literature and clinical data, extracting valuable information such as nanodrug mechanisms, efficacy, and side effects. This facilitates rapid screening of nanodrugs with potential therapeutic effects, discovering their side effects and safety, and designing more precise and effective treatment plans. Additionally, sentiment analysis techniques can be used to analyze the emotional responses and attitudes of patients toward nanodrugs, allowing healthcare providers to better understand patient needs and expectations and improve treatment efficacy and patient quality of life.

16.4 Artificial Intelligence Facilitates Precise and Personalized Nanomedicine Based on Genomics

In the process of advancing nanomedicine, an increasing number of challenges are constraining the translation of nanomedical theoretical therapies into clinical practice. The primary limitation of nanomedical therapies in clinical applications is the lack of standardized quantitative methods for detecting and analyzing patient responses to treatments. This is because doctors cannot effectively analyze the data generated during treatment in real-time to inform their subsequent decisions, thus hindering the clear therapeutic benefits for patients. Additionally, the lack of access to extensive genomic data and information when selecting candidate small molecule targets hampers the development of nanodrug-based therapies. To advance the field of nanomedicine, interdisciplinary development is essential, which can be facilitated through AI. Figure 16.3 shows a diagram of AI's disuse for auxiliary computing functions in nanomedicine.

The application of AI in the field of nanodrug design holds great potential. Through virtual screening and molecular design, it can efficiently predict the interactions between nanodrug molecules and targets, thereby identifying the most promising candidate drugs. AI can also extend the possibilities in drug metabolism and pharmacology prediction, enabling the evaluation of nanodrug metabolism pathways and safety, optimizing nanodrug design and dosage. Furthermore, AI can discover synergistic effects between different drugs in nanodrug combination therapies, enhancing treatment efficacy and reducing resistance risks. In the realm of rapid clinical trial design, AI optimizes trial protocols, expediting the introduction of new drugs to the market. It can also predict early side effects of nanodrugs, thus reducing clinical trial failure rates. Additionally, in patent and literature research related to drugs, AI can mine vast amounts of information to guide future nanodrug development. These applications of AI significantly enhance the efficiency, accuracy, and progress of nanodrug development, offering more opportunities and challenges for new, precise, personalized nanodrug discovery.

AI successfully overcomes the bottlenecks in translating nanomedicine by harnessing its significant computational power and throughput. AI has found extensive application in fields like treatment imaging and cancer nanodrug development, promoting the growth of nanomedicine. In this context, we mainly elucidate the synergy between the nanomedical paradigm and AI through examples in cell replacement therapy for diabetes and precision medicine in cancer treatment.

16.4.1 Cell Replacement Therapy for Diabetes

Type 1 diabetes (T1D) is characterized by the inability of a patient's pancreatic cells to produce and secrete insulin, leading to a range of metabolic and systemic issues. Current treatments for T1D include short-term therapy, primarily involving insulin injections, which do not cure the disease. Long-term therapy typically involves clinical pancreatic transplantation but is limited by high rates of post-transplant pancreatic cell loss and inadequate longitudinal monitoring of the graft. In the field of nanomedicine, there are new molecular imaging techniques, such as magnetic particle imaging, for tracking cell transplants and the development of novel nanodrugs for genetic modification and imaging of transplanted cells.

However, these studies face limitations due to a lack of standardized and quantifiable methods. Current research in nanomedicine has begun to employ various machine learning algorithms, such as K-means, an unsupervised machine learning clustering algorithm, in fields like biomedical imaging, for monitoring in vivo pancreatic islet cell transplants. K-means effectively and automatically segments and analyzes images from pancreatic islet MPI scans with high accuracy. This research extends beyond the boundaries of AI and nanomedicine, enabling the quantitative and real-time monitoring of pancreatic islet cell transplants for diabetes patients. This can empower physicians and researchers to make faster, more informed decisions and to automatically screen for early graft loss and potential signs thereof. Furthermore, in the field of pancreatic islet cell transplantation, many deep learning algorithms find widespread applications and offer greater predictive capacity on dynamic datasets.

Apart from providing an image analysis platform for monitoring pancreatic islet cell transplants and other cell therapies, AI can also assist nanoscience and technology-based TID treatment by monitoring glucose levels and triggering insulin release through optogenetic stimulation. Optogenetics involves transfecting insulin promoter-conjugated channel rhodopsin proteins into β cells, which open and allow the expression of functional insulin upon exposure to light. Deep learning algorithms, specifically those involving recurrent high-order neural networks, can monitor glucose levels and regulate insulin release triggered by timed optogenetic stimulation to maintain a stable state. The key distinction of recurrent high-order neural networks from convolutional neural networks lies in their ability to classify and predict dynamic systems using "dynamic neurons." By training recurrent high-order neural networks on dynamic protein responses under different conditions, the algorithm can make predictions and classifications for future instances, possessing high spatiotemporal specificity for measuring protein responses.

16.4.2 Precision Medicine in Cancer Treatment

Precision medicine is an emerging approach to disease prevention and management that considers individual genetic, environmental, and lifestyle variations. It relies heavily on targeting specific molecular targets with diagnostic probes. However, the histological characteristics of a patient's diseased tissue, as well as their interactions with other molecular processes and lifestyle factors, make the formulation and optimization of targeted nanoparticles or nanodrugs complex. Effective reduction of unnecessary binding and interactions with nontarget cells in healthy hosts can significantly improve treatment outcomes. Negative reactions can be caused by potential nonspecificity of nanoparticle probes and similar therapeutic agents. Therefore, AI algorithms for predicting biocompatibility, membrane-binding ligand interactions, and other surface-binding characteristics can advance the progress and development of nanomedicine.

Currently, research on gene screening and the selection of nanodrug targets for optimization using AI algorithms is limited. One study used a random forest classifier algorithm to determine candidate drugs based on specific genomic spectra of patients. The random forest algorithm, trained on previously labeled data, predicts the effectiveness of specific genomes by assessing selected features. In this study, researchers sequenced and analyzed tumor samples from 48 individuals, identifying mutated cancer drivers and categorizing them based on the different ways these drivers acted. However, this approach did not account for tumor heterogeneity and the complex interactions of multiple drugs within the body. Nonetheless, this research offers a promising direction for developing patient-specific therapies and provides hope for genome sequencing and drug design through machine learning.

AI can be uniquely applied to the development of nanodrugs in cancer treatment, for example, in designing and optimizing spherical nucleic acids for immune modulation using machine learning algorithms. Spherical nucleic acids are nanostructures with immune-modulating properties that are fundamentally different from the short linear oligonucleotides that make up traditional nucleic acids. They are 80 times more potent and show a 700-fold increase in antibody titer compared to conventional linear oligonucleotides, leading to enhanced treatment of lymphoma in mice. Spherical nucleic acids can actively traverse mammalian cell membranes without the need for transfection agents, resist nucleases, and carry a wide variety of complex payloads into many types of cells. Researchers proposed using supervised machine learning models to aid in the selection and development of spherical nucleic acids targeting the TLR9 pathway. TLR9 is composed of unmethylated cytosine-phosphate-guanine (CpG) dinucleotides, which can activate the secretion of cytokines and trigger an inflammatory response. In machine learning models used for nanodrug screening and development, supervised machine learning algorithms can accurately predict the TLR9 stimulatory activity of spherical nucleic acids based on structural features.

As the field of nanomedicine continues to advance, the modification of drug compounds and imaging agents with nanoparticles has significantly enhanced the efficacy of cellular therapies and imaging agents. Various imaging methods have been applied to track nanoparticles in vivo, facilitate drug delivery, and monitor treatment outcomes. Similar to traditional combination therapies, molecular imaging can further improve treatment outcomes by targeted delivery of multiple drugs, thereby maintaining their synergistic effects.

In the realm of AI, deep learning algorithms such as convolutional neural networks have found widespread use in tasks related to image segmentation and classification. They have been applied in the field of nanomedicine for monitoring tumor responses to chemotherapy. In a study, a low glycosylated Mucin-1 (uMUC1)-targeting peptide, EPPT, was combined with superparamagnetic iron oxide nanoparticles for targeting ovarian cancer cells in a mouse model. This approach quantified the interference of the probe with weighted signals to evaluate changes in the expression of biomarkers after chemotherapy. Deep learning algorithms were employed to segment and quantify regions of interest. The convolutional neural network demonstrated satisfactory results in classifying tumor treatment responses based on established criteria for solid tumors. However, the limited training dataset in this study included only one type and location of cancer, which presents scalability limitations.

These examples, along with many others in the fields of precision medicine and molecular imaging, illustrate that AI offers the necessary analytical capabilities for advancing the forefront of nanomedicine through standardization and automation.

The importance of AI in the development of nanodrugs cannot be overstated. Whether through machine learning or deep learning algorithms for analyzing genomics data or employing computer simulations for the design of more effective nanodrugs, AI contributes to the progress of nanomedicine and the treatment of human diseases. In the future, with the continuous development and application of AI technologies, the design and development of nanodrugs will become more efficient and precise. Additionally, AI can assist scientists in better understanding the interactions between genomics, proteomics, and cell lines, leading to the creation of more personalized and precise treatment strategies. In summary, AI's application in precision medicine and nanodrug development holds significant importance, with ample room for further development in the future.

References

1 Huang, S., Cai, N., Pacheco, P.P. et al. (2018). Applications of support vector machine (SVM) learning in cancer genomics. *Cancer Genom. Proteom.* 15 (1): 41–51.
2 Williams, A.M., Liu, Y., Regner, K.R. et al. (2018). Artificial intelligence, physiological genomics, and precision medicine. *Physiol. Genom.* 50 (4): 237–243.
3 Kabanov, A.V. (2006). Polymer genomics: an insight into pharmacology and toxicology of nanomedicines. *Adv. Drug Deliv. Rev.* 58 (15): 1597–1621.
4 Cano-Gamez, E. and Trynka, G. (2020). From GWAS to function: using functional genomics to identify the mechanisms underlying complex diseases. *Front. Genet.* 11: 424.
5 Hayat, H., Nukala, A., Nyamira, A. et al. (2021). A concise review: the synergy between artificial intelligence and biomedical nanomaterials that empowers nanomedicine. *Biomed. Mater.* 16 (5): 052001.

6 Ho, D., Wang, P., and Kee, T. (2019). Artificial intelligence in nanomedicine. *Nanoscale Horiz.* 4 (2): 365–377.

7 Serov, N. and Vinogradov, V. (2022). Artificial intelligence to bring nanomedicine to life. *Adv. Drug Deliv. Rev.* 184: 114194.

8 Wang, J., Li, Y., Nie, G., and Zhao, Y. (2019). Precise design of nanomedicines: perspectives for cancer treatment. *Nat. Sci. Rev.* 6 (6): 1107–1110.

9 Zaslavsky, J., Bannigan, P., and Allen, C. (2023). Re-envisioning the design of nanomedicines: harnessing automation and artificial intelligence. *Expert Opin. Drug Deliv.* 20 (2): 241–257.

10 Mattson, D.L. and Liang, M. (2017). From GWAS to functional genomics-based precision medicine. *Nat. Rev. Nephrol.* 13 (4): 195–196.

17

Microfluidic Conjugating AI Platform for High-throughput Nanomedicine Screening

Xing Huang[1], Wenya Liao[1,2], Zhongbin Xu[1,2], and Yujun Song[3,4,5]

[1]*Department of Mechanical Engineering, Zhejiang Provincial Engineering Center of Integrated Manufacturing Technology and Intelligent Equipment, Hangzhou City University, 50/51 Huzhou Street, Hangzhou 310005, China*
[2]*College of Energy Engineering, Institute of Process Equipment, Zhejiang University, 38 Zheda Rd, Hangzhou 310027, China*
[3]*University of Science and Technology Beijing, Center for Modern Physics Technology, School of Mathematics and Physics, 30 Xueyuan Road, Haidian District, Beijing 100083, China*
[4]*Zhengzhou Tianzhao Biomedical Technology Company Ltd., Zhengzhou New Technology Industrial Development Zone, 7B-1209 Dongqing Street, Zhengzhou 451450, China*
[5]*Key Laboratory of Pulsed Power Translational Medicine of Zhejiang Province, Hangzhou Ruidi Biotechnology Company Ltd., Room 803, Bldg. 4, 4959 Yuhangtang Road, Cangqian Street, Hangzhou 310023, China*

17.1 Introduction

With the rapid advancement of nanoscience, nanomedicines are assuming an increasingly vital role in modern medicine, spanning early diagnosis, disease prevention, and treatment, among other applications. Nanomedicines can be broadly categorized into two primary groups: nanodrug carriers and nanodrug particles [1]. In contrast to conventional drug preparation methods, nanomedicines leverage the unique effects associated with their finite size and surface characteristics. These attributes can enhance the solubility of otherwise insoluble drugs, improve the pharmacokinetics and safety profiles of nanomedicines [2], bolster drug targeting capabilities [3, 4], enhance the stability of drug delivery processes, and afford better control over drug release kinetics [5]. These distinctive properties make nanomedicines particularly well-suited for applications in the realm of targeted drug therapy, especially in the context of cancer treatment.

However, the development of novel nanodrugs typically entails a protracted timeline and substantial financial investments. Consequently, the expedient and efficient execution of drug screening poses a critical challenge in the domain of new drug discovery. To date, only a few nanomedicines have been approved by the US Food and Drug Administration and the European Medicines Agency, including liposomes (Doxil) encapsulating anticancer drugs for chemotherapy [6],

Xing Huang and Wenya Liao contribute equally to this chapter.

Nanomedicine: Fundamentals, Synthesis, and Applications, First Edition. Edited by Yujun Song.
© 2025 WILEY-VCH GmbH. Published 2025 by WILEY-VCH GmbH.

iron oxide NPs (Feridex [7] and Resovist [8]) for magnetic resonance imaging, and polymer NPs (Copaxone and Neulasta) for drug delivery and diagnostic imaging [9]. On one hand, development of a novel nanomedicine relies on the discovery of appropriate target-based drugs, which takes a lot of time to screen out suitable lead compounds from hundreds of thousands to millions of compounds. On the other hand, due to ethical issues, limited animal models and human models are available for drug screening. The traditional 2D cell culture cannot simulate the human body environment precisely and thus fails to obtain credible results on effect of drugs. Although the emergence of 3D cell models has partially filled the gap between oversimplified 2D models and nonrepresentative animal models [10–12] to achieve better simulation of the in vivo environment, they still face problems such as low throughput, low efficiency and high cost for nanodrug screening.

Emerging microfluidic technologies and artificial intelligence (AI) provide a new concept for developing efficient platforms for nanomedicine screening. On the one hand, in recent years, computer-aided drug design and screening based on AI have shown powerful abilities in accelerating virtual screening and structural design of nanomedicines [13–15]. On the other hand, microfluidic devices provide precisely controlled and well-simulated in vivo microenvironments for cells [16, 17]. By virtue of simple design, high integration degree and high economic efficiency, microfluidic methods enable highly monodisperse drug preparation and high-throughput screening [18]. Microfluidic drug screening generates a large amount of data, which is naturally suitable for training of AI models. Therefore, nanomedicine screening systems combining AI and microfluidic technology are expected to further promote the high-throughput screening of nanomedicine. In this chapter, the application of microfluidic technology in the synthesis and screening of nanomedicine and nanodrug screening based on AI and microfluidic technology are reviewed. Finally, the synergistic effect of the two technologies on nanodrug screening and future development trends are prospected.

17.2 Microfluidic Technologies for Medicine Development

17.2.1 Basics of Microfluidics

Microfluidic technologies, also known as lab on a chip (LOC) or micrototal analysis system (μTAS), accurately control and manipulate fluids on microscale [19]. It integrates basic functions of biology, chemistry, and other laboratory operations on a chip of several square centimeters, such as sample preparation, reaction, separation, and detection. The flexible combination and integration of the multiple basic operations allow microfluidics to be widely used in microprocessing, material synthesis, and especially biomedical engineering.

The microfluidic technique originated from capillary electrophoresis. In the early 1990s, Manz et al. first proposed the concept of microfluidic chip and then put forward the concept of μTAS through early chip electrophoresis research. The initial microfluidic technologies are mainly used in analysis, which achieves high-precision separation and detection in the case of a small number of samples and reagents, and meanwhile has the advantages of low cost and short analysis time. This is because the reduction of the microchannel size not only makes the device size miniaturized and lightweight but also makes it easy to carry for onsite analysis. More importantly, its performance has been significantly improved benefiting from the distinguished fluid flow behaviors in the microsized channels. Compared to macroscale flows, a laminar flow state with a low Reynolds number dominated by viscous force is performed in the microchannels, which facilitates precise control of the fluid. Besides, small channels have high specific surface area and thus high heat transfer efficiency, leading to higher reaction speed.

Microfluidic systems are usually composed of basic functional components such as micropumps, microvalves, microfluidic sensors, micromixers, and microreactors. Micropumps play the role of fluid transportation as the power source of microfluidic systems. The microvalves play the role of fluid control. The micromixers and microreactors are the output components of the microfluidic system, which execute operations such as accelerating fluid mixing and controlling biological or chemical reactions, respectively. In microfluidic systems, the fluid operation schemes are usually divided into active mode and passive mode according to whether the fluid control mode has participation of external energy. For passive type, it makes the fluid deform at interface of the flow field by means of the significant physical phenomena in the microscopic field of the fluid, such as surface tension, capillary action, geometric structure, and fluid characteristics of the fluid channel. For active type, microfluidic control is mainly achieved by external force stimulation or local drive, such as electric field [20], magnetic field [21], sound field [22], and light field [23].

17.2.2 Fabrication of Microfluidic Chips

The analytical performance of microfluidic chips largely depends on the forming quality of the microchannels. Different materials have different processing technologies. With the continuously expansion of the application fields, higher processing accuracy, more complex structure, and lower price requirements are proposed for the microfluidic chips. The materials used to fabricate microfluidic chips include silicon materials (such as monocrystalline silicon wafers, quartz, and glass), organic polymer materials (such as polymethylmethacrylate [PMMA], polydimethylsiloxane [PDMS], polycarbonate [PC], and hydrogels), and paper microfluidic chips.

Siliceous materials are the main materials of early microfluidic chips, which mainly adopts microprocessing technologies such as lithography, wet etching and novel flow lithography, and soft lithography. Photolithography and wet etching technology are usually composed of three processes: film precipitation, lithography, and etching [24]. Firstly, a thin photoresist film is coated on the substrate uniformly. The pattern on the mask is transferred to the photoresist by ultraviolet light irradiation and developed. For photolithography, the developed photoresist is ready for use as a mold. For etching methods, the substrate is further etched by specific chemical environments, followed by wash-off of photoresist to form a microchannel with the etched substrate as a micromold or directly using it as a chip.

The microchannel structure fabricated by lithography process has high precision and can realize the fabrication of complex channel structures. However, the mask must be manufactured by lithography in advance. Once multiple lithography is required to obtain channels with different depths, processing cost increases rapidly. Therefore, this method is mostly suitable for mass production of simple patterns [25].

Organic polymer chips are mainly fabricated by molding methods (including hot embossing, imprinting, soft lithography, injection molding, etc.) and direct fabrication methods (including laser ablation, 3D printing and micromilling, etc.) [26–28].

Molding methods use positive molds made by lithography or etching methods as micropatterning templates. Representatively, soft lithography, developed from the standard lithography for processing glass and silicon chips by the Whitesides group, is now one of the most popular methods in lab-scale fabrication of microfluidic chips (as shown in Figure 17.1a). The soft lithography method directly casts the PDMS precursor on the microstructures formed by the photoresist. After the PDMS is thermally polymerized, it is separated from the template to obtain the microchannels. It does not need expensive equipment but can obtain channels with good optical properties and biocompatibility, which makes it widely used in cell analysis and other fields. Considering the limited solvent resistance and gas permeability of PDMS, plastic materials are more suitable for applications such as organic microreactors or polymerase chain reactions (PCR). Hot embossing and imprinting methods melt plastic pellets or partially melt flat plastic substrates to mold microstructures onto plastic surface. To further improve productivity, mold injection is adopted for massive manufacturing of plastic microfluidic chips. LIGA technology (as shown in Figure 17.1b), which is short for Lithographie, Galvanoformung, and Abformung, is a typical method based on mold injection. Micropatterns are usually produced by X-ray-deep lithography or substitutive ultraviolet light lithography. Metal mold is then formed by microelectroforming. After that, micromold injection is conducted to produce plastic chips. Due to the need for lithography mask production and mold injection equipment, the production cost is usually very high. But it is still one of the best methods to process polymer chips with a high aspect ratio, high resolution, and high efficiency [29].

Compared to molding methods, direct fabrication methods process polymer chips by selectively removing or adding materials on substrates. For example, laser processing technology is a noncontact processing method, which can be directly implemented according to CAD data. Microchannels are engraved by burning the polymers with high-energy lasers. Similarly, micromilling removes materials by mechanical milling to create microchannels. Compared with methods mentioned above, laser processing and micromilling save the process of molding, shorten the production time and improve the production efficiency for prototyping. 3D printing technology is a processing method that additively 3D packs precursors layer-by-layer to form designed structures. It directly generates the final product quickly and accurately through the computer. Due to its low cost, simple processing, and high efficiency, 3D printing technology has gradually become one of the most attractive microfluidic chip processing methods. The use of 3D printing technology

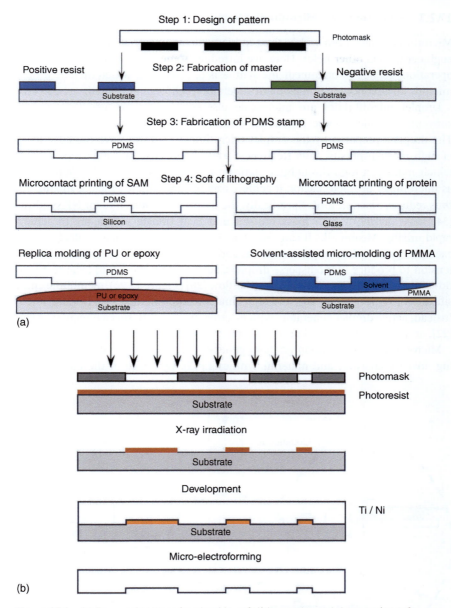

Figure 17.1 (a) Four major steps involved in soft lithography and three major soft lithographic techniques. (b) Simplified fabrication process for LIGA technology.

can significantly simplify the processing of microfluidic chips [30, 31], and the choice of printing materials is also very flexible. Compared with other microprocessing technologies, it greatly reduces the technical requirement and processing cost of microfluidic chips, which has a very significant impact on the application of microfluidic technology.

17.2.3 Representative Microfluidic Units

Microfluidics have been widely used in biomedicine, material synthesis, chemical engineering and other fields. Here, we introduce some representative microfluidic operations related to nanomedicine synthesis or screening, which include droplet microfluidics, concentration gradient generators, cell/organ-on-a-chip, and so on.

Droplet microfluidics are microscale discretization techniques that transfer conventional microfluidic functional elements into independent microdroplets. The droplets prepared by the droplet microfluidic platform have the advantages of high monodispersity, high frequency, controllable size, and high repeatability, which make it a high-throughput platform for experiments [32, 33]. When used as micromixer or microreactor, it achieves significantly more efficient mixing and thus reduces reaction time. Specifically, the droplet microreactors for NP synthesis [34] have merits of high mass transfer and heat transfer efficiency to achieve rapid mixing. Meanwhile, it can also avoid cross-contamination and blockage. For example, passive droplet generators (as shown in Figure 17.2 [35]) utilize surface tension to produce monodisperse droplets, such as cross-flow [36] (Figure 17.2a), flow-focusing [37] (Figure 17.2b), coaxial-flow [38] (Figure 17.2c), membrane emulsification [39] (Figure 17.2d), and step emulsification method [40] (Figure 17.2e). Instead, active droplet generators with electrospray [41, 42], surface acoustic waves [22], or magnetic forces [21] provide better control over droplet size and dispersity.

Microfluidic concentration gradient generators are widely used in drug screening, toxicity analysis, and material synthesis [43]. By virtue of the laminar flow

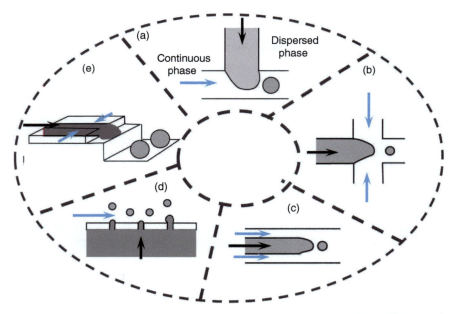

Figure 17.2 Schematic of various microfluidic device geometries (a) T-junction device, (b) flow-focusing, (c) coaxial-flow, (d) membrane emulsification, and (e) step emulsification method.

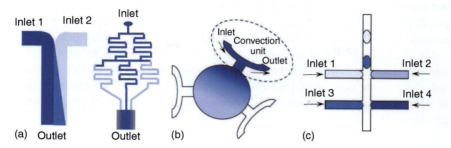

Figure 17.3 (a) Convection-based method: (left) (right) Y junction, (right) "Christmas tree" structure microfluidic concentration gradient chip. (b) diffusion-based method, and (c) droplet-based method.

behavior, the microfluidic concentration gradient generators create precisely controlled concentration gradient fields in microchannels, while reducing the use of chemical reagents and costs. In 2000, Jeon et al. [44] (Figure 17.3a) first proposed the concept of microfluidic concentration gradient and designed a classic Christmas tree structure based on the laminar diffusion principle of low Reynolds number to achieve a continuous change of concentration gradient. Up to now, the concentration gradient generation methods based on microfluidics mainly include convection-based [44, 45], diffusion-based [46], and droplet-based methods [47] (as shown in Figure 17.3).

Organ-on-a-chip is another major sort of microfluidic applications. Microfluidic chips are used to control the fluids, cells, and their microenvironment to construct systems that simulate the function of human tissues and organs. They serve as low-cost research models and screening platforms for evaluation of effectiveness and biosafety of drugs and vaccines under similar environment with real physiological and pathological conditions of human body. Implementation of verification of candidate drugs, drug toxicology, and pharmacological effects on organ-on-a-chip is used to realize individualized treatment. Current organ-on-a-chip mainly includes lung-on-a-chip [48–50], kidney-on-a-chip [51–53], brain-on-a-chip [54, 55], heart-on-a-chip [56, 57], and vessel-on-a-chip [58, 59], which mimic corresponding organs or tissues. Multiorgan-on-a-chip [60] combines more than one kind of organ-on-a-chip to study coordination effects between different organs.

17.3 Microfluidic Preparation of Nanomedicines

Nanomedicines are mainly divided into two categories: nanodrug particles and nanodrug carriers. Making drugs as NPs increases solubility by increasing the surface area, and thus avoiding toxicity brought by cosolvents that are frequently used in insoluble compounds [61–63]. However, production of nanodrug particles is relatively complex and costly. In contrast, nanocarrier drugs utilize nanocarriers, such as liposomes, polymer NPs, and polymer micelles, to improve curative effects. However, conventional nanomedicine preparation methods have limited precision but meanwhile complex fabrication process, which easily leads to formation of polydisperse particles and also aggregation of particles. Therefore, the prepared nanomedicines are likely to fail to proceed to drug tests on in vitro models [64].

The clinical translation of nanomedicines remains slow due to difficulties in assessing and predicting their behavior in complex organisms. Microfluidic technologies provide good platforms for preparation of high-quality nanodrugs, such as NP processing, nanostructure assembling, and high-throughput preparation of nanodrugs [65, 66].

17.3.1 Drug Nanoparticle Preparation

Nanonization of drugs can change physical and chemical properties of drugs, which thus affects absorption and metabolism of drugs and ultimately achieves the purpose of enhancing drug efficacy and reducing adverse drug reactions. For example, in 1995, Liversidge and Cundy [62] successfully increased oral bioavailability of Danazol (Danazol is a hydrophobic compound [$10\,\mu g\,ml^{-1}$] with poor bioavailability) from 5.1% of conventional suspension to 82.3% of nanosuspension (NS) by reducing diameter of drugs.

The main methods for producing drug NPs can be classified as top-down and bottom-up methods. In top-down approaches, large drug particles are converted into nanosized particles by external force. These methods have high repeatability and are easy for industrial production, but meanwhile have drawbacks like contamination, uncontrolled particle growth, wide particle size distribution, and high equipment costs. Bottom-up approaches are to keep the size of drugs at nanoscale during preparation. These methods are usually used to produce amorphous particles, thereby improving the solubility and bioavailability of drugs. Bottom-up approaches are simple, rapid, energy-efficient, and cost-effective, but often lead to rapid particle growth and poor size distribution [64, 67, 68].

Emerging microfluidic nanoprecipitation, as a recent, simple and cost-effective bottom-up technique of drug nanonization, provides novel platforms for preparing NPs with stable physical and chemical properties, uniform size, and high repeatability. For example, to further improve drug therapy of Danazol, Zhao et al. [69] presented a microfluidic-assisted liquid antisolvent precipitation (LASP) method for the direct synthesis of danazol NPs without additives. In this method, danazol solution and deionized water were, respectively, pumped into two feed-in channels and then mixed at junction of channels to prepare precipitation. Chemical composition and physical properties of Danazol NPs prepared by this method remained unchanged, while the dissolution rate of drugs increased significantly due to reduced particle size and expanded specific surface area. Danazol NPs obtained by this method achieved 100% dissolution in five minutes, while the dissolution rate of original Danazol particles was 35%. Similarly, Ali et al. [70] developed hydrocortisone (HC) NSs for ophthalmic delivery using microfluidic nanoprecipitation. They show that NSs prolonged effects with comparable significant improvement in ocular bioavailability in rabbits relative to the drug solution. Wang et al. [71] combined Y-shaped microchannels and nanoprecipitation to obtain amorphous cefuroxime axetil (CFA) NPs with controllable size. In this method, acetone and isopropyl ether were used as solvents and antisolvents of CFA. Further, size of NPs can be reduced by reducing the temperature, CFA concentration, and flow rate of the CFA solution in acetone or by increasing the flow rate of the antisolvent. The NPs precipitated by microfluidic method are more uniform and monodisperse than the NPs obtained by spray drying method. Shrimal et al. [72] used a serpentine microfluidic channel to improve quality of nanoscale particles of poorly water-soluble drug telmisartan (TEL), thereby increasing its solubility and bioavailability. In this method, solvents and antisolvents mixed at T-junction are further infused into a serpentine microchannel to enhance mixing. TEL crystal size decreased to 423.65 nm and polydispersity index (PDI) value decreased to 0.115. The above studies prove the microfluidic nanoprecipitation technology as a promising technology for reducing the size and polydispersity of drug nanocrystals (NCs).

17.3.2 Nanocarrier Preparation

Nanomaterials like polymer NPs, lipid NPs (LNPs) and inorganic NPs have been widely used as nanocarriers for drug delivery due to their low toxicity and high bioavailability. In contrast to drug NPs, nanocarriers usually do not have curing effect but instead carry drug molecules to the desired position for release. Up to now, more than 10 nanocarrier drugs have been approved such as AmBisome, Doxil, Onpattro, Comirnaty, and Spikeva [73–77].

17.3.2.1 Polymer Nanoparticles

Various polymeric NPs have been widely considered as nanocarriers, including natural (gelatin, albumin, etc.) and synthetic polymers (polylactic acid, polyacrylate, etc.). The preparation methods of polymeric NPs are generally divided into two categories: dispersion of polymers and polymerization of monomers. Conventional polymeric NPs preparation methods have drawbacks of large diameter, high toxicity, complex preparation process, and low efficiency. To solve these problems, microfluidic technologies have been gradually adopted to prepare polymeric NPs in recent years.

Frequently used natural polymers for preparation of NP carriers include chitosan, alginate, shellac, etc. Majedi et al. [78] presented a T-shaped microfluidic chip for the synthesis of monodisperse chitosan NPs via self-assembly at physiological pH. Obtained chitosan NPs by this method are shown to encapsulate hydrophobic anticancer drugs and provide a sustainable release profile with high tunability. Alginate NPs can be formed by polyelectrolyte complexation in a mild aqueous environment. However, alginate NPs formed by conventional methods aggregate easily. To minimize aggregation and narrow size distribution, Kim et al. [79] combined polyelectrolyte complexation and microfluidic mixing devices to prepare alginate NPs. By changing the flow rates of the Ca-alginate pre-gel and PLL solutions, the size of alginate NPs could be controlled (380–520 nm). El-Sayed et al. [80] used another method to prepare alginate NPs by combining T-shaped microfluidic devices and bubble bursting. In this method, alginate NPs can be produced with different sizes (80–200 nm) depending on the viscosity of the solution used to prepare alginate microbubbles. Kong et al. [81] used flow-focusing microfluidic devices to achieve rapid mixing of curcumin and shellac solutions with water in ethanol, increasing NP drug encapsulation efficiency and drug loading. In a study conducted by Baby et al., a microfluidic nanoprecipitation approach was proposed to make curcumin-loaded shellac NPs with tunable drug loading up to 50%. These NP nanocarriers are highly stable and pH-responsive, having great potential in field of oral nanodrugs [82, 83].

Synthetic polymer nanocarriers frequently adopt poly(ethylene glycol) (PEG), polyvinylpyrrolidone (PVP), and poly(lactic-co-glycolic) acid, etc., as matrix materials. Here, to briefly introduce the synthesis methods, we take PLGA as an example. In a study conducted by Karnik et al. [83], poly(lactide-co-glycolide)-b-poly(ethylene glycol) (PLGA-PEG) NPs were produced on a flow-focusing microfluidic device. They found that microfluidic preparation favors smaller particle sizes (34 nm) at a high PLGA: PLGA-PEG content. They further changed flow rate and polymer concentration to optimize sizes of NPs, which is beneficial to controlling NP drug load and release. To solve problem of channel blockage during the NP manufacturing process of two-dimensional hydrodynamic flow focusing (HFF) devices, a 3D HFF device [84] was proposed. This microfluidic chip consists of three continuous inlets and horizontal focusing cross-junctions. The polymers are sequentially focused in the vertical and horizontal directions to achieve 3D focusing, and self-assembly of NPs is achieved downstream of the channel. To further improve the high-throughput preparation of NPs, Lim et al. [85] proposed a parallelized 3D HFF device with eight identical HFF channels.

17.3.2.2 Liposomes

Liposomes represent one of the earliest and most established nanodrug carriers, characterized by their exceptional stability, remarkable drug loading efficiency, substantial surface area, and superior biocompatibility when compared to polymers. They are often used to encapsulate hydrophilic, hydrophobic drugs and ribonucleic acids. Traditional lipid nanocarrier preparation methods include solvent emulsification diffusion, high-pressure homogenization, microemulsion, solvent emulsification evaporation, solvent injection/solvent displacement, membrane contraction, phase inversion techniques, microemulsion cooling, supercritical fluid method, gas-assisted melting atomization, and ultrasonication [86]. However, the LNP particles obtained by these methods are large and polydisperse, which easily leads to reduction of tissue penetration and activity of LNPs. Studies have revealed that average particle size/diameter and the PDI are key parameters of lipid-based nanocarriers, where obtaining lipid nanocarrier preparations with constant and narrow size distribution is one of the key problems in achieving optimal clinical results. Microfluidic technologies have emerged as novel approaches for the preparation of lipid NPs (LNPs). In contrast to conventional methods, these technologies offer enhanced precision in regulating particle size and facilitate concurrent high-throughput production. In a study conducted by Jahn et al. [87], microfluidic flow-focusing device was used to achieve spontaneous self-assembly of liposomes in phosphorus liposome solution. Through manipulation of parameters such as flow rate, channel dimensions, and lipid composition, precise control over the size range of liposomes, maintained within the 100–300 nm range, was achieved. Krzyszton et al. [88] used microfluidic flow-focusing device to prepare small nucleic acid/lipid particles (38 nm). Compared with vortex mixing method, it increases the encapsulation efficiency by 20%. Hood et al. [86] employed 3D microfluidic flow-focusing devices as an alternative to their 2D counterparts. This transition resulted in a significant increase in the production yield of lipid particles, by up to four orders of magnitude, primarily attributable to the heightened operational efficiency of the 3D system. Microfluidic systems have the capacity to precisely mix solvents within the temporal range of microseconds to milliseconds, which notably outpaces the characteristic time scale associated with lipid aggregation, typically falling within the range of 10–100 ms. Thus, smaller particles with uniform size can be produced using this microfluidic device [89]. Chen et al. [90] harnessed a microfluidic device featuring a crossed herringbone micromixer to achieve the swift generation of LNPs incorporating siRNA. The resultant LNPs exhibited a size approximately threefold smaller (ranging from 60 to 90 nm) and showcased reduced size heterogeneity in comparison to LNPs produced through conventional methodologies.

17.3.2.3 Inorganic Nanoparticles

Compared to organic NPs, inorganic NPs, such as quantum dots (QDs), metal oxide NPs, and silica NPs hold unique characteristics like superior mechanical and chemical stability, easy surface functionalization, and large surface area [91]. Nonetheless, NPs of inorganic origin, synthesized using conventional techniques, tend to be prone to aggregation. Microfluidic synthesis methods can not only reduce preparation time of inorganic NPs but also improve physical properties of inorganic NPs [92, 93]. Also, microfluidics provide convenience for high-throughput production of inorganic NPs.

QDs are 5–20 nm-sized semiconductor NCs with outstanding electronic and optical performances such as high fluorescence brightness, long fluorescence lifetime, good stability, etc. The nucleation and growth of QDs can be controlled by changing the microfluidic structure and adjusting the microfluidic parameters, thus controlling size and structure of QDs [93]. In 2002, Yen et al. [94] successfully prepared a series of colloidal CdSe NCs in a microfluidic capillary. Two precursor liquids: cadmium oleate and tri-n-octyl phosphine selenide in a high-boiling solvent system were delivered into a continuous flow microreactor and combined in a mixing chamber before the mixture was led to a heat glass reaction channel (180–320 °C) to form CdSe QDs. By changing the precursor concentration, it was able to control the nucleation rate of the CdSe NCs. Carbon QDs, another kind of important QDs have also been synthesized using microfluidic approaches to improve the synthetic efficiency and quantum yields. Shao et al. [95] synthesized carbon QDs with full-spectrum emission fluorescence from citric acid and urea solutions as precursors using a continuous flow microreactor. The microflow-based reaction greatly reduced the conventional reaction time from 12 to 24 hours to

20 min, while the uniform and rapid mass transfer facilitated the production of smaller (average diameter of 2.88 nm) and more uniform carbon QDs than those synthesized in the autoclave.

Silica NPs, including mesoporous silica NPs, represent another crucial category of inorganic NPs, notable for their chemical stability and biodegradability [96]. Silica NPs prepared by microfluidic methods have high monodispersity and quality compared to other methods. Yan et al. [97] proposed a microfluidic-based platform to prepare pH/redox-triggered mesoporous silica NPs for doxorubicin/paclitaxel (DOX/PTX) codelivery. Drug ratio was precisely controlled by microfluidics to improve stability, dispersion, and loading of the hydrophobic drug PTX. Other studies showed that porous structures can be easily and precisely adjusted by microfluidic technologies. For example, Zhao et al. [98] proposed a capillary microfluidic device with three cylindrical glass capillary tubes (one inner, one middle, and one outer) nested within a square glass tube. The porous structure adjustment of NPs is realized by flow regulation.

17.4 Microfluidic High-throughput Drug Screening

Drug screening, an indispensable phase in the drug development process, involves the identification of biologically active lead drug compounds from a vast pool of drug molecules. The scope and complexity of drug screening have grown progressively challenging in light of the ever-expanding drug libraries and the diversification of drug action targets. In comparison to conventional drug screening methods reliant on Petri dishes and animal models, microfluidic platforms offer the advantage of closely emulating the microenvironment of human cells and tissues. This characteristic circumvents ethical concerns linked to animal models and mannequins, reduces reagent wastage, and minimizes costs. Moreover, the high degree of system integration within microfluidic setups facilitates efficient high-throughput drug screening.

17.4.1 Microfluidic Drug Screening Based on Cell Assays

Cell-based drug screening is a commonly employed approach for sifting through extensive drug databases in search of potential candidate drugs. These models, known for their ability to faithfully mimic human physiological conditions and enable high-throughput drug screening, play a pivotal role in expediting the drug discovery process. Initial cell-based high-throughput screening methods mainly use expensive and time-consuming microplates, which are costly and complicated to operate. Microfluidic technologies have brought miniaturization and high-throughput development to cell-based drug screening technologies. At present, microfluidic-based cell high-throughput technologies mainly include perfusion flow mode [99], droplet-based mode [100, 101], and microarray mode [102, 103].

17.4 Microfluidic High-throughput Drug Screening

Wang et al. [104] devised a microfluidic chip with the objective of facilitating high-throughput cytotoxicity screening. The chip boasts a grid structure, comprising 24 channels in both horizontal and vertical orientations, thereby yielding a total of 576 individual chambers. The cell chambers are strategically positioned at the junction points of these channels, and the channels themselves are demarcated by pneumatic elastic microvalves. This innovative device enables concurrent high-density parallel cytotoxicity screening, resulting in a substantial reduction in reagent consumption. Clausell-Tormos et al. [105] employed a droplet microfluidic system for cell cultivation, which exhibited a remarkable 500-fold enhancement in cell growth compared to the traditional orifice plate structure. Within this system, cells thrived for extended periods, with cell viability remaining at a commendable 80% even after three days of incubation. Besides, how to achieve precise and complicated control of droplets is of great significance for drug screening of droplet cells. Kim et al. [106] introduced a fully automated and programmable microfluidic array system comprising 64 autonomous cell culture chambers, tailored for high-throughput drug screening. Varied drug concentration inputs were achieved through meticulous control of microfluidic diffusion mixers. Cells were adept at processing a series of 64 drug concentrations either sequentially or concurrently. To streamline the system's operations, LabVIEW software was harnessed for comprehensive automation and control.

Due to limitations of 2D cell models in simulating complex environment in vivo, the effect of screening drugs in clinical practice is greatly reduced. The emergence of 3D cell models (such as tumor spheres, organoids) makes up for these limitations and can better simulate complex tissue structures and human physiological environments [107]. In recent years, a combination of 3D tumor sphere models and microfluidic technologies has attracted more and more attention in the field of high-throughput drug screening.

Chen et al. [108] realized culture of 3D tumor cells in sodium alginate hydrogel solution based on a microfluidic system. This device allows simultaneous testing of concentrations of several drugs to increase throughput of drug screening. Similarly, Prince et al. [109] successfully cultured breast tumor sphere arrays in biomimetic hydrogels on a microfluidic device. The effects of drug dose and delivery speed on breast tumor spheroids were studied to demonstrate the correlation between in vivo drug efficacy and on-chip drug response. This device greatly saves time and cost and is expected to be applied to personalized tumor research. Fan et al. [110] demonstrated a brain cancer chip for high-throughput drug screening, which was integrated by microporous arrays and microchannels. By culturing GBM cells in a microporous array to form 3D brain cancer tissues and applying pitavastatin and irinotecan treatment on the chip, the possibility of high-throughput GBM cancer spheroid formation and drug parallel screening based on the chip was demonstrated. Albanese et al. [111] proposed a melanoma tumor-on-a-chip device that enables real-time analysis of NP accumulation under physiological flow conditions. The device uses gravity-driven fluids to load MDA-MB-435 melanoma cell spheres into a microfluidic chip, allowing the microtissue to remain stationary in a moving solution, producing a controllable flow strip that also provides an optical window for uninterrupted real-time analysis using confocal microscopy imaging. The fluorescently labeled gold NPs were further loaded into the system to study their accumulation, distribution, and penetration depth in the tumor sphere. The results showed that the penetration of NPs into tissues was limited by their diameter (<110 nm). These findings were confirmed in mouse xenograft models, indicating that the tumor on the chip is a promising platform for screening the best NP design before in vivo research. In order to promote the complex combination of multiple drugs, Li et al. [112] developed a high-throughput, multilayer PMMA microfluidic chip for combination drug screening on tumor spheres. The device consists of an inlet layer and multiple dispersion layers, which can simultaneously load different samples into the chip to achieve drug combination and screening.

The organoid formation method, despite its advantages, has certain limitations. Precise control over organoids and their local microenvironment is challenging, making it difficult to faithfully replicate the dynamic and intricate microenvironment encountered during natural organ development [113]. In order to overcome these problems, researchers proposed to combine organoids with organ chip technology. Wang et al. [114, 115] developed a microfluidic brain organoid chip model to generate hiPSCs-derived 3D brain organoids on the organ chip. Through the integration of 3D matrigel, fluid flow, and a multicellular tissue structure, an attempt was made to replicate the brain's microenvironment. As a result, the generated organoids exhibited distinct characteristics, including neural differentiation, regionalization, and the development of cortical tissue. The model was further used to study the effects of prenatal nicotine exposure on brain development.

17.4.2 Microfluidic Drug Screening Based on Organ-on-a-Chip

In recent years, the advent of microfluidic organ-on-a-chip devices, capable of replicating the microarchitecture and functionality of living human organs, has significantly advanced drug screening technology. The primary advantage of organ chips lies in their capacity to emulate the intricate structures and functions of organs in vitro, as evidenced by various studies. The drug screening process utilizing organ chips, as depicted in Figure 17.4, encompasses several crucial steps. In the first step, microfluidic chips are designed and fabricated, specifically tailored to human organs or tissues. The second step involves cell cultivation directly on the microfluidic chip, simulating the microenvironment of human organs or tissues. In the third step, test drugs are introduced, and the interaction between drugs and cells is monitored, concluding with data collection. The final step entails drug evaluation, assessing the efficacy and safety of the tested drugs, thus completing the screening process. To date, a range of organ chips has been developed for various organs, encompassing vessel-on-a-chip, lung-on-a-chip, liver-on-a-chip, kidney-on-a-chip, and multiorgans-on-a-chip systems.

The progression of vessel-on-chip technology has yielded a potent in vitro model for drug screening and assessment of vascular function. The microchannel design within microfluidic devices closely mimics the architecture of the microvascular system, enabling the replication of vascular conditions by precise control of

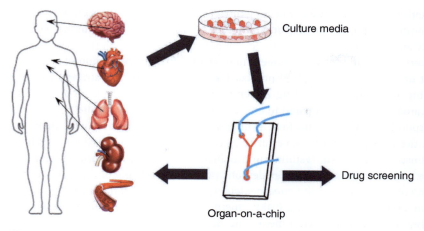

Figure 17.4 Process of microfluidic drug screening based on organ-on-a-chip.

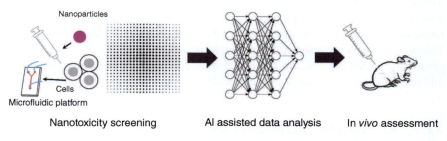

Figure 17.5 AI assisted microfluidic development of nanomedicine.

shear rate distribution and fluid flow patterns within the channels [116]. Sobrino et al. [117] developed a vascularized 3D micro-organ model for anticancer drug screening. They found that the microvasculature in the platform was sensitive to antiangiogenesis (pazopanib, linifenib, cabotantinib, axitinib, etc.) and vascular disrupting agents (vincristine, paclitaxel). It is proved that this vascularized tumor model can be used to identify drugs that directly target cancer cells. On this basis, Phan et al. [118] further developed an arrayed vascularized microtumor platform for larger-scale drug screening applications, and successfully demonstrated its ability to identify antiangiogenesis and anticancer drugs from small compound libraries.

Huh et al. [50] designed a lung-on-a-chip device, which reproduced the key structure, function, and mechanical properties of human alveolar-capillary interface. The lung model chip consists of two tightly connected microchannels separated by a thin (10 μm) porous and flexible PDMS membrane. In order to simulate the physiological respiratory movement, two large transverse microchambers were integrated into the device, in which the air pressure (controlled by a small vacuum pump) was periodically reduced and increased, inducing periodic stretching and relaxation of the membrane. The lung model chip can be used to study the complex organ-level responses in lung inflammation and cell-level responses to bacterial infection.

Heart-on-a-chip provides a screening model for toxicity evaluation of cardiovascular drugs. Zhang et al. [119] proposed a new 3D bioprinting strategy to create on-chip heart models for drug screening and disease modeling. Their findings indicated that when compared to conventional drugs, the myocardial chip model, incorporating endothelialization, exhibited analogous outcomes upon treatment with the widely used anticancer drug doxorubicin. Specifically, a dose-dependent response was observed, resulting in a reduction in beating rate and an elevation in toxicity. Moreover, the authors demonstrated the versatility of their platform by testing interactions between nanomedicines, such as NPs and heart cells.

Jang et al. [53] developed a PDMS microfluidic device designed to replicate the in vivo renal tubular environment. The device comprises a "fluid cavity" and a "chamber" separated by a porous structure. Experiments were conducted under both static and dynamic flow conditions, revealing a noteworthy disparity in drug toxicity. This innovative device holds promise as a valuable in vitro model for investigating renal physiology, renal diseases, and nephrotoxicity. Weber et al. [120] harnessed a microfluidic system to reconstruct the proximal tubules of the kidney. This in vitro system closely mirrors the secretion and reabsorption functions of proximal tubules in vivo, demonstrating its applicability in renal drug clearance and research on drug-induced nephrotoxicity.

17.5 AI-assisted Microfluidic Development of Nanomedicine

The development of innovative nanomedicines typically begins with the identification of a promising drug candidate that targets a specific disease or condition. This candidate is subsequently subjected to rigorous safety and efficacy screenings to eliminate compounds that do not demonstrate therapeutic potential. The drug development process proceeds with additional steps, including cell culture analysis, animal model experiments, and patient trials. This entire process demands significant financial investments and time resources.

However, with advancements in technology, the integration of computer-aided drug design and screening, bolstered by AI and microfluidic technologies, has revolutionized the development of nanomedicines. These innovations are streamlining the drug development process, potentially reducing costs and accelerating the path to effective therapies.

For example, machine learning has emerged as a powerful tool for predicting the interactions between drug compounds and their target molecules, thereby playing a pivotal role in several critical aspects of drug discovery and development. It aids in the identification of potential drug targets, the discovery of new drugs, and the selection of the most promising drug candidates. By leveraging vast datasets and complex algorithms, machine learning can expedite the screening and evaluation of drug–target interactions, facilitating more efficient and effective drug development processes.

Yamankurt et al. [121] reported a method for determining structure–activity relationships and utilized AI to help screen and develop SNAs that target Toll-like receptor 9 (TLR9) pathways. A high-throughput device was engineered to enable the production of SNAs at the picomolar scale. Approximately 1000 candidate SNAs were first identified on the basis of a reasonable range of 11 design parameters. Further, machine learning was used to quantitatively model SNA immune activation, and to determine the minimum number of SNAs needed to capture the optimal structure–activity relationship for a given SNA library. The method can reduce the number of NPs that need to be tested by an order of magnitude, rendering it a valuable screening tool in the development of therapeutic NP-based drugs.

In another study by Damiati et al. [122], a novel approach is presented, involving machine learning to forecast the size of poly(lactic-co-glycolic acid) (PLGA) particles generated by diverse microfluidic systems, either alone or in combination.

Furthermore, AI is also employed to investigate various classes of polymer particles, with the ultimate objective of facilitating production of precisely controlled polymer particles.

Boehnke et al. [123] developed an AI-based high-throughput screening method for systematically assessing the interaction of 35 different NP types with 488 cancer cell lines, which can determine some characteristics of cancer cells and NPs to help successful delivery of NPs. Beyond virtual screening and high-throughput screening, machine learning classification methods also find valuable application in the categorization of effective or ineffective drug compounds, as well as in the assessment and prediction of toxicity [124, 125]. George et al. [126] utilized computer data processing to develop an in vitro screening assessment method for high-throughput NP toxicity based on animal models, and proposed to further achieve NP toxicity screening. Lin et al. [127] have used AI image recognition platform to achieve a larger NP library (24 types) screening. Concu et al. [128] used artificial neural network (ANN) to develop a quantitative structure–toxicity relationship (QSTR)-perturbation model to jointly probe the toxicological characteristics of metal and metal oxide NPs in different experimental environments, which showed >97% accuracy in both training and validation. The toxicity of several types of NPs is predicted based on this method.

Building upon AI-assisted drug screening as described previously, microfluidic techniques serve to enhance and refine the overall performance of the screening process. Due to the large gap between experimental results and clinical results, most of the nanomedicines approved for clinical use are not as effective as previously expected. A suitable preclinical model for drug screening is of great significance for drug screening and development. The microfluidic platform can overcome the problems of low flux, high cost, and low efficiency of traditional animal models and cell models. It serves as a good platform for high-throughput screening of nanomedicines because it provides a large amount of data, including biomedical images, videos, cell behavior measurements, and sequence reading data. At present, microfluidic chips integrating AI and machine learning have been gradually applied to the screening and research of nanomedicines. AI-assisted microfluidic nanodrug screening is illustrated in Figure 17.5, and we will also introduce some examples below.

Similarly, Zhang et al. [129] proposed a fast, low-cost label-free cell evaluation method based on neural network and microfluidic technology, which can be used to evaluate tumor spheroids for large-scale drug screening in microfluidics. A microfluidic chip with 1920 chambers was developed for drug testing, which was supported by automated image collection and cropping system. A convolutional neural network was trained to estimate the viability of the spheroids, which was based on the fact of the traditional live/dead staining sphere vitality. The experiment accurately estimated the efficacy of three chemotherapeutic drugs (doxorubicin, oxaliplatin, and irinotecan) and cross-validated the performance of doxorubicin and oxaliplatin by trained networks. The finding suggests it is possible to use a number of representative drugs to train the generic drug network and apply it to many different drugs in large-scale screening.

In 2018, Moore et al. [130] integrated microfluidic technology with AI to create an in vitro model that replicates the dynamic immune–tumor microenvironment for tumor biopsy. In this model, live tumor slices (up to 12 slices) were incorporated to dynamically simulate the interactions between tumor cells and tumor-infiltrating lymphocytes (TILs). Importantly, this setup enabled the acquisition of high-resolution real-time images, facilitating detailed observations of these interactions. Building upon this foundation, they utilized machine-learning image analysis algorithms to quantitatively assess the experimental results. This approach allowed for the demonstration of the interactions between mouse tumor cells and anti-PD-1 immune checkpoint inhibitors, as well as the evaluation of the response of human tumor cells to immune checkpoint inhibitors.

Kingston et al. [131] designed a novel machine-learning analysis method based on 3D imaging for the quantitative evaluation of gold NP-tumor cell interactions in local and liver metastases. The analysis of the physiological and NP delivery conditions of 1301 liver microtransferases revealed that NPs were more easily incorporated into microtransferases (50% NP-positive cells) than into primary tumors (17% NP-positive cells).

These studies have shown that the development of in vitro microfluidic modeling of tumor microenvironment provides a powerful platform for in vitro experiments of cancer and other diseases, enabling researchers to better develop drugs and treat cancer more effectively, while machine learning has been used to evaluate the effect of nanomedicine on metastatic tumors and predict NP delivery.

17.6 Conclusions and Perspectives

In recent years, the utilization of nanomedicines has gained widespread recognition across various therapeutic domains, notably in the realm of cancer treatment.

Nevertheless, the development of new drugs continues to be a resource-intensive and time-consuming endeavor. The challenge lies in finding ways to mitigate the costs and enhance the overall efficiency of drug development in the field of nanomedicine.

The synergy of microfluidic technology with the advancements in AI, particularly the implementation of deep learning methods, offers a promising platform for addressing these challenges in nanomedicine development. AI harnesses machine learning models to predict the interactions between target molecules and drugs, thereby playing a pivotal role in target identification, novel drug discovery, and the selection of the most promising drug candidates. This fusion of microfluidics and AI holds the potential to revolutionize the field of nanomedicine by streamlining drug development processes and improving the cost-effectiveness of research and innovation. Microfluidic cell assays and organ-on-a-chips offer superior platforms for predicting candidate drug reactions and assessing drug toxicity in the process of drug screening. These chips facilitate the verification of potential drug candidates and generate substantial image data. In conjunction with microfluidic technology, deep learning methods in AI excel at rapidly processing vast quantities of data, leading to improved efficiency and accuracy in the screening of nanodrugs. This integration of microfluidic systems and AI contributes to enhanced drug screening capabilities and the acceleration of drug development processes.

However, it is true that the high-throughput screening of nanomedicines through the amalgamation of AI and microfluidic technology is still relatively nascent. While AI has indeed unveiled numerous potential targets and novel compounds for various diseases, the success rate of compounds generated solely by AI remains limited. Further advancements and refinement of this innovative approach are essential to fully realize its potential in the development of effective nanomedicines.

The effective utilization of AI in drug screening hinges on access to extensive data sources, including scientific literature, patents, and clinical trial results, for model construction and validation. Addressing the challenges of aggregating, curating, and ensuring data quality in these diverse sources is crucial. Establishing a reliable and comprehensive database that provides high-quality data is essential for enhancing the drug screening process.

Incorporating a microfluidic system into the equation presents additional challenges, primarily in generating stable and substantial drug screening data through this technology. Overcoming these technical hurdles is key to realizing the full potential of the drug screening platform based on AI and microfluidic technology.

Despite these challenges, there is a strong belief that such a drug screening platform has the potential to usher in revolutionary changes in the drug discovery and development process, ultimately leading to more efficient and effective drug development. This integration of AI and microfluidics has the capability to streamline drug screening, reduce costs, and accelerate the pace of innovation in the field of medicine.

References

1 Agrawal, Y.K. and Patel, V. (2011). Nanosuspension: an approach to enhance solubility of drugs. *Journal of Advanced Pharmaceutical Technology & Research* 2 (2): 81.
2 Onoue, S., Yamada, S., and Chan, K. (2014). Nanodrugs: pharmacokinetics and safety. *International Journal of Nanomedicine* 20: 1025–1037.
3 Petros, R.A. and Desimone, J.M. (2010). Strategies in the design of nanoparticles for therapeutic applications. *Nature Reviews. Drug Discovery* 9 (8): 615–627.
4 Xin, Y., Huang, Q., Tang, J. et al. (2016). Nanoscale drug delivery for targeted chemotherapy. *Cancer Letters* 379 (1): 24–31.
5 Lu, Y., Aimetti, A.A., Langer, R. et al. (2017). Bioresponsive materials. *Nature Reviews Materials* 2: 1.
6 Barenholz, Y.C. (2012). Doxil® – the first FDA-approved nano-drug: lessons learned. *Journal of Controlled Release* 160 (2): 117–134.
7 Bu, L., Xie, J., Chen, K. et al. (2012). Assessment and comparison of magnetic nanoparticles as MRI contrast agents in a rodent model of human hepatocellular carcinoma. *Contrast Media & Molecular Imaging* 7 (4): 363–372.
8 Reimer, P. and Balzer, T. (2003). Ferucarbotran (Resovist): a new clinically approved RES-specific contrast agent for contrast-enhanced MRI of the liver: properties, clinical development, and applications. *European Radiology* 13 (6): 1266–1276.
9 Ventola, C.L. (2017). Progress in nanomedicine: approved and investigational nanodrugs. *P T* 42 (12): 742–755.
10 Castro, F., Leite Pereira, C., Helena Macedo, M. et al. (2021). Advances on colorectal cancer 3D models: the needed translational technology for nanomedicine screening. *Advanced Drug Delivery Reviews* 175: 113824.
11 Ravi, M., Paramesh, V., Kaviya, S.R. et al. (2015). 3D cell culture systems: advantages and applications. *Journal of Cellular Physiology* 230 (1): 16–26.
12 Kapałczyńska, M., Kolenda, T., Przybyła, W. et al. (2018). 2D and 3D cell cultures – a comparison of different types of cancer cell cultures. *Archives of Medical Science* 14 (4): 910–919.
13 Serov, N. and Vinogradov, V. (2022). Artificial intelligence to bring nanomedicine to life. *Advanced Drug Delivery Reviews* 184: 114194.
14 Julkunen, H., Cichonska, A., Gautam, P. et al. (2020). Leveraging multi-way interactions for systematic prediction of pre-clinical drug combination effects. *Nature Communications* 11 (1): 6136.
15 Gupta, R., Srivastava, D., Sahu, M. et al. (2021). Artificial intelligence to deep learning: machine intelligence approach for drug discovery. *Molecular Diversity* 25 (3): 1315–1360.
16 Gao, D., Liu, H., Jiang, Y. et al. (2012). Recent developments in microfluidic devices for in vitro cell culture for cell-biology research. *Trends in Analytical Chemistry* 35: 150–164.
17 Hashimoto, M., Tong, R., and Kohane, D.S. (2013). Microdevices for nanomedicine. *Molecular Pharmaceutics* 10 (6): 2127–2144.

18 Du, G., Fang, Q., and den Toonder, J.M. (2016). Microfluidics for cell-based high throughput screening platforms – a review. *Analytica Chimica Acta* 903: 36–50.
19 Whitesides, G.M. (2006). The origins and the future of microfluidics. *Nature* 442 (7101): 368–373.
20 Zhou, J. and Pei, Z. (2020). Experimental study of the piezoelectric drop-on-demand drop formation in a coaxial airflow. *Chemical Engineering and Processing: Process Intensification* 147: 107778.
21 Ghaderi, A., Kayhani, M.H., Nazari, M. et al. (2018). Drop formation of ferrofluid at co-flowing microcahnnel under uniform magnetic field. *European Journal of Mechanics – B/Fluids* 67: 87–96.
22 Lee, C., Lee, J., Kim, H.H. et al. (2012). Microfluidic droplet sorting with a high frequency ultrasound beam. *Lab on a Chip* 12 (15): 2736–2742.
23 Park, S.Y., Wu, T.H., Chen, Y. et al. (2011). High-speed droplet generation on demand driven by pulse laser-induced cavitation. *Lab on a Chip* 11 (6): 1010–1012.
24 Lin, C., Lee, G., Lin, Y. et al. (2001). A fast prototyping process for fabrication of microfluidic systems on soda-lime glass. *Journal of Micromechanics and Microengineering* 11 (6): 726–732.
25 Nuzaihan, M.N.M., Hashim, U., Rahim Ruslinda, A. et al. (2015). Fabrication of silicon nanowires array using e-beam lithography integrated with microfluidic channel for pH sensing. *Current Nanoscience* 11: 239–244.
26 Goral, V.N., Hsieh, Y.C., Petzold, O.N. et al. (2010). Hot embossing of plastic microfluidic devices using poly (dimethylsiloxane) molds. *Journal of Micromechanics and Microengineering* 21 (1): 017002.
27 Sahli, M., Millot, C., Gelin, J.C. et al. (2013). The manufacturing and replication of microfluidic mould inserts by the hot embossing process. *Journal of Materials Processing Technology* 213 (6): 913–925.
28 Becker, H. and Gärtner, C. (2000). Polymer microfabrication methods for microfluidic analytical applications. *Electrophoresis* 21 (1): 12–26.
29 Malek, C.K. and Saile, V. (2004). Applications of LIGA technology to precision manufacturing of high-aspect-ratio micro-components and -systems: a review. *Microelectronics Journal* 35 (2): 131–143.
30 Horn, T.J. and Harrysson, O.L. (2012). Overview of current additive manufacturing technologies and selected applications. *Science Progress* 95 (Pt 3): 255–282.
31 Gross, B.C., Erkal, J.L., Lockwood, S.Y. et al. (2014). Evaluation of 3D printing and its potential impact on biotechnology and the chemical sciences. *Analytical Chemistry* 86 (7): 3240–3253.
32 Guo, M.T., Rotem, A., Heyman, J.A. et al. (2012). Droplet microfluidics for high-throughput biological assays. *Lab on a Chip* 12 (12): 2146–2155.
33 Shang, L., Cheng, Y., and Zhao, Y. (2017). Emerging droplet microfluidics. *Chemical Reviews* 117 (12): 7964–8040.
34 Teh, S., Lin, R., Hung, L. et al. (2008). Droplet microfluidics. *Lab on a Chip* 8 (2): 198.

35 Zhu, P. and Wang, L. (2016). Passive and active droplet generation with microfluidics: a review. *Lab on a Chip* 17 (1): 34–75.

36 Thorsen, T., Roberts, R.W., Arnold, F.H. et al. (2001). Dynamic pattern formation in a vesicle-generating microfluidic device. *Physical Review Letters* 86 (18): 4163.

37 Ganán-Calvo, A.M. and Gordillo, J.M. (2001). Perfectly monodisperse microbubbling by capillary flow focusing. *Physical Review Letters* 87 (27): 274501.

38 Umbanhowar, P.B., Prasad, V., and Weitz, D.A. (2000). Monodisperse emulsion generation via drop break off in a coflowing stream. *Langmuir* 16 (2): 347–351.

39 Nakashima, T., Shimizu, M., and Kukizaki, M. (2000). Particle control of emulsion by membrane emulsification and its applications. *Advanced Drug Delivery Reviews* 45 (1): 47–56.

40 Huang, X., Eggersdorfer, M., Wu, J. et al. (2017). Collective generation of milliemulsions by step-emulsification. *RSC Advances* 7 (24): 14932–14938.

41 He, B., Huang, X., Xu, H. et al. (2018). Creating monodispersed droplets with electrowetting-on-dielectric step emulsification. *AIP Advances* 8 (7).

42 Huang, X., He, B., Xu, Z. et al. (2020). Electro-coalescence in step emulsification: dynamics and applications. *Lab on a Chip* 20 (3): 592–600.

43 Toh, A.G.G., Wang, Z.P., Yang, C. et al. (2014). Engineering microfluidic concentration gradient generators for biological applications. *Microfluidics and Nanofluidics* 16 (1–2): 1–18.

44 Jeon, N.L., Dertinger, S.K.W., Chiu, D.T. et al. (2000). Generation of solution and surface gradients using microfluidic systems. *Langmuir* 16 (22): 8311–8316.

45 Long, T. and Ford, R.M. (2009). Enhanced transverse migration of bacteria by chemotaxis in a porous T-sensor. *Environmental Science & Technology* 43 (5): 1546–1552.

46 Atencia, J., Morrow, J., and Locascio, L.E. (2009). The microfluidic palette: a diffusive gradient generator with spatio-temporal control. *Lab on a Chip* 9 (18): 2707.

47 Zeng, S., Li, B., Su, X.O. et al. (2009). Microvalve-actuated precise control of individual droplets in microfluidic devices. *Lab on a Chip* 9 (10): 1340.

48 Zhang, M., Xu, C., Jiang, L. et al. (2018). A 3D human lung-on-a-chip model for nanotoxicity testing. *Toxicology Research* 7 (6): 1048–1060.

49 Doryab, A., Amoabediny, G., and Salehi-Najafabadi, A. (2016). Advances in pulmonary therapy and drug development: lung tissue engineering to lung-on-a-chip. *Biotechnology Advances* 34 (5): 588–596.

50 Huh, D., Matthews, B.D., Mammoto, A. et al. (2010). Reconstituting organ-level lung functions on a chip. *Science* 328 (5986): 1662–1668.

51 Theobald, J., Abu El Maaty, M.A., Kusterer, N. et al. (2019). In vitro metabolic activation of vitamin D3 by using a multi-compartment microfluidic liver-kidney organ on chip platform. *Scientific Reports* 9 (1): 4616.

52 Kim, S. and Takayama, S. (2015). Organ-on-a-chip and the kidney. *Kidney Research and Clinical Practice* 34 (3): 165–169.

53 Jang, K.J., Mehr, A.P., Hamilton, G.A. et al. (2013). Human kidney proximal tubule-on-a-chip for drug transport and nephrotoxicity assessment. *Integrative Biology (Camb)* 5 (9): 1119–1129.

54 Park, J., Lee, B.K., Jeong, G.S. et al. (2015). Three-dimensional brain-on-a-chip with an interstitial level of flow and its application as an in vitro model of Alzheimer's disease. *Lab on a Chip* 15 (1): 141–150.

55 Pamies, D., Hartung, T., and Hogberg, H.T. (2014). Biological and medical applications of a brain-on-a-chip. *Experimental Biology and Medicine (Maywood, N.J.)* 239 (9): 1096–1107.

56 Grosberg, A., Alford, P.W., ML, M.C. et al. (2011). Ensembles of engineered cardiac tissues for physiological and pharmacological study: heart on a chip. *Lab on a Chip* 11 (24): 4165–4173.

57 Jastrzebska, E., Tomecka, E., and Jesion, I. (2016). Heart-on-a-chip based on stem cell biology. *Biosensors and Bioelectronics* 75: 67–81.

58 Zhang, W., Zhang, Y.S., Bakht, S.M. et al. (2016). Elastomeric free-form blood vessels for interconnecting organs on chip systems. *Lab on a Chip* 16 (9): 1579–1586.

59 Caballero, D., Blackburn, S.M., De Pablo, M. et al. (2017). Tumour-vessel-on-a-chip models for drug delivery. *Lab on a Chip* 17 (22): 3760–3771.

60 Schimek, K., Busek, M., Brincker, S. et al. (2013). Integrating biological vasculature into a multi-organ-chip microsystem. *Lab on a Chip* 18: 3588–3598.

61 Rabinow, B.E. (2004). Nanosuspensions in drug delivery. *Nature Reviews Drug Discovery* 3 (9): 785–796.

62 Liversidge, G.G. and Cundy, K.C. (1995). Particle size reduction for improvement of oral bioavailability of hydrophobic drugs: I. Absolute oral bioavailability of nanocrystalline danazol in beagle dogs. *International Journal of Pharmaceutics* 125 (1): 91–97.

63 Moghimi, S.M., Hunter, A.C., and Murray, J.C. (2004). Nanomedicine: current status and future prospects. *The FASEB Journal* 19 (3): 311–330.

64 Gaumet, M., Vargas, A., Gurny, R. et al. (2008). Nanoparticles for drug delivery: the need for precision in reporting particle size parameters. *European Journal of Pharmaceutics and Biopharmaceutics* 69 (1): 1–9.

65 He, Z., Ranganathan, N., and Li, P. (2018). Evaluating nanomedicine with microfluidics. *Nanotechnology* 29 (49): 492001.

66 Ahn, J., Ko, J., Lee, S. et al. (2018). Microfluidics in nanoparticle drug delivery; from synthesis to pre-clinical screening. *Advanced Drug Delivery Reviews* 128: 29–53.

67 Hubbell, J.A. and Chilkoti, A. (2012). Nanomaterials for drug delivery. *Science* 337: 303–305.

68 Raemdonck, K., Braeckmans, K., Demeester, J. et al. (2014). Merging the best of both worlds: hybrid lipid-enveloped matrix nanocomposites in drug delivery. *Chemical Society Reviews* 43 (1): 444–472.

69 Zhao, H., Wang, J., Wang, Q. et al. (2007). Controlled liquid antisolvent precipitation of hydrophobic pharmaceutical nanoparticles in a microchannel reactor. *Industrial & Engineering Chemistry Research* 46 (24): 8229–8235.

70 Ali, H.S.M., York, P., Ali, A.M.A. et al. (2011). Hydrocortisone nanosuspensions for ophthalmic delivery: a comparative study between microfluidic nanoprecipitation and wet milling. *Journal of Controlled Release* 149 (2): 175–181.

71 Wang, J., Zhang, Q., Zhou, Y. et al. (2010). Microfluidic synthesis of amorphous cefuroxime axetil nanoparticles with size-dependent and enhanced dissolution rate. *Chemical Engineering Journal* 162 (2): 844–851.

72 Shrimal, P., Jadeja, G., and Patel, S. (2021). Microfluidics nanoprecipitation of telmisartan nanoparticles: effect of process and formulation parameters. *Chemical Papers* 75 (1): 205–214.

73 Stone, N.R.H., Bicanic, T., Salim, R. et al. (2016). Liposomal amphotericin B (AmBisome®): a review of the pharmacokinetics, pharmacodynamics. *Clinical Experience and Future Directions. Drugs* 76 (4): 485–500.

74 Ickenstein, L.M. and Garidel, P. (2019). Lipid-based nanoparticle formulations for small molecules and RNA drugs. *Expert Opinion on Drug Delivery* 16 (11): 1205–1226.

75 O'Brien, M.E.R., Wigler, N., Inbar, M. et al. (2004). Reduced cardiotoxicity and comparable efficacy in a phase III trial of pegylated liposomal doxorubicin HCl(CAELYX™/Doxil®) versus conventional doxorubicin for first-line treatment of metastatic breast cancer. *Annals of Oncology* 15 (3): 440–449.

76 Zhang, X., Goel, V., and Robbie, G.J. (2020). Pharmacokinetics of Patisiran, the first approved RNA interference therapy in patients with hereditary transthyretin-mediated amyloidosis. *The Journal of Clinical Pharmacology* 60 (5): 573–585.

77 Schoenmaker, L., Witzigmann, D., Kulkarni, J.A. et al. (2021). mRNA-lipid nanoparticle COVID-19 vaccines: structure and stability. *International Journal of Pharmaceutics* 601: 120586.

78 Majedi, F.S., Hasani-Sadrabadi, M.M., Emami, S.H. et al. (2013). Microfluidic assisted self-assembly of chitosan based nanoparticles as drug delivery agents. *Lab on a Chip* 13: 204–207.

79 Kim, K., Kang, D., Kim, M. et al. (2015). Generation of alginate nanoparticles through microfluidics-aided polyelectrolyte complexation. *Colloids and Surfaces A: Physicochemical and Engineering Aspects* 471: 86–92.

80 El-Sayed, M., Huang, J., and Edirisinghe, M. (2015). Bioinspired preparation of alginate nanoparticles using microbubble bursting. *Materials Science and Engineering C* 46: 132–139.

81 Kong, L., Chen, R., Wang, X. et al. (2019). Controlled co-precipitation of biocompatible colorant-loaded nanoparticles by microfluidics for natural color drinks. *Lab on a Chip* 19: 2089–2095.

82 Baby, T., Liu, Y., Yang, G. et al. (2021). Microfluidic synthesis of curcumin loaded polymer nanoparticles with tunable drug loading and pH-triggered release. *Journal of Colloid and Interface Science* 594: 474–484.

83 Karnik, R., Gu, F., Basto, P. et al. (2008). Microfluidic platform for controlled synthesis of polymeric nanoparticles. *Nano Letters* 8 (9): 2906–2912.

84 Rhee, M., Valencia, P.M., Rodriguez, M.I. et al. (2011). Synthesis of size-tunable polymeric nanoparticles enabled by 3D hydrodynamic flow focusing in single-layer microchannels. *Advanced Materials* 23 (12): H79–H83.

85 Lim, J., Bertrand, N., Valencia, P.M. et al. (2014). Parallel microfluidic synthesis of size-tunable polymeric nanoparticles using 3D flow focusing towards in vivo study. *Nanomedicine: Nanotechnology, Biology and Medicine* 10 (2): 401–409.

86 Hood, R.R., DL, D.V., Atencia, J. et al. (2014). A facile route to the synthesis of monodisperse nanoscale liposomes using 3D microfluidic hydrodynamic focusing in a concentric capillary array. *Lab on a Chip* 14: 2403–2409.

87 Jahn, A., Vreeland, W.N., Gaitan, M. et al. (2004). Controlled vesicle self-assembly in microfluidic channels with hydrodynamic focusing. *Journal of the American Chemical Society* 126 (9): 2674–2675.

88 Krzysztoń, R., Salem, B., Lee, D.J. et al. (2017). Microfluidic self-assembly of folate-targeted monomolecular siRNA-lipid nanoparticles. *Nanoscale* 9 (22): 7442–7453.

89 Valencia, P.M., Farokhzad, O.C., Karnik, R. et al. (2012). Microfluidic technologies for accelerating the clinical translation of nanoparticles. *Nature Nanotechnology* 7 (10): 623–629.

90 Chen, D., Love, K.T., Chen, Y. et al. (2012). Rapid discovery of potent siRNA-containing lipid nanoparticles enabled by controlled microfluidic formulation. *Journal of the American Chemical Society* 134 (16): 6948–6951.

91 Wang, F., Li, C., Cheng, J. et al. (2016). Recent advances on inorganic nanoparticle-based cancer therapeutic agents. *International Journal of Environmental Research and Public Health* 13 (12): 1182.

92 Abou-Hassan, A., Sandre, O., and Cabuil, V. (2010). Microfluidics in inorganic chemistry. *Angewandte Chemie International Edition* 49 (36): 6268–6286.

93 Liu, Y., Yang, G., Hui, Y. et al. (2022). Microfluidic nanoparticles for drug delivery. *Small* 18 (36): 2106580.

94 Yen, B.K.H., Stott, N.E., Jensen, K.F. et al. (2003). A continuous-flow microcapillary reactor for the preparation of a size series of CdSe. *Advanced Materials* 15 (21): 1858–1862.

95 Shao, M., Yu, Q.b., Jing, N. et al. (2019). Continuous synthesis of carbon dots with full spectrum fluorescence and the mechanism of their multiple color emission. *Lab on a Chip* 19: 3974–3978.

96 Trofimov, A.D., Ivanova, A.A., Zyuzin, M.V. et al. (2018). Porous inorganic carriers based on silica, calcium carbonate and calcium phosphate for controlled/modulated drug delivery: fresh outlook and future perspectives. *Pharmaceutics* 10 (4): 167.

97 Yan, J., Xu, X., Zhou, J. et al. (2020). Fabrication of a pH/redox-triggered mesoporous silica-based nanoparticle with microfluidics for anticancer drugs doxorubicin and paclitaxel codelivery. *ACS Applied Biomaterials* 3 (2): 1216–1225.

98 Zhao, X., Liu, Y., Yu, Y. et al. (2018). Hierarchically porous composite microparticles from microfluidics for controllable drug delivery. *Nanoscale* 10 (26): 12595–12604.

99 Zhu, X., Yi Chu, L., Chueh, B. et al. (2004). Arrays of horizontally-oriented mini-reservoirs generate steady microfluidic flows for continuous perfusion cell culture and gradient generation. *The Analyst* 129 (11): 1026.

100 Kimura, H., Yamamoto, T., Sakai, H. et al. (2008). An integrated microfluidic system for long-term perfusion culture and on-line monitoring of intestinal tissue models. *Lab on a Chip* 8 (5): 741.

101 Trivedi, V., Doshi, A., Kurup, G.K. et al. (2010). A modular approach for the generation, storage, mixing, and detection of droplet libraries for high throughput screening. *Lab on a Chip* 10 (18): 2433.

102 Wlodkowic, D., Faley, S., Zagnoni, M. et al. (2009). Microfluidic single-cell array cytometry for the analysis of tumor apoptosis. *Analytical Chemistry* 81 (13): 5517–5523.

103 Wei, C.W. (2005). Using a microfluidic device for DNA microarray hybridization in 500s. *Nucleic Acids Research* 33 (8): e78.

104 Wang, Z., Kim, M., Marquez, M. et al. (2007). High-density microfluidic arrays for cell cytotoxicity analysis. *Lab on a Chip* 7 (6): 740.

105 Clausell-Tormos, J., Lieber, D., Baret, J.C. et al. (2008). Droplet-based microfluidic platforms for the encapsulation and screening of mammalian cells and multicellular organisms. *Chemistry & Biology* 15 (5): 427–437.

106 Kim, J., Taylor, D., Agrawal, N. et al. (2012). A programmable microfluidic cell array for combinatorial drug screening. *Lab on a Chip* 12 (10): 1813.

107 Thoma, C.R., Zimmermann, M., Agarkova, I. et al. (2014). 3D cell culture systems modeling tumor growth determinants in cancer target discovery. *Advanced Drug Delivery Reviews* 69–70: 29–41.

108 Chen, M.C.W., Gupta, M., and Cheung, K.C. (2010). Alginate-based microfluidic system for tumor spheroid formation and anticancer agent screening. *Biomedical Microdevices* 12 (4): 647–654.

109 Prince, E., Kheiri, S., Wang, Y. et al. (2022). Microfluidic arrays of breast tumor spheroids for drug screening and personalized cancer therapies. *Advanced Healthcare Materials* 11 (1): 2101085.

110 Fan, Y., Nguyen, D.T., Akay, Y. et al. (2016). Engineering a brain cancer chip for high-throughput drug screening. *Scientific Reports* 6 (1): 25062.

111 Albanese, A., Lam, A.K., Sykes, E.A. et al. (2013). Tumour-on-a-chip provides an optical window into nanoparticle tissue transport. *Nature Communications* 4 (1).

112 Li, L., Chen, Y., Wang, H. et al. (2021). A high-throughput, open-space and reusable microfluidic chip for combinational drug screening on tumor spheroids. *Lab on a Chip* 21 (20): 3924–3932.

113 Mccauley, H.A. and Wells, J.M. (2017). Pluripotent stem cell-derived organoids: using principles of developmental biology to grow human tissues in a dish. *Development* 144 (6): 958–962.

114 Wang, Y., Wang, L., Guo, Y. et al. (2018). Engineering stem cell-derived 3D brain organoids in a perfusable organ-on-a-chip system. *RSC Advances* 8 (3): 1677–1685.

115 Wang, Y., Wang, L., Zhu, Y. et al. (2018). Human brain organoid-on-a-chip to model prenatal nicotine exposure. *Lab on a Chip* 18 (6): 851–860.

116 Wang, X., Sun, Q., and Pei, J. (2018). Microfluidic-based 3D engineered microvascular networks and their applications in vascularized microtumor models. *Micromachines* 9 (10): 493.

117 Sobrino, A., Phan, D.T.T., Datta, R. et al. (2016). 3D microtumors in vitro supported by perfused vascular networks. *Scientific Reports* 6 (1): 31589.

118 Phan, D., Wang, X., Craver, B.M. et al. (2017). A vascularized and perfused organ-on-a-chip platform for large-scale drug screening applications. *Lab on a Chip* 17 (3): 511–520.

119 Zhang, Y.S., Arneri, A., Bersini, S. et al. (2016). Bioprinting 3D microfibrous scaffolds for engineering endothelialized myocardium and heart-on-a-chip. *Biomaterials* 110: 45–59.

120 Weber, E.J., Chapron, A., Chapron, B.D. et al. (2016). Development of a microphysiological model of human kidney proximal tubule function. *Kidney International* 90 (3): 627–637.

121 Yamankurt, G., Berns, E.J., Xue, A. et al. (2019). Exploration of the nanomedicine-design space with high-throughput screening and machine learning. *Nature Biomedical Engineering* 3 (4): 318–327.

122 Damiati, S.A., Rossi, D., Joensson, H.N. et al. (2020). Artificial intelligence application for rapid fabrication of size-tunable PLGA microparticles in microfluidics. *Scientific Reports* 10 (1).

123 Boehnke, N., Straehla, J.P., Safford, H.C. et al. (2022). Massively parallel pooled screening reveals genomic determinants of nanoparticle delivery. *Science* 377 (6604): eabm5551.

124 Zhao, C., Zhang, H., Zhang, X. et al. (2006). Application of support vector machine (SVM) for prediction toxic activity of different data sets. *Toxicology* 217 (2–3): 105–119.

125 Korkmaz, S., Zararsiz, G., and Goksuluk, D. (2014). Drug/nondrug classification using Support Vector Machines with various feature selection strategies. *Computer Methods and Programs in Biomedicine* 117 (2): 51–60.

126 George, S., Xia, T., Rallo, R. et al. (2011). Use of a high-throughput screening approach coupled with in vivo zebrafish embryo screening to develop hazard ranking for engineered nanomaterials. *ACS Nano* 5 (3): 1805–1817.

127 Lin, S., Zhao, Y., Ji, Z. et al. (2013). Zebrafish high-throughput screening to study the impact of dissolvable metal oxide nanoparticles on the hatching enzyme, ZHE1. *Small* 9 (9–10): 1776–1785.

128 Concu, R., Kleandrova, V.V., Speck-Planche, A. et al. (2017). Probing the toxicity of nanoparticles: a unified in silico machine learning model based on perturbation theory. *Nanotoxicology* 11 (7): 891–906.

129 Zhang, Z., Chen, L., Wang, Y. et al. (2019). Label-free estimation of therapeutic efficacy on 3D cancer spheres using convolutional neural network image analysis. *Analytical Chemistry* 91 (21): 14093–14100.

130 Moore, N., Doty, D., Zielstorff, M. et al. (2018). A multiplexed microfluidic system for evaluation of dynamics of immune–tumor interactions. *Lab on a Chip* 18: 1844–1858.

131 Kingston, B.R., Syed, A.M., Ngai, J. et al. (2019). Assessing micrometastases as a target for nanoparticles using 3D microscopy and machine learning. *Proceedings of the National Academy of Sciences* 116 (30): 14937–14946.

Index

a

ablation therapy 3, 12–27, 299–323
acoustic impulse and acoustic chemical effects, ultrasound 271–272
acridine orange 280
active targeting 9, 13, 49, 74, 84, 137, 140, 142, 152, 198, 199, 201, 252, 255, 300, 363
  function design of nanoparticles
    antibody-directed active targeting design 52–53
    protein and peptide-directed active targeting function design 52
    small molecule and aptamer-directed active targeting function design 51–52
  of nanomedicine 364
Ag NPs 284, 287–288
agonists 18, 114, 255–257
AGuIX (activation and guidance of irradiation by X-ray) 352
alloys materials 335
Alzheimer's disease (AD) treatment 202, 299, 311
angiogenic nano-biomaterials
  delivery vehicles 175
  intrinsic angiogenic agents 176
antibacterial nano-biomaterials
  inorganic 164–165
  organic 165–166
antibody-directed active targeting design 52–53
antibody-drug conjugates (ADCs) 364
anti-CD19 chimeric antigen receptor (CAR-T 19) 257
anti-HIV
  concept of 132–142
  protease inhibitors 133–141
  reverse transcriptase inhibitors 133
anti-SARS-CoV-2 257
anti-smoking measures 2
anti-tumor drugs 2
artificial antibody 13, 252–253
artificial intelligence (AI)
  application of 379–380
  cell replacement therapy for diabetes 380
  computer vision technology 378
  deep learning 372
  definition 371
  emerging microfluidic technologies 386–389
  machine learning 372, 378
  Natural Language Processing (NLP) Technology 378–379
  precision medicine in cancer treatment 380–382
  risk assessment models 373
artificial intelligence generation content (AIGC) 28
artificial neural network (ANN) 374, 401
Au@CoFeB nanoparticles 90
Au@CoFeB-Rg3 nanomedicines 90, 131
Au-based nanomaterials 350
Au nanoparticles (AuNPs) 13, 82, 283–284, 287, 289, 352, 358

Nanomedicine: Fundamentals, Synthesis, and Applications, First Edition. Edited by Yujun Song.
© 2025 WILEY-VCH GmbH. Published 2025 by WILEY-VCH GmbH.

Index

auto-targeting design
  exosomes and supramolecular cell membrane vesicle targeting design 54
  macrophage targeting function design 53–54
  platelets targeting function design 53
  RBC targeting function design 53

b

ball milling 55–56, 131, 335
bioconjugated QDs 89
BiOI QDs 356
bio-safety of nano-biomaterials 182–183
blood–brain barrier (BBB) 3, 5, 9, 18, 61, 62, 126, 127, 132, 142, 202, 323, 347
Blumlein pulse generator 304, 305

c

Cabiliv 13
CA-Gd@C-dots 356
cancer radiotherapy, milestones in 344–345
carbon-based nanomaterials 288
carbon dots (CDs) 285, 288, 356
carbon nanotubes (CNT) 83, 92, 117, 120, 131, 223, 288, 289
cavitation effect, ultrasound 270–271
cefuroxime axetil (CFA) nanoparticles 393
cell death 10, 11, 255, 257, 282–284, 286, 287, 289, 300, 302, 307, 331, 342, 359
cell-based drug screening 396
cell membrane coating nanotechnology
  electroporation 94
  extrusion 93
  graphene nanoplatform-mediated cell membrane coating 94
  in situ encapsulation with cell-derived vesicles 94
  sonication 93
cerebrovascular diseases 1, 2, 5, 6, 301, 302

chemiluminescence biochemical reaction excitation (CLE) 14, 22
chimeric antigen receptor (CAR) T cell therapy 248, 257–258
chronic non-healing wounds 157, 158
Co/Co_3O_4 MNPs 236
coherent scattering 347, 348
cold tumors 250, 251
Coley toxin 248
colloidal gold nanoparticles (GNPs) 83–84
combination therapy, for cancer treatment 321
complex cohesive phase separation 144
Compton effect 347
computational simulation and machine learning 183
computer vision technology 378
condensed phase separation methods 143, 144
convection-based method 391
conventional polymer micelles 145
copper/carbon quantum dot 288
co-precipitation method 149, 218
core/shell magnetic nanoparticles 235–237, 240
coronary artery disease (CAD) 2, 388
CT nanoprobe 213
  application of 223–224
cubic Fe_3O_4 MNPs 233
cutaneous wounds
  definition of 157
  management of 158
  nano-biomaterials-based approaches 158, 159
  nano-biomaterials-based therapeutics, wound healing
    antibacterial 164–166
    enabled biophysical regulation 169–175
    engineering of the wound microenvironment 166–169
    hemostasis 160–163
  nano-biomaterials for imaging and monitoring

wound infection monitoring 177–179
wound parameters and markers 179–181
nanomaterial-based approaches 158
therapeutic and diagnostic management 158
CY-PSMA-1-Au$_{25}$NCs radiosensitizer 352–353
cytokine interferon-α (IFN-α) 248
cytokines 12, 24, 53, 172, 175, 245–247, 250, 253, 258, 259, 261–263, 332, 381
cytotoxicity of Au@CoFe-Rg3 320

d

danger/damage-associated molecular patterns (DAMPs) 282
deep brain stimulation (DBS) technology 115, 310, 311
deep learning 372, 373, 380, 382, 403
dendrimers 74, 131, 138, 139, 272
dendritic macromolecules 143, 145, 272
diagnostic ultrasound 269
Diels–Alder reaction 334
DLVO theory 60, 61
doxorubicin/paclitaxel (DOX/PTX) 396
droplet microfluidics 390
drug genomics 374–376
drug therapy 114, 117, 132, 153, 299, 300, 385, 392

e

efavirenz 133, 139
electric field responsive nanomedicine 312
  drug loading 312
  efficacy 312
  surface modification 312
  synthesis process 312
emulsification method 116, 143, 312, 390
emulsion polymerization 133, 143
engineering of the wound microenvironment

MMPs 167–169
pro-inflammatory mediators 169
redox modulation 167
enhanced permeability and retention (EPR) effect 4, 49, 73, 74, 199, 250, 272, 280, 352
epibromohydrin functionalized TiO$_2$@γ-Fe$_2$O$_3$ nanoparticles 240
epithelial barrier 52, 137
exchange coupling interaction 235–237, 239
external beam radiotherapy (EBRT) 342, 346, 362

f

facile freeze-annealing process 27
FA-LNPs 52
Fenton nanoparticles 284
Fe$_{3-x}$O$_4$/CoFe$_2$O$_4$ core–shell MNPs 235
Fe$_3$O$_4$ nanoparticles 149, 211, 227, 262, 333, 334, 336
Fe$_3$O$_4$@SiO$_2$@PNG-Hy MNPs 239, 240
ferrite materials 334–335
focused ultrasound (FUS) 14, 269, 275, 280
frequently-used natural polymers 393

g

gadolinium chelates coated gold nanoparticles Au@DTDTPA-Gd$_{50}$ 352
Gd-based theranostic nano-agent 352
gene-based therapies 197
genomics of nanomedicine
  definition 375
  genomic responses 377
  modifications 376
  pharmacokinetic characteristics 375
  toxicity 376–377
genomics, definition 372
gold-levonorgestrel nanoclusters (Au$_8$NCs) 350
  structure and size characterization 351

gold nanoclusters (AuNCs)   350
gold nanoparticles for bioimaging   54
gold-nanoparticles-reduced graphene oxide (NAu-rGO) nanocomposite   289
gold nanorods (GNRs)   54, 94, 287, 288, 346, 350
gold nanoshells   287
Goldbody   13
graphene nanoplatform-mediated cell membrane coating   94
graphene oxide (GO)   85, 119, 120, 165, 288–289
graphene, synthesis and application of   85–86

h

hard magnetic nanoparticles   229, 230
hard/soft ($CoFe_2O_4/Fe_3O_4$) nanocomposites   238
hematoporphyrin   279, 280
heterogeneous MNPs   229, 230
high-intensity focused ultrasound (HIFU)   8, 14, 18, 23, 28, 269, 270, 273
highly fluorescent QDs   88
high-pressure steep pulse third-generation machine   308
high-throughput drug screening
  based on cell assays   396–398
  organ-on-a-chip   398–400
high-Z-element-based drugs   347
high-Z-metal-based nanomaterials
  physical and biochemical radio-sensitization mechanism of   349
  physicochemical and biochemical radio-sensitization mechanism of   349
hot tumors   250, 251
hydrodynamic flow focusing (HFF) devices   394
hydrophilic polymers   79, 81
hydrophobic QDs   88
hydrothermal synthesis method   149, 219
hyperbaric oxygen therapy (HBO)   158
hyperthermia   22, 54, 83, 86, 94, 227, 228, 329
  magnetic hyperthermia   227, 228, 236, 238–239, 329–332, 334–339
  tumor hyperthermia method   329
hypoxia-inducible factor (HIF)   175, 176, 293

i

^{131}I-based RIT   359
immune agonist   12
immune checkpoints   248, 250
immune system and immune response   245–247
indoleamine 2,3-dioxygenase (IDO)   248
inorganic antibacterial nano-biomaterials   164–165
inorganic-based nanomedicines   146, 148
  magnetic material   120–121
  nano-mesoporous silicon   122
  nanocalcium   122
  nanocarbon   120
  nanogold   119–120
  quantum dots (QDs)   117–118
inorganic nanoparticles   395
  colloidal gold nanoparticles (GNPs)   83–84
  graphene   85–86
  layered double hydroxides (LDHs)   89
  magnetic nanoparticles (MNPs)   86–88
  mesoporous silica nanoparticles (MSNPs)   84–85
  multifunctional composite nanoparticles   90
  quantum dots (QDs)   88–89
Instance 1   337
Instance 2   337–338
Instance 3   338–339
intelligent nanomedicine   6, 358
interleukin-2 (IL-2)   248
internal radioisotope therapy (RIT)   343, 358
  advantages   358
  alpha (α) particle emitters   359, 360

Auger particle emitters 360, 362
beta (β) particle emitters 359, 361
drawbacks 358
intracellular magnetic hyperthermia
    therapy 332
intractable diseases 1–3, 6, 10, 14, 23, 27,
    28, 49, 263, 302, 323
intrinsic angiogenic agents 176
ionizing radiation 341
    response of normal cells to 342–343

k

knee osteoarthritis (KOA) 338, 339

l

lab on a chip (LOC) 386
$La_{1-x}Sr_xMnO_3$ (LSMO) 336, 337
layered double hydroxides (LDHs) 89
linear energy transfer (LET) 342, 359,
    360
lipid nanoparticles (LNPs) 52, 113, 131,
    140, 200–202, 393–395
liposomal doxorubicin (Doxil) nanodrugs
    9
liposomal nanoparticles
    in drug delivery 77–78
    extrusion technique 76
    heating method 77
    microfluidization 76
    sonication 76
liposomes 138, 394
    nanocarrier preparation 394–395
    preparation methods 138
liquid antisolvent precipitation (LASP)
    method 392
low-intensity focused ultrasound (LIPUS)
    269, 270

m

machine learning 28, 182, 183, 372, 373,
    378, 380–382, 400–403
macrophage targeting function design
    53–54
macroscopic dose enhancement factors
    (MDEF) 350

magnetic field responsive nanomedicine
    synthesis
    alloys materials 335
    composite materials 335–336
    Fe_3O_4 materials 333–334
    ferrite materials 334–335
    other materials 336–337
magnetic heat therapy apparatus 330
magnetic hyperthermia 227, 228, 236,
    238–239, 329–332, 334–339
    on tumor cells 331–332
magnetic mediated hyperthermia (MMH)
    applications
        Instance 1 337
        Instance 2 337–338
        Instance 3 338–339
    cytotoxicity 332–333
    killing effect 331–332
    magnetic field responsive
        nanomedicine synthesis
        alloys materials 335
        composite materials 335–336
        Fe_3O_4 materials 333–334
        ferrite materials 334–335
        other materials 336–337
    magnetic hyperthermia on tumor cells
        331–332
    magneto-thermal conversion
        mechanism 330–331
    overview 329–330
    thermal effects 329
magnetic nano-biomaterials 210, 212
magnetic nanoparticles (MNPs) 86
    biomedical applications
        magnetic hyperthermia 238–239
        magnetic resonance imaging (MRI)
            236–237
        neuromodulation 240–242
        targeted drug delivery 239–240
    in biomedical science 228
    exchange coupling interaction
        235–236
    hard magnetic nanoparticles 230
    magnetic moment structure of
        228–229

magnetic nanoparticles (MNPs) (contd.)
  schematic representation 230
  soft magnetic nanoparticles 228–229
magnetic nanoprobe
  application of 222–223
  co-precipitation method 218–219
  hydrothermal synthesis method 219
  thermal decomposition method 219
magnetic regulation
  composition 233–234
  shape 232–233
  size 230–232
magnetic resonance imaging (MRI) 22, 87, 88, 90, 94, 120, 212, 214, 222, 224, 227, 236–237, 242, 275, 313, 318–320, 333, 334, 344, 351–353, 356, 364, 378, 386
magnetoelectric nanoparticles (MENPs) 241, 242
magnetoplasma nanocore–shell structure 208
magneto-thermal conversion mechanism 330–331
mechanical effects, ultrasound 270
mechanism of nanomedicine enhancing immunotherapy 248–252
mechanism of tumor immunotherapy 247–248
MedPulser® system 307
mesoporous silica nanoparticles (MSNPs) 84–85
metallic-based nanomedicines 122–125, 148–150
metal nanoclusters (NCs) 284
metal-organic frameworks (MOFs)
  preparation and characterization of 123
  tumor therapy 123–125
methylene blue (MB) 280
methyl methacrylate (MMA) 336
micelle 9, 22, 51, 55, 114, 131, 138–140, 143–146, 148, 150, 217, 272, 300, 301, 391
microfluidic chips fabrication 387–389

microfluidic concentration gradient generators 390, 391
microfluidic synthesis 56–58, 395
microfluidic systems 387, 395, 401, 403
microfluidic technology 386
  AI assisted nanomedicine 400–402
  high-throughput drug screening 396–400
    based on cell assays 396–398
    organ-on-a-chip 398–400
  medicine development 386–391
  microfluidic chips fabrication 387–389
  microfluidic units 390–391
  nanomedicines 391–396
    drug nanoparticles preparation 392–393
    nanocarrier preparation 393–396
micro total analysis system (µTAS) 386
mixture (hybrid) nanoparticles 90–92
molecular imaging probe 208, 219, 224
monoclonal antibody ipilimumab 248
monodisperse Fe_3O_4 nanoparticles 227
mononuclear phagocytosis system (MPS) 50, 62
multi-drug resistant (MDR) 11, 12, 164, 166
multifunctional composite nanoparticles 90
multifunctional integrated theranostic nanoparticle platforms design 363
multifunctional/multimodal nanoprobes 214
multi-functional (target) nanomedicines 150–152
multimodal iron-based nanodrug (Au@CoFe-Rg3)
  design and synthesis 313–315
  in situ anticancer effects, in liver cancer 320–321
  physical imaging applications 318–320
  physical properties and structural analysis 315–318

multi-modal theraunostics instruments 27
multi-mode nanomedicines 6, 14, 18, 24, 25, 27, 28

n

nano-biomaterials, cutaneous wound assessment and therapy 158
nano-biomaterials-enabled biophysical regulation of wound healing
  angiogenic nano-biomaterials 175
  bio-electrical stimulation 173–175
  mechanical cues 171–173
  surface nanotopographical features 171
nanocalcium 117, 122
nanocarbon 117, 120
nanocarriers, characteristics 73
nanocrystals (NCs) 55, 90, 149, 161, 167, 209, 290, 292, 354, 395
nano-dosage forms or nano-carriers 138
nano drug carriers 141, 251
nanodrug mediated PEF combined therapy 322
  bacterial infection treatment 323
  cancer treatment 322–323
  enhancing therapeutic effects 303
  instruments and parameters
    capacitor discharge device 305
    electrode system 305
    pulse generator 304–305
    pulse parameter adjustment 305–306
    pulse power amplifier 305
  neurological disease treatment 323
  overcoming drug resistance 303
  personalized treatment 303
  reduced side effects 303
nanoemulsion 6, 52, 138, 140, 141, 152
nanoenzymes 7, 11, 24, 27
nanogold 117, 119–120
nano-immunotherapy 24
nanoknife system 307, 308
nanomaterial-based PSs
  Ag NPs 284
  AuNPs 283–284
  carbon-based nanomaterials 288–289
  carbon dots (CDs) 285
  metal nanoclusters (NCs) 284–285
  noble metal-based nanomaterials 286
  photothermal agent 286
  photothermal therapy 286
  semiconductor-based nanomaterials 289
  transition metal carbides (TMCs) 285
nanomaterials
  ball milling 55–56
  biological methods 58–59
  manufacturing issues 96
  microfluidic synthesis 56–58
  nanoprecipitation 56
  personalized protein nanoparticles (PNP) 58
  physico-chemical characterization of 95
  regulatory issues 95–96
  safety concerns 95
nanomedicine(s)
  active targeting function design of nanoparticles 51–53
  application examples 275–276
  artificial intelligence 377–379
  auto-targeting design
    exosomes and supramolecular cell membrane vesicle targeting design 54
    macrophage targeting function design 53–54
    platelets targeting function design 53
    RBC targeting function design 53
  concept of 272
  future development of 27–28
  genomics of 374–377
  historical timeline of major developments of 6–7
  interdisciplinary features of 12–27
  in vitro or in vivo 3D traceable function gold nanoparticles for bioimaging 54

nanomedicine(s) (contd.)
    quantum dots for Bioimaging 54–56
    key features of 4–6
    microfluidic technology 391–396
    nanotechnology translational 6–12
    passive targeting function design 49, 51
    surface modification
        charged materials 60–62
        conformation change 62–63
        multi-ligand modified nano-preparations 66
        single targeting ligand 63–65
        steric resistance 62
    targeting delivery of gene
        based nanomedicines to the brain 202
        based nanomedicines to the liver 200, 201
        based nanomedicines to the lung 200–202
        based nanovaccines to the LNs 200
        based nanovaccines to the spleen 199–200
    therapies to the tumors 198–199
    therapy theory of 272–273
    ultrasonic atomization pyrolysis 274
    ultrasonic chemical precipitation 274
    ultrasonic electrochemical technology 274–275
    ultrasonic reduction 275
nanomedicine mediated immunotherapy
    immunological applications
        agonists 255–257
        artificial antibody 252–253
        cytokines 261–263
        nanomedicines combined CAR-T therapy 257–258
        nanovaccines 258–261
        reprogrammed immunity 253–255
    mechanism of nanomedicine enhancing immunotherapy 248–252
    mechanism of tumor immunotherapy 247–248
nanomedicine mediated pulsed electric field (PEF) therapy 24, 299–323
    advantages 302
nanomedicine mediated ultrasound therapy, equipment for 273–274
nanomedicine mediating immunotherapy
    immune system and immune response 245–247
nanomedicines combined CAR-T therapy 257–258
nanomedicines for nervous system diseases
    concepts and types of 113–114
    inorganic-based nanomedicines
        magnetic material 120–121
        nanocalcium 122
        nanocarbon 120
        nanogold 119–120
        nano-mesoporous silicon 122
        quantum dots (QDs) 117–118
    metallic-based nanomedicines 122–125
    multi-functional nanomedicines for 125–128
        chemotherapy 125–126
        combination treatment strategy 127
        perspectives of 127–128
    polymer-based nanomedicine
        emulsification method 116
        reductive reaction reduction method 116
        solvent evaporation method 116
    polymer-based nanomedicine typical application
        inclusion of nanomaterials 117
        targeted drug delivery 116–117
        tissue repair and regenerative medicine 117
        vaccines and immunotherapy 117
    therapeutical methods for 114–116
nanomedicines, microfluidic technology
    drug nanoparticles preparation 392–393
    nanocarrier preparation 393–396

nanomedicine synthesis with radio/nuclear radiation response 362–364
nano-mesoporous silicon 117, 122
nanoparticle carriers 135, 312
nanoprecipitation 56, 80, 81, 131, 143, 144, 392–394
nanoprobes
  CT imaging 213
  magnetic 210–212
  multifunctional/multimodal 214
  nuclide imaging 213–214
  optical nanoprobe 208–210
  photoacoustic imaging 212–213
  synthesis methods
    magnetic nanoprobe 218–219
    noble metal 216–218
nanoprobes for wound infection monitoring 177
nanoscale coordination polymers (NCP) 11, 12
nanosecond pulsed electric fields (nsPEFs) 300
  tumor treatment 300
nanosecond pulse tumor ablation system 308, 309
nanospheres 54, 114, 119, 131, 138, 139, 141, 152, 283, 287
nanotechnology 371
  mediated radiation therapy 346
  translational nanomedicine 6–12
nanotheranostics 6, 10, 14, 24
nanotherapeutics 6, 49
nanovaccines 7, 199–200, 257–261
Natural Language Processing (NLP) Technology 378–379
needle shaped electrode 305
neuromodulation 227, 228, 236, 240–242
noble metal-based nanomaterials 286–288
noble metal nanoprobes
  chemical synthesis method 216–218
  electrochemical process 218
  photochemical method 217
  template-based method 217–218

non-porphyrinoid-based PSs 280
non-specific immune response 245
non-viral vectors 197
NP albumin-bound paclitaxel 9
nucleolin-targeted AS1411 aptamer 240
nuclide imaging nanoprobe 208, 213–214

o

oil-in-water (O/W) emulsion 140
optical nanoprobe 208–210, 219–222
organic antibacterial nano-biomaterials 164–166
organic polymer chips 388
organ-on-a-chip 390, 391, 398–400, 403
overhealing 157, 180

p

parallel electrode 305
Parkinson's disease 13, 114–116, 242, 299, 310, 311, 323
passive targeting 49–51, 54, 135, 137, 142, 151, 152, 198–199, 220, 223, 252, 255, 280
passive targeting function design
  shape dependent biological effect of nanoparticles 50–51
  size dependent biological functions 50
  surface charge and PEGylation of nanoparticles 50
pathogen-associated molecular patterns (PAMPs) 246
PEGylated FePt–Fe$_2$O$_3$ core–shell nanoparticles 240
personalized protein nanoparticles (PNP) 58
phagocytosis 6, 50, 62, 66, 113, 135–137, 140, 152, 223, 261
photoacoustic imaging nanoprobe 208, 212–213, 222–223
  application of 222–224
photodynamic and photothermal synergistic therapy 291–292

photodynamic therapy (PDT)   3, 14, 18, 22, 54, 123–125, 128, 153, 164, 220, 259, 279–293, 322, 323, 346
photosensitizers (PSs)/photothermal agent   279
photoshock ablation   22
photothermal ablation (PTA)   22
photothermal agent   176, 279, 280, 286, 289, 290, 292, 293
photothermal therapy   286
photothermal treatment (PTT)   94, 164, 165, 177, 178, 279–286, 288–293, 364
physical sensitization mechanism   349
pilot microfluidic process   27
plant mediated green synthesis of nanoparticles   59
platelets targeting function design   53
PLGA$_{SS}$D@BPQDs nanosystem   356–358
polydimethylsiloxane (PDMS)   92, 171, 172, 387, 388, 399, 400
polydispersity index (PDI)   393, 394
poly (ethylene glycol)-poly (lactide-co-glycolic acid) (PEG-PLGA)   56
poly(lactic-co-glycolic acid) (PLGA)   52, 80, 82, 93, 139, 141, 143, 253, 356, 357, 394, 401
polymerase chain reaction (PCR)   170, 172, 388
polymer-based nanomedicines   116–117, 142–146
  emulsification method   116
  reductive reaction reduction method   116
  solvent evaporation method   116
polymer-based nanomedicine typical application
  inclusion of nanomaterials   117
  targeted drug delivery   116–117
  tissue repair and regenerative medicine   117
  vaccines and immunotherapy   117
polymeric micelle paclitaxel   9

polymeric nanoparticles (NPs)   74, 78–82, 133, 393
  controlled drug release   73
  hydrophilic polymers   81
  nanoprecipitation/solvent displacement method   80
  polymerization methods   81
  salting out method   79–80
  solvent evaporation method   79
  spontaneous emulsification/solvent diffusion method   79
polymerization method   81, 143, 336
polymer micelles   143, 145, 146, 272, 300, 301, 391
polymer nanocarriers, characteristics of   141–145
polymethacrylic acid (PMMA)   336, 387, 398
polyvinylpyrrolidone (PVP)   62, 80, 289, 313, 394
porphyrinoid-based PSs   280
pPD-Gd@C-dots   356
precision medicine in cancer treatment   380–382
protein and peptide-directed active targeting function design   52
pulse count   306
pulsed laser photogenesis   22
pulse frequency   305, 306
pulse voltage   305, 306
pulse waveform generators, advantages and disadvantages of   305, 306
pulse width   305–307

q

quantitative structure–toxicity relationship (QSTR)   401
quantum dots (QDs)   88, 117, 353, 395
  for Bioimaging   54–55
  copper/carbon   288
  inorganic-based nanomedicines   117–118
quantum drugs   24

r

radiation sensitizers 347
radiation therapy/radiotherapy (RT)
   for cancer treatment 341–364
   side effects 344
radio-enhancers 347
radio/nuclear radiation therapy,
     nanomedicine mediated 341–364
radiosensitizers 347–358
   for radiation therapy 347–358
Raman nanoprobe 209
RBC targeting function design 53
RD® steep nanosecond pulse generator 309
redox modulation of the wound microenvironment 167
reductive reaction reduction method 116
reprogrammed immunity 253–255
Rg3-sheltered dynamic nanocatalysts 91–92
rose Bengal (RB) 280

s

selective organ targeting (SORT) strategies 200, 201
self-assembled core-shell nanoscale coordination polymer (NCP) nanoparticles 11, 12
semiconductor-based nanomaterials 289
   copper sulfide (CuS) 290
   molybdenum disulfide 290
   WS_2 290
semiconductor QDs, as photo-and radio-sensitizers 356
siliceous materials 387
single or multi walled carbon nanotube (SWCNT/MWCNT) 289
size-dependent magnetic moments 231
small molecule and aptamer-directed active targeting function design 51–52
smart nanomedicine 6, 10, 24, 27, 28
sodium iodide transporter (NIS) 359

soft/hard ($Fe_3O_4/CoFe_2O_4$) nanocomposites 238
soft magnetic nanoparticles 229
solid lipid nanoparticles (SLN) 113, 131, 140
solid-phase sintering and vapor–liquid–solid growth (SS–VLS-like) method 27
solvent evaporation method 79, 116
sonochemical dynamics (SDT) 14, 18
specific immune response 245, 259
spray drying method 143, 145, 393
standardized preparation protocols 182
stealth effect 140
stimuli-responsive drug delivery 51
superatom cluster drugs 24
superparamagnetic iron oxide nanoparticles (SPIONs) 222, 338, 382
supervised learning 372, 373
synergistic anticancer research 312–321
synthetic polymer nanocarriers 394

t

targeted drug delivery 64, 78, 113, 116–117, 140, 151, 152, 199, 227, 228, 239–240, 300
targeted drug delivery system (TDDS) 78, 140, 151, 152
targeted siRNA polymeric NPs (CALAA-01) 9
targeting delivery of gene based nanomedicines to the brain 202
targeting delivery of gene-based nanomedicines to the liver 200, 201
targeting delivery of gene based nanomedicines to the lung 200–202
targeting delivery of gene based nanovaccines to the LNs 200
targeting delivery of gene based nanovaccines to the spleen 199–200

targeting delivery of gene therapies to the tumors 198–199
targeting the pro-inflammatory mediators 169
telmisartan (TEL) 393
theranostic nano-biomaterials for wound infection control 177–179
theranostic nano-platforms 363
therapeutic α-emitters 359–360
therapeutic β-emitters 359, 361
therapeutic Auger-emitters 362
thermal decomposition method 219, 235
thermal effect, ultrasound 270
thixotropic effect, ultrasound 271
tissue Repair and Regenerative Medicine 117
Toll-like receptor 9 (TLR9) pathways 401
transient receptor potential vanilloid 4 (TRPV4) 241
transition metal based nanomedicines 11
transition metal carbides (TMCs) 285
transition-metalloptosis 11
tricarboxylic acid (TCA) cycle 10
tumor hyperthermia method 329
tumor immune editing hypothesis 254
tumor immune editing process 255
tumor immune monitoring hypothesis 253, 254
tumor immune stimulation hypothesis 254
tumor infiltrating lymphocytes (TILs) 250, 402
tumor treatment
  nanomedicine advantages 302
  nanosecond pulsed electric fields (nsPEFs) 300
two-pore domain potassium channels (TREK-1) 241
type 1 diabetes (T1D) 380

u

ultrasmall Au_{25} nanoclusters (NCs) for selective prostate cancer targeting 352
ultrasonic atomization coupling pyrolysis process 27
ultrasonic atomization pyrolysis 274
ultrasonic chemical precipitation 274
ultrasonic electrochemical technology 274
ultrasonic reduction 275
ultrasound therapy and therapy theory 269–270
ultrathin Ti_3C_2 MXene nanosheets 292
underhealing 157
unsupervised learning 372

v

vaccines and immunotherapy 117
vascular cell adhesion molecule-1(VCAM-1) 52
vibrating sample magnetometer (VSM) method 335, 336
viral vectors 197, 257
VivaGel™ 139

w

water-in-oil (W/O) 140, 141, 150
water-insoluble organic solvents 144
wound infection monitoring
  nanoprobes for 177
  theranostic nano-biomaterials 177–179
wound parameters and markers
  physiological and pathological 179–180
  stem cell-based wound therapy 180
  wound scarring markers 180–181

x

X-rays interaction with high-Z materials 348